WORLD ENERGY INVESTMENT OUTLOOK

2003
INSIGHTS

INTERNATIONAL ENERGY AGENCY

INTERNATIONAL ENERGY AGENCY
9, rue de la Fédération,
75739 Paris, cedex 15, France

The International Energy Agency (IEA) is an autonomous body which was established in November 1974 within the framework of the Organisation for Economic Co-operation and Development (OECD) to implement an international energy programme.

It carries out a comprehensive programme of energy co-operation among twenty-six* of the OECD's thirty member countries. The basic aims of the IEA are:

- to maintain and improve systems for coping with oil supply disruptions;
- to promote rational energy policies in a global context through co-operative relations with non-member countries, industry and international organisations;
- to operate a permanent information system on the international oil market;
- to improve the world's energy supply and demand structure by developing alternative energy sources and increasing the efficiency of energy use;
- to assist in the integration of environmental and energy policies.

** IEA member countries: Australia, Austria, Belgium, Canada, the Czech Republic, Denmark, Finland, France, Germany, Greece, Hungary, Ireland, Italy, Japan, the Republic of Korea, Luxembourg, the Netherlands, New Zealand, Norway, Portugal, Spain, Sweden, Switzerland, Turkey, the United Kingdom, the United States. The European Commission also takes part in the work of the IEA.*

ORGANISATION FOR ECONOMIC CO-OPERATION AND DEVELOPMENT

Pursuant to Article 1 of the Convention signed in Paris on 14th December 1960, and which came into force on 30th September 1961, the Organisation for Economic Co-operation and Development (OECD) shall promote policies designed:

- to achieve the highest sustainable economic growth and employment and a rising standard of living in member countries, while maintaining financial stability, and thus to contribute to the development of the world economy;
- to contribute to sound economic expansion in member as well as non-member countries in the process of economic development; and
- to contribute to the expansion of world trade on a multilateral, non-discriminatory basis in accordance with international obligations.

The original member countries of the OECD are Austria, Belgium, Canada, Denmark, France, Germany, Greece, Iceland, Ireland, Italy, Luxembourg, the Netherlands, Norway, Portugal, Spain, Sweden, Switzerland, Turkey, the United Kingdom and the United States. The following countries became members subsequently through accession at the dates indicated hereafter: Japan (28th April 1964), Finland (28th January 1969), Australia (7th June 1971), New Zealand (29th May 1973), Mexico (18th May 1994), the Czech Republic (21st December 1995), Hungary (7th May 1996), Poland (22nd November 1996), the Republic of Korea (12th December 1996) and Slovakia (28th September 2000). The Commission of the European Communities takes part in the work of the OECD (Article 13 of the OECD Convention).

FOREWORD

To the best of my knowledge, no previous attempt has been made by any organisation to build a comprehensive and authoritative picture of future investment needs, worldwide, in all parts of the energy-supply chain. This is the task which has been undertaken by the team, under Fatih Birol, responsible for the *World Energy Investment Outlook 2003*. It is my pleasure to acknowledge the work of this team and the contribution made by a wide circle of collaborators, wider than ever before, drawn from international organisations, energy companies and the financial community.

The total figure which emerges for the required global investment over thirty years is enough to make anyone pause: $16 trillion. No one claims that this figure will be precisely validated in thirty years time. Indeed, policy-makers will have failed if the energy economy has not been reshaped to make it more sustainable. But the figure is well enough founded to establish an order of magnitude for the task ahead.

Fortunately, no individual nor organisation has sole responsibility for mobilising that level of finance. But all of us who have a role need to recognise that the task is daunting. We all need to contribute constructively and sagely to easing the way. If the problem goes unsolved, someone, somewhere in the world, will go without the energy he (or, more likely, she) needs.

It is in the detail that most readers will find the insights they seek. Fuel by fuel and region by region, the global picture is built up. Each chapter presents an overall analysis, a regional breakdown and, in most cases, the consequences of making one or more changes in key assumptions, such as more intensive efforts in OECD countries to limit emissions of greenhouse gases.

Quantifying global investment needs and then analysing the obstacles to funding leads, inevitably, to the question: "Will the required sums be forthcoming?". Rather than attempt to answer that question definitively, we shall be satisfied if we promote an urgent and serious debate, leading governments, in particular, to adopt policies which will help overcome the obstacles. It is clear that the greatest part of the challenge lies in financing investment in developing countries for domestic consumption. I would venture the personal judgement that finance will not be available on the required scale if the circumstances described in the analysis persist unchanged. We should turn our attention now to defining the changes of policy needed to invalidate that judgement and, to take up the theme on which my predecessor concluded this foreword last year, to go further and raise the funds necessary to bring electricity to every world citizen before the year 2030.

This work is published under my authority as Executive Director of the IEA and does not necessarily reflect the views or policies of the IEA member countries.

Claude Mandil, Executive Director

Comments and questions are welcome and should be addressed as follows:

Fatih Birol
Chief Economist
Head, Economic Analysis Division
International Energy Agency
9, rue de la Fédération
75739 Paris Cedex 15
France

Telephone: (33-1) 4057 6670
Fax: (33-1) 4057 6659
Email: Fatih.Birol@iea.org

World Energy Outlook Series

World Energy Outlook - 1993
World Energy Outlook - 1994
World Energy Outlook - 1995
Oil, Gas & Coal Supply Outlook - 1995
World Energy Outlook - 1996
World Energy Outlook - 1998
World Energy Outlook - 1999 / Insights
  Looking at Energy Subsidies: Getting the Prices Right
World Energy Outlook - 2000
World Energy Outlook - 2001 / Insights
  Assessing Today's Supplies to Fuel Tomorrow's Growth
World Energy Outlook - 2002
World Energy Investment Outlook - 2003 / Insights
World Energy Outlook - 2004 (forthcoming)

ACKNOWLEDGEMENTS

This study was prepared by the Economic Analysis Division (EAD) of the International Energy Agency (IEA) in co-operation with other divisions of the IEA. The study was designed and managed by Fatih Birol, Head of the Economic Analysis Division. Other members of the EAD who were responsible for bringing this study to completion include: Maria Argiri, Marco Baroni, Amos Bromhead, François Cattier, Laura Cozzi, Lisa Guarrera, Hiroyuki Kato, Armelle Lecarpentier, Trevor Morgan and Michael Taylor. Claudia Jones, Emeline Tisseyre and Sally Bogle provided essential support.

Carmen Difiglio (Energy Technology Policy Division) and Peter Fraser (Energy Diversification Division) were part of the Outlook team.

The study also benefited from input provided by other IEA colleagues, namely: John Cameron, Dunia Chalabi, Doug Cooke, Sylvie Cornot, Ralf Dickel, David Fyfe, Rebecca Gaghen, Dolf Gielen, Mitsuhide Hoshino, Nicolas Lefevre, Lawrence Metzroth, Isabel Murray, Rick Sellers, Nancy Turck, Fridtjof Unander, Christof van Agt and Noé van Hulst. Pierre Lefevre, Fiona Davies, Loretta Ravera, Muriel Custodio and Bertrand Sadin of the Public Information Office provided substantial help in producing this book.

Robert Priddle carried editorial responsibility.

The work could not have been achieved without the substantial support provided by many government bodies, international organisations and energy companies worldwide, notably the US Department of Energy, the Ministry of Petroleum and Energy of Norway, the World Bank, the Organization of Petroleum Exporting Countries (OPEC), the World Energy Council, the World Coal Institute, BP and Electricité de France (EDF).

Overall guidance for the work, shaping the analysis and drawing out key messages, was contributed by an Advisory Panel, whose members were:

Advisory Panel Members

Olivier Appert, President and CEO, Institut Français du Pétrole, France

Gerald Doucet, Secretary-General, World Energy Council, United Kingdom

Andrei Konoplyanik, Deputy Secretary-General, Energy Charter Secretariat, Belgium

Peter Nicol, Global Sector Director, Oil and Gas, ABN-AMRO, UK

Michael Oppenheimer, President, Energy Coal, BHP Billiton, Australia

Rajendra Pachauri, Director-General, The Energy and Resources Institute (TERI) and Chair, Intergovernmental Panel on Climate Change (IPCC), India
John Paffenbarger, Vice-President, Constellation Energy, United States
Jamal Saghir, Director, Energy and Water, World Bank, United States
Yoshiro Sakamoto, Former Vice-Minister, Ministry of Economy, Trade and Industry of Japan; CEO Arabian Oil, Japan
Tore Sandvold, Former Director-General, Ministry of Petroleum and Energy of Norway; Consultant, Norway
Adnan Shihab-Eldin, Director, Research Division, OPEC Secretariat, Austria
Matt Simmons, President, Simmons & Company International, United States

Many other experts commented on the underlying analytical work and reviewed early drafts of each chapter. Their comments and suggestions were of great value. Prominent contributors include:

General and Finance

Robin Baker, Société Générale, United Kingdom
John E. Besant-Jones, World Bank, United States
Michael Hamilton, Arab Petroleum Investment Corperation (APICORP), Saudi Arabia
Kokichi Ito, Institute of Energy Economics of Japan, Japan
Nobuyuki Higashi, Japan Bank for International Cooperation, France
Michelle Michot Foss, Institute for Energy, Law & Enterprise, United States
Martin Raiser, EBRD, United Kingdom
Alastair Syme, Merrill Lynch, United Kingdom

Oil and Gas

Nathalie Alazard-Toux, Institut Français du Pétrole, France
Debbie Armstrong, Alberta Department of Energy, Canada
Guy Caruso, Department of Energy, United States
Marie Francoise Chabrelie, CEDIGAZ, France
Xavier Chen, BP China, People's Republic of China
Kevin Cliffe, Natural Resources Canada, Canada
William Davie, Schlumberger, France
Herman Franssen, International Energy Associates, United States
Donald Gardner, ExxonMobil, United States
Nadir Gürer, OPEC Secretariat, Austria
Jim Jensen, Jensen Associates, United States
Nancy Johnson, Department of Energy, United States
Jostein Dahl Karlsen, Ministry of Petroleum and Energy, Norway / IEA Advisory Group on Oil and Gas Technologies, France

Alex Kemp, Department of Economics, University of Aberdeen, United Kingdom
Pankaj Khanna, Teekay Shipping Ltd, Canada
David Knapp, Energy Intelligence Group, United States
Alessandro Lanza, Eni S.p.A., Italy
Jean-François Larive, CONCAWE, Belgium
Stephen O'Sullivan, UFG, Russia
Stephan Ressl, Gas Transmission Europe (GTE), Belgium
Robert Skinner, Oxford Institute for Energy Studies, United Kingdom
Michael D. Smith, BP, United Kingdom
Jonathan Stern, Royal Institute of International Affairs, United Kingdom
Klaus Otto-Wene, Consultant, Sweden

Coal

William A. Bruno, CONSOL Energy Inc., United States
Christopher Cosack, German Hard Coal Association, Germany
Hans Gruss, Consultant, Germany
Brian Heath, Coal Industry Advisory Board to IEA, United Kingdom
Malcolm Keay, World Coal Institute, United Kingdom
Mike Mellish, Department of Energy, United States
Karen Schneider, ABARE, Australia
P V Sridharan, TERI, India
Mustafa Yorukoglu, Turkish Coal Enterprises, Turkey
Hu Yuhong, China Coal Information Institute, People's Republic of China
Keith Welham, Rio Tinto plc, United Kingdom
Ross Willims, BHP Billiton, Australia

Electricity and Renewables

Laurent Corbier, World Business Council for Sustainable Development, Switzerland
Edmilson dos Santos, University of Sao Paulo, Brazil
Gurcan Gulen, Institute for Energy, Law & Enterprise, United States
Cahit Gurkok, United Nations Industrial Development Organization (UNIDO), Austria
Reinhard Haas, Technical University of Vienna, Austria
Mark Hammonds, BP, United Kingdom
Gudrun Lammers, EDF, France
Leonidas Mantzos, National Technical University of Athens, Greece
Pietro Menna, European Commission, DG-Tren, Belgium
Dean Travers, Electrabel, Belgium
Francois Verneyre, EDF, France

Yufeng Yang, Center National Development and Reform Commission, People's Republic of China

Advanced Technologies
Henrik S. Andersen, Norsk Hydro, Norway
Graham Campbell, Natural Resources Canada, Canada
John Davison, IEA Greenhouse Gas and R&D Programme, United Kingdom
Barbara McKee, US Department of Energy, United States
Olav Kaarstad, Statoil ASA, Norway
Dale Simbeck, SFA Pacific, Inc., United States
John Topper, IEA Greenhouse Gas and R&D Programme, United Kingdom

All errors and omissions are solely the responsibility of the IEA.

TABLE OF CONTENTS

List of Figures in Text

Chapter 4: Oil

Chapter 5: Natural Gas

Chapter 7: Electricity

List of Tables in Text

Chapter 6: Coal

Chapter 7: Electricity

Chapter 8: Advanced Technologies

Annex 2: Methodology

EXECUTIVE SUMMARY

The total investment requirement for energy-supply infrastructure worldwide over the period 2001-2030 is $16 trillion. This investment is needed to expand supply capacity and to replace existing and future supply facilities that will be exhausted or become obsolete during the projection period. These estimates are based on the Reference Scenario of the *World Energy Outlook 2002*, in which the global energy market is projected to grow by two-thirds over the next three decades, equal to annual demand growth of 1.7% per year. Although the total sum of investment needs is large in absolute terms, it is modest relative to the size of the world economy, amounting to only about 1% of global GDP. But the extent of the challenge differs among regions. Russia's investment requirement will amount to 5% of GDP, Africa's to 4%. The share is much lower in OECD countries.

The world's energy resources are sufficient to meet projected demand, but mobilising the investment required to convert resources into available supplies depends on the ability of the energy sector to compete against other sectors of the economy for capital. The energy-investment challenge is heightened by the fact that capital needs in the next thirty years will be much bigger, in real terms, than over the past thirty years. In the case of electricity, the investment requirement will be nearly three times greater. This makes it all the more important that investment conditions in the energy sector are right to attract the required amounts of capital.

The electricity sector dominates the investment picture: power generation, transmission and distribution will absorb almost $10 trillion, or 60%, of total energy investment. This share is more than 70% if investment in the fuel chain to meet power station fuel requirements is included. Total investments in the oil and gas sectors will each amount to more than $3 trillion, or around 19% of global energy investment. The coal industry requires only $400 billion, or 2%: supply of one unit of energy from coal is only about one-sixth as capital-intensive as producing and transporting the same unit from gas. Renewables will capture nearly a third of investment in new power plants in the OECD.

Developing countries, where production and demand increase most rapidly, will require almost half of global investment in the energy sector as a whole, even though the unit cost of capacity additions is generally lower than in the OECD. China alone will need to invest $2.3 trillion, or

14% of the world total. Capital needs will be almost as big in the rest of Asia, including India and Indonesia. Investment needs amount to $1.2 trillion in Africa and $1 trillion in the Middle East, where upstream oil and gas developments account for more than half of total investment. Russia and other transition economies will account for 10% of global investment and OECD countries for the remaining 41%. Investment needs will remain the largest in the United States and Canada — $3.2 trillion. Over 40% of total non-OECD investments in the oil, gas and coal supply chains will be devoted to projects to export those fuels to OECD countries, because most reserves are located outside the OECD. These investments will be more readily financed than investment in projects to supply domestic markets, where payments are made in local currencies.

A substantial proportion of all this energy investment is required simply to maintain the present level of supply. Oil and gas wells are depleting, power stations are becoming obsolescent and transmission and distribution lines need replacing. Much of the new production capacity brought on stream in the early years of the projection period will itself need to be replaced before 2030. In total, 51% of investment in energy production will be needed simply to replace or maintain existing and future capacity. The remaining 49% will be in capacity to meet rising demand. Primary demand for natural gas will grow fastest among the fossil fuels, at 2.4% per year. Oil demand is expected to rise by 1.6% per year and coal use by 1.4%. Electricity demand will also grow at a brisk annual rate of 2.4%, driving much of the demand for gas and coal as an input to power generation.

Extraction costs, including those incurred in exploring for reserves, will account for most of the investment in the fossil-fuel industry, though the proportion differs among fuels. Mining will absorb 88% of total coal investment — despite the fact that international trade in coal, requiring investment in port facilities and shipping, will increase faster than global demand. Similarly, exploration and development will take nearly three-quarters of total investment in oil. The share is lower for gas, at 55%, because of the higher cost of transportation. For electricity, the share of generation is even lower, at 46%. Indeed, global electricity investment in transmission and distribution — driven by the growing number of household connections in developing countries and the need to replace infrastructure in OECD countries and transition economies — will be almost as large as the total capital needed for the oil and gas industries combined.

Financing Energy Investment Cannot Be Taken for Granted

Just as global energy resources are not believed to be a constraint in absolute terms, so financial resources at a global level are sufficient to finance this projected energy investment, though the conditions need

to be right. Domestic savings — the single most important source of capital for investment in infrastructure projects — exceed by a large margin total energy-financing requirements. But in some regions, energy-capital needs are very large relative to total savings. In Africa, the share is half. And energy investment has to compete for funds which might equally well be devoted to other sectors. More important than the absolute amount of finance available worldwide, or even locally, is the question of whether conditions in the energy sector are right to attract the necessary capital. Most investors require a return related to their perceived risk. If they do not see that being achieved in the energy sector, they will invest elsewhere.

The risks faced by investors in energy projects are formidable and are changing. Those risks, which include those of a geological, technical, geopolitical, market, fiscal and regulatory nature, vary by fuel, by the stage of the fuel chain in question and by region. But the energy sector has, in most cases, been able to mobilise the required financing in the past. It will be able to do so in the future only if financing mechanisms are in place, investment returns are high enough and investment conditions are appealing.

More of the capital needed for energy projects will have to come from private and foreign sources than in the past. There has already been a marked trend away from financing energy investments from public budgets. Many governments have privatised energy businesses, both to raise money and to limit the future call on the public budget, and have opened up their markets to foreign involvement. Foreign direct investment is expected to remain an important source of private capital in non-OECD regions, particularly for oil and gas projects. Private capital flows are very sensitive to macroeconomic conditions and to the nature and stability of government policies.

Financing the required investments in developing countries is the biggest challenge revealed by this analysis. The financial needs for energy developments in transition economies and developing nations are much greater relative to the size of their economies than in OECD countries. In general, investment risks are also greater, particularly for domestic electricity and downstream gas projects. Few of these governments could fund fully the necessary investment, even if they wanted to. Poorly developed financial markets often limit opportunities for borrowing from domestic private lenders. Exchange-rate risks, economic and political instability and uncertain legal and regulatory regimes impede inward capital flows. Governments with heavy demands on their domestic budgets may be tempted to overtax exploitation of national natural resources, inhibiting investment. It is especially pressing for non-OECD countries to create an investment framework and climate that will enable them to mobilise the necessary capital.

The Power Sector Will Dominate Energy Investment

Almost $10 trillion of the total $16 trillion of capital needed in the energy industry will be for the power sector, because of relatively rapid growth in demand and the much higher capital cost of electricity per unit of energy supplied compared to fossil fuels. Around $4.5 trillion will be needed for power generation. The construction of 4,700 GW of new generating capacity, 2,000 GW of which will be gas-fired, will cost over $4 trillion. Developing countries will account for the larger part of both new capacity and investment. Over $400 billion will be spent on refurbishing existing power stations, mostly coal-fired plants located in OECD countries and the transition economies. Transmission and distribution together will call for $5.3 trillion of capital, 55% of which will be spent in the developing countries.

OECD countries will need to invest $4 trillion in their power sectors, half of it for transmission and distribution grids. Replacement of old power plants will account for much of the investment in power generation in those countries, since more than a third of today's capacity is likely to be retired over the next thirty years. More than 40% of OECD electricity investment will occur in the United States and Canada, which will remain the largest electricity market in the world. Despite the relative maturity of the system, its investment requirements will also be larger than in any region except China.

Whereas financing the electricity sector in the OECD has not been a problem until now, new doubts have now been introduced by the transition to fully competitive markets. Liberalisation increases risks to investors in power generation, especially peaking capacity. There are also uncertainties about prospective investment in transmission networks, which has lagged behind investment in generation in some OECD countries, for example in the United States and some European countries. Recent events in North America and Europe have demonstrated the importance of transmission and distribution reliability. Liberalised electricity markets require increased levels of investment in transmission to accommodate greater volumes of electricity trade. Higher investments in transmission will also be required because of increased use of intermittent renewables. The owner, operator and generator are increasingly distinct, complicating the allocation of responsibilities and network planning. Long-established siting problems persist in many places, while uncertainty about future environmental regulation is also a growing constraint on investment in electricity.

The five largest countries in the world outside the OECD — China, Russia, India, Indonesia and Brazil — will need about a third of global electricity investment. The transition economies and developing countries, together, will account for about 60%. There will be no guarantee that the developing countries will be able to finance the $5 trillion of investment in electricity which is needed to meet their projected demand, two-thirds of it in developing Asian countries. The challenge is particularly stiff in Africa. India has very difficult circumstances, too. It needs to raise $665 billion over the thirty years to 2030, equivalent to 2% of GDP every year. This will not be achievable without major reforms: the State Electricity Boards currently earn on average a negative rate of return on capital of 35% and revenues from electricity sales meet only 70% of costs. A crucial element in the reform process in India and many other countries will be to make tariff structures more cost-reflective.

More private sector involvement in developing countries will be required. How successful those countries will be in attracting private capital is one of the biggest uncertainties about future electricity investment. In fact, private investment has been declining since 1997. There are major uncertainties about when and to what extent private investment will rise again and where the new investors will come from. Renewed expansion of private sector participation will take time and will call for appropriate policies.

The rate of growth in investment and supply projected here will leave 1.4 billion people without access to electricity in 2030, a mere 200 million fewer than now. Boosting global electricity investment by just 7% would be sufficient to bring a minimal level of supply to these marginalised people. But that would mean raising another $665 billion in regions that are already struggling to raise capital. The international community will be called upon to take on some responsibility for financing the provision of basic electricity services to the very poor.

Oil Investment Will Shift away from OECD Countries

Total investment in the global oil industry will amount to almost $3.1 trillion over the projection period — $2.2 trillion, or 72%, of this devoted to exploration and development for conventional oil. Investment in non-conventional oil (including gas-to-liquids) will amount to $205 billion, or 7% of total oil investment. Tankers and pipelines will absorb investment of $260 billion (8%), driven by an 80% increase in trade between now and 2030. Investment in crude oil refining will amount to around $410 billion, or 13% of total oil investment. This will be needed to boost refinery capacity and to upgrade refineries so that their output meets the shift in the product-demand mix towards lighter and cleaner products. These estimates are derived from a projected 45 mb/d increase in global oil demand to 120 mb/d in 2030. Around one-third of global oil investment will occur in the OECD regions. But 45% of the investment outside the OECD will be in projects to supply oil to OECD countries.

About a quarter of upstream oil investment will be needed to meet rising demand. The rest will be needed to counter the natural decline in production from wells already in production and those that will start producing in the future. At a global level, upstream investment needs are, in fact, far more sensitive to changes in natural decline rates – the decline in production that would be observed in the absence of additional investment to sustain production – than to the rate of growth of oil demand. Estimated decline rates vary among regions, ranging from 4% per year in some Middle East countries to 11% in the North Sea in Europe. Offshore fields will account for almost a third of the increase in production from now to 2030, but they will take a bigger share of investment because they cost more to develop.

The share of the Middle East in total upstream spending, at less than 20%, is small relative to its contribution to the increase in global production capacity, because exploration and development costs in the region are very low. With half the world's remaining conventional oil reserves, the Middle East is projected to meet almost two-thirds of the increase in global oil demand between now and 2030. Non-conventional oil will win a significant and growing share of the market over that period, coming largely from Canada and Venezuela and accounting for about 5% of global oil investment (excluding gas-to-liquids). Capital and operating costs for such projects are high compared with most conventional oil projects, though their exploration costs are negligible.

If the projected amount of investment in the Middle East is not forthcoming and production does not, therefore, increase as rapidly as expected, more capital would need to be spent in other more costly regions. Under a Restricted Middle East Investment Scenario, in which those countries adopt policies to restrict their production growth and investments, global oil-investment requirements are 8% higher than in the Reference Scenario. World oil demand would be 8% lower because of the higher prices that would result. Nonetheless, oil revenues in Middle East OPEC and other OPEC countries would be lower, together with global economic growth. These findings imply that it will be in the interests of both consumer and producer countries to facilitate capital flows to the Middle East upstream oil sector.

Gas Investment Will Continue to Grow, though Bottlenecks Could Emerge

Cumulative investment in the natural gas supply chain over the projection period will be $3.1 trillion, more than half of it in exploration and development. This investment will be needed to compensate for the natural decline in production capacity and to meet a near-doubling of gas demand over the projection period. On average, an additional 300 bcm of new gas-production capacity will be needed each year – the equivalent of the total current gas production capacity of OECD Europe. Annual spending will increase from an average of less than $80 billion in the 1990s to $95 billion during the current decade and close to $120 billion in the third decade of the projection period. The OECD will account for half of total investment in natural gas and North America alone for well over a quarter. Outside the OECD, the transition economies will need to attract the largest amounts of capital, much of it for projects to produce and export gas to Europe and Asia.

Global investment in transmission and distribution networks, underground storage and LNG liquefaction plants, ships and regasification terminals will amount to $1.4 trillion. LNG investments will be higher than in the past, because a sixfold increase in inter-regional LNG trade will more than offset further falls in unit costs. By 2030, half of inter-regional gas trade is projected to be in the form of LNG.

Energy-market reforms, more complex supply chains and the growing share of international trade in global gas supply will give rise to profound shifts in gas-investment risks, required returns and financing costs. Securing financing for large-scale greenfield projects – especially in developing countries – will be difficult, time-consuming and, therefore, uncertain. The private sector will have to provide a growing share of investment needs, because state companies will not be able to raise adequate public finance. In many cases, only the largest international oil and gas companies, with strong balance sheets, will be able to take on the required multi-billion dollar investments. Long-term take-or-pay contracts in some form will remain necessary to underpin most large-scale projects. The lifting of restrictions on foreign investment and the design of fiscal policies will be crucial to capital flows and production prospects, especially in the Middle East, Africa and Russia, where much of the increase in global production and exports is expected to occur.

As a result of these factors, there is a great risk that investment in some regions and parts of the supply chain might not always occur quickly enough. In that event, supply bottlenecks could emerge and persist because of the physical inflexibility of gas-supply infrastructure and the long lead times in developing gas projects. Such investment shortfalls would drive up prices and accentuate short-term price volatility, which would, in turn, signal the need for more investment.

Coal Investment Will Hinge on Relative Prices and Environmental Policies

Coal investment needs, at only $400 billion over the projection period, will be much smaller than for the other fossil fuels, but will, similarly, be centred outside the OECD. Investment in coal rises to $1.9 trillion if coal-fired power stations are included. China will account for 34% of global coal investment, excluding shipping, amounting to $123 billion. The overall figure for the OECD in total is only just in excess of this, despite the importance of the North American coal market and of production in Australia. Overall, developing countries will account for more than half of the investment in coal, with the OECD and transition economies together accounting for the rest. The relatively low capital intensity of investment in the coal chain – six times less than that for gas – means that, despite the capital advantage of gas over coal in power station construction costs, the capital advantage of gas over coal in power generation diminishes and may even disappear where gas prices are high.

Tougher government action to address environmental problems could counter coal's price advantage and reduce coal demand and the need for investment. Uncertainty about future environmental policies is already pushing up required rates of return for new projects and creating a barrier to coal investment. However, continued research into clean coal technologies and carbon sequestration offers the potential for further improvements in the environmental performance of coal-fired power plants.

Environmental Policies under Consideration in the OECD Would Change Investment Patterns Dramatically

More intensive efforts than those currently made in OECD countries to cut greenhouse-gas emissions and save energy would change significantly the level and pattern of energy investment. The actions envisaged in the OECD Alternative Policy Scenario, which assumes implementation of the policies that OECD countries are currently considering, would lead to a dramatic shift in the pattern of energy investment and reduce overall energy needs. Investment in the OECD power-generation sector remains about the same, as the higher cost of renewables offsets the reduced need for new capacity, but investment in electricity transmission and distribution is reduced by almost 40%. Renewables capture one-half of power-generation investment, at a cost of $720 billion, compared to $480 billion in the Reference Scenario. Coal-mining and transportation investment falls by $25 billion, with almost half of this occurring outside the OECD because of lower demand for exports from major non-OECD producers. Demand for natural gas – a fuel with low carbon-intensity – also falls, as many new power stations which would have burned gas are not built. Demand-side investments, not covered by this analysis, would be higher.

Technology Could Dramatically Alter the Long-term Investment Outlook

Environmental considerations are a major driver of new energy technology. Advanced technologies being developed today could dramatically alter energy investment patterns and requirements in the

longer term. Carbon-sequestration technologies could increase investment in the OECD power-generation sector by up to a quarter, but they face unresolved environmental, safety, legal and public acceptance issues. These investments would come on top of the $16 trillion of total energy investment in the Reference Scenario. A small contribution by hydrogen fuel cells to electricity generation – 100 GW – towards the end of the projection period is included in that scenario, but widespread application of fuel cells in motor vehicles is not: large reductions in the cost of fuel-cell vehicles would first have to be achieved, even though competitive fuel costs can already be envisaged in an established hydrogen economy. Advanced nuclear generation systems and improved electricity transmission and distribution are other areas in which technical achievement can be expected in the long term.

Governments Action to Lower Potential Barriers to Energy Investment Will Be Vital

The role of governments in securing energy investment will continue to change, with greater emphasis being given to creating the right enabling conditions. Most governments will continue to seek greater private participation in the energy sector. Some governments will continue to finance oil and gas investment directly or through their national companies, but they will often have to pay more for their capital than major international companies. Governments everywhere will have to pay attention to how the policy, legal and regulatory framework affects investment risks and how barriers to investment can be lowered.

Governments that have promoted competitive energy markets have introduced new investment risks – alongside benefits to consumers. Many uncertainties remain about how to make competitive markets function in such a way that security of supply is ensured in a cost-effective manner, so governments need to monitor developments closely and assess the need for changes to market rules and regulations. They also need to create more stable, transparent and predictable regulatory conditions in order to enable players in competitive markets to evaluate those risks and to ensure that market structures do not impede investments that are economically viable. Some compromises will be necessary, for example on long-term take-or-pay contracts for natural gas.

Governments in many non-OECD countries continue to intervene in energy markets more directly, with potentially adverse implications for investment. The decisions of energy-producing countries, for example in relation to oil-production quotas or the terms for access to their resources, will affect greatly the attractiveness to foreign investors of investment opportunities. In many non-OECD regions, there is still a long way to go to ensure that basic principles of good governance, both in the energy sector and more generally, are applied properly and respected.

CHAPTER 1: INTRODUCTION

Objectives and Scope

The objective of this study is to assess quantitatively the prospects for investment in the global energy sector through to 2030. Specifically, it attempts to answer two questions of major interest to the energy industry, the financial community and policy-makers:

- How much capital will be needed to finance the construction of energy-supply infrastructure, including the exploration and development of resources, and the transformation, transportation and distribution of energy sources?
- What obstacles will the energy sector need to overcome to attract this investment?

A further question inevitably arises: will the obstacles be overcome to secure all the funds needed to finance fully the projected investment needs worldwide? The answer to this question varies from fuel to fuel, from region to region and according to the perspective of the reader. One purpose of this study is to stimulate debate among policy-makers, investors and the energy industry about the extent of the challenge and the actions that will need to flow from it.

Securing a reliable supply of affordable energy is crucial to the economic health of all countries, members of the International Energy Agency and non-members alike. Ensuring long-term security of energy supply requires continuing large-scale investment in every step of the supply chain. The scale of the challenge is getting larger as energy needs grow: the 2002 edition of the *World Energy Outlook* projects that global primary energy demand will increase by two-thirds over the next three decades, reaching 15.3 billion tonnes of oil equivalent in 2030. The earth's energy resources are judged adequate to meet this growth in demand, but the amount of investment needed for these resources to be exploited will be significantly larger than in the past – a central finding of *WEO-2002*. Quantifying just how much additional capital will be required in each major region, for each fuel and for each stage of the fuel-supply chain, is the first task of this study. To our knowledge, this is the first time any attempt has been made to quantify future global energy-investment needs in a detailed, fuel-by-fuel manner.

There are grounds for concern about whether all this investment will be forthcoming, both because of the extent of the increase in capital requirements and because of potential barriers to investment. In many cases, the scale of energy projects is increasing. And responsibility for investment is shifting. Traditionally, large-scale industrial projects have been dominated by state-owned firms, but private investors are playing a larger role in energy investment as countries liberalise their economies. But political and economic constraints will limit access to this alternative source of capital in a number of countries. Investment will be forthcoming only where the investment climate is sufficiently attractive to generate competitive financial returns. The investment challenge will be toughest in developing countries and in emerging market economies — where most of the increase in energy demand and almost all the increase in production are expected to occur (Figure 1.1). More generally, market reforms and new policies to address environmental concerns, together with technological developments, are adding to the uncertainties that inevitably face the energy investor. Public resistance to certain technologies and local objections to the siting of energy facilities can also impede investment. Identifying the potential barriers to investment and the uncertainties surrounding future investment flows is this study's second main goal.

Figure 1.1: **Increase in World Energy Production and Consumption**

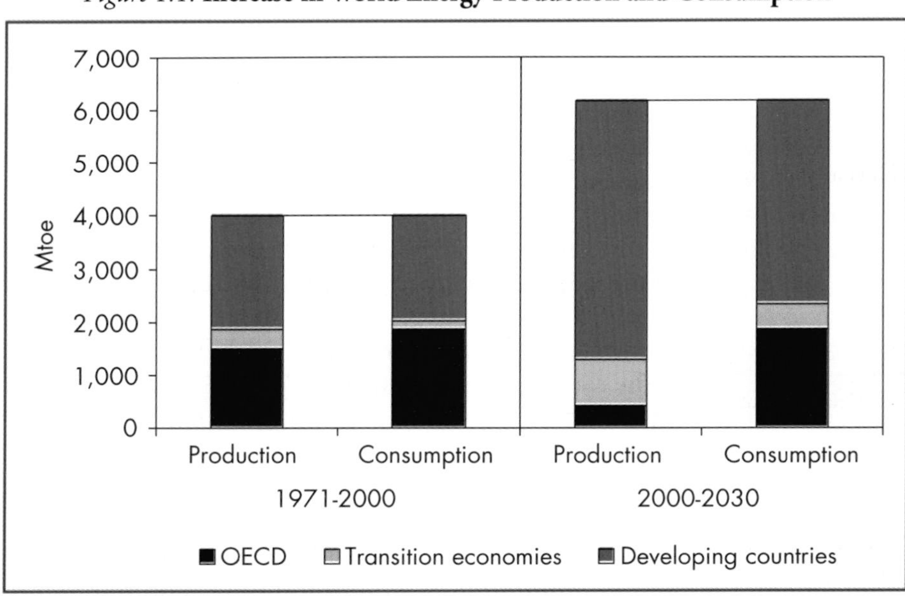

Source: IEA (2002).

The time frame of the analysis is to 2030. The study is limited to supply-side investments in the oil, gas, coal and electricity sectors. Investments in renewable sources of energy and nuclear power are included in electricity generation. But investments in the equipment and infrastructure involved in the *use* of final energy, such as boilers, machines, cars or appliances, are *not* considered. The study covers each of the *World Energy Outlook* regions (Table 1.1). Some regional markets are analysed in more detail, either because they play an important role in world energy supply or because their investment prospects are especially uncertain. These include the oil and gas sectors in the Middle East, Russia and North America, and the coal and electricity industries in China and India.

Table 1.1: **World Energy Outlook Regions**

OECD	North America	Canada and United States Mexico
	Europe	European Union-15 Other
	Pacific	Japan, Australia and New Zealand Korea
Transition economies	Russia	
	Other (former Soviet Union and Eastern Europe)	
Developing countries	China	
	East Asia	Indonesia Other
	South Asia	India Other
	Middle East	
	Africa	
	Latin America	Brazil Other

Note: Detailed regional definitions are provided in Annex 3.

The aggregate projections presented in this report do not include investments in advanced technologies such as carbon storage, hydrogen production, advanced nuclear reactors and advanced electricity transmission and distribution networks. The outlook for these technologies and their costs is nonetheless discussed in Chapter 8.

Methodological Approach

The central estimates of investment requirements in this study cover the period 2001-2030 and are derived from the projections of energy supply and demand of the *World Energy Outlook 2002* Reference Scenario. This scenario is based on a consistent set of assumptions about macroeconomic conditions, population growth, energy prices, government policies and technology.[1] Crucially, it takes into account only those government policies and measures that had been enacted as of mid-2002. Later, or potential, policy initiatives, including those aimed at reducing greenhouse-gas emissions and energy imports, are not taken into account in the Reference Scenario. However, new policies adopted since mid-2002, though locally significant, do not appreciably change the global energy outlook.

The impact on investment of the introduction by OECD countries of new policies related to climate change and energy security is analysed separately, based on the OECD Alternative Policy Scenario projections in *WEO-2002*. Basic assumptions on macroeconomic conditions and population are the same as for the Reference Scenario. However, energy prices change as they respond to the new energy supply and demand balance. The Alternative Policy Scenario differs from the Reference Scenario by assuming that OECD countries adopt a range of new policies on environmental problems, notably climate change, and on energy security.[2] It also assumes that there will be faster deployment of new energy technologies.

The quantification of capital requirements for this study, notably the compilation of unit capital cost estimates and capacity needs for each fuel, component and region, involved compiling and processing large quantities of data. A significant contribution to this work has been made by a number of organisations in the energy sector and in the financial community, acknowledged at the beginning of this report. The component cost and capacity estimates may prove to equal in value the final investment numbers. A detailed description of the methodology used for each fuel can be found in Annex 2.

The calculation of the investment requirements involved the following steps for each fuel and region:

- New capacity needs for production, transportation and (where appropriate) transformation were calculated on the basis of projected supply trends, estimated rates of retirement of the existing supply infrastructure and decline rates for oil and gas production.
- Unit capital cost estimates were compiled for each component in the supply chain. These costs were then adjusted for each year of the

1. For details, see IEA (2002). The assumptions take account of feedbacks between the different factors driving energy supply and demand.
2. See IEA (2002).

projection period using projected rates of change based on a detailed analysis of the potential for technology-driven cost reductions and on country-specific factors.[3]

- Incremental capacity needs were multiplied by unit costs to yield the amount of investment needed.

The results are presented by decade in year 2000 dollars. The estimates of investment in the current decade take account of projects that have already been decided and expenditures that have already been incurred. The convention of attributing capital expenditures to the year in which the plant in question becomes operational has been adopted. In other words, no attempt has been made to estimate the lead times for each category of project. This is because of the difficulties in estimating lead times and how they might evolve in the future.

For the purposes of this study, investment is defined as capital expenditure only. It does not include spending that is usually classified as operation and maintenance.

The final investment projections, like the underlying supply projections, are subject to a wide range of uncertainties, the most important of which are discussed at the end of Chapter 2. To analyse the sensitivity of the investment projections to particular uncertainties, alternative scenarios are presented for certain fuels and regions. These include the following:

- A scenario in which investment in **Middle East oil-production capacity** is restricted, to test the effect on global oil supply of lower production in that region and to assess the implications for investment needs.
- The investment implications of the *WEO-2002* Alternative Policy Scenario for the **electricity and coal** markets in the European, Pacific and North American regions of the OECD and for the **gas** market in the European Union.
- The additional investment that would be needed in order to provide all households worldwide with **access to electricity** by 2030. *WEO-2002* projects that, in the absence of new measures to promote electrification in poor developing countries, 1.4 billion people will still lack access to electricity in 2030 — only 200 million fewer than today.

3. Cost trends are described for each fuel in Chapters 4 to 7.

CHAPTER 2: GLOBAL ENERGY INVESTMENT NEEDS TO 2030

HIGHLIGHTS

- More than $16 trillion, or $550 billion a year, needs to be invested in energy-supply infrastructure worldwide over the three decades to 2030, an amount equal to 1% of projected gross domestic product. This investment is needed to expand supply capacity and to replace existing and future supply facilities that will be exhausted or become obsolete during the projection period. More than half of investment in energy production will be needed simply to replace or maintain existing and future capacity.
- The electricity sector alone will need to spend almost $10 trillion to meet a projected doubling of world electricity demand, accounting for 60% of total energy investment. If the investments in the oil, gas and coal industries that are needed to supply fuel to power stations are included, this share reaches more than 70%. Total investments in the oil and gas sectors will each amount to more than $3 trillion, or 19% of global energy investment. Coal investment will be almost $400 billion, or 2%.
- Almost half of total energy investment will take place in developing countries, where production and demand are expected to increase most. Russia and other transition economies will account for 10% and OECD countries for 41%. Over 40% of non-OECD investments in the oil, gas and coal supply chains will be devoted to projects to export those fuels to OECD countries.
- China alone will need to invest $2.3 trillion, or 14% of the world total. Capital needs will be almost as big in the rest of Asia, including India and Indonesia. Africa has investment needs of $1.2 trillion. OECD North American investment needs remain greatest — $3.5 trillion.
- The weight of energy investment in the economy varies considerably across regions. It will amount to a mere 0.5% of GDP in the OECD. Its share is much larger in non-OECD regions. In Russia, the annual average investment requirement will exceed 5% of GDP. Africa needs to allocate 4% of GDP to energy investment on average each year.
- Investment in energy infrastructure is a key driver of economic growth. Energy-investment requirements will account for a significant share of total domestic investment in Russia and other transition economies, Africa and the hydrocarbon-rich Middle East.

- Between primary fossil fuels, the capital intensity of investment varies considerably. Natural gas is about six times more capital-intensive than coal for the unit of equivalent energy supplied. This is a powerful consideration for capital-constrained countries.
- Future investment needs are subject to many uncertainties, including macroeconomic conditions, energy prices, environmental policies, geopolitical factors, technological developments and the pace and impact of market reforms.

Overview

To meet projected demand growth of 1.7% per year over the next three decades,[1] $16 trillion, or almost $550 billion a year, will need to be invested in global energy-supply infrastructure.[2] This is equal to around 1% of projected global GDP and 4.5% of total investment on average. Capital needs will grow steadily through the projection period. The average annual rate of investment is projected to rise from $455 billion in the decade 2001-2010 to $632 billion in 2021-2030 (Table 2.1). This compares with estimated energy investment of $413 billion in 2000. Actual capital flows will fluctuate around these levels according to project and business cycles.

Table 2.1: **World Energy Investment** ($ billion in year 2000 dollars)

	2000	2001-2010	2011-2020	2021-2030	Total 2001-2030	Share of total 2001-2030 (%)
Oil	87	916	1,045	1,136	3,096	19
Gas	80	948	1,041	1,157	3,145	19
Coal	11	125	129	144	398	2
Electricity	235	2,562	3,396	3,883	9,841	60
Total	**413**	**4,551**	**5,610**	**6,320**	**16,481**	**100**
Annual average	**413**	**455**	**561**	**632**	**549**	**100**

1. See IEA (2002) for the detailed demand and supply projections that underlie the analysis of future investment needs in this study.
2. Investment in this study includes capital expenditure on the following: exploration and development, refining, tankers, pipelines and non-conventional oil production facilities for the oil sector; exploration and development, liquefied natural gas facilities, transmission and distribution pipelines and underground storage facilities for the gas sector; mining, shipping and ports for the coal sector; and power stations and transmission and distribution networks for the electricity sector.

This investment will be needed to replace hydrocarbon reserves and existing and future supply facilities that will be exhausted or retired during the projection period and to expand production and transport capacity to meet demand growth. For the energy sector as a whole, 51% of investment in production will be simply to replace existing and future capacity. The rest will be needed to meet the increase in demand. The share of investment to replace existing and future production capacity is highest for oil upstream, at 78%, followed by gas upstream (70%) and coal mining (65%). In the electricity sector, investment to replace power plants will be 30%. Table 2.2 details developments in key aspects of the global supply infrastructure for each fuel.

Table 2.2: **Global Energy Supply and Infrastructure**

		Units	**2000**	**2030**	**Annual average rate of growth** (%)
Oil	Production*	mb/d	75	120	1.6
	Refining capacity	mb/d	82	121	1.3
	Tanker capacity	million DWT	271	522	2.2
Gas	Production	bcm	2,513	5,280	2.5
	Transmission pipelines	thousand km	1,139	2,058	2.0
	Distribution pipelines	thousand km	5,007	8,523	1.8
	Underground storage working volume	bcm	328	685	2.5
Coal	Production	Mt	4,595	6,954	1.4
	Port capacity**	Mt	2,212	2,879	0.9
Electricity	Generating capacity	GW	3,498	7,157	2.4
	Transmission network	thousand km	3,550	7,231	2.4

* Oil production includes processing gains.
** The port capacity figures in this table do not include the capacity required for internal coal trade in China, India and Indonesia (cumulative additions for internal trade are 192 Mt of capacity).

Investment Outlook by Fuel

The electricity sector will account for 60% of total energy investment over the projection period. If those investments in the oil, gas and coal industries that are needed to supply fuel to power stations are included, the share of electricity reaches more than 70% (Figure 2.1). Total oil investment will make up 19% of global energy investment, gas a further 19% and coal a mere 2%.

Figure 2.1: **Cumulative World Energy Investment by Energy Use, 2001-2030**

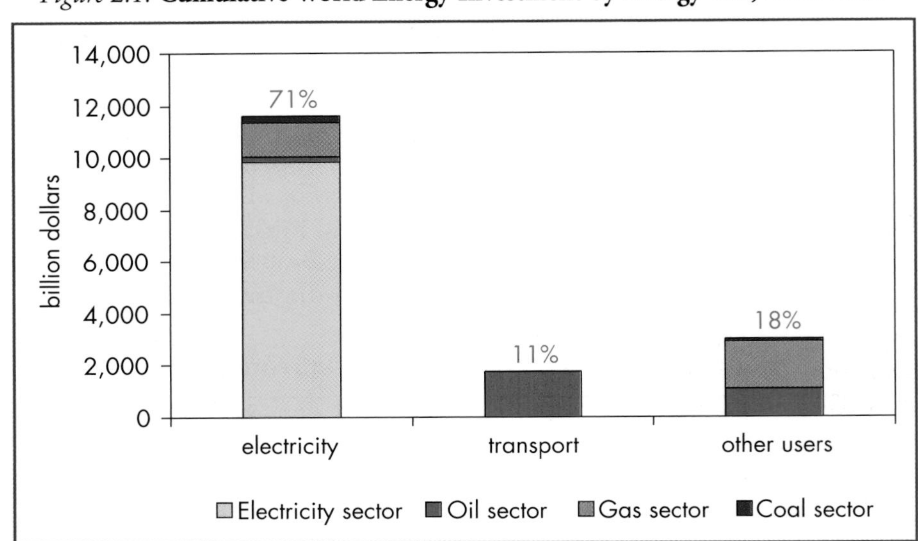

Of the primary fossil fuels, natural gas is the most capital-intensive. Each tonne of oil equivalent of gas supply will call for an average of almost $28 of investment in exploration and development, transportation and storage, compared to $22 for oil and less than $5 for coal (Figure 2.2).[3]

Figure 2.2: **Capital Intensity of Fossil Fuel Supply, 2001-2030**

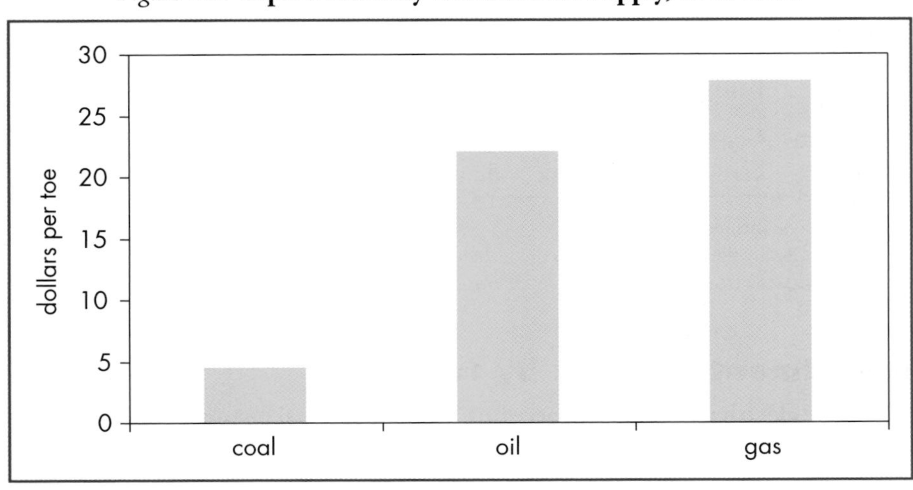

3. Total cumulative investment in all stages of the energy supply chain divided by total production increase between 2001 and 2030.

Oil

The projected increase in world oil demand from 77 mb/d in 2002 to 120 mb/d in 2030 will require more than 200 mb/d of new production capacity to be brought on stream. Most of the new capacity will be needed to replace depleted wells that are already producing or wells that will be brought into production and subsequently depleted during the next three decades. Bringing all this capacity on stream will entail upstream investment of $2.2 trillion. Inter-regional trade in crude oil and refined products is projected to increase by around 80% over the period 2001-2030.[4] The construction of tankers and pipelines will cost $257 billion. Investment in refining, with new capacity concentrated in the Middle East and Asia, will be $412 billion dollars. Development of non-conventional oil projects, which are expected to contribute over 8% to total world oil supply by 2030, will cost $205 billion, mainly in Canada and Venezuela.

OECD countries will account for the largest share of total oil investment (31%), followed by the Middle East (18%), the transition economies (16%) and Africa (13%). Nonetheless, the largest addition to production capacity will take place in the Middle East, where the unit capital cost of exploration and development per barrel is only a quarter that of OECD countries. The impact on global oil supply and investment needs of potential constraints in Middle East oil production is analysed in a Restricted Investment Scenario in Chapter 4.

Natural Gas

Production of natural gas, the fastest growing fossil fuel in the future, will need to rise from 2.5 tcm/year in 2000 to 5.3 tcm/year in 2030. This will mean adding a cumulative total of 9 tcm of capacity over that period, since much of the additional capacity will be needed to replace existing and future wells that will be depleted during the next three decades. Only a third will be needed to meet rising demand. Total investment requirements in the gas sector will reach $3.1 trillion. Exploration and development spending will account for 55%, costing $1.7 trillion. More than tripling of physical inter-regional gas trade by 2030 will call for rapid growth in cross-border supply infrastructure. Investment in high-pressure transmission, the liquefied natural gas (LNG) supply chain and local distribution networks to supply gas to power plants and final consumers will also grow in all regions, totalling $1.4 trillion worldwide over the projection period.

4. Total international trade will be even larger because of trade between countries within each region and re-exports.

Almost 30% of global gas investment will take place in OECD North America, which has the largest and most mature gas market. As a result, OECD countries in total will account for nearly half of gas sector investment. The share of the transition economies will be 16%, and that of China and East Asia together, 9%. Investment requirements in each of these three regions are higher than those in the Middle East. This is because unit exploration and development costs are higher in these regions and because transmission and distribution pipelines are longer.

Coal

Total investment in the coal production chain over the next three decades will be much smaller at $398 billion, or $13 billion per year, despite the fact that coal will account for 24% of world primary energy supply in 2030. This reflects the higher labour intensity of the coal sector, the much lower cost of extracting coal compared to oil and gas, and the expectation that increasing market competition and consolidation in the main producing countries will result in lower unit capital costs. Mining will account for the bulk of investment requirements (88%). China alone will absorb 34% of world coal investment, followed by the OECD countries — mainly North America and Australia. The main uncertainty facing coal investment is the impact of additional environmental policies on demand, which is analysed in an Alternative Policy Scenario in Chapter 6.

Electricity

Global electricity-sector investment will be nearly $10 trillion. World electricity demand will double by 2030, and installed power generation capacity will increase from 3,498 GW in 2000 to 7,157 GW in 2030. As over 1,000 GW of existing plants will be retired over the next three decades, a total of some 4,700 GW of generating capacity needs to be built, costing $4.1 trillion. In many countries, investment in transmission and distribution will need to be even greater than that in power generation, in contrast to past patterns. Transmission networks will be extended by 3.7 million kilometres worldwide. The global electricity sector needs investment of $1.6 trillion in transmission and $3.8 trillion in distribution networks.

Some 60% of electricity sector investment needs to arise outside the OECD. Asia will account for 36% of global electricity investment, more than half of which will be in China. OECD North America has the second-largest investment requirements, followed by OECD Europe. Replacement of old plants is a major feature of investment in OECD countries. The investment implications for OECD electricity markets of a range of policies and measures under consideration in order to curb greenhouse-gas emissions and save energy are assessed in an Alternative Policy Scenario in Chapter 7. The additional investment that would be needed in order to provide universal electricity access by 2030 is also quantified.

Investment Outlook by Region

As nearly 70% of the increase in world primary energy demand and almost all the growth in energy production between 2001 and 2030 will occur in developing countries and the transition economies, energy-investment needs will be greatest and increase most rapidly in those regions (Table 2.3). Almost half of total energy investment, or $7.9 trillion, will take place in developing countries, and 10% ($1.7 trillion) in the transition economies.[5] The OECD countries will account for the remaining 41% ($6.6 trillion), of which more than half will go to North America. The unit cost of capacity additions in developing countries and transition economies, especially for oil and gas production, is generally lower than in the OECD.

Table 2.3: **Cumulative Energy Investment by Region, 2001-2030**
($ billion in year 2000 dollars)

	2001-2010	2011-2020	2021-2030	Total 2001-2030
OECD North America	1,062	1,179	1,247	3,488
OECD Europe	650	717	697	2,064
OECD Pacific	381	333	287	1,000
Total OECD	**2,093**	**2,228**	**2,231**	**6,552**
Russia	269	391	389	1,050
Other transition economies	168	221	233	622
Total transition economies	**438**	**612**	**622**	**1,672**
China	578	787	888	2,253
Other Asia (including India)	489	689	876	2,055
Middle East	268	332	444	1,044
Africa	248	393	567	1,208
Latin America	339	440	558	1,337
Total developing countries	**1,923**	**2,641**	**3,332**	**7,897**
Inter-regional transportation	97	129	134	360
Total world	**4,551**	**5,610**	**6,320**	**16,481**
Annual average	**455**	**561**	**632**	**549**

5. The regional investment requirements exclude seaborne coal carriers, oil tankers and LNG carriers, as it is not clear where the investments in those assets will occur. This exclusion extends to inter-regional oil pipelines.

Although the bulk of energy investment will occur in developing countries and the transition economies, over 40% of total non-OECD investments in the oil, gas and coal supply chains will be devoted to projects to export those fuels to OECD countries (Table 2.4). This is because most fossil-fuel reserves are located outside the OECD. About a third of the investment in the oil sector will be related to exports from non-OECD countries to the OECD. Indeed, that investment will be larger than the investment within OECD countries themselves. The share in global investment of export investments in non-OECD countries is 21% for natural gas and 8% for coal.

Table 2.4: **Cumulative Energy Investment by Fuel and Market, 2001-2030**
($ billion in year 2000 dollars, % share by market)

	In OECD countries		In non-OECD countries for OECD markets		In non-OECD countries for domestic and other non-OECD markets		Total	
	$ billion	%	$ billion	%	$ billion	%	$ billion	%
Coal	147	37	31	8	220	55	398	100
Oil	892	29	1,001	32	1,203	39	3,096	100
Gas	1,523	48	646	21	976	31	3,145	100
Electricity	4,036	41	0	0	5,806	59	9,841	100
Total	**6,598**	**40**	**1,678**	**10**	**8,205**	**50**	**16,481**	**100**

Note: Investment in shipping is included.

Investment requirements in China, which are projected to account for nearly 20% of world incremental energy demand, will amount to $2.3 trillion, equivalent to 14% of world energy-investment needs (Figure 2.3). The share of investment is lower than that of demand because much of the increase in primary energy demand will be met by indigenous coal, which is less capital-intensive than other fuels, and imported oil and gas. Investment in other Asian countries together is almost as high, driven mainly by the power sector. India and Indonesia account for much of the region's investment requirements. African energy investment, at $1.2 trillion, is the third-largest of the non-OECD regions, after China and other Asian countries (excluding India), because of the relatively high cost of developing oil and gas reserves and of continuing electrification. Both Russia and the Middle East will need to invest around $1 trillion, or 6.5% of total world energy investment.

Figure 2.3: **Cumulative Energy Investment by Region, 2001-2030**

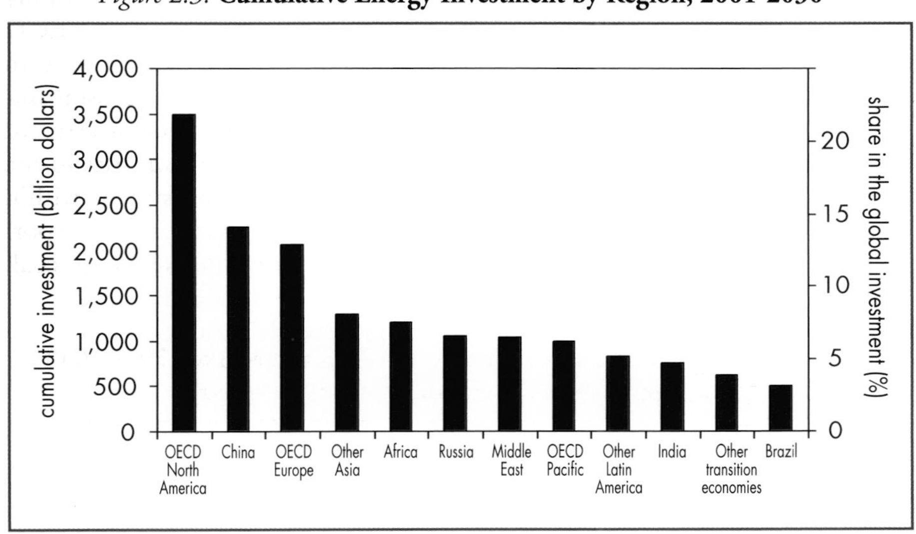

The breakdown of projected investment by fuel varies considerably among the regions according to their resource endowment and the expected evolution of the fuel mix (Figure 2.4). The electricity sector will account for 50% or more of energy investment in all regions except the Middle East and Russia.

Figure 2.4: **Fuel Shares in Cumulative Energy Investment by Region, 2001-2030**

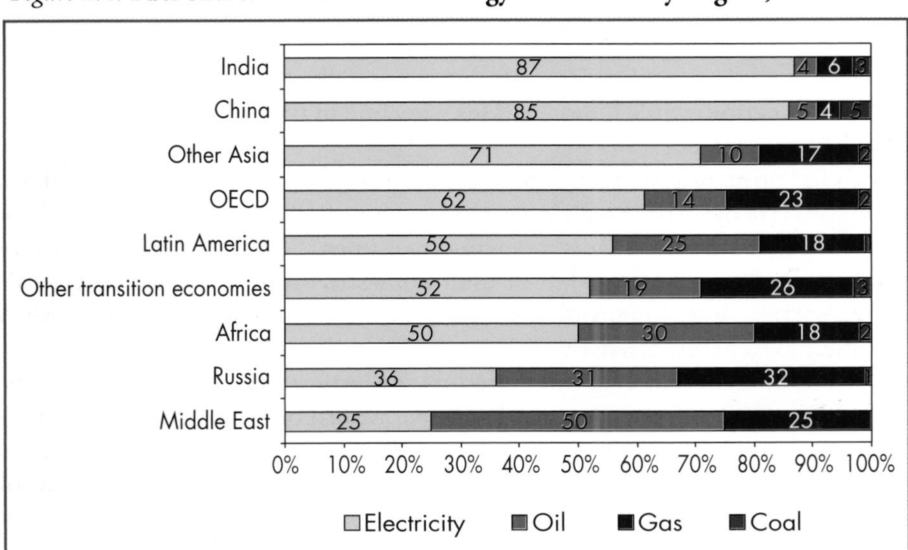

In India and China, 85% or more of total energy investment will go to this sector. The equivalent share is 71% in other Asia, 56% in Latin America and 62% in the OECD as a whole. Rapid growth in electricity demand and the high capital cost of power generation, transmission and distribution explain this pattern.

Investment in the oil and gas sectors combined is larger than that in the power sector in the Middle East and Russia, where the bulk of the world's hydrocarbon resources are located and where much of the increase in global production will occur. In the Middle East, 50% of energy investment will go to the oil sector and another 25% to the gas sector. In Russia, the share of oil and gas investment combined is also high, at 63%, partly because of the higher unit capital costs of new wells.

Share of Energy Investment in the Economy

Projected global energy investment of $16 trillion equates to 1% of global GDP on average over the next thirty years.[6] The proportion is expected to fall slightly over the projection period, from 1.1% in the current decade to 0.9% in the decade 2021-2030. The share of energy investment in GDP varies significantly across regions. It will remain much higher in developing countries and the transition economies than in the OECD countries. The share is highest in Russia, averaging more than 5% between 2001 and 2030, followed by Africa (4.1%), other transition economies (3.6%), and the Middle East (3%). Energy investment will amount to only 0.5% of GDP in the OECD (Figure 2.5). Although global energy investment is projected to grow during the next three decades, its share of GDP will evolve differently in different countries. For example, it is expected to increase in Africa, which needs to accelerate investment in electrification, especially towards the end of the projection period, while China will see a decline in the share of GDP devoted to energy investment.

The projected share of energy in total global domestic investment averages 4.5% over the projection period.[7] There is also a wide variation across regions (Figure 2.6). In Russia, energy investment in the period 2001-2030 needs to be equivalent to 31% of total domestic investment. In Africa, the figure is 21%. Investment needs in the energy sector in the hydrocarbon-rich Middle East and the other transition economies represent more than 15% of total investment. These figures underline the importance of the energy sector in the future economic growth and national wealth of these regions.

6. Total cumulative investment divided by cumulative world GDP (in year 2000 dollars at market exchange rate) between 2001 and 2030.

7. It is assumed that total domestic investment will increase in line with GDP.

Figure 2.5: **Share of Energy Investment in GDP by Region, 2001-2030**

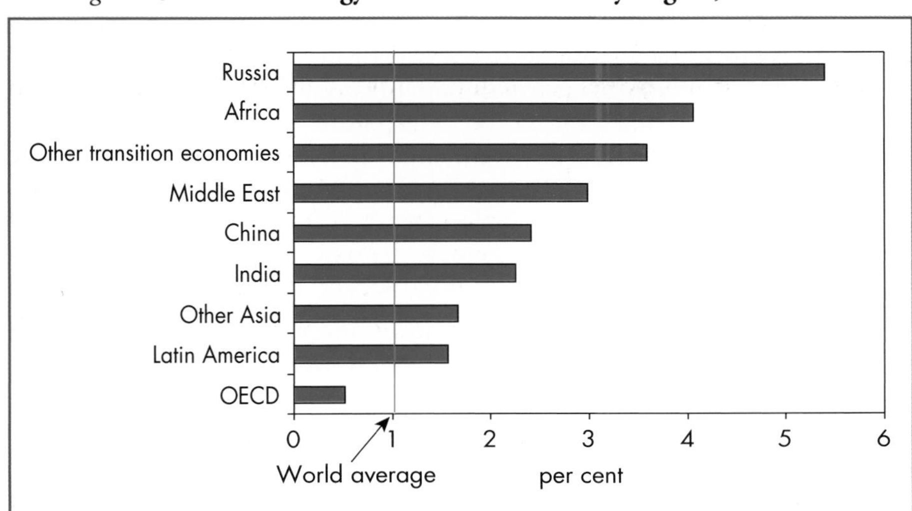

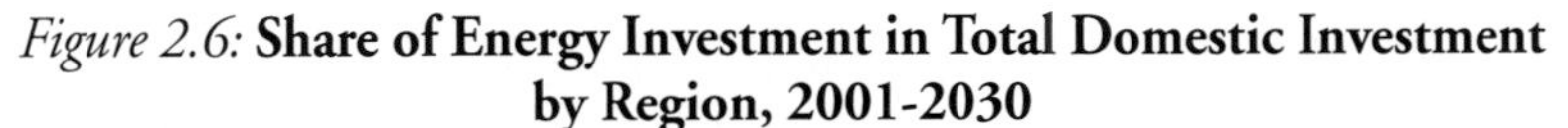

Figure 2.6: **Share of Energy Investment in Total Domestic Investment by Region, 2001-2030**

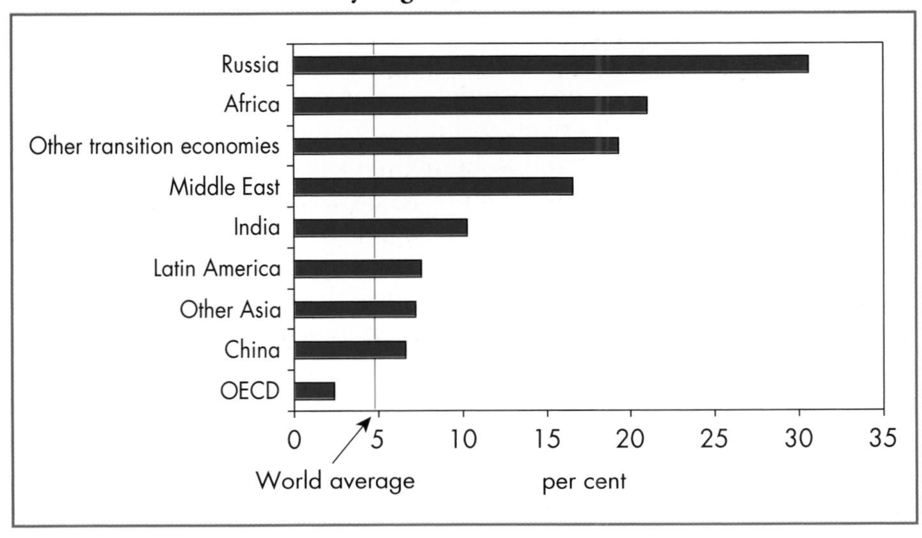

Principal Uncertainties

Future investment needs will be determined by the rate of growth in energy demand and the amount and cost of supply capacity needed to meet the increase in demand and to replace old plants. These variables depend on a wide range of inter-related factors, many of which are very uncertain.[8] There are also doubts about the return on energy-sector investments compared to other types of investment, affecting the ability of energy companies to compete with other companies to raise the necessary finance for future investments (see Chapter 3). The major sources of uncertainty include macroeconomic conditions, energy prices, environmental policies, geopolitical factors, technological developments and government policies towards the energy sector. These factors are discussed briefly below.

Macroeconomic Conditions

The pace of economic activity is clearly the most important driver of energy demand. If global GDP grows faster or slower than the 3% per annum rate assumed in the *WEO-2002* Reference Scenario, demand too will grow faster or slower. Rates of economic growth at the regional and country levels could differ substantially from those assumed, especially over short periods. Growth prospects for the transition economies, China and Africa are perhaps the least certain. The impact of structural economic changes, including the gradual shift from manufacturing to service activities, and the effects of technological advances on demand for energy are also very uncertain.

Energy Prices

Energy prices both affect and are determined by the demand for energy and the economics of investing in energy supply. Oil prices continue to influence the prices of gas and, to a lesser extent, coal through inter-fuel competition in end uses. The interplay of oil demand and supply remains hard to predict, not least because of uncertainties about future OPEC production and pricing policies. Higher oil prices would choke off demand to some extent, while making investments in production capacity more attractive, at least temporarily. The Middle East and the former Soviet Union, where the greater part of the world's remaining reserves is located, are expected to meet most of the increase in demand for oil over the coming three decades. Since production costs in the Middle East are still the lowest in the world, that region is particularly well placed to meet much of the increase in oil demand. But the increasing market dominance of the biggest Middle East producers could lead to a shift in their production and investment policies in pursuit of higher crude oil prices. Taken too far, such a policy could depress demand for oil while also encouraging investment in other countries both in conventional and non-conventional oil production and alternative forms of energy. On the other hand, lower oil prices would curb such investments and enable the lowest-cost producers to augment their market shares.

8. A detailed discussion of demand-side factors can be found in *WEO-2002*.

Short- and medium-term volatility in energy prices also plays a role in energy-investment decisions. In principle, investors will be more reluctant to invest at times of volatile prices, since the uncertainty associated with future returns is higher. There is empirical evidence that the recent increase in oil-price volatility has indeed raised the cost of capital and curtailed marginal investments.[9] The growing importance of short-term trading of gas and electricity, together with more volatile spot and futures prices, is also raising the cost of capital for utilities. The development of inter-regional trade in LNG is expected to lead to convergence of regional gas prices, which may help reduce price volatility.

The impact on pricing and investment of liberalisation of gas and electricity markets, typically involving the privatisation of utilities and the introduction of competition in production and supply, is another source of uncertainty. In competitive markets, private investors will invest only if prices are high enough and predictable enough to recover and reward the initial investment taking account of risk. Under past monopoly conditions, price risks were usually carried largely by consumers. The new circumstances may raise the cost of capital. Lower prices that may result from deregulation would also constrain utilities' operating cash flows and their ability to borrow capital for future investment. Political intervention to cap prices below full-cost levels can make matters worse.

Environmental Policies

The production, transformation and consumption of energy give rise to environmental problems, such as the emission of CO_2 and local pollutants like SO_2, which are of growing concern worldwide. Environmental policies and regulations affect energy investment directly, by requiring or encouraging the installation of cleaner technologies and, indirectly, by altering energy demand and changing the fuel mix.

New measures to mitigate greenhouse-gas emissions could have a major impact on the level and pattern of energy investment in many countries. The Alternative Policy Scenario in *WEO-2002* showed that the new policies aimed at curbing CO_2 emissions that OECD countries are considering would reduce their oil demand by 9.5%, gas demand by 12.5% and coal demand by 26% in 2030, compared to the Reference Scenario. Such demand reductions would significantly affect energy supply, international energy-trade flows and, therefore, investment in energy-supply infrastructure, especially in non-OECD countries. The impact on investment trends of these additional policies is discussed in the relevant chapters.

9. The results of the empirical analysis reported in IEA (2001) point to a strong inverse relationship between investment and volatility, implying that an increase in volatility results in a decline in investments and *vice versa*.

Geopolitics

Political instability can be a major barrier to investment. The oil and gas industries are subject to acute geopolitical risks and uncertainties because of the concentration of resources in regions susceptible to instability and the growing role of trade in global energy supply. Geopolitical uncertainties in the Middle East have increased further following the events in Iraq in early 2003 and the continuing Israel-Palestine conflict. Civil unrest in other areas of the world, including Nigeria, Venezuela and Indonesia, has led to disruptions in oil and gas exports and raised new doubts about future investment.

Given a reasonable degree of stability, the Middle East, the transition economies and Africa will account for 47% of oil investment and more than 30% of gas investment over the next 30 years. Geopolitical tensions in these regions could delay or halt investment in some projects, as higher risk premia push up the cost of capital. Under-investment there would drive up prices and increase the viability of higher-cost projects in other regions. The impact of slower oil investment in the Middle East than projected in the Reference Scenario is analysed in Chapter 4. Electricity-sector investment is also sensitive to geopolitical uncertainty, especially in developing countries, where private and foreign capital needs to play a growing role.

Technological Developments

Technological advances, by lowering unit capital or operating costs, typically make investments more attractive. But the degree and pace of unit capital cost reductions due to technology development are very hard to predict. Improvements in one sector can adversely affect the economics or investment in another. Demand-side technologies also affect supply-side infrastructure investments, through their impact on demand growth and the fuel mix. Supply-side technological developments, which may be of particular significance, include the following:

- New seismic techniques are improving the success rates of oil and gas drilling. New production and engineering technologies, including ultra-deepwater offshore and non-conventional production techniques, are expected to reduce development costs.
- High-pressure technology is expected to play a major role in reducing the unit capital cost of large-scale, long-distance pipeline projects.
- Further advances in combined cycle gas turbine (CCGT) technology will increase the thermal efficiency of new gas-fired power plants. CCGT plants are smaller and quicker to build than coal-fired plants and nuclear reactors, making it easier to attract private financing. Small-scale, highly efficient generating plants will also boost the role of distributed generation.

- Advances in coal technology will also increase the thermal efficiency of new coal-fired power plants, and reduce their capital costs, particularly those of integrated gasification combined-cycle (IGCC) plants.
- New LNG production and shipping technologies are expected to continue to drive down unit costs.
- Further technological advances and unit cost reductions could increase the economic viability of gas-to-liquids and coal-to-liquids production.
- The use of advanced coal-mining technology will continue to lower the capital and operating costs of coal extraction and preparation.
- The capital costs of renewable energy technologies are expected to fall substantially, but the range of projected unit cost reductions is wide and depends heavily on government policies and the pace of deployment (see Chapter 7).
- Hydrogen-based fuel cells, carbon sequestration and storage technologies, and advanced nuclear reactors could radically change the energy supply outlook and investment patterns in the longer term (see Chapter 8).

These trends have been taken into account in the core investment analysis in this study.

Government Energy Policies

Governments affect energy investments in a multitude of ways, directly and indirectly. Major sources of uncertainty, other than environmental policies, include the following:

- The general legal and regulatory framework governing trade and investment affects the opportunities for and attractiveness of energy investment. Policies on foreign direct investment can be a key uncertainty for energy-investment prospects, particularly for power generation in developing countries and upstream oil and gas development in the Middle East, Russia and Africa. Improvements in governance would encourage capital flows into developing and emerging market economies. Borrowing constraints may mean that countries with coal reserves rely more heavily on indigenous coal, the extraction of which is much less capital-intensive than oil and gas.
- Energy-sector reforms, bringing increased competition and efficiency, may reduce unit investment needs. Privatisation can expand sources of finance. But the reform process, if poorly managed, can add to uncertainty and raise the cost of capital. Where the state remains the sole owner of energy companies, investment may be constrained by the availability of public funds.

- Energy taxation and subsidy policies can be a critical uncertainty for demand prospects and for potential investment returns. The removal of subsidies could result in lower demand for fuels and energy services, but may also build confidence that markets are becoming freer and less distorted. The stability of the fiscal regime is a major issue for investors in the upstream oil and gas industry.
- International co-operation, including the development of bilateral and multilateral legal instruments aimed at minimising investment risk and the cost of capital, can play an important role in stimulating investment flows to the transition economies and developing countries.

Although the Reference Scenario assumes that current energy and environmental policies remain unchanged at both national and regional levels throughout the projection period, the pace of implementation of those policies and the approaches adopted are nonetheless assumed to vary by fuel and by region. For example, electricity and gas market reforms are assumed to move ahead at varying speeds among countries and regions. Similarly, progress will be made in liberalising energy investment and reforming energy subsidies, but will be faster in OECD countries than in others. In all cases, energy taxes are assumed to remain unchanged.

CHAPTER 3: FINANCING GLOBAL ENERGY INVESTMENT

HIGHLIGHTS

- Global financial resources are ample to cover energy investments of more than $16 trillion over the next thirty years. But conditions in the energy sector have to be right to capture the required share of that finance in a competitive market for capital.
- Ease of access to capital will vary among countries, energy sub-sectors and types of investment. Raising capital for energy projects in OECD countries has not been a problem in the past, but uncertainties arising from deregulation and probable greater price volatility create a new hurdle.
- The situation in developing countries and the transition economies is less secure. Risks are higher there and capital needs are greater relative to the size of the economy. Export-oriented projects will be more readily financed than projects to supply domestic markets in non-OECD countries.
- Energy investment needs relative to domestic savings — the main source of financial resources — are particularly high in Africa.
- The private sector will be called upon to play an increasing role in financing energy investments in all regions, as governments retreat from direct ownership and intervention. Foreign direct investment (FDI) is expected to remain an important source of private capital in non-OECD regions. All types of private capital flows are very sensitive to macroeconomic conditions and to the nature and stability of government policies.
- Financial markets in non-OECD countries are often poorly developed. This limits access to capital, especially long-term capital, and is likely to be a more important constraint than the absolute level of available funds.
- Particularly because income is generated in hard currencies, the oil and upstream gas industries worldwide will be able to raise capital relatively easily, unless abnormal risks arise. Major international oil companies have strong balance sheets to underpin efforts to raise capital and should have few difficulties. Independent upstream companies face greater difficulties; and a shift to deepwater and emerging areas, such as West Africa, will tend to push up the price of capital. Major oil-producing countries and their national companies can expect to have to pay more for their funds than the international oil companies.

- Few governments in developing countries and the transition economies will be able to finance the large-scale investments required in domestic electricity and downstream gas, while borrowing from domestic lenders will be constrained by poorly developed financial markets. Access to international capital will be limited by the exchange-rate risk, deficiencies in the legal and regulatory systems and more general fears of economic or political instability.
- OECD countries will face few such problems, but liberalisation and deregulation are creating new uncertainties. Prospects in the short term have been damaged by the difficulties experienced by merchant power companies.
- Consolidation in the coal industry and improved operational and financial performance where markets are competitive will ease the financing of coal projects, provided environmental fears are not reinforced.
- Governments play a big role in determining whether or not energy projects can be financed, even where they are not directly involved in the financing. They set the general conditions which determine the extent of economic, political and legal risks.

This chapter discusses how the energy sector as a whole, and its different sub-sectors in different parts of the world, might fare in securing capital for investment purposes. It reviews how energy investments are currently financed and considers in qualitative terms how financing sources and mechanisms might change in the future. The first section reviews the availability of financial resources for energy investment, and considers the particular characteristics of energy investment, both generally and in developing countries and the transition economies specifically. The second and third sections analyse recent trends in and prospects for energy-investment financing. The last section looks at the role of governments in promoting energy investment.

The Energy Sector's Access to Capital

How is the $16 trillion of capital that is expected to be needed for energy investment over the period 2001-2030 to be mobilised? Globally, capital is available on the necessary scale. But ease of access to capital varies among sectors and regions. Access to capital for particular energy projects might be constrained by an absolute insufficiency of financial resources, underdeveloped financial markets or the expectation of inadequate returns on investment relative to the risks, when compared to alternative use of the capital.

These potential constraints are of particular significance for developing countries and the transition economies. Financial resources in those regions are more limited, yet, as Chapter 2 demonstrates, their energy-investment needs will be especially large compared with OECD countries — both in absolute terms and relative to the size of their economies. The overall question posed by this chapter is whether there will be *sufficient* capital available, *on appropriate terms*, to finance all the identified energy investment needs *in a timely fashion*.

Financial Resources

Domestic Savings

Domestic savings are the main source of capital for infrastructure projects, including energy, in most countries. The share of domestic savings in gross domestic product (GDP), therefore, provides one important measure of the overall size of capital available domestically in the economy. In 2000, savings accounted for nearly 23% of global GDP. By comparison, energy-investment needs are projected at only 1% of projected GDP. The total amount of capital available worldwide from this source will, therefore, be some twenty times larger than energy-investment needs. But the ratio of domestic savings to energy investment varies considerably among regions (Figure 3.1). Given competition from other economic sectors, there could be local problems in attracting capital from domestic savings into energy investment on the required scale.

In OECD countries, domestic savings, at 23% of GDP, are more than 40 times higher than their energy-investment needs. Outside the OECD, countries in Asia, especially East Asia, have high domestic savings rates, which have underpinned high investment and rapid economic development over the past decades. The share of savings in China, at nearly 40%, is among the highest in the world, and is 15 times the average annual energy-investment requirement (2.4% of GDP).[1] Russia, which needs to spend 5.4% of its GDP on energy investment, also has a high domestic savings rate, 37%. India has a savings rate of 20%, but energy-investment needs are relatively high at 2.2% of GDP. Thanks to abnormally high oil prices in recent years, the Middle Eastern countries on average have the third highest savings rate worldwide. Domestic savings are below average in the transition economies (other than Russia),

1. Given that China lacks a well-established social security system, the savings actually available for investment could be much smaller than the figures suggest.

Figure 3.1: **Energy Investment and Domestic Savings as a Percentage of GDP by Region**

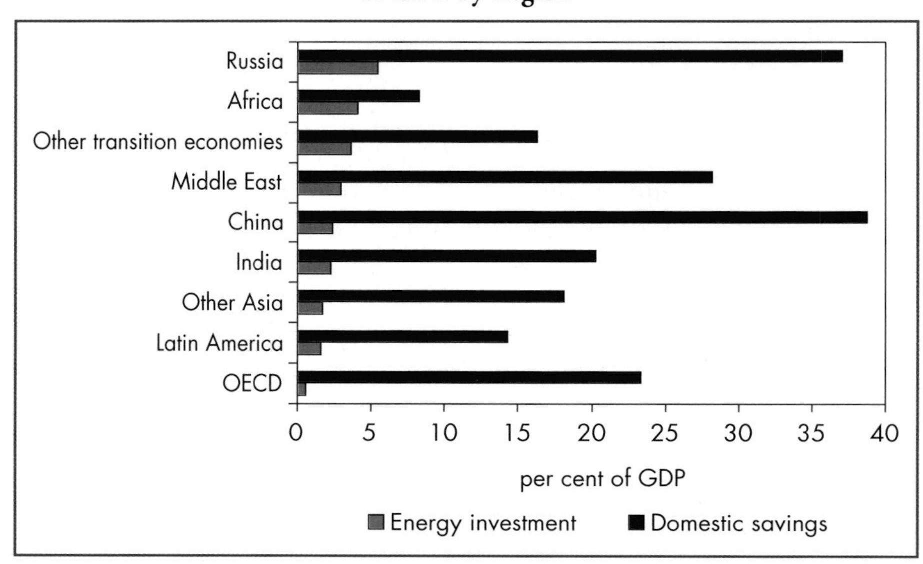

Note: Domestic savings as a percentage of GDP is based on 2000 data. The share of energy investment in GDP is based on projected averages for the period 2001-2030.
Source: World Bank (2003a); IEA analysis.

where energy-investment needs will be equal to one-fifth of domestic savings, and in Latin America. Availability of domestic capital will be most constrained in Africa, where the demand for capital for energy investment alone will amount to almost half of total savings. This casts doubts over whether it will be possible to expand the region's energy infrastructure as rapidly as projected, even if international capital inflows and savings rates increase.

Even where domestic savings comfortably exceed the energy sector's thirst for capital, energy companies will still have to compete with other sectors for domestic financial resources. In addition, energy investment can involve occasional mega-projects and the excess of domestic savings over energy-investment requirements could be much smaller at times. Furthermore, domestic savings need to be mobilised through financial markets.

Figure 3.2 shows the balance between total domestic investment and domestic savings, the so-called IS balance, which is a common indicator of the availability of financial resources for *all* types of investment in the economy. There will certainly be strong competition for financial resources among sectors in all countries. In addition, capital outflows could limit the resources available for domestic use, intensifying competition for capital, unless there are offsetting inflows of foreign capital.

Figure 3.2: **Domestic Investment and Savings as a Percentage of GDP by Region, 2000**

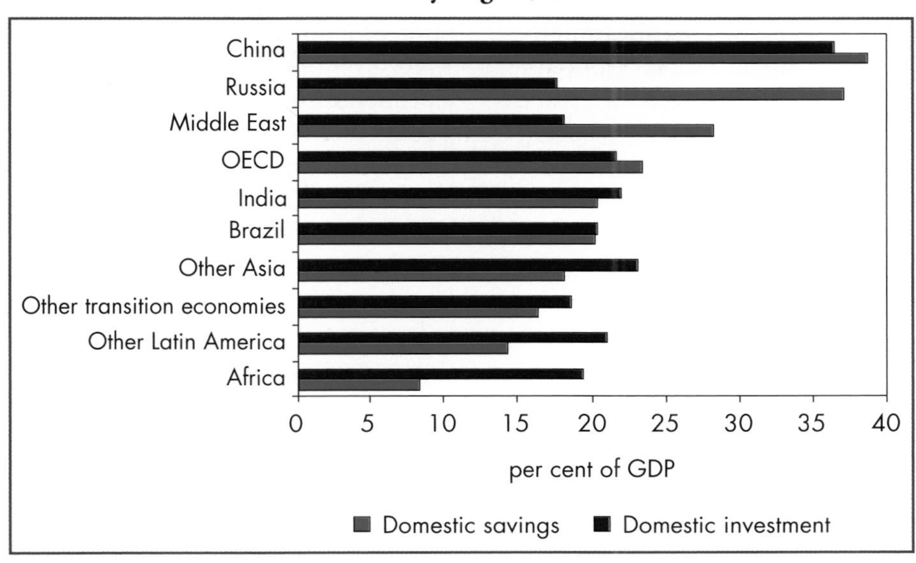

Source: World Bank (2003a).

Domestic savings are lower than total domestic investment in many developing countries and the transition economies. Among the five biggest non-OECD countries, India and Brazil have such a deficit. The IS gap is over 2 percentage points in the transition economies (excluding Russia) and nearly 7 percentage points in Latin America (excluding Brazil). Domestic savings are less than half of total domestic investment in Africa, where the need to expand access to modern energy is most pressing. In all of these countries or regions, the electricity sector, which will face greater difficulties in raising international capital than the oil and gas industry, will account for most energy-investment needs.

China, with strong support from the government, makes use of its abundant domestic savings to finance very high rates of domestic investment (36% of GDP), which have been a basis of its rapid economic growth in recent years. In Russia, there appears to be under-investment in manufacturing sectors and non-energy infrastructure, and the energy sector could be exposed to greater competition for capital once investment in those sectors recovers. Countries in the Middle East, which have experienced high domestic savings rates in recent years as a result of high oil prices, will face a shortage of domestic savings in the future if oil prices decline. Their strong growth in population is unlikely to be matched by employment opportunities, and increasing social expenditure will pre-empt an increasing portion of government expenditure.

External Finance

The shortfall between investment requirements and domestic savings in some developing countries and the transition economies highlights the need to mobilise capital inflows from abroad, especially from OECD regions, where domestic savings exceed investment. Dependence on external finance brings both benefits and risks. Financing from abroad often reduces the cost of capital and provides longer debt maturity, since international financial markets are usually better organised, more competitive, and have a large base of investors and lenders. At the same time over-dependence on foreign investment flows can destabilise an economy. Overseas capital inflows can be volatile and currency depreciation can increase the debt burden of borrower countries if the revenues generated by the investment are mainly in local currency, which is generally the case of investment in the electricity and downstream gas sectors.

Countries that already have a large external debt will face difficulties in securing additional foreign capital. Among non-OECD regions, the ratio of total external debt[2] to gross national product (GNP),[3] which assesses the sustainability of a country's debt-service obligations, is highest in Africa, followed by the transition economies and some Asian countries (Figure 3.3). Although there are no

Figure 3.3: **Total External Debt as a Percentage of GNP in Non-OECD Regions, 2000**

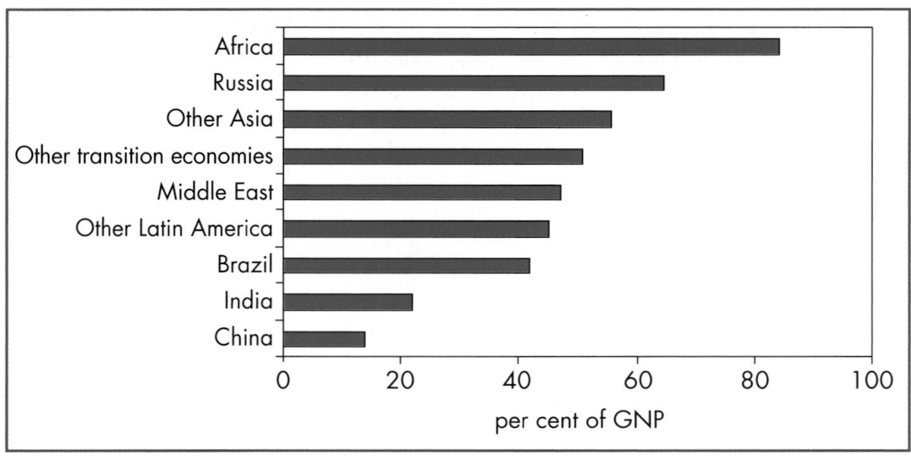

Source: World Bank (2003a).

2. Total external debt is debt owed to non-residents repayable in foreign currency, goods, or services. It is the sum of public, publicly guaranteed, and private non-guaranteed long-term debt, use of International Monetary Fund credit and short-term debt.

3. A major difference between GDP and GNP is that GNP includes the dividends and workers' remittances that the country receives from abroad, which can be used to service external debt, and excludes those paid by businesses operating in the country, which cannot. Thus, GNP is a more appropriate variable to measure the country's ability to service external debt.

hard and fast rules on the acceptable level of external indebtedness, only those countries with rapidly growing economies and/or exports are normally able to sustain rising levels of debt.

Competition for foreign capital has increased since the early 1990s, as market reforms, including privatisation, have increased opportunities for private investors to participate in energy projects. This has exacerbated the task of securing funding for new investment in developing countries. The role of foreign direct investment in energy-investment financing is discussed in more detail later in the chapter.

Financial Markets

The lack of appropriate mechanisms in domestic financial markets tailored to the needs of energy projects is likely to be more important than the absolute level of funds as a constraint on financing energy investment, especially in many non-OECD countries. Access to financial resources and, consequently, the capital structure of companies are strongly affected by the stage of development of a country's financial market.[4] Countries with more developed financial sectors provide better access for companies to equity, bonds and borrowing. Long-term debt, which is more suited to energy and other capital-intensive infrastructure projects, is usually available only in deep and sophisticated financial markets. The ratio of short-term debt to total debt is typically much higher in countries with underdeveloped financial markets. Large companies in countries with active stock markets and small companies in countries with large banking systems have longer debt maturity. Companies with access to international markets can make themselves more attractive to investors and lenders by signalling their commitment to higher standards of corporate governance, enabling them to lengthen their debt maturity structure.

The absolute size of financial institutions is usually correlated with income, as is the level of their annual financing activities. The main exception is China, where the banking sector relative to GDP is larger and more active than in the OECD. The activity of the banking sector, measured as the level of domestic credit provided by banks relative to energy-investment needs, is highest in the OECD, China and Brazil (Figure 3.4).[5] Banks are much less active in the transition economies and Africa, where the share of energy-investment requirements in the economy is expected to be higher. Further development of

4. See, for example, IMF (2002); Demirguc-Kunt and Maksimovic (1996); Schmukler and Vesperoni (2001); and World Bank (2003b).

5. Domestic credit includes that provided by foreign banks, so the actual lending capacity of domestic banks is often weaker than indicated by this measure, especially in non-OECD regions. For example, local banks in Saudi Arabia, whose banking sector is the largest in the Middle East, can lend only up to about $500 million to any one hydrocarbon project.

Figure 3.4: **Energy Investment Needs and Domestic Bank Credit by Region**

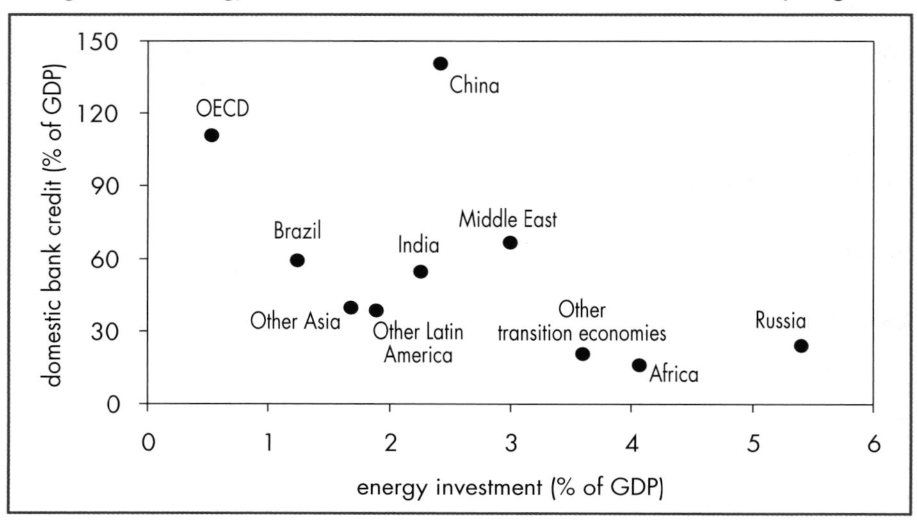

Note: Bank credit as a percentage of GDP is based on 2000 data, while energy investment as a percentage of GDP is the average of projected investment during the period 2001-2030.
Sources: World Bank (2003a); and IEA analysis.

the banking sector in these countries will be vital. The degree of concentration of the banking sector, defined as the ratio of the assets of the three largest banks to total banking sector assets, is much higher in the transition economies, Africa and South Asia, indicating a lack of competition among banks.

Stock and bond markets are also much less developed in non-OECD regions, though their size and scope have increased over the last three decades.[6] In Russia, India, other Asia, and Latin America (excluding Brazil), the size of the stock market, measured as the value of listed shares divided by GDP, is between 30% and 40% that of the OECD in aggregate. China and Brazil have relatively large stock markets. The size of Africa's stock markets relative to GDP is only 20% that of the OECD. In developing countries and the transition economies (excluding China, Brazil, and other Asian countries), stock market activity, measured as total shares traded divided by GDP, is less than one-tenth that of the OECD. The size of markets for bonds (particularly private ones) in non-OECD regions is also smaller. Partly because of this, long-term private debt issues are limited.

6. Financial sector development usually starts with the development of the banking sector, followed by that of capital markets. However, the pace and path of development depend on government policy, as well as historical factors. For example, among OECD regions, the financial market in North America is oriented to equity and bonds, while the Pacific region has a bank-dominated financial market. European financial markets lie somewhere between the two.

Less developed financial markets in developing countries and the transition economies mean that financial services to energy investors are of lower overall quality in those countries. Long-term local currency financing, which is often sought by electricity and gas utilities, is not always available. Short-term financing might be the only available option, often in foreign currencies, exposing electricity and gas utility companies to high refinancing and foreign exchange rate risks. Moreover, since there are few instruments in less developed financial markets to mitigate such risks, the supply of private financing is limited and the cost of capital tends to be higher.

Characteristics of Energy Investment

Access to capital depends on the risk and reward profile of the investment concerned, as well as on the availability of financial resources and mechanisms. For the energy sector to attract adequate funding for investment, it must offer terms and rates of return which compare favourably with those offered in other sectors, taking into account the different risk profiles. The amount and type of investment and the financing mechanisms used vary enormously according to the energy sub-sector, the stage of the supply chain within the same sub-sector (for example, upstream and downstream oil and gas), the choice of technology and the location of investment.

Nature of Energy Investment

Energy projects are usually more capital-intensive than projects in most other industries, involving large initial investments before production or supply can begin. The electricity sector is the most capital-intensive of all the major industrial sectors, measured by capital investment per unit of value added (Figure 7.4 in Chapter 7). On average, the electricity sector requires two to three times as much investment as manufacturing industries, such as automobile manufacturing, in order to generate one dollar of added value. Oil and gas extraction, processing and refining are also relatively capital-intensive.

The more capital-intensive an industry, the more exposed it is to financial risks such as changes in interest rates and other events in financial markets. High capital intensity also gives rise to investment cycles and the associated risk of over- or under-building of capacity. High energy prices usually bring forth more investment until such time as excess capacity emerges, pushing down prices and, in turn, discouraging investment.

Energy assets also have a relatively high degree of specificity: that is, once built, they cannot be moved to different locations, and, in the case of electricity and gas utility companies, their services are bound to specific domestic markets. They often have very long operating lives, more than 30 years in many cases, whereas revenue streams in the long term may be highly unpredictable. This is especially true for electricity and gas transmission and distribution networks. In addition, some energy investments, such as nuclear power plants and hydroelectric plants, have long lead times, requiring a substantial grace period before starting to repay debt.

Energy assets are usually part of a complex supply chain, with interdependencies between energy projects and between energy and other projects. Risks add up along this chain. For example, a gas pipeline and associated upstream facilities may be built to supply a gas-fired power plant. Obtaining financing for the gas infrastructure is normally contingent on there being a long-term gas supply contract with the power project. Investment in coal mining usually requires investment in railway or port infrastructure. If producers cannot be sure that the transport network has sufficient capacity for their production, they will prefer to invest in old fields equipped with adequate infrastructure, rather than in development of new fields or mines.

Given these characteristics, energy investments are normally financed with long-term capital. The basic logic behind this is the investor's need to match as closely as possible the maturities of both the assets and liabilities sides of the balance sheet. In general, longer-term debt reduces the risk that a project's cash flow might fall short of the amounts required to service debt obligations.[7] Debt maturity is more important to project finance, where debts have to be repaid from the project's cash flow and lenders have no, or limited, recourse to the assets of the sponsoring company.

Risks Associated with Energy Investment

Energy investments are exposed to differing types and degrees of risk, with consequences for the cost and allocation of capital. The higher the risk associated with an investment, the higher the cost of capital and the higher the return required by investors and lenders. The investment projections in this *Outlook* are based on the assumption that the risk profiles of energy projects will not change dramatically in the future.

The amount of risk involved in any energy project and its significance vary, depending on the scope of the project: planning, construction, start-up, and operation. Risks arise not only from the project itself but also from changes in the domestic and international investment environment, such as economic conditions, political circumstances and energy policies. The identification, evaluation and mitigation of risks are key steps in securing financing for energy projects. Broadly speaking, there are four types of risks associated with energy investment: economic, political, legal and *force majeure* (Table 3.1).

7. Dailami and Leipziger (1997) show how the probability of default increases when debt maturity is shortened.

Table 3.1: **Risks in Energy Investment**

	Type of risk	Examples
Economic risk	Market risk	• Inadequate price and/or demand to cover investment and production costs • Increase in input cost
	Construction risk	• Cost overruns • Project completion delays
	Operation risk	• Insufficient reserves • Unsatisfactory plant performance • Lack of capacity of operating entities • Cost of environmental degradation
	Macroeconomic risk	• Abrupt depreciation or appreciation of exchange rates • Changes in inflation and interest rates
Political risk	Regulatory risk	• Changes in price controls and environmental obligations • Cumbersome administrative procedures
	Transfer-of-profit risk	• Foreign exchange convertibility • Restrictions on transferring funds
	Expropriation/ nationalisation risk	• Changing title of ownership of the assets
Legal risk	Documentation/ contract risk	• Terms and validity of contracts, such as purchase/supply, credit facilities, lending agreements and security/collateral agreements
	Jurisdictional risk	• Choice of jurisdiction • Enforcement risk • Lack of a dispute-settlement mechanism
Force majeure risk		• Natural disaster • Civil unrest/war • Strikes

Profitability or return on investment is the key determinant of any investment and financing decision. A shortfall in project revenues is always a major risk. If oil and gas prices fall, oil and gas companies, reacting to reduced cash flow, typically cut upstream investments, shifting the available capital to projects in regions where investment and operation risks are low enough to offset lower prices and revenues. Increasing energy price volatility is likely to lead to an increase in the cost of capital (see Chapter 2).

Geological factors, such as the rate of depletion of the resource base, add complexity to investment in the energy sector, as capital is often secured against oil and gas reserves. Large sums must be spent at the outset, to compensate for uncertain returns over a long period. If oil-well productivity is lower than expected or gas reserves fail to come up to expectations, the investor's ability to recover his capital may be compromised.

While project developers, investors and lenders can reduce economic risks, political and legal risks are often outside their control. With increased reliance on foreign capital and private participation in energy investment, political risk is attracting more attention. Governments may change the regulatory and economic framework at any time in a way that substantially affects a project's financial viability. Domestic and foreign investors may lack confidence in commercial contracts if policies, laws and regulations (including those concerning taxation, foreign exchange, energy pricing and state ownership of energy assets) are unstable. Uncertainty about market reforms, such as changes in subsidies and taxes, and the unbundling and privatisation of state companies, and doubts about whether a level competitive playing-field will be established impose additional risks for energy investment. In China and Russia, mining, oil and gas companies have to deal with uncertain distribution of authority between central and local governments, both of which may influence the project and the revenue distribution.[8] Even contract terminology can be a risk: it differs from country to country. For example, reserves need to be carefully defined to avoid disputes. Servicing foreign debt or equity depends on broader government policies, such as attitudes to capital mobility and currency convertibility.

Various techniques to mitigate energy project risks have been developed. Risks can be identified and allocated to the different parties through agreements and contracts such as fixed price contracts and performance guarantees. Contractors and equipment suppliers are usually bound to pay compensation if they fail to fulfil their obligations, such as completing construction work on time and to budgeted cost. Fuel-supply and power-purchase contracts in the power generation sector and take-or-pay agreements in the gas sector are standard ways of reducing market risk. Production-

8. Asian Development Bank (1999).

Box 3.1: **Environmental Risk and Energy Investment**

Environmental risk, a type of regulatory risk, is becoming much more important in energy-investment decisions, as concerns about the damage to the local and global environment caused by energy production and consumption are growing among investors and residents. Energy companies have been facing increasing costs to meet more stringent environmental regulations in both OECD and some developing countries, such as China. Cleaning up oil spills and gas leakages can be very expensive. Investors may hesitate to provide funds to energy projects which operate to low environmental standards, because of concerns about public opinion and potential litigation.[9]

sharing agreements have become a common basis for large-scale oil and gas investments in some emerging market economies. While formal guarantees can be provided by host governments and by multilateral/bilateral development institutions, financial contracts, such as forward contracts, futures contracts and option contracts, are increasingly used to reallocate price risks.

Financial Performance of the Energy Sector[10]

The financial track record of any company affects its access to financial markets and cost of capital and, therefore, the attractiveness of a given investment. In addition to qualitative factors, such as management capability and experience of project development, investors and lenders look at financial variables, such as the prospective return on investment and the company's capital structure, in evaluating investment risks. Cross-sector comparisons of financial variables provide insights into the relative attractiveness of investment in energy.

9. Ten large international banks have recently adopted "Equator Principles", social and environmental guidelines for project financing in emerging markets, which are applied by the International Finance Corporation to its loans and investment.

10. Data for this analysis are based primarily on Standard & Poor's Compustat Global Database. It should be noted that the data capture the profitability and capital structure of the firm's *overall* activities, including non-energy activities. Also, the database covers only those companies listed on stock markets. Many state-owned companies and special purpose companies, including independent power producers (IPPs) and joint ventures for oil and gas projects, are not included.

Box 3.2: **Definitions of Financial Variables Used in this Outlook**

The measures of financial performance particularly discussed in this chapter in order to evaluate the relative attractiveness of the energy sector to lenders and investors are the following:

Return on investment (ROI) is one of the major profitability variables and is defined here as operating income divided by invested capital (all sources for the long-term financing of a company, the majority of which is shareholders' equity and long-term debt). As operating income is calculated before the deduction of interest and dividend payments, this definition of ROI measures the total return from a company's business in relation to the total money invested in it in the form of borrowing or equity.[11]

Debt equity ratio, often called leverage, is the ratio of total debt to the sum of shareholders' equity and total debt. A company with a higher debt equity ratio tends to be perceived to be riskier, though there is no definite level beyond which the ratio should not go.

Debt maturity is the ratio of short-term debt, including long-term debt due in one year, to total debt, and measures the maturity structure of a company's debt – the lower the ratio, the more the company is dependent on long-term debt.

Further relevant definitions are contained in Annex 3.

Profitability

Profitability is the key factor in a company's ability to raise finance for investment, whether on a corporate or a project basis. If the capital employed in a company is not generating an adequate return, the company will have limited access to new capital, as investors and lenders seek more profitable opportunities elsewhere.

Figure 3.5 shows, by industry, the average ROI achieved over the period 1993-2002.[12] Companies whose stocks are publicly traded are included here, both in the OECD and in non-OECD regions.[13] The ROI achieved in the

11. Earnings before interest, tax, depreciation and amortisation (EBITDA), which are closer to cash flows generated from a company's operation, are sometimes used as a numerator instead of operating income. Comparisons of EBITDA-based ROI yield similar results to those derived from operating-income-based ROI that are presented here.

12. Investors and lenders take investment decisions based on the *future* expected return, while profitability/risk indicators such as ROI capture the *actual* return.

13. In non-OECD countries, where state-owned energy companies are common, a large part of investment is financed through official government borrowing and government budgets. However, as policies shift towards more private sector involvement, state companies are increasingly seeking private capital in the same way as listed companies.

Figure 3.5: **Average Return on Investment* by Industry, 1993-2002**

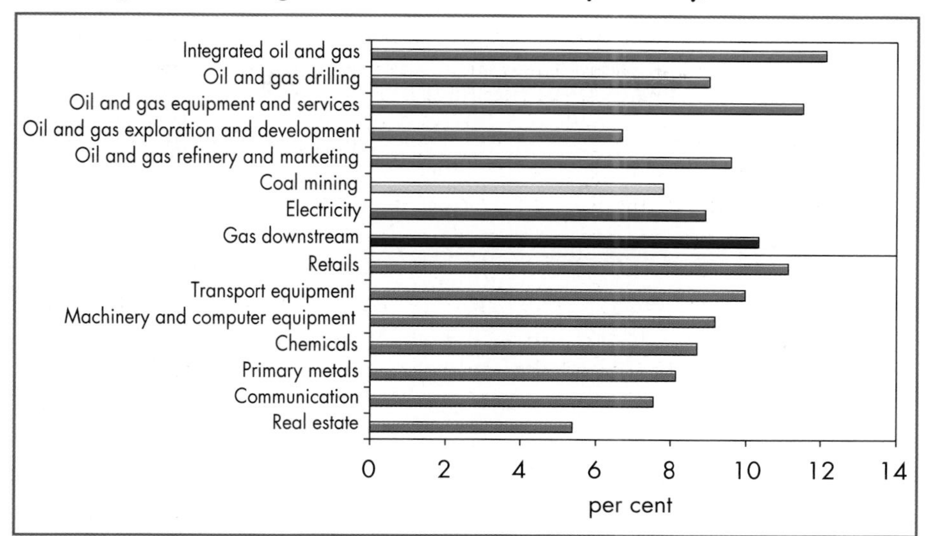

* Operating income divided by invested capital.

energy industry, while varying widely, was often higher than that achieved in other industries. Among the different segments of the oil and upstream gas sectors, integrated companies, including the major international companies, had the highest ROI, 12%, while the ROI of independent companies in exploration and production, which were less risk-averse than integrated companies, was little more than half that, at the low end of the range for all industries.[14] Coal companies' average ROI was also lower, but a wide variation among coal companies emerged in the late 1990s, when restructuring of industry got underway in several countries. Coal companies in countries with less competitive markets tend to have a lower ROI.

The volatility of investment returns is another factor taken into consideration by investors and lenders: the higher the volatility, the higher the cost of capital. Electricity and downstream gas companies have had the most stable ROI of all industries: the volatility of their returns has been, on average, just one-quarter to 40% of that of major manufacturing industries (Figure 3.6).[15]

14. Prices of goods such as oil and real estate show spikes and plunges, making it difficult to measure profitability under "normal conditions". For example, the ROI of oil and gas companies jumped after 2000 as a result of higher oil prices.

15. Gas downstream includes transmission and distribution pipelines, underground gas storage and LNG liquefaction plants, regasification terminals and ships.

On the other hand, the ROI of oil and upstream gas companies, especially exploration and development companies and oilfield equipment and service companies, is relatively volatile, owing to swings in international oil prices. This complicates their investment planning and tends to lead to more conservative investment policies. There is evidence that reduced capital spending by oil-services companies, resulting in less availability of oilfield equipment, has at times forced oil and gas companies to scale back their investment programmes.[16] The volatility of oil prices and, therefore, that of investment returns is expected to remain high and might even rise in the medium term, which would drive up the cost of capital.[17]

Figure 3.6: **Volatility* of Return on Investment by Industry, 1993-2002**

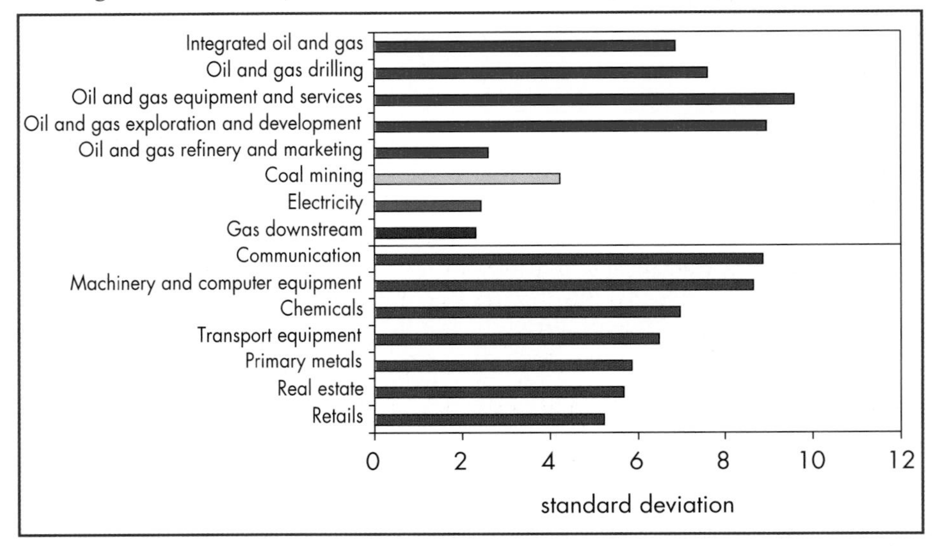

* Measured by the standard deviation of ROI over the period 1993-2002. Standard deviation measures how widely actual values are dispersed from the average.

The higher profitability of integrated oil and gas companies compared with independent oil and gas exploration and development companies and service companies reflects the nature and diversity of their assets and operations. With geographically diverse assets in all parts of the supply chain, integrated oil and gas companies are markedly less vulnerable to specific events than independent companies. Breadth of operations also allows integrated companies to reduce their cost of capital, as the overall risk to investors and lenders is an average of all

16. IEA (2001).
17. See Chapter 2.

the different investments.[18] Independent companies sometimes find that project finance gives them better access to capital, by apportioning risks precisely among stakeholders through complex and well-defined agreements.

Like major international oil and gas companies, vertically integrated electricity companies, with several power stations and large networks, are often better placed financially than independent power producers. The returns on investment of electricity and downstream gas companies are implicitly or explicitly guaranteed by regulations in many countries, keeping investment risk low. But competition in supply has tended to drive down the ROI of electricity companies in the United States, the United Kingdom and Australia (see Chapter 7), as productivity gains have not kept pace with falling electricity prices, squeezing profit margins. This has reduced the ability of companies in those markets to raise funds to finance investment. A similar trend is developing for electricity companies in non-OECD regions, as the initial cost-cutting associated with privatisation and deregulation has run its course and increased competition has driven down prices and profit margins.

Data relating to merchant power companies are not included in the data set behind Figures 3.5 and 3.6. Many such companies in liberalised electricity markets in OECD countries are facing severe financial difficulties due to large operating losses and debts. Deficiencies in the design and implementation of market reforms as well as flaws in their business model have contributed to these problems. The collapse of Enron has led the markets to favour asset-based companies once again. Many electricity companies in non-OECD countries, especially state-owned companies, are also in trouble. In India, for example, the State Electricity Boards (which are not listed and therefore not included in the data set underlying this analysis) continue to make heavy losses, since tariffs are set too low to cover the cost of generation. As a result, they find it very difficult to service debt and finance new investment (see Chapter 7).

The financial performance of coal companies has improved in recent years thanks to restructuring. The benefits of diversification would be expected to apply to coal companies whose primary business is not coal production, though this is not captured by this financial analysis owing to the data constraints. Non-coal operations, such as steel and electricity, often allow these companies to raise funds for coal investments more cheaply than their coal operations alone could support (see Chapter 6).

Capital Structure

Capital structure affects a company's exposure to financial risks, such as swings in interest rates. In practice, capital structure variables reflect both the

18. Jechoutek and Lamech (1995).

company's past performance and the degree of development of financial markets. Where financial markets are poorly developed, the availability of debt, especially long-term, may be limited, forcing companies to rely more on short-term debt and equity.

The debt equity ratio has been rising slowly in recent years in the energy sector. For electricity and downstream gas companies, total debt exceeds equity. The opposite is the norm in other industries (Figure 3.7). This is partly due to the capital-intensive nature of the electricity and gas industries. Low interest rates in recent years have also induced electricity and downstream gas companies to increase debt levels relative to equity. The ratio for integrated oil and gas companies is also higher than for other industries. It increased over the past ten years, reflecting greater reliance on bank borrowing and bonds and a shift away from reliance on internal cash flows.[19] Coal companies have a low proportion of debt in their capital structure on average, but the picture varies widely over time and across companies. This partly reflects the recent consolidation process. Increased reliance on debt, with its associated interest payments, will tend to reduce energy companies' ability to deal with financial shocks. And the cost of capital could increase once a company's indebtedness has reached a certain level.

Figure 3.7: **Debt Equity Ratio* by Industry, 1992-2001**

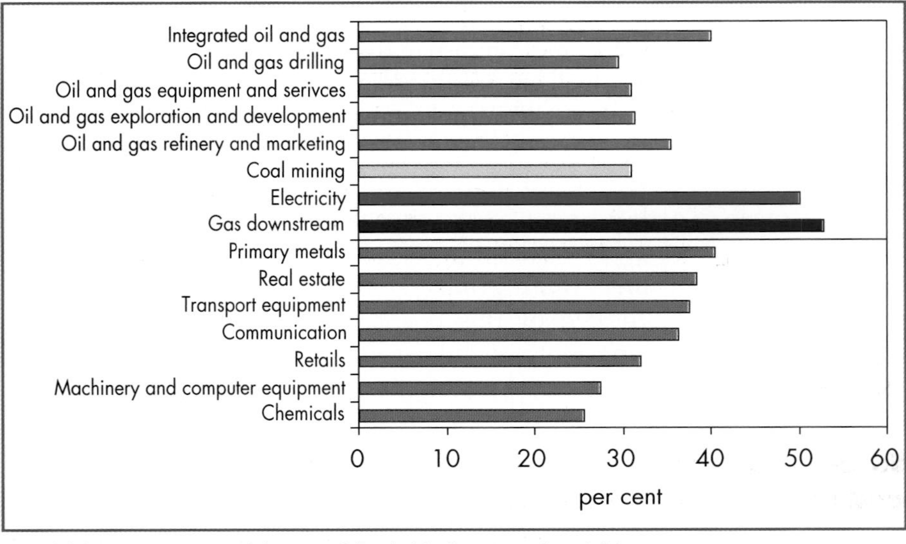

* Total debt as a percentage of the sum of shareholders' equity and total debt.

19. During the last two to three years, some major oil companies reduced their debt and repurchased their shares using strong cash flow, causing their debt equity ratios to decline.

One major determinant of future energy-investment financing will be the availability of long-term debt. Debt maturity — the ratio of short-term debt to total debt — is generally low for energy companies, reflecting the need to balance the maturity of their debt with the lives of their assets (Figure 3.8). Long-term debt accounts for more than 80% of the total debt of electricity and downstream gas companies, a higher share than in most other industries. Long-term debt accounts for 50% to 60% of total debt in most manufacturing industries. Among energy companies, integrated oil and gas companies have the highest reliance on short-term debt, which they use to supplement their internal cash flows. Independent exploration and development companies have much lower debt maturity. The debt maturity of coal companies lies between electricity and gas utilities and manufacturing industries, reflecting their greater capital intensity compared to non-energy sectors and their higher labour intensity than other energy sub-sectors. Longer debt maturity reduces the demand for early cash flow from a project and refinancing risks associated with short-term debt.

Figure 3.8: **Debt Maturity* by Industry, 1992-2001**

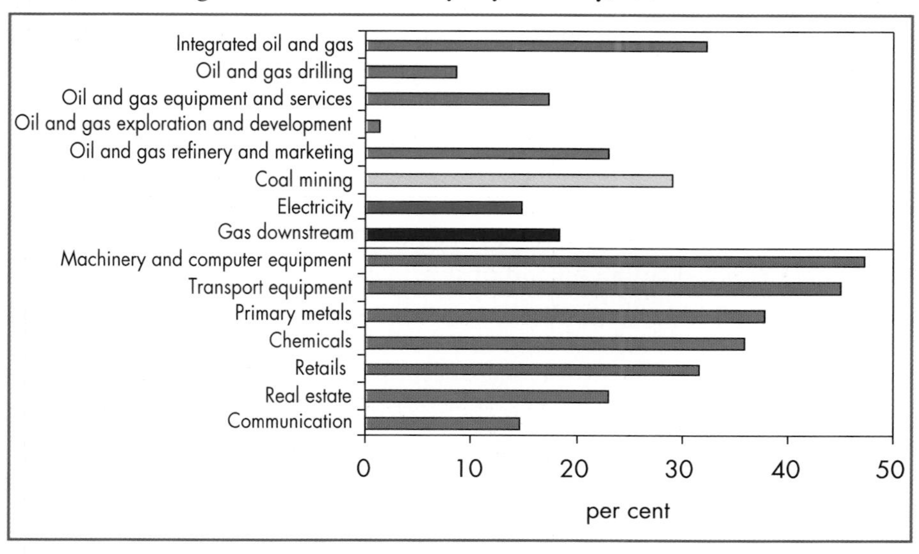

* Short-term debt as a percentage of total debt.

Energy Investment in Developing Countries and the Transition Economies

Access to capital is likely to be a major issue for non-OECD countries. Some 60% of global energy investment needs over the next 30 years will arise in these regions. But their domestic savings are smaller relative to domestic investment and financial markets are less developed.

Risk and Return

Investing in energy projects in developing countries and the transition economies is generally riskier than in OECD countries because of less well-developed institutional and organisational structures, lack of clear and transparent energy, legal and regulatory frameworks, and poorer economic and political management. The availability of capital goods and skilled labour is also limited in those countries, which can increase construction and operation risks. Therefore, even though the unit cost of investment in developing countries and the transition economies is lower, a higher risk premium usually applies in determining the minimum acceptable rate of return on investment.

There is a significant difference in the risk and return profile between an export project and a project for the domestic market in non-OECD regions. A project whose earnings come primarily from the domestic market, where the economic and political environment is unstable, may be subject to greater risk than an export-oriented project. The exchange rate risk is an important factor in the electricity and downstream gas industries, because they often need to purchase fuel in foreign currency, typically in US dollars, but revenues are usually generated in local currencies, while reliance on foreign borrowing is high, because local currency funds are limited. Moreover, in many cases, much of this borrowing is short-term, thus involving refinancing risks. Under these circumstances, investors hesitate before providing financial resources to energy projects.

As a result of these factors, returns on energy investment have to be significantly higher in non-OECD regions. Over the ten years to 2002, returns in the non-OECD oil and upstream gas industries were more than a fifth higher than in OECD countries. In the downstream gas industry, they were more than 40% higher (Figure 3.9). The risk premium accounts for the greater part of this sharp difference. The picture is very different for the electricity sector. The ROI of electricity companies in non-OECD regions was 10% lower than in the OECD. The ROI was not high relative to prevailing lending interest rates in these countries, but it is very likely that the actual cost of borrowing of these companies was kept lower than the level that those interest rates suggest, thanks to government intervention such as guarantees and interest ceilings on loans to electricity projects. Moreover, the picture is probably much worse than the data suggest, since many electricity companies are state-owned companies and are, therefore, not listed. The financial performance of state power companies tends to be worse than that of listed ones.

Rating agencies attempt to quantify country and political investment risks by means of so-called sovereign credit ratings. These ratings are used to help establish the cost of capital for each country. The risk premium demanded by

Figure 3.9: **Return on Investment* of Energy Companies by Sub-sector and Region, 1993-2002**

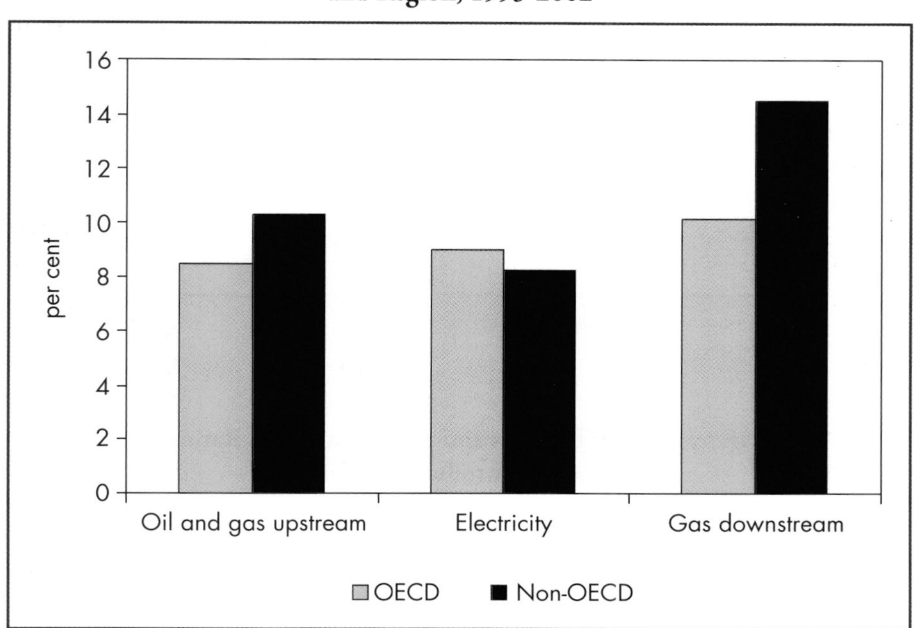

* Operating income divided by total capital invested.

investors and lenders is a function of many variables, and is ideally calculated on a project-by-project basis. However, in practice, country ratings often serve as a benchmark for risk premia, especially when sovereign guarantees are arranged. Countries in Asia and Latin America generally have lower (that is, less favourable) ratings than OECD countries (Table 3.2). Ratings for African countries are better than those for Latin American countries, but this is because only a limited number of African countries are actually rated, as most have no access to international financial markets.

Figure 3.10 plots the sovereign ratings of 14 major oil-producing countries outside the OECD against their proven oil reserves. Saudi Arabia and Russia together hold nearly 40% of the world's remaining proven oil reserves, yet they have lower ratings than countries with much smaller reserves, like the United Arab Emirates (UAE) and Qatar. Countries struggling to deal with severe economic, social and political problems, such as Brazil, Indonesia and Venezuela, have the lowest ratings. State-owned oil companies in these countries, whose credit ratings are at best equivalent to sovereign ratings, can procure funds from foreign investors and lenders only on relatively unfavourable terms. They also bear a higher cost of capital than international

Table 3.2: **Distribution of Sovereign Credit Ratings* by Region, April 2003**

	Aaa	Aa	A	Baa	Ba	B	Caa	Ca	C
OECD	18	5	4			1			
Transition economies		1	2	2	1	4		4	
Middle East		5	3	1	1				
East Asia	1	1	3	2	2	3			
South Asia					1	1			
Latin America		1	2	5	7	5	3		
Africa		1	3	2					

* Ratings for long-term government bonds in foreign currency.
Source: Moody's Investors Service website.

Figure 3.10: **Oil Reserves and Sovereign Credit Ratings of Major Oil-producing Countries**

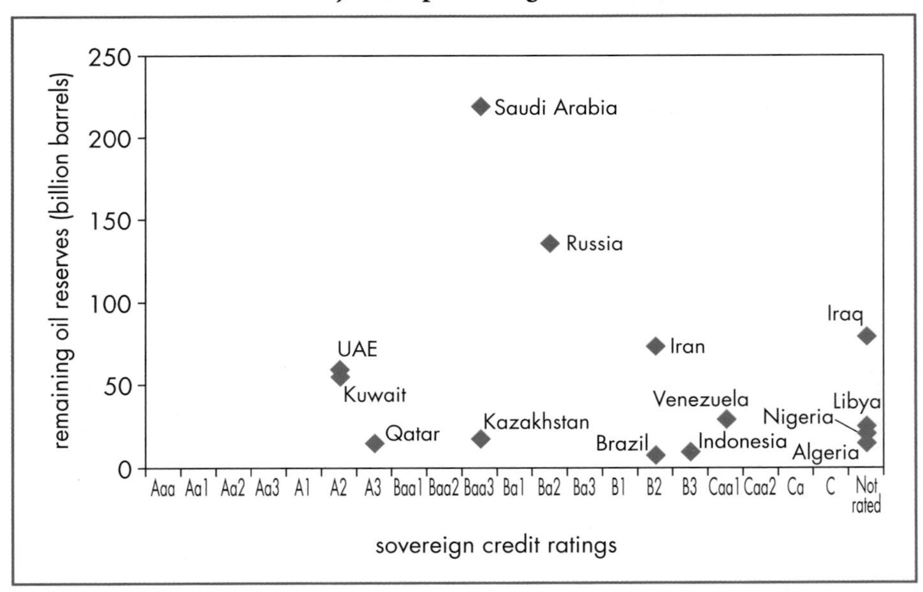

Note: Ratings are those for long-term government bonds in foreign currency as of April 2003. Reserves are effective as of January 1, 1996.
Sources: IEA databases; Moody's Investors Service website.

oil companies, which enjoy higher credit ratings (Aa3 or higher). The United States, the United Kingdom and Norway, the leading OECD oil producers, have the best sovereign credit ratings among oil-producing countries.

Capital Structure

For the reasons already indicated, there are major differences in capital structure between companies based in OECD regions and those in non-OECD regions. Energy companies incorporated in non-OECD regions are much less leveraged compared to their counterparts in OECD regions (Figure 3.11). The average debt equity ratio of electricity and downstream gas companies in non-OECD regions is just over 60% of that of OECD-based companies. This should not be interpreted to mean that they enjoy lower exposure to financial risks. Rather, their capital structure reflects borrowing constraints due to less developed domestic financial markets. These constraints have forced energy companies to rely more on equity, which was traditionally held directly or indirectly by governments.

The debt maturity of energy companies based in non-OECD regions is shorter too. Compared to OECD-based companies, their dependence on short-term debt is 1.7 times higher in the electricity sector, 2.5 times higher for downstream gas companies, and six times higher for oil and upstream gas companies, compared to OECD-based companies. This suggests that banks

Figure 3.11: **Capital Structure of Energy Companies by Sub-sector and Region, 1992-2001**

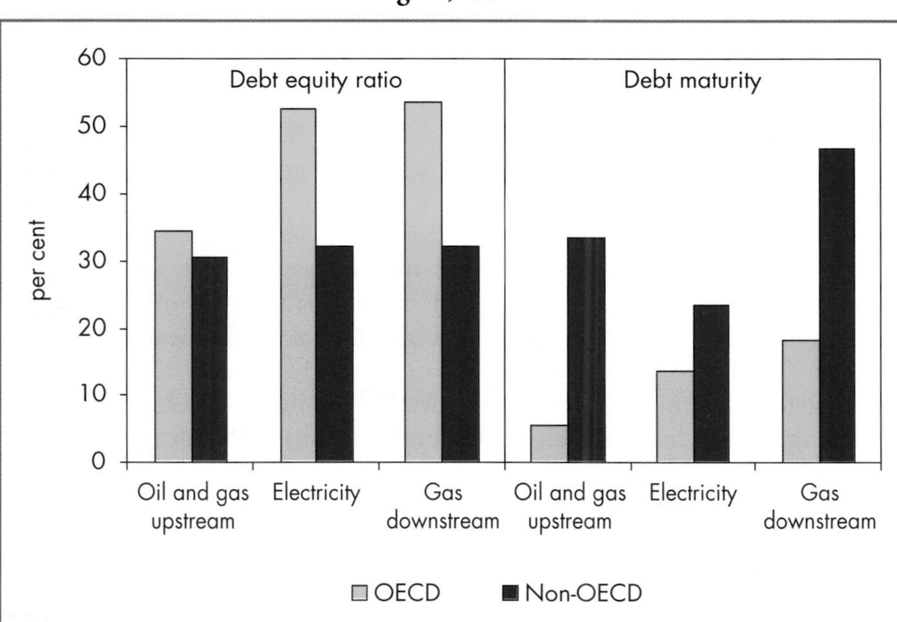

Note: The debt equity ratio is total debt as a percentage of the sum of shareholder's equity and total debt. Debt maturity is short-term debt as a percentage of total debt.

are their primary source of funds, since financial markets, especially capital markets, are not deep enough to provide adequate long-term debt instruments. Macroeconomic instability, such as high inflation, also contributes to shorter debt maturity. Heavier reliance on short-term debt can make financing riskier, especially if such sort-term debt is in foreign currency.

Trends in Energy Investment Financing

Changing Structure of Energy Investment Financing

Governments have traditionally intervened heavily in the energy sector, often through ownership of energy companies. As a result, the government budget and government-sponsored borrowing were often the main sources of financing for energy investment, especially in natural gas and electricity projects. From the late 1980s, countries began to privatise and introduce competition into energy markets. Private sector participation was expected to improve economic efficiency, while reducing pressure on central government budgets. These developments have forced the energy sector to take advantage of growing international financial markets and find new investment financing mechanisms, which mitigate investment risks caused by deregulation.

Oil and Gas

Until the 1970s, most investments in the oil sector were financed from the internal cash flows of the international oil companies. Their balance sheets were sound, so they had little trouble in raising additional funds from financial markets. The situation changed in the 1970s and 1980s as producing countries asserted their sovereignty over oil resources and concern grew in consuming countries about the security of oil supply and rising prices. Governments became heavily involved in the upstream oil and gas sector through the establishment of national oil companies and other forms of intervention. Rising oil prices offered substantial cash flows and almost unlimited credit to oil-rich countries. Consequently, an increasing amount of funding for oil projects came from government budgets. Outside OPEC, financing came from official government-sponsored borrowings from multilateral financial institutions and bilateral donors. Emerging independent oil companies became major users of commercial debt.

By the late 1980s, the emphasis had shifted back towards private financing, as many governments redefined the role of national oil companies as a result of the sovereign debt crisis and government-funding constraints as oil prices fell. This was particularly the case in non-OPEC oil-producing countries, where finding and development costs are generally high. Several countries commercialised their oil industry and opened it up to foreign investment. International oil companies applied more stringent criteria to their investment decisions. In many cases, they now seek to share risks through complex financing arrangements and local participation, typically by setting up joint ventures with national oil companies. As their investments have grown, companies have also tended to try to keep debt off their balance sheet, isolating the impact of large-scale stand-alone projects on overall corporate performance by using project finance on a non-recourse or limited-recourse basis[20].

The important difference between gas and oil projects is that upstream gas investment is normally contingent on long-term contracts with downstream gas companies and large end-users, mainly electricity companies. The financing of investments in gas transmission and distribution is similar to that in the electricity sector. The specific characteristics of liquefied natural gas project financing are discussed in Box 5.3 in Chapter 5.

Coal

Coal has a long history of extensive government involvement, often in the form of ownership. Governments have used numerous measures, including trade restrictions and subsidies, to support coal mining. Many governments used to finance investment either directly through budget allocations or through indirect subsidies. In non-OECD countries, such as India, Russia and China, governments frequently kept prices to consumers below the cost of supply. Since artificially low prices led to poor financial performance, investment by state-owned producers was heavily dependent on government budget allocations.

As the competitiveness of domestic coal against other fuels, including internationally traded coal, declined, governments became less willing to bear the burden of rising subsidies and launched restructuring programmes. This often involved the closure or consolidation of small unprofitable mining companies, diversification of their activities and privatisation of state-owned companies. Large diversified companies now dominate coal production in many parts of the world. This development has been particularly strong in the United States, Australia and South Africa. The largest companies are able to raise funds from domestic and international financial markets through traditional corporate finance mechanisms, often backed by the financial performance of non-coal operations.

20. In project financing, investment risks need to be well identified and future cash flow reasonably predictable. This is one of the reasons why lenders seek take-or-pay contracts in gas-production and IPP projects. The main disadvantage of project finance is the complexity and time involved in arranging the finance and higher transaction costs, since it involves intensive negotiations, detailed contracts and often the establishment of a so-called special purpose company.

The coal industry has become increasingly international over the past 30 years, with the expansion of international coal trade and capital flows. The shares of large companies are now traded in various stock markets, giving access to additional funds. Several major non-OECD producing countries have opened the sector to foreign investors in order to fill the gap between available domestic financial resources and investment needs.

Electricity

Because of the industry's natural monopoly characteristics and the treatment of power supply as a public service, state-owned utilities or government departments were traditionally responsible for running the electricity industry in most countries. The absence of well-developed domestic financial markets also obliged governments to take on responsibility for financing electricity-sector investment in many non-OECD countries. But government borrowing on behalf of electricity companies reduced the availability of credit for other sectors and programmes, especially in those countries. Where companies were privately-owned, most often in OECD countries, governments regulated them tightly. Private companies were able to take on some commercial debt, sometimes backed by government guarantees.

In the 1990s, a number of countries privatised their electricity utilities and opened up the sector to competition. Public utilities started borrowing from commercial banks and issuing equities and bonds in capital markets. Although most private investment in the electricity sector was made through corporate financing, the introduction of independent power producers (IPPs) accelerated the use of project finance. In non-OECD regions, foreign capital contributes most to this type of finance. The typical structure of project finance for IPPs involves around 30% equity. The rest of the capital is borrowed from various sources, such as commercial banks, infrastructure funds, equipment suppliers, international development banks and export credit agencies. IPP projects in developing countries are usually secured by power purchasing agreements with publicly-owned distribution companies, which bear most of the market risk.

Private Financing

Although the private sector has traditionally dominated the oil sector and the upstream part of the gas sector, it emerged as an important investor in and lender to the electricity and downstream gas industries only in the 1990s. This reflected a general shift in government policy towards greater reliance on market forces and an increased role for the private sector in the economy. The private sector can expect to be called upon to play an increasingly important role in financing energy investments in the coming decades. Foreign investment will need to make up a large part of these flows.

Investment by the private sector in electricity and natural gas transmission and distribution in developing countries and the transition economies (in year 2002 dollars) increased substantially in the early to mid-1990s, from around $1 billion in 1990 to $52 billion in 1997 (Figure 3.12).[21] But it has since fallen sharply, to $17 billion in 2002, equal to the level of 1994. This largely reflects the economic crises in East Asia and Latin America, the completion of some large investment programmes launched in the 1990s and disillusionment with the outcomes of some earlier investments. In East Asia, energy investment with private-sector participation slumped from $14 billion in 1997 to $2 billion in 2002. In Latin America, that investment dropped from $26 billion to $5 billion. Some investors have withdrawn from these sectors because of macroeconomic instability and uncertainties about market reforms. Difficult conditions in international capital markets and falls in the market capitalisation of potential investors have also made it harder to raise capital for new investment.[22]

Figure 3.12: **Private Participation in Electricity and Gas Investment in Developing Countries and the Transition Economies** (in year 2002 dollars)

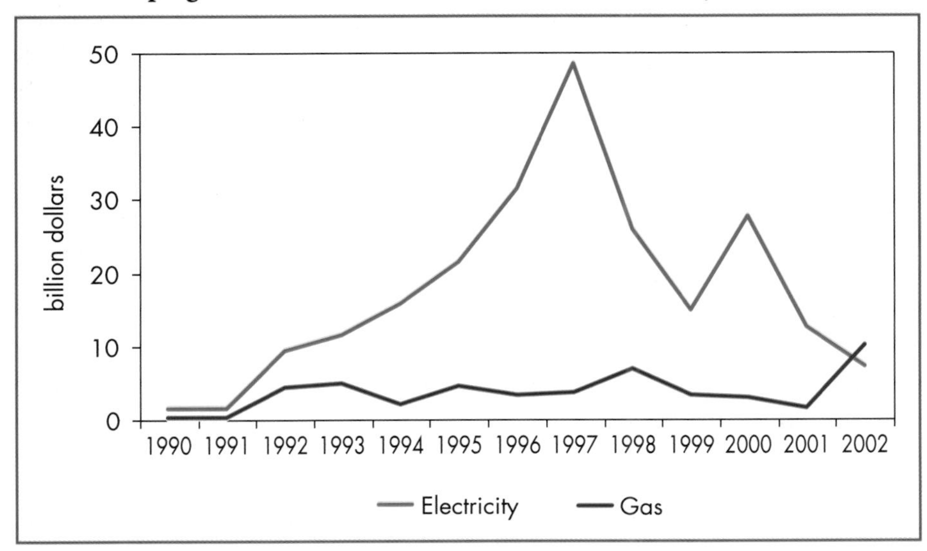

Source: World Bank (2003c).

21. The definition of developing countries and the transition economies used here differs from that used in the *Outlook*. For example, the World Bank's definition of developing countries excludes Hong Kong, Singapore, Chinese Taipei, Kuwait, Qatar and UAE, while its definition of transition economies includes the Czech Republic, Hungary and Poland, which are all OECD countries.
22. Lamech and Saeed (2003).

Over the period 1990-2002, Latin America accounted for 43% of all private investment in electricity and 48% in natural gas transmission and distribution. East Asia accounted for a further 31% of such investment in electricity and 12% in natural gas. Investment has flowed mainly to a few large countries, led by Argentina, Brazil, Chile, China and India, where the government has shown a strong commitment to promoting private-sector participation in energy and where energy demand has grown strongly. The form of private participation varies across regions, reflecting differences in approaches to reform.[23] In Latin America, privatisation has been the main driving force, accounting for 67% of total investment in the region during 1990-2002. By contrast, greenfield projects accounted for 82% of private electricity sector investment in East Asia and 96% in South Asia. Governments in these regions have put the emphasis on IPPs. Private investment in Africa in electricity has changed very little.

International Capital Flows

Steady growth in energy demand, further moves towards competitive energy markets and public sector budget constraints will continue to provide scope for private capital in energy-investment financing. But, as the recent financial crises in Asia and Latin America have demonstrated, private capital flows, especially international flows, are very sensitive to macroeconomic conditions and the stability of government policy.

Net long-term capital flows to *all sectors* in developing countries and the transition economies (in current year dollars) almost tripled from $124 billion in 1991 to $337 billion in 1998, but fell back to $207 billion in 2002 — below the level of 1993 (Table 3.3).[24] Private debt flows have fallen much more than foreign direct investment (FDI). FDI increased fivefold from $35 billion in 1991 to $196 billion in 1999, before falling back to $152 billion in 2002. Its share in total flows has reached more than 70% (see the latter part of this section for a further discussion of FDI). This shifting pattern of private flows from debt to equity is mainly due to reduced borrower demand and a change in investors' preferences.[25] In Asia, a major recipient of international capital flows, persistent current account surpluses and relatively steady inflows of FDI since the 1997 economic crisis have reduced the region's need for external debt. Worldwide, investors now tend to focus more on credit risk than rates of return, and have become more reluctant to hold debt, regardless of the destination of capital flows. This trend is expected to continue in the short term. The long-term picture will depend on developments in the global economy.

23. Izaguirre (2002).
24. World Bank's definition of developing countries and the transition economies (see footnote 21).
25. World Bank (2003b).

Table 3.3: **Net Long-term Capital Flows to Developing Countries and the Transition Economies** ($ billion in current year dollars)

	1991	1995	1997	1998	1999	2000	2001	2002
Total flows	**124**	**261**	**327**	**337**	**276**	**238**	**234**	**207**
Official flows	**62**	**54**	**40**	**62**	**43**	**23**	**58**	**49**
Aid	35	33	27	28	29	30	30	33
Debt	27	21	13	34	14	-6	28	16
Private flows	**62**	**207**	**287**	**274**	**233**	**214**	**177**	**158**
Equity flows	*43*	*144*	*203*	*187*	*211*	*200*	*185*	*161*
FDI	35	108	176	180	196	174	179	152
Portfolio	8	36	27	7	15	26	6	9
Debt flows	*19*	*63*	*84*	*87*	*22*	*15*	*-9*	*3*
Bonds	11	31	38	40	30	17	10	19
Bank lending	5	31	43	51	-6	3	-12	-16
Others	3	2	3	-4	-2	-6	-7	-6

Sources: World Bank (2000 and 2003b); UNCTAD (2003).

Net official flows have traditionally been an important source of external capital in developing countries and the transition economies. Their decline in recent years is due, on the one hand, to fewer financial rescue packages by multilateral institutions and, on the other, to repayments to bilateral export credit agencies under debt-restructuring agreements. The scale of official development assistance (ODA) has fallen relative to economic activity in most recipient countries over the past decade.[26] The World Bank and other regional development banks started lending to the oil and gas sector after the oil crises of the 1970s in order to help countries develop their indigenous energy resources. The emphasis has since shifted to development of infrastructure and privatisation of state-owned oil and gas companies. By the 1990s, the power sector had also become a key area of funding, though the development banks have a long tradition of lending to hydropower plants. However, traditional World Bank lending to infrastructure projects — especially energy — has declined since the early 1990s. The International Finance Corporation (IFC) used to be very active in promoting energy investment on a more commercial basis than the World Bank itself, providing loans, equity and quasi-equity finance. Unlike the World Bank, the IFC can finance private projects. Financing by multinational institutions often generates a "cow bell" effect,

26. ODA is included in official flows.

attracting private capital in its wake, because of their capacity to evaluate the investment opportunities and risks by country, sector and project. The recent decline in energy-investment lending by those institutions could, in consequence, adversely affect the private capital flows into the sector.[27]

One component of official flows that has been growing is export credit. A significant number of export credit agencies provide loans, guarantees and insurance to support energy investment. They increasingly provide support to project finance and are assuming a wider range of risks, as hydrocarbon-rich countries shift some of the burden of financing energy projects to the private sector.

Among the several channels open to energy companies to tap international financial markets, FDI has attracted much attention. This source of funding can ease a company's financing constraints by bringing in scarce capital, but it may also exacerbate financing constraints for domestic companies if foreign-invested companies borrow heavily from domestic financial markets. Studies have confirmed that FDI flows tend to reduce overall financing constraints, although crowding out can occur.[28] FDI also brings state-of-the-art technologies, which are usually very expensive, to host countries.

Global FDI inflows to all sectors in 2002 (in current year dollars) were $651 billion, down by 22% from 2001 or only half the peak in 2000. OECD regions have been the main recipients of global FDI flows for the last 20 years, accounting for 75% of global FDI inflows in 2002. But the OECD's share has declined over much of that period, from 87% in 1989 to 60% in 1997. FDI inflows into developing countries and the transition economies have not yet returned to pre-Asian crisis levels, but have been more stable than other capital flows since then. Weak economic growth, falling market capitalisation, fewer privatisation programmes and a slump in large, cross-border mergers and acquisitions have contributed to the recent decline in FDI flows both in OECD and non-OECD regions.[29]

Among developing countries and the transition economies, China has been the largest recipient of FDI inflows during the period 1991-2002 (Figure 3.13). It received 10% of global FDI inflows in 2002, compared to 3% in 1991. Over the past two decades, most Latin American countries have opened various sectors to foreign capital, making the region another key recipient of FDI. Inflows to this region have gone mainly to Argentina, Brazil (the second-largest recipient of FDI among developing countries and the transition economies), Chile, Colombia and Peru. The recent decline in FDI in the region is

27. The World Bank Group is currently considering revitalising its support to infrastructure projects, including those in the energy sector. This could lead to a reversal of the recent trend.
28. Harrison, Love and McMillan (2002); and UNCTAD (1999).
29. UNCTAD (2003).

Figure 3.13: **Net Foreign Direct Investment Inflows to Developing Countries and the Transition Economies by Region** (in current year dollars)

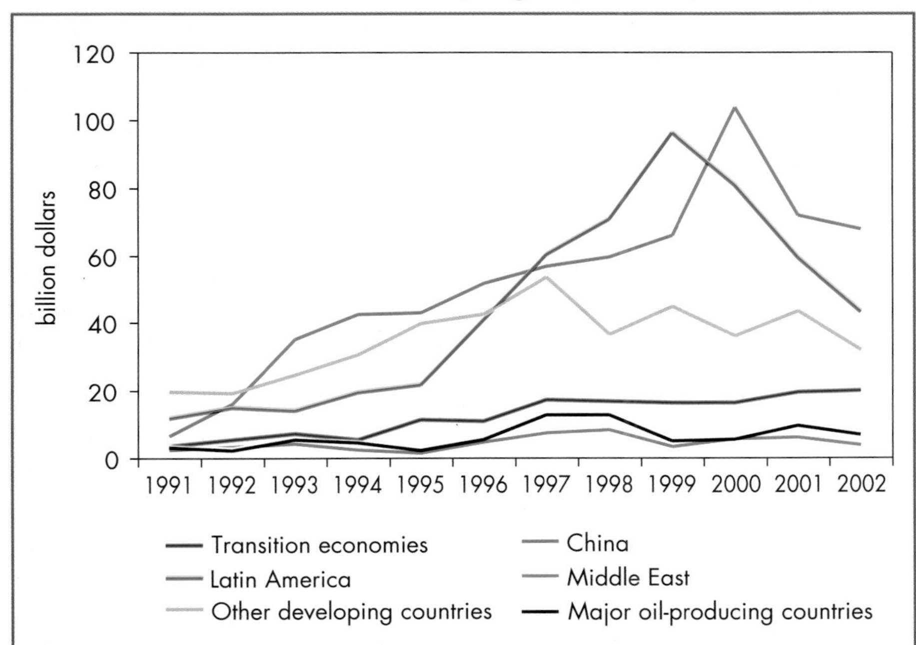

Source: UNCTAD (2003).

primarily due to the economic downturn in the region triggered by the economic crisis in Argentina and the postponement or cancellation of privatisation projects in the first three countries.

FDI inflows in the Middle East, most of which go to the oil and gas upstream sectors, grew until 1998 and then declined, but their share in global FDI has remained at around 1%. Inflows to major oil-producing countries, a proxy for global FDI flows in the oil and gas upstream sectors, have also been rising, accounting for 2% to 7% of global FDI through the 1990s.[30] Saudi Arabia and the UAE experienced negative inflows in some years, implying that international investors did not reinvest earnings and preferred to repatriate them, or that the recipient enterprises paid back their debts. This may have been because of a lack of investment opportunities in those countries. Another explanation is that FDI, for definitional reasons, does not capture the increasing capital flows through production-sharing agreements and similar contractual arrangements.

30. Only the oil producers with relatively undiversified economies were included in this analysis: Iran, Iraq, Kuwait, Qatar, Saudi Arabia, UAE, Algeria, Nigeria, Venezuela, and Kazakhstan.

Only limited data are available on FDI inflows by sector, particularly for developing countries and the transition economies.[31] In the OECD regions, the share of the energy sector in total FDI inflows averaged around 10% during 1995-2001, peaking at 20% in 1998.[32] Among the energy sub-sectors, oil production and refining received much more FDI than the electricity and gas utility sectors. The share of the energy sector in total FDI inflows to most developing countries, where data by industry are available, has declined in recent years (Figure 3.14). This is largely because of a surge of direct investment in the service sector and the information technology and communication industry up to 2000.[33]

Figure 3.14: **Share of the Energy Sector in Foreign Direct Investment Inflows by Region**

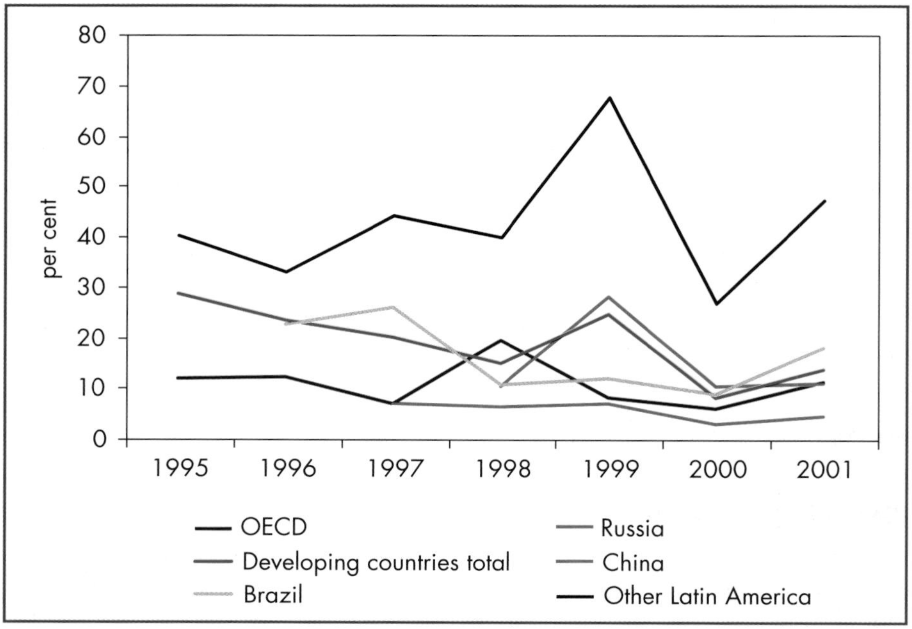

Note: See footnote 32 for the definition of the energy sector. Data are not available for China in 1995 and 1996, Brazil in 1995, and Russia in 1995, 1996 and 1997 and this chart reflects these limitations.
Source: UNCTAD (2003).

31. The UNCTAD database covers 50 countries, including most OECD countries, for which data by industry are available for 1999, 2000 and 2001. Those countries account for 89% of global FDI inflows.
32. The energy sector here is defined as mining, quarrying and petroleum; coke, petroleum product and nuclear fuel; and electricity, gas and water utilities. Mining, quarrying and petroleum can be used as a proxy of upstream activities of the hydrocarbon sector, while the coke, petroleum product and nuclear fuel category includes oil refining and other downstream oil activities.
33. World Bank (2003b).

The energy sector was not the largest recipient of FDI inflows in China. The manufacturing sector attracted 60% of total FDI on the mainland, with another 35% going to the service sector.[34] In Hong Kong, services account for most of the large FDI inflows. The oil and gas sectors have received about 10% of Russia's FDI inflows in recent years, but the figure ballooned to 28% in 1998, when the financial crisis hit non-energy investment particularly hard.[35] In Latin America, the energy sector has accounted for a much higher share of total FDI inflows, thanks to the privatisation of state-owned electricity and gas utility companies and the opening up of oil and gas reserves to foreign capital. In Brazil, FDI in the electricity sector peaked in the late 1990s, while FDI in the oil and gas industries has grown in recent years. In the other parts of the continent, the energy sector has been the largest recipient sector of FDI. The upstream oil and gas industries, mainly in Venezuela, Ecuador, Colombia, Chile and Peru, accounted for 70% to 80% of FDI inflows into the energy sector. Almost all of the rest went to the electricity and gas utility sectors, especially in Argentina, Chile and Colombia.

The prospects for FDI in the energy sector will depend on the economic environment, incomes, market growth, the quality of labour and government policies. Openness to foreign capital and protection of foreign investors' rights over earnings and technologies, in addition to the adequacy of returns on investment, will be key factors. For example, the liberalisation of foreign investment policies and ongoing reforms in the energy sector are expected to help China to attract more foreign investment, particularly in developing its western gas resources and in new electricity projects.

Prospects for Financing Energy Investment

Near-term Outlook

The shares of electricity and gas utilities on the world's major stock markets have been outperformed on average by other sectors over the past eight years. But oil and gas companies have performed better since late 2000, owing to higher oil prices (Figure 3.15). The energy sector tends to fall in and out of favour with investors, according to expectations of oil prices and earnings, and the outlook for other sectors.[36] The relatively poor performance of energy stocks in the late 1990s resulted partly from the boom in technology stocks.

34. See IEA (2002).
35. UNCTAD (2003).
36. While ROI and other financial variables measure the actual performance of companies/industries and, to a limited extent, their prospects, the performance of shares can reflect additional elements, notably investors' perceptions about risk and, even, ethical considerations.

Figure 3.15: **Performance of Standard and Poor's Global 1200 Indices**

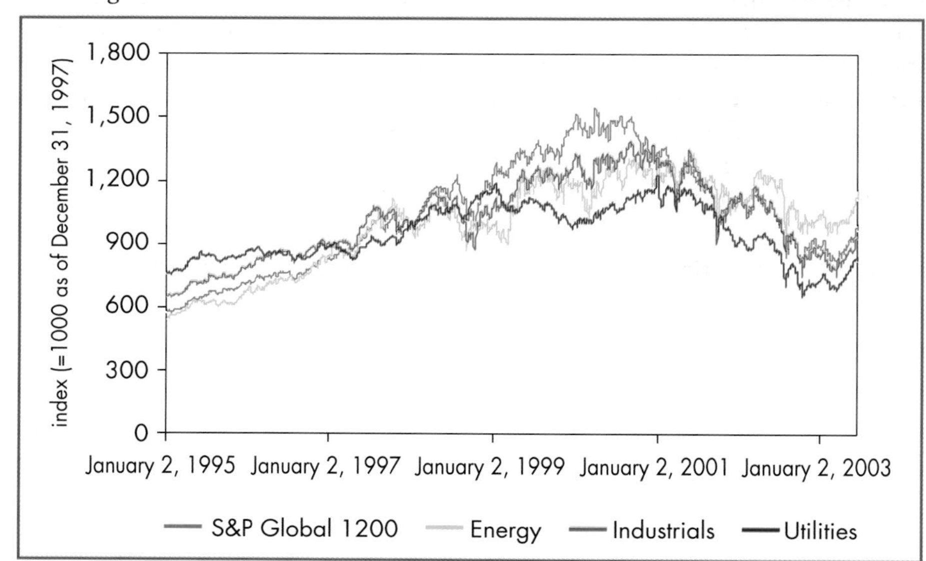

Source: Standard and Poor's (2003b).

Returns on new upstream investments in the oil sector currently comfortably exceed the cost of capital, because of relatively high oil prices and low interest rates. The average return in mid-2003 was believed to be around 15%, while the cost of capital for international oil companies is estimated at between 8% and 10%. Even so, investment by international oil companies remains low relative to cash flow and capital flows to other sectors where margins between investment return and the cost of capital are significantly lower (see Chapter 4). This is probably because the oil companies and the financial markets expect oil prices to decline, or because the "risk-adjusted" return on investment is not high enough at current prices.[37] Further, some oil companies have recently been buying back their own shares, while FDI inflows in some Middle East countries have turned negative (see the previous section). These trends suggest that there is a lack of new investment opportunities that can generate returns high enough to satisfy shareholders. Oil companies may also have been opting to improve their financial health by using cash flows to reduce debt, in the wake of recent mergers and acquisitions.

The situation is rather different in the electricity sector. Banks were once keen to exploit lending opportunities thrown up by market reforms in many countries. But they have become very reluctant to extend loans to power

37. Merrill Lynch (2003).

companies, whether corporate or project finance, unless a high "risk-adjusted" return is expected. Debt issuance by power projects in 2002 plummeted by more than 40% (Figure 3.16). This was largely caused by the severe financial problems experienced by merchant power companies in the shift to competitive markets in several OECD countries. A rebound in lending to the power sector will depend on improved market conditions and renewed confidence on the part of the commercial banks.

Figure 3.16: **Global Debt Issued by Power Projects** (in current year dollars)

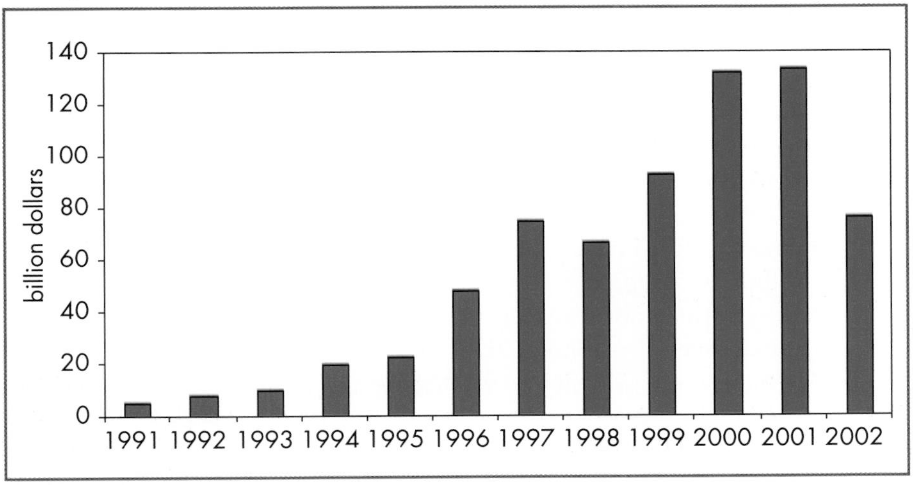

Source: Société Générale (2003).

Looking Forward

Ease of access to capital for energy investment is expected to continue to vary widely according to the sub-sector concerned and the risk-return profiles of individual projects. Current prospects are as follows:

- Major international oil and gas companies are unlikely to face any real shortage of capital, especially for upstream investment. Their healthy balance sheets and strong creditworthiness, based on higher profitability and abundant cash flow, will ensure that they will be able to attract capital for projects which show a high risk-adjusted return. The policies of major oil-producing countries towards foreign access to reserves will be a more important determinant of their investment.
- Independent oil and gas companies, including those involved in exploration and development in the most mature producing areas, where investment returns could deteriorate quickly, are more likely to face difficulties in raising funds for future investment.

- National oil companies in oil-producing countries will need to find new sources of capital on international financial markets. However, they will face less favourable terms than international oil companies, as indicated by sovereign credit ratings.
- A shift in oil and gas investment towards higher-cost deepwater and/or emerging areas such as West Africa, Russia, and the Caspian region would push up the cost of capital, as economic, political and legal risks in those areas will be higher, even though the unit capital cost of such investment is expected to decline further in the future. The sheer size of some projects will induce companies to opt for project financing, which is sometimes more costly and time-consuming to arrange.
- Price volatility, if it continues to increase, could raise significantly the cost of capital and the required risk-adjusted returns.
- Electricity and gas downstream companies will remain more debt-oriented to fund their heavy capital needs. Returns on investment by regulated and vertically integrated companies (listed on major stock markets), which have generally been fairly high and stable, are likely to decline as competition intensifies with market reforms. High debt equity ratios and lower profitability could constrain their access to capital, especially in non-OECD regions.
- Financing investment in renewable electricity plants, which are expected to account for 33% of total OECD investment in power generation over the next 30 years, will be very much subject to the pace of the future decline in their capital costs, relative to those of fossil fuel-based power plants, and to the electricity price. Governments almost certainly have to intervene to ensure adequate returns to investment, through pricing measures and the provision of various incentives, if renewables are to play their expected role in climate change abatement and energy security.
- The merchant power companies' financial difficulties are likely to continue in the short term, making it hard for them to meet their financing needs both for investment and general operations, including debt repayment. Mid- and long-term prospects for their access to capital at reasonable rates will depend very much on government policies towards the electricity sector.
- Transmission and distribution of electricity and gas will remain relatively low-risk businesses in OECD countries, with returns protected to a large degree by the regulator. Their access to capital will depend partly on the future regulatory framework, and — in the case of state firms — the ability and readiness of the governments to finance investment themselves.

- Financing will be more difficult for electricity and downstream gas companies in many non-OECD regions, especially in Africa, the transition economies, Latin America and South Asia, because of poorly-developed domestic financial markets and the higher cost of capital due to perceived higher risks. To the extent that these companies serve domestic markets, the viability of new investment and the companies' access to capital will depend on the economic, political, regulatory and legal environment in those countries. Export-oriented oil, gas and coal projects will, in many cases, be easier to finance, as their markets are more secure and there are fewer financial risks. In particular, they are less exposed to exchange rate risks, since their revenues are usually generated in foreign currencies.
- Coal companies' access to capital will tend to improve, as consolidation continues, assuming that the non-coal operations of diversified companies remain profitable and that environmental fears are not reinforced.
- The share of private equity and debt in total energy investment will continue to increase. Although state-owned companies in the Middle East and Russia still own a large share of the world's oil and gas reserves, their reliance on government financing is expected to decline as they seek further private finance through concessions or other means, such as production-sharing contracts. Public shareholders and institutional investors may contribute new forms of equity.
- Energy-related FDI will continue to grow. Major oil- and gas-producing countries in Africa, the Middle East, Russia and Latin America are expected to continue to modify their upstream policies and practices in order to attract investment by international oil companies. Major coal-producing countries will also seek greater private investment to improve productivity. The benefits of FDI will be significant in the power sector in non-OECD countries, which needs long-term financing and modern technology. Liberalisation and restructuring should provide more opportunities to participants in international financial markets. But no country will be able to finance its energy-investment needs through foreign capital alone. Foreign capital is by nature more volatile than domestic capital and carries exchange rate risks. A large part of energy-investment financing in the future will still need to come from domestic sources, underpinned by domestic private savings. This will call for additional efforts to develop the domestic banking sector and capital markets in many developing countries and the transition economies.

- The use of non-traditional financing, including off-balance sheet vehicles and derivatives, will grow in the long term, despite the current loss of confidence in such instruments in the wake of a series of accounting scandals. The energy sector will need to tap the ample resources of institutional investors, such as pension funds and insurance companies, which have long-term liabilities and steady and predictable cash flow. In rural areas in developing countries, where financial markets will require time to grow, the development of community-based micro-credit programmes could play an important role in financing small-scale, distributed power generation projects, including solar photovoltaic and mini-hydro (see Chapter 7 for investment to provide universal electricity access by 2030).
- In general, investors in the energy sector, especially large, vertically integrated companies, are likely to continue to make more use of traditional corporate finance than project finance, since corporate finance tends to be cheaper. Project finance will continue to be deployed selectively to supplement corporate finance, particularly where the investment and the associated risks are too large to be absorbed into a company's balance sheet. New entrants investing in renewable electricity plants, typically wind power in the liberalised OECD markets, are likely to draw upon project finance, while there will be a greater sectorwide appetite for such finance in non-OECD countries. The energy sector may also make more use of project finance backed by export credit agencies. However, governments will hesitate to increase their exposure to contractual liabilities such as financial guarantees to power projects or obligations to repay debt in foreign currencies. Such contingent liabilities do not usually show up in governments' budget statements, but they can prove extremely costly (see Box.3.3 for a discussion of the impact of the 1997 economic crisis on Indonesia's energy sector).

Government Policies

Though they will be less directly involved in financing the energy sector in the future as direct investors and lenders, governments will continue to play a major role in ensuring that adequate finance for energy infrastructure is mobilised in a timely fashion. This is warranted by the size of the challenge and by the importance to national economic and social well-being of meeting that challenge successfully. There is much governments can do to lower the barriers to private investment, notably by managing the economy effectively in order to improve the overall investment climate, by promoting the development of domestic financial markets and by establishing an effective regulatory and policy framework for the energy sector. This chapter concludes by discussing these responsibilities of government.

Macroeconomic Management

The macroeconomic environment is the single most important driver of energy demand and, therefore, energy-investment needs. As one of the main sources of risk, it affects not only the amount of investment needed but also the access of the energy sector to capital. Low sovereign ratings due to poor macroeconomic performance lead to higher risk premia in the cost of capital, which can jeopardise the viability of energy projects in the country concerned. Investors and lenders are more reluctant to provide funds to projects in lower-rated countries.

Box 3.3: **Impact of Asian Economic Crisis on Energy Investment in Indonesia**

The 1997 Asian economic crisis wrought an unprecedented shock to many of the region's economies, with far-reaching effects on energy investment. Indonesia's GDP fell by 13% in 1998 (in 1995 US dollars and purchasing power parity terms), driving down total energy demand by 1.7% and oil demand by nearly 4%. Electricity consumption grew by only 1.4%, compared to average annual growth of 11% in the 1990-1997 period. IPPs in the country sold power to the national utility under long-term contracts denominated in dollars, while customers of the national utility paid their bills in the local currency, the rupiah, at regulated prices. As a result, the collapse of the rupiah squeezed the finances of the national utility.

The fall in oil and gas demand in Japan and Korea, as well as the fall in oil prices after the crisis, further exacerbated the macroeconomic environment, as the hydrocarbon sector generated nearly 30% of the government's revenue and 5% of Indonesia's GDP at that time. Together with the unstable political situation, the crisis dented investor confidence and many energy projects that were to be financed by foreign capital were postponed or cancelled.[38]

Korea, where GDP fell by 6.7% in 1998, also experienced an abrupt decline in energy demand and delays to energy investment, but both recovered strongly in the following years.

38. The crisis resulted in temporary overcapacity in the power sector, but the long lead time for investment and potential for rapid electricity demand growth have raised concerns about the long-

Investors and lenders often respond to macroeconomic turbulence by rationing credit. They do this by increasing interest rates or by withdrawing money from the country. High inflation is often a reason for the absence of long-term capital, since it penalises future returns. Negative real interest rates due to high inflation also discourage domestic savings, which are the main source of funds for investment. High domestic interest rates relative to international rates, especially if combined with fixed exchange rates, create strong incentives to borrow abroad and lend domestically. But this carries considerable risks. The mismatched and unhedged currency positions of financial institutions and companies were a major factor behind the 1997 Asian economic crisis, which had a devastating impact on energy investment in the region (Box 3.3).[39] The recent crisis in Argentina also led to a downturn in energy investment, especially that financed by foreign private capital.

Financial Sector Development

The development of financial markets in developing countries and the transition economies will be vital to securing sufficient amounts of private capital to meet the projected growth in energy investment. Financial markets will need to become deeper and more efficient, drawing on domestic savings and tapping into international financial resources.

The underdevelopment of financial markets has to some extent been a result of inappropriate government policies: governments have often concentrated energy investment financing and ownership of assets in public hands. In addition, interest rate and credit controls have discouraged savers from contributing to the energy sector in particular, and the capital market in general. Financial institutions operating in less developed financial markets rely much more on short-term credit, while the use of derivatives, a useful tool to reduce the risks surrounding energy investment, is much more limited.

The participation of the energy sector in the domestic bond and equity markets can contribute to deepening the capital markets, as some developed countries have shown in the past. Harmonising policies towards the energy sector and the financial sector can help ensure that enough financial resources are channelled to energy investment in appropriate ways.

Although foreign capital flows are often volatile, they can ease energy companies' financial constraints in non-OECD countries. FDI often helps deepen liquidity and hasten the maturity of domestic financial markets. Foreign investors bring efficiency gains through the provision of advanced

39. Bosworth and Collins (2000).

financial services and good management practices. Restrictions on capital account transactions can exacerbate corporate financing constraints, though the Asian economic crisis confirmed the need for careful design in the removal of such restrictions.

The development of financial markets requires adequate disclosure of information, allowing investors and lenders to monitor debt issuers and to exert corporate control. Lack of transparency and poor information about companies and the financial sector lead to a higher cost of capital and vulnerability to external shocks. If accounting and auditing are underdeveloped or if it is difficult or expensive to enforce loan contracts, financial institutions will hesitate to lend on a long-term basis. Several studies have confirmed that the benefits of lender protection are large.[40] In general, countries with lower standards of investor and lender protection display more risky financing patterns and lower rates of return on assets and equity. Weak judicial and legal systems also limit the development of long-term finance.

Energy Policy and Governance

The difficulties that many countries will face in mobilising financial resources for energy investment in the future will be exacerbated by poor and unpredictable energy policies. Governments still have an important role to play in creating and maintaining an enabling environment for investment. By minimising policy-induced risk and clarifying economic risk, reforms can reassure equity investors that energy companies will be able to generate a reasonable rate of return. Bankers have to be sure that debts will be serviced.

The development and implementation of energy reforms inevitably generate uncertainties. Governments can minimise these uncertainties by communicating consistently and clearly its policy goals, strategy and details of planned reforms, and by implementing them in an orderly and programmed manner.

Financial resources flow to sectors and countries that have established sound and predictable systems of corporate governance. The relationship between the quality of governance and the level of FDI flows is particularly clear and positive.[41] A well-governed energy sector is characterised by stable and enforceable legal and regulatory systems, with companies operating under the best commercial practices by international standards. The issue is often not whether the law and regulations exist but whether they are enforced in a fair and transparent manner. The risk-reward profile of a project can be substantially improved by clarifying the rules of the game and assuring the stability and enforcement of relevant policies. Imperfections in the design of

40. Claessens, Djankov and Nenova (2000); and Caprio, Jr. and Demirguc-Kunt (1997).
41. OECD (2002).

competitive electricity and gas markets and the financial collapse of some major energy companies, notably Enron, have severely shaken investor confidence in some countries.

A survey by the World Bank of international investors' views of the power sector in developing countries identifies tariff levels and payment discipline, maintenance of stable and enforceable laws and contracts, improvement of host governments' administrative efficiency and minimisation of government interference as key priorities for governments seeking to attract international investment into the power sector.[42] Improved corporate governance and practice would certainly contribute to increasing the flow and lowering the costs of capital, especially from private investors and lenders abroad. The energy sector can attract capital, even if taxes are high, as long as the policy and business environment is credible and predictable. Important factors include the following:

- *Establishment of a comprehensive, fair and transparent sectoral framework and a system of enforcement:* This issue remains to be addressed by many countries. Many countries lack an independent regulatory authority to supervise competition in the energy markets. The ability of authorities in the reformed markets to collect and disclose information about the performance of the energy sector is an important concern to investors and lenders. Efforts to promote public understanding of energy reforms are often weak.
- *Stamping out corruption:* In some countries, privatisation has led to a high degree of concentration of vested interests in and around the energy sector. A weak jurisdictional system can allow insider-dealing and corruption to develop, generating extra investment and production costs.
- *Application of international accounting standards:* Energy companies do not always employ accounting systems to international standards. Some state-owned energy companies in developing countries and the transition economies do not provide detailed information about their assets and operations, or a breakdown of revenues and costs. Uncollected inter-enterprise debts, barter trade, lack of payment discipline and inadequate and non-transparent financial information create uncertainty, which discourages investors and lenders.
- *Fair policies towards foreign capital:* Fair and consistent provisions with respect to foreign companies' repatriation of earnings, procurement and property are critical to attracting foreign investment. The fiscal regime for upstream activities, which has a major impact on actual and expected returns from investment in oil and gas exploration and

42. Lamech and Saeed (2003).

production, must balance the needs of encouraging private capital inflows and retaining for the state an appropriate share of the rent from hydrocarbon assets. Lack of an appropriate tax code brought investment to a virtual halt in Russia in the late 1990s.

- *Enforcement of laws and contracts:* Reliable enforcement of the legal and regulatory framework is central to bringing energy projects to financial closure. In many developing countries and the transition economies, the lack of enforcement, rather than the lack of a framework *per se*, is the more serious concern. Enforcement is particularly critical in project finance, where risks are allocated precisely through agreements and contracts among equity holders, lenders, input suppliers and buyers. An effective mechanism for resolving disputes, usually in a manner that permits enforcement of a court judgement or arbitral award outside the host country's jurisdiction, is an important element. Further development of bilateral and multilateral legal instruments would stimulate international capital flows by reducing investors' risk and thus cost of capital.

CHAPTER 4: OIL

HIGHLIGHTS

- Investment of over $3 trillion, or $103 billion per year, will be needed in the oil sector through to 2030. Capital spending will have to increase steadily through the period as capacity becomes obsolete and demand increases. Investment in OECD countries will be high relative to their production capacity because of higher unit costs compared to other regions.
- Exploration and development will dominate oil-sector investment, accounting for over 70% of the total over the period 2001-2030. About a quarter of upstream investment will be needed to meet rising demand. The rest will be needed to counter the natural decline in production from wells already in production and those that will start producing in the future. At a global level, investment needs are, in fact, far more sensitive to changes in decline rates than to the rate of growth of oil demand.
- The projected investment will permit an increase in world oil supply from 77 mb/d in 2002 to 120 mb/d in 2030. Offshore fields will account for almost a third of the increase in production from now to 2030, but they will take a bigger share of investment because they cost more to develop.
- The share of the Middle East in total upstream spending, at less than 20%, is small relative to its contribution to the increase in global production, because exploration and development costs in the region are very low. Investment in non-conventional oil projects, mostly in Canada and Venezuela, will account for a growing share of total upstream spending. Capital and operating costs for such projects are high compared with most conventional oil projects, though their exploration costs are negligible.
- Production prospects in Iraq are highly uncertain. The pace of production growth is linked to the pace of political recovery. It is estimated that raising capacity to around 3.7 mb/d by 2010 will cost about $5 billion. Further increases are possible but will depend on the future Iraqi government's production strategy.

- Cumulative investment in crude oil refining will total $412 billion, or close to $14 billion per year. This will be needed to boost refining capacity to 121 mb/d and to upgrade refineries so that their output meets the shift in the product demand mix towards lighter and cleaner products.
- Investment in oil tankers and oil pipelines for international trade will amount to $257 billion through to 2030. Supply chains will lengthen, so most of the investment in transport will go towards oil-tanker capacity rather than pipelines.
- Although 69% of total oil sector investment, excluding transportation, will occur outside the OECD, more than 40% of this investment will be in projects to supply crude oil and products to OECD countries.
- Sufficient capital exists to meet projected investment requirements. But the extent to which capital will be invested in the oil sector will vary between countries, depending on a number of perceived risk factors, including oil prices, fiscal terms, political conditions and technical issues such as geological risk.
- Financing may be a constraint where government policies or perceived geopolitical factors prevent or discourage foreign involvement – especially in the Middle East and Africa. State budgetary pressures could squeeze the amount of earnings that national oil companies are allowed to retain for investment purposes and, therefore, increase their need to borrow. Although most oil-producing countries now allow some form of private and foreign investment, the commercial terms on offer may not always be sufficiently attractive to investors and lenders.
- If the projected amount of investment in the Middle East is not forthcoming and production does not, therefore, increase as rapidly as expected, larger amounts of capital would need to be spent in other more costly regions. Under a Restricted Investment Scenario, in which Middle East countries adopt policies to restrict their production growth and investments, global oil-investment requirements are 8% higher than in the Reference Scenario. Global oil demand would be 8% lower because of the higher prices that would result. Overall oil revenues in Middle East OPEC countries, and in OPEC countries in general, would be lower. These findings imply that it will be in the interests of both consumer and producer countries to facilitate capital flows to the Middle East upstream oil sector. This is a key issue that will need to be addressed in the context of the consumer-producer dialogue.

This chapter summarises the outlook for oil investment globally and then by sub-sector: conventional crude oil exploration and development, non-conventional oil production, refining and transportation. A further section assesses the main drivers of investment and uncertainties about future needs and whether they will be financed. This is followed by an analysis of oil investment prospects in each major region. The chapter ends with an analysis of a scenario in which investment in Middle East oil production capacity is restricted, to test the effect of lower production in that region on global oil supply and assess the implications for investment needs.[1]

Global Investment Outlook

A little over $3 trillion of investment will be needed in the oil sector through to 2030 (Figure 4.1). Investment needs will average $103 billion per year, but will increase steadily through the period as demand increases. Annual capital spending will rise from $92 billion in the current decade to $114 billion in the last decade of the projection period. The share of investment spending in OECD countries is high relative to their production capacity because unit costs are higher, particularly in the upstream segment of the supply chain, compared to other regions.

Figure 4.1: **World Cumulative Oil Investment by Decade**

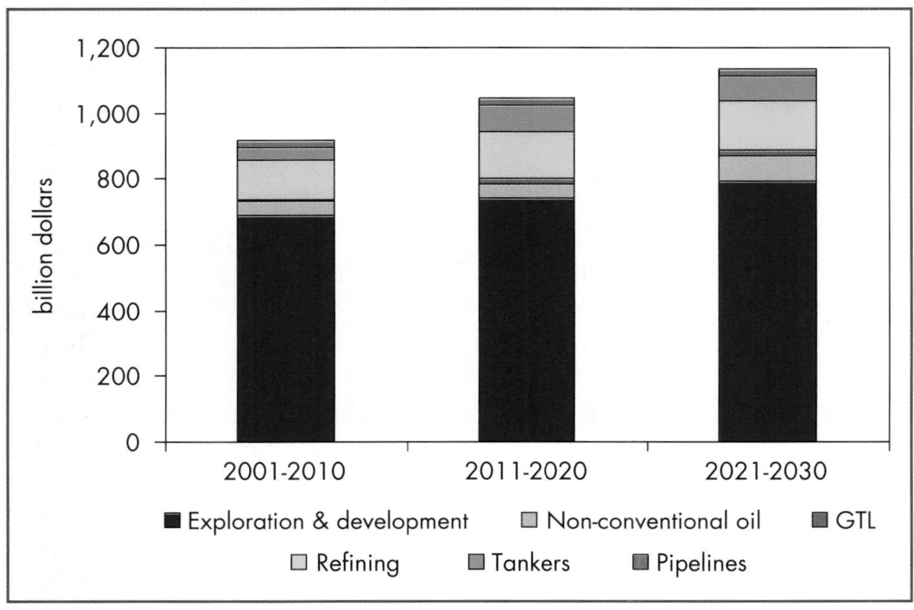

Note: Non-conventional oil excluding GTL

1. The preparation of this chapter benefited from the joint OPEC/IEA Workshop on Oil Investment Prospects held at the OPEC Secretariat in Vienna on 25 June 2003. See OPEC (2003).

Box 4.1: **World Oil Production Outlook**

In the Reference Scenario of the *World Energy Outlook 2002 (WEO-2002)*, oil demand is projected to grow from 77 mb/d in 2002 to 120 mb/d in 2030. To meet this increase in demand, production in OPEC countries[2] will have to grow continuously from 29 mb/d in 2002 to 65 mb/d by 2030 (Figure 4.2). Their output will grow most rapidly in the second and third decades, when they will account for most of the increase in world crude oil production. Supported by recent high oil prices, non-OPEC supply will remain around the current level until 2010. Then it will decline slowly as production in non-OPEC regions, notably the transition economies, Africa and Latin America, will no longer compensate for output declines in mature areas, such as North America and the North Sea. Non-OPEC production will fall to 42 mb/d in 2030. Non-conventional oil production and processing gains provide the balance with world oil demand.[3]

Uncertainties associated with the above projections are discussed throughout this chapter. In particular, the Restricted Middle East Investment Scenario highlights the consequences on the global oil market of insufficient investment in the Middle East. Demand-side uncertainties are discussed in the OECD Alternative Policy Scenario of *WEO-2002*. Under this scenario, new government policies and measures aimed at saving energy and cutting greenhouse-gas emissions would reduce OECD oil demand by 4.6 mb/d in 2030, lowering the call on OPEC by 7%.

Capital spending on exploration and development will dominate oil-sector investment, accounting for 72% of the total over the period 2001-2030 (Table 4.1). The share of upstream spending in total oil investment will be highest in the transition economies and Latin America (Figure 4.3). The bulk of this investment will be needed to maintain production levels at existing fields and in new fields that will produce in the future. The rest will be needed to meet projected growth in demand. Investment in non-conventional oil projects, mainly in Canada and Venezuela, will represent an important and growing share of total spending. Investment in crude oil tankers and oil pipelines will amount to $257 billion. This will be driven by rapid growth in inter-regional trade. Most of this investment will go towards expanding the

2. OPEC production in the IEA's World Energy Model is assumed to be the residual supplier to the world market. See IEA (2002a) for a description of the methodology for projecting oil production.
3. In the *WEO-2002*, Canadian raw bitumen was not classified as non-conventional oil but is now included in this category.

Figure 4.2: **World Oil Production**

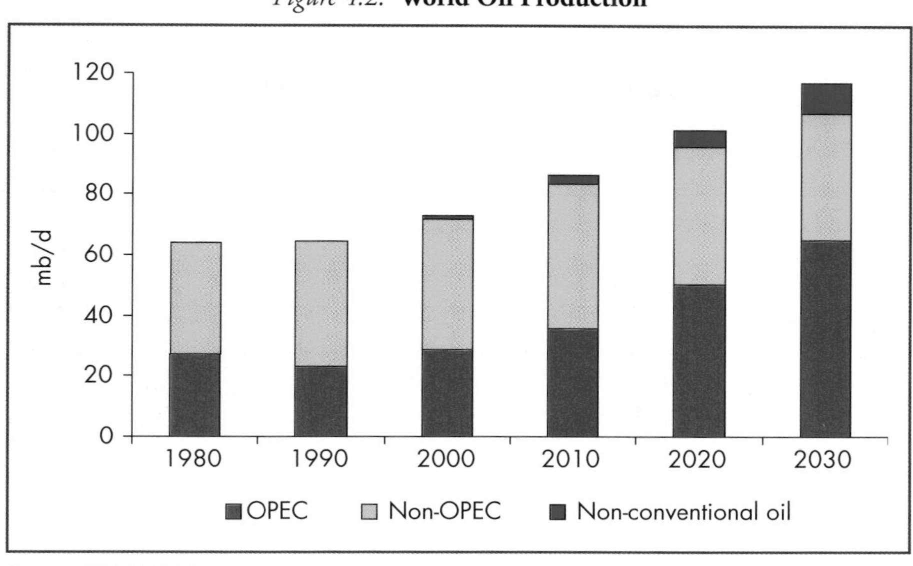

Source: IEA (2002a).

capacity of the oil-tanker fleet. Global investment in pipeline capacity will be relatively small, but will be important in some locations where large deposits are situated far from the coast.

Figure 4.3: **Oil Investment by Region, 2001-2030**

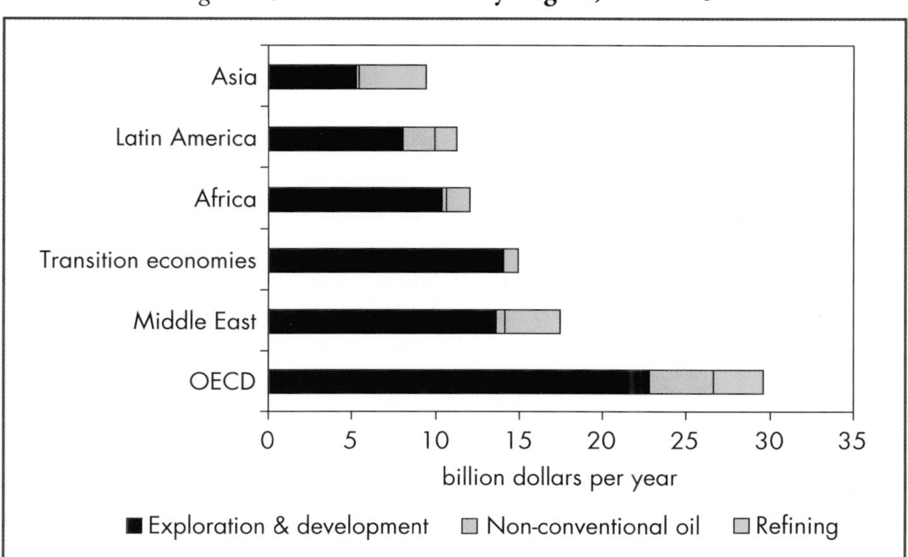

Table 4.1: **Global Oil Cumulative Investment by Region and Activity, 2001-2030*** (billion dollars)

	Exploration & development	Non-conventional oil	Refining	Total
OECD North America	466	114	43	622
OECD Europe	199	1	22	222
OECD Pacific	19	1	24	44
Total OECD	**684**	**115**	**89**	**888**
Russia	308	0	20	328
Other transition economies	113	0	7	120
Total transition economies	**422**	**0**	**26**	**448**
China	69	0	50	119
South and East Asia	87	7	69	163
Middle East	408	16	99	523
Africa	311	7	42	360
Latin America	241	59	37	336
Total developing countries	**1,116**	**89**	**297**	**1,502**
Total non-OECD	**1,538**	**89**	**323**	**1,950**
Total world	**2,222**	**205**	**412**	**2,839**

* Not including global transportation investment of $257 billion.

Cumulative investment in crude oil refining will total $412 billion, or $14 billion per year. This will go to increasing refining capacity to 121 mb/d and to upgrading refineries to match output to the changing product-demand slate. The greater part of refinery investment will be in Asia in response to strong growth in demand for transport fuels in the region. Large investments will also be needed in the Middle East and Africa as these regions seek to increase returns on their indigenous oil production by establishing export refining centres.

Although 69% of all oil-sector investment, excluding transportation, will occur outside the OECD countries, more than 40% of non-OECD investment will be in projects to supply crude and products to OECD countries (Table 4.2). In the Middle East, over half of oil investment is for exports to the OECD.

Table 4.2: **Oil Investment* in Non-OECD Countries by Supply Destination, 2001-2030**

	For supply to OECD markets		For supply to domestic and other non-OECD markets		Total
	$ billion	%	$ billion	%	$ billion
Total non-OECD	**792**	**41**	**1,159**	**59**	**1,950**
Of which :					
Russia	138	42	190	58	328
Middle East	237	45	286	55	523
Africa	166	46	194	54	360
OPEC	360	45	440	55	800

* Exploration and development, refining and GTL.

Exploration and Development

Investment Outlook

Cumulative global investment in exploration and development from 2001 to 2030 will amount to $2.2 trillion, or around $74 billion per year. Investment will grow over the projection period, from an estimated $69 billion per year in the first decade to $79 billion per year in the third. In comparison, spending was somewhat lower, at $64 billion, in 2000 as a consequence of low oil price in the previous year. Although capacity additions will be substantial, average upstream investment costs per unit of output will fall as more oil is expected to come from the Middle East, which is by far the world's lowest cost region (Figure 4.4).

Estimated investment requirements for each region depend on projected production, production decline rates, and exploration and development costs. Developing countries are expected to account for nearly 55% of global upstream investment. Investments in OECD countries will remain large despite the small and declining share of these countries in world oil production, which is projected to fall from 30% in 2002 to only 11% in 2030. In contrast, investment in Middle East OPEC countries represents only 18% of total investment, because of low unit costs in this region. This region will account for 43% of world oil supply in 2030, up from 29% at present.

Conventional Oil Production Prospects and Capacity Requirements

Some 470 billion barrels of oil will have to be found in the next three decades, to replace reserves in existing producing fields and to meet the growth in demand. This implies a decline in the global proven reserves to production ratio from around 40 years at present to under 20 years in 2030 (Figure 4.5).

Figure 4.4: **Average Development Costs and Proven Reserves by Region**

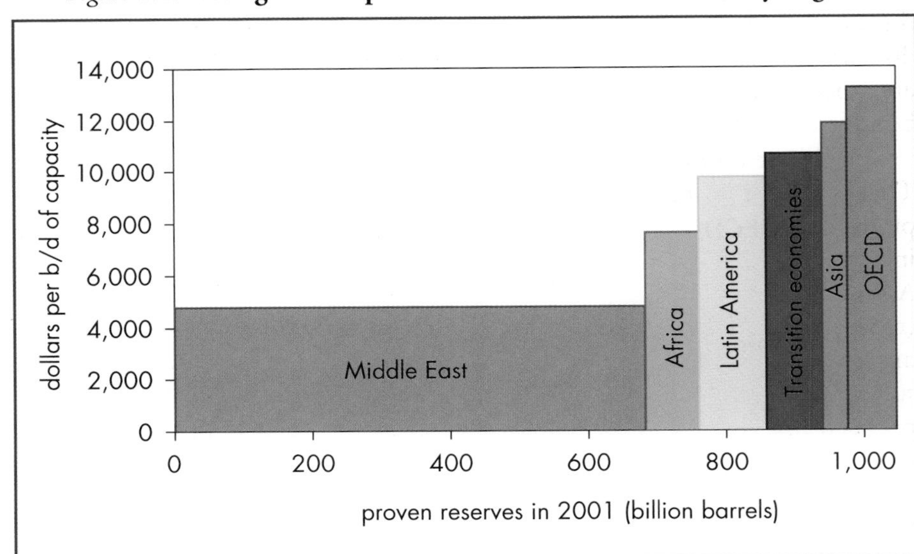

Note: The size of the area for each region provides an indication of the relative amounts of investment that would be notionally required to develop all existing proven reserves.
Source: IEA, *Oil and Gas Journal* (24 Dec 2001).

Figure 4.5: **World Reserves to Production Ratio**

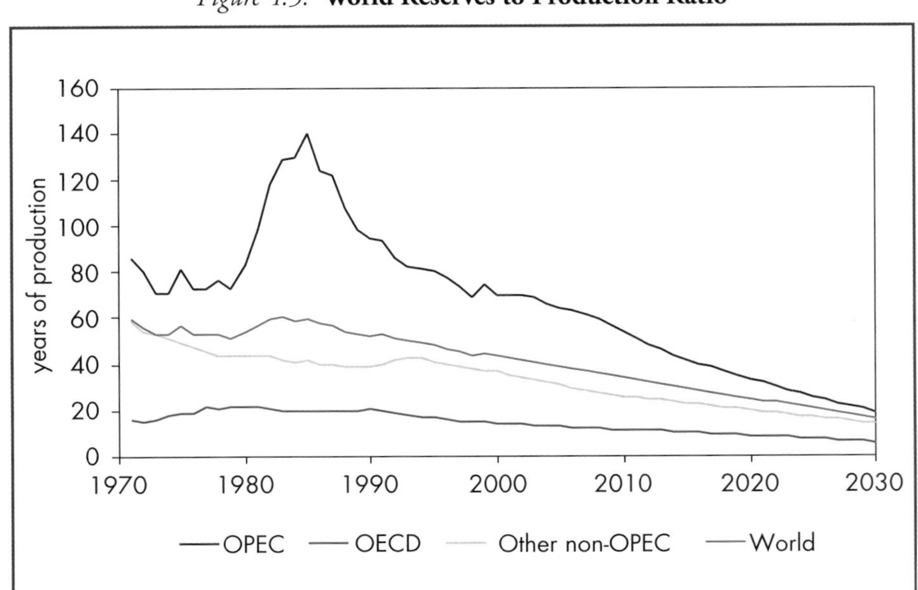

The ratio will remain much higher in OPEC countries, despite a rapid decline during the projection period as their production increases sharply. In the past 30 years, 768 billion barrels of oil have been proved to exist. Offshore discoveries are expected to account for close to half of the total amount of reserves that will be added over the projection period.

Offshore production currently represents around 30% of world oil supply. The Gulf of Mexico, Brazil, West Africa and the North Sea are the main producing areas at present. There is significant potential for offshore production increases in the Chinese Sea, Gulf of Oman, the Caspian Sea and South-East Asia. Potential but undiscovered offshore resources, especially in ultra-deep water (more than 1,500 metres), are believed to be nearly as large as undiscovered onshore reserves (Figure 4.6). Offshore developments are projected to increase substantially, reaching 34 mb/d in 2030. They will be the main source of new production in non-OPEC countries, where offshore production will represent more than 50% of total output in 2030.

In total, more than 200 mb/d of new production capacity will have to be added during the next three decades. This will be required mainly to replace progressive declines in production capacity from wells already in production or that come onstream during the projection period, as well as to meet demand

Figure 4.6: **Undiscovered Onshore and Offshore Oil by Region**

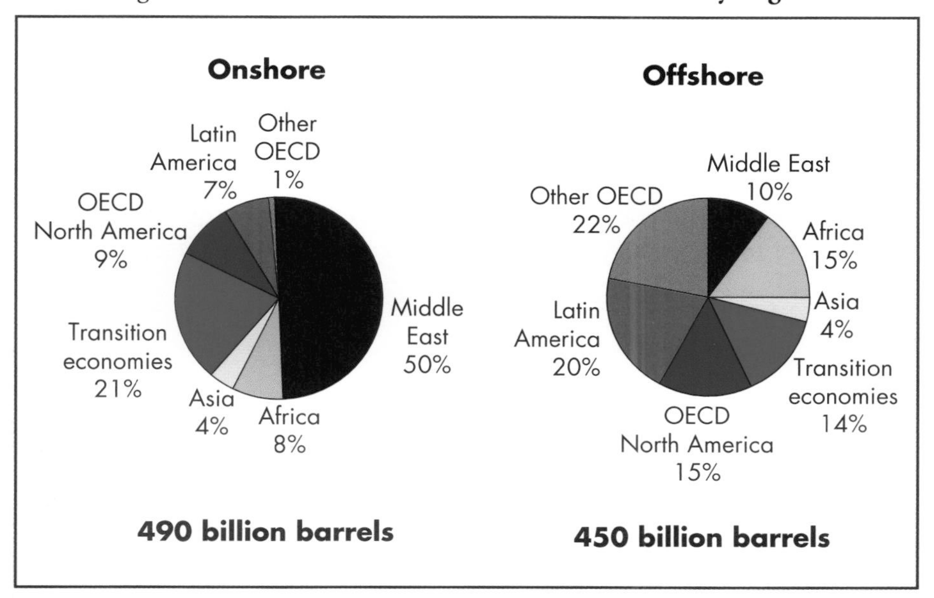

Source: USGS (2000).

growth (Box 4.2). Replacement capacity of 175 mb/d will be more than five times larger than the 33 mb/d of capacity additions required to meet demand growth (Figure 4.7). Of this replacement capacity, 37 mb/d will be needed simply to maintain capacity related to the increase in demand over the projection period. The rest will be needed to maintain *current* capacity.

Box 4.2: **Oil Production Decline Rates**

The term *decline rate* corresponds to the rate of decline in production over a given period from a given oil well or field, or averaged over a country or region. A distinction is made between the observed decline rate and the "natural" or "cashless" decline rate. The natural decline rate is the decline in production that would be observed in the absence of additional investment to sustain production.

For the purposes of projecting future investment needs, our analysis uses estimates of year-on-year natural decline rates averaged over all producing fields in a given country or region. These estimates were derived from information on observed decline rates, from which the estimated effects of investment were stripped out. In reality, there are rarely cases where no investment occurs to sustain production, so estimates of natural decline rates are inevitably uncertain. In addition, there is only partial information available on either observed or natural decline rates, partly because data on oilfield production and investments to sustain production from specific fields are confidential in many cases.

A typical production profile for an oil well or field shows production rapidly increasing to reach its peak and then declining relatively rapidly, before the rate of decline slows. The decline rate is, therefore, usually higher just after the production peak and falls progressively as the field ages. At a country or regional level, the average decline rate will reflect the production profiles of many fields with different characteristics and production start dates.

The amount of new capacity that will be needed to replace wells that will be phased out during the projection period is highly dependent on the assumed rate of natural production decline. At a global level, investment needs are, in fact, far more sensitive to the decline rate than to the rate of growth in oil demand. For example, for a given cost of development, an average decline rate of 10% per year would mean that more than twice as much investment would be needed compared to a rate of 5%.

Figure 4.7: **Conventional Oil Production Capacity**

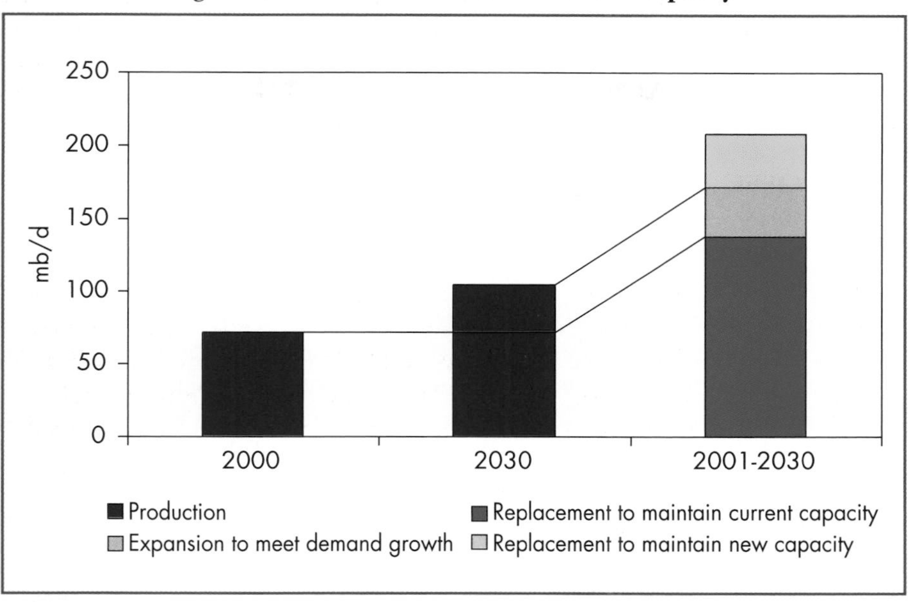

In practice, natural decline rates vary widely among regions and over time according to several factors, including geology, extraction technology, field age and production policies. Decline rates are usually lower for larger fields, because initial production rates are low relative to reserves and the fall in oilfield pressure is slower. Decline rates are affected by technology, which allows new fields that would have been too small in the past to be developed. In addition, more efficient oil extraction techniques, including horizontal wells, enable a higher peak to be reached more quickly, thereby sharpening the production profile of a field (see Northwest Europe Continental Shelf, below). However, the deployment of technologies that increase the recovery of reserves may mitigate production declines by increasing a field's recoverable reserves and prolonging its life.

The age of fields in a given country or region affects the average year-on-year decline rate. Production declines at fields that have already reached their peak could be partially offset by rising output at new fields, which need on average two to four years to reach their peak. The average decline rate for a region will also depend on the average age of fields in production. The stage of maturity of a field also affects the decline rate. After the peak is reached, the decline rate is often high in the first year or two, but then tends to fall progressively.

Natural decline rates are affected by the way in which an oilfield is developed and the production policies adopted once it comes on stream. In general, the more wells that are drilled in an oilfield, the quicker the drop in pressure in the field and the higher the eventual decline rate. The operator of the field must seek a balance between maximising production and cash flow early in the life of the field and maximising the amount of oil that can be recovered. Over-rapid exploitation can damage the long-term productivity of a field and reduce the proportion of the reserves that can be ultimately recovered. Decline rates at many fields in OPEC countries, especially in the Middle East, have been artificially reduced by production caps, which hold actual production rates well below capacity. The decline rate for some fields is negligible, reducing the aggregate decline rate for the country.

The decline rates assumed in our analysis vary over time and range from 5% per year to 11% per year. Rates are generally lowest in regions with the best production prospects and the highest reserves/production ratios, such as the Middle East, where they range from 4% to 6%. This study's assumed decline rates are highest in mature OECD producing areas, including onshore North America (where they average 9%) and Europe (11%). In Asia, rates range from 6% in Indonesia to 9% in India. In Latin America, the lowest rates are in Venezuela (5%) and highest in Argentina (9%). In African countries with large onshore reserves, such as Algeria and Libya, rates are typically around 6%. They average around 8% in West Africa.

Technology and Cost Developments[4]

Exploration costs, which include all investment that is needed before a discovery is confirmed, include mainly geophysical and geological analysis and drilling of exploration wells. Development costs cover spending after a discovery is confirmed, and mainly involve drilling of production wells and the installation of surface equipment. To calculate investment needs for exploration and development, this study drew on cost estimates from a wide range of sources, including commercial databases, private and national oil companies, international organisations, including OPEC, and literature surveys. The resulting database includes costs for different types of onshore and offshore locations and different sizes of field on a regional and, in some cases, country-by-country basis. Figure 4.8 summarises these cost estimates.

Average exploration and development costs per barrel fell sharply in the 1980s, continued to decline (though more slowly) in the early to mid-1990s and then started to rise in the second half of the 1990s.[5] The earlier fall in these costs can be explained primarily by rapid technological advances driven by the crude oil price spikes of 1973 and 1979. The application of vastly increased computing power to geophysical and geological interpretation has stimulated the development and interpretation of geophysical data. Developments in geophysical data

4. The following analysis also applies largely to the upstream gas sector.
5. DOE/EIA (2003).

Figure 4.8: **Current Exploration and Development Cost Estimates by Region**

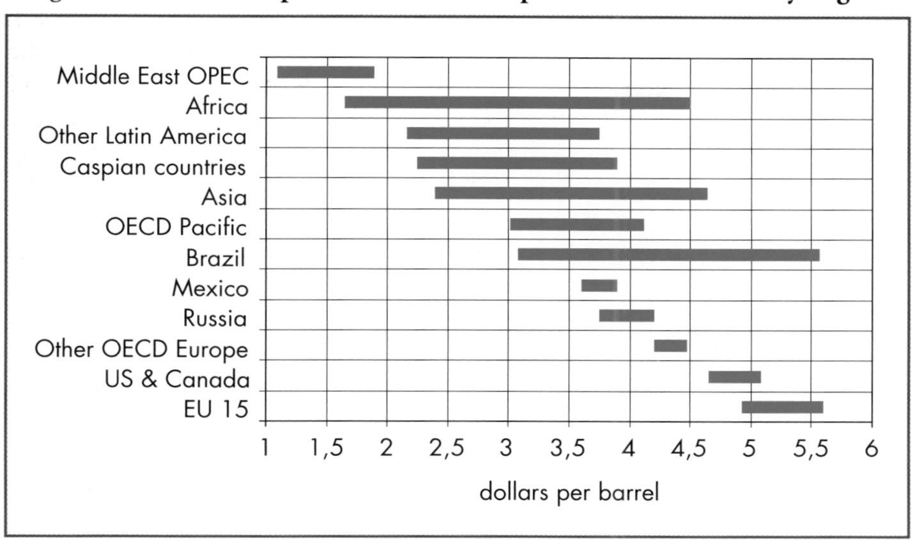

acquisition and interpretation, including 3-D and 4-D modelling and reservoir simulation, have led to a much better understanding of reservoir characteristics. As a result, drilling success rates have improved markedly and the number of dry holes has dropped (Figure 4.9), while exploration, development and operating (lifting) costs have all fallen.

Continuing improvements are expected in upstream technology, especially in the area of seismic techniques to better delineate reservoir characteristics. The deployment of underground sensors, for example, is expected to bring further cost reductions, but how big they will be is uncertain. Improvements in recovery rates would increase the overall recoverable resources and reduce unit production costs and the need for exploration. Enhanced oil recovery (EOR) or improved oil recovery[6] techniques typically allow the recovery of an additional 4% to 11% of the original oil in place, depending on the characteristics of the reservoir. Further technological improvements that boost EOR would help maintain production at existing fields for longer, so delaying the need for new field developments. Future unit cost reductions through technological improvements may be partly offset by a decline in the average size of fields. Development costs are very sensitive to field size, especially for very small fields (Figure 4.10).

6. Improved oil recovery techniques refer to processes designed to recover oil remaining after primary and secondary recovery techniques have been used. They include thermal and gas flooding and chemical and microbial methods.

Figure 4.9: **Oil Drilling Success Rate, Number of Well Completions Worldwide and International Oil Price, 1977-2001**

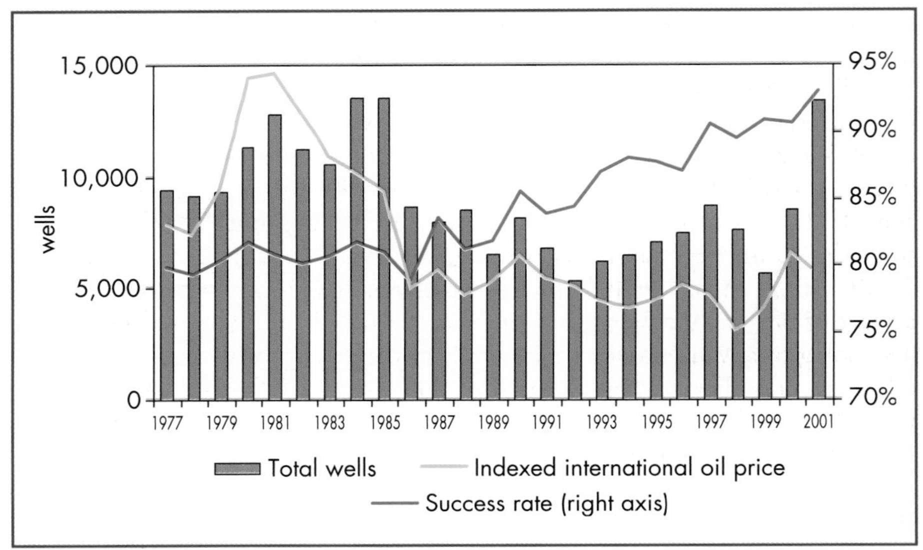

Source: IEA databases, US DOE EIA (2001).

Figure 4.10: **Field Size and Development Costs for OECD Europe, Onshore and Offshore**

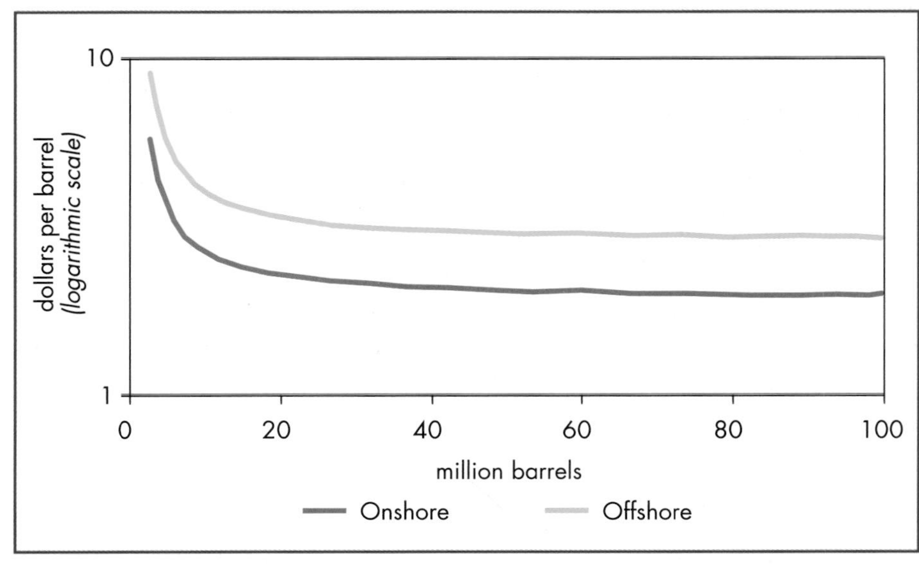

Source: IHS Energy (2003); IEA analysis.

Progress during the 1990s in oil exploration, development and production techniques in deep water (water depth exceeding the economic limits of fixed platforms – around 500 metres) and in ultra-deepwater (more than 1,500 metres) has led to reserve and capacity additions in regions that were not previously accessible. Progress has been led by innovations in 3-D and 4-D seismic techniques, horizontal drilling, multiphase pumps and floating production storage and off-take vessels. The record depth for offshore production has risen from 752m in 1993 (the Marlim field in Brazil) to 1,852m (the Roncador field in Brazil). This trend is expected to continue as exploration wells are continuing to be drilled in deeper and deeper water, the record currently stands at around 3,000 metres.[7] Advances in deepwater exploration have increased substantially the acreage of offshore sedimentary basins that can now be explored and developed. Technological advances have also drastically improved the economics of deep water oil production. For example, capital costs for the first Gulf of Mexico deep water fields developed in the 1970s were high, at around $25 per barrel. Fields in this region can now be brought on line for less than $10 per barrel.

Non Conventional Oil[8]

Investment Outlook

Global cumulative investment in non-conventional oil projects, including oil sands, raw bitumen, extra-heavy oil and gas-to-liquids, will amount to $205 billion over the period 2001-2030, or $6.8 billion per annum. The greater part of this investment will go towards developing Canada's oil sands resources and Venezuela's extra-heavy oil deposits, both of which are enormous. Cumulative investment is projected to amount to $92 billion in Canada and $52 billion in Venezuela. Global cumulative investment in gas-to-liquids (GTL) projects will amount to $40 billion over the period 2001-2030. Development of other non-conventional oil projects, including heavy oil plants in California and oil-shale plants in Australia, will make a small contribution to investment in the sector.

Non-conventional oil, excluding GTL, is expected to contribute around 8% to total world oil supplies by 2030. Production by that time is projected to reach 9.5 mb/d, most of it in Canada and Venezuela. Official estimates of proven reserves in the two regions total 580 billion barrels of recoverable oil – more than the entire proven reserves of conventional crude oil in the Middle East. The outlook for investment in Canada and Venezuela is discussed in the regional analysis (OECD North America and Latin America) below.

7. Unocal in the Gulf of Mexico.

8. See Annex 3 for a definition of non-conventional oil. The analysis of non-conventional oil in this study benefited from the results of an IEA workshop in Calgary in November 2002 (see www.worldenergyoutlook.org/weo/papers.asp for the proceedings).

International oil prices are the main source of uncertainty for investment in all non-conventional oil projects. Although technological improvements have dramatically reduced supply costs, significant expansion of the industry will depend on the expectation that oil prices will exceed $20 per barrel for a long period. Once projects are in place, and capital costs are sunk, they can continue to operate at relatively low oil prices.

Gas-to-Liquids (GTL)

Advances in technology, tighter environmental regulations and higher oil prices have led to a surge in interest in developing GTL projects based on low-cost gas reserves located far from markets. GTL plants produce high-quality oil products, as well as specialist products such as lubricants, by converting natural gas into synthetic gas and catalytic reforming or synthesising into liquids. Although the only large-scale commercial plants currently in operation are located in South Africa and Malaysia, much of the growth in GTL production is likely to occur in the Middle East, centred initially at least on Qatar. Investment in GTL projects is expected to be concentrated in the second and third decades of the projection period (Figure 4.11).

Figure 4.11: **GTL Cumulative Investment by Region, 2001-2030**

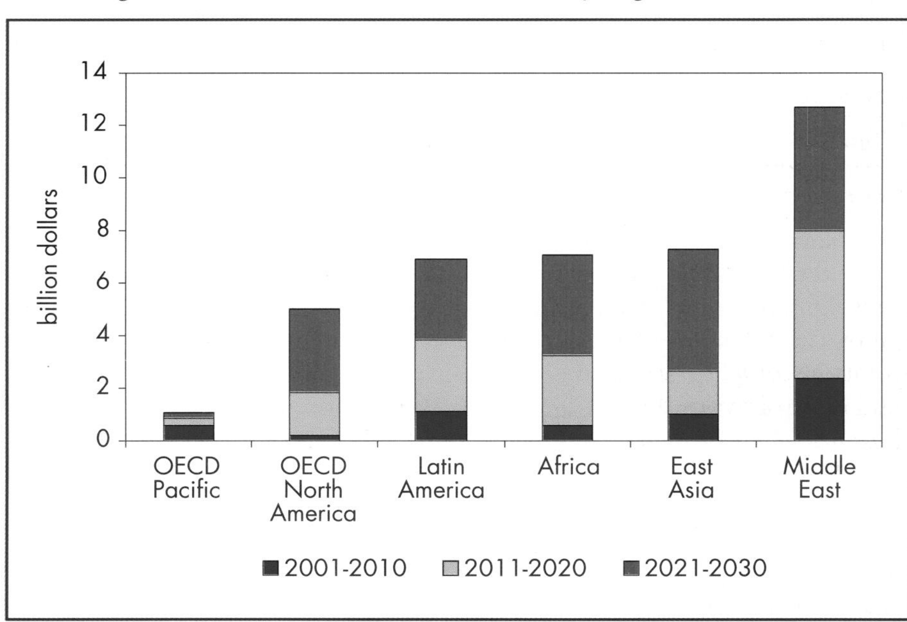

Output of liquids from GTL plants in 2002 was 42,500 b/d (Table 4.3). This is projected to increase to around 300,000 b/d by 2010 and 2.3 mb/d by 2030. The steeper rise in GTL production after 2010 is due to rising oil prices which are assumed to increase by $8/barrel (in year 2000 dollars) between 2010 and 2030. Global GTL demand for gas is projected to increase from 4 bcm in 2000 to 29 bcm in 2010 and 233 bcm in 2030. Much of this gas will be consumed in the conversion process, which is very energy-intensive.

Table 4.3: **Existing and Planned Commercial Scale GTL Plants**

Operator	**Location**	**Capacity** (b/d)	**Projected start-up date**
In operation			
Mossgas	Mossel Bay, S.Africa	30,000	1990
Shell	Bintulu, Malaysia	12,500	1993
Sub-total		**42,500**	
Commercial plants under construction or planned			
Sasol Qatar	Qatar	34,000	2005
Shell	Qatar	140,000	2008
Chevron	Nigeria	34,000	2005
Shell International Gas	Egypt	75,000	2005
Iran National Petroleum Company	Iran	110,000	-
Sub-total		**393,000**	
Total		**435,500**	

Source: IEA database; *Oil and Gas Journal* (25 November 2002).

The economics of GTL processing are highly dependent on plant construction costs, product types and yields, the market prices of the liquids produced and the gas feedstock, and the cost of carbon dioxide emissions. GTL plants are complex and capital-intensive. They require large sites and construction lead times of two-and-a-half to three years. Capital costs typically account for at least half of total levelised[9] costs for an integrated plant with power production on site. Syngas production accounts for about 30% and the Fischer-Tropsch

9. The present value of a cost, including capital, financing and operating costs, expressed as a stream of equal annual payments.

synthesis process itself about 15% of capital costs, with other processing units, power generation and other services making up the rest.

The latest GTL technologies being developed by Shell and Sasol, a South African energy company, are thought to involve capital costs of around $20,000 per barrel per day of capacity. A 75,000-b/d plant would, therefore, cost about $1.5 billion. Capital costs for particular GTL projects vary regionally because of differing labour and construction costs and on the basis of the desired mix of oil product outputs from the plant. As around 10 MBtu of natural gas is required per barrel of liquid produced, access to low-cost gas feedstock is crucial to the economics of GTL projects. For example, if natural gas can be secured for $1/MBtu, the cost of the gas feedstock alone will be $10 per barrel of liquid produced. Assuming a gas feedstock price of $0.75/MBtu, the breakeven price is around $15 per barrel of crude oil.

As technology improves, GTL is likely to compete for investment funds against both oil refining and alternative ways of exploiting gas reserves. GTL may be the preferred option for "geographically stranded" gas reserves where the costs of piping or shipping the gas as LNG to markets are prohibitive. Gas that is stranded owing to quality concerns is unlikely to be favoured as a GTL feedstock in the absence of technological breakthroughs to either clean the gas at low cost or to process it as is. GTL products are generally able to command a premium price because of their very high quality.

Oil Refining

Investment Outlook

Global cumulative investment in the refining sector will amount to $412 billion over the period 2001-2030, or an average of $14 billion per annum.[10] This comprises investment to increase refining capacity to meet demand growth and investment to increase conversion and quality-treatment capability so that refinery output continues to match changes in the mix of oil product demand.

The bulk of refinery investment will occur in Asia in response to the region's strong growth in demand for refined products, particularly in China and India (Figure 4.12). Refinery investment in OECD regions will be moderate and will be dominated by spending to improve product quality and increase conversion capability. OECD refiners will rely largely on capacity creep[11] at existing refineries and increased imports to meet their demand growth. Environmental and planning regulations, and in some cases high land costs, have

10. This does not include investment in refinery maintenance which is typically around 1-2% of refinery replacement value per year. Investment in product quality is limited to road transport fuels.
11. Capacity creep is the increase in a refinery's capacity arising from conventional expansion and/or debottlenecking investments.

made it increasingly difficult to build new refineries in OECD regions over the past decade. Refinery investment in the transition economies will be concentrated on updating existing facilities rather than expanding capacity. Technologies in this region, especially for product conversion, are generally outdated in comparison to international norms.

Figure 4.12: **Refinery Sector Cumulative Investment by Region, 2001-2030**

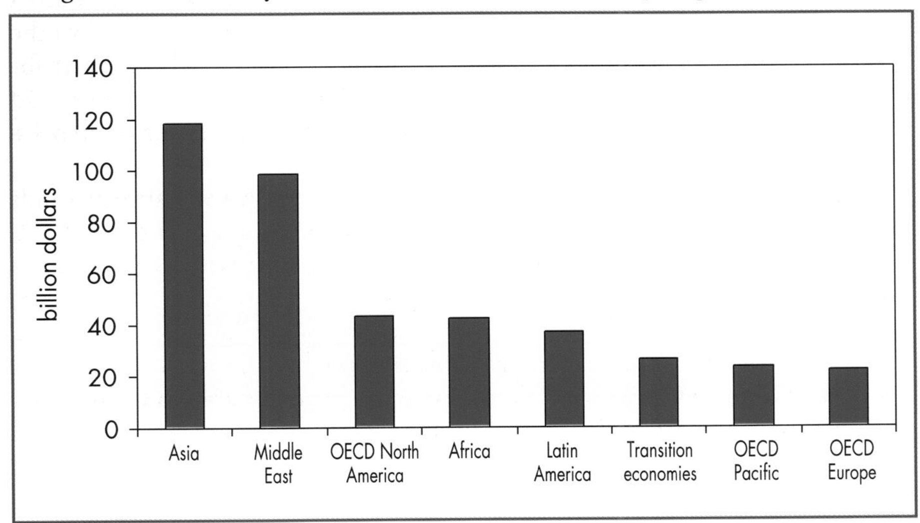

Refinery investment will be considerable in the Middle East and Africa. Much of this will be to build new refineries to export products to expanding markets in the Asia-Pacific region as well as to Europe and North America. For this trade to be possible, export refiners will need to invest to match the high fuel quality requirements of their target markets. Development of export refining centres in producer countries is usually motivated by the goal of capturing the rent in the supply chain. The optimal location for a refinery is usually close to markets, as transportation costs for refined products are higher than for crude oil.

Refining Capacity Requirements

To meet demand for refined products of 114 mb/d in 2030, global refining capacity will need to increase by an average 1.3% a year, reaching 121 mb/d in 2030. This represents expansion of close to 50% of current capacity. Demand for transportation fuels – essentially gasoline, diesel and aviation turbine fuel – will grow at more than twice the rate of heavier refined products. In response, refinery conversion capability will have to be increased to ensure refinery output continues to match the changing product demand mix. The rate of growth in refinery conversion capacity has been derived from the projected decline in heavy fuel oil demand. If, as expected, global crude oil supply becomes gradually heavier, conversion-capacity requirements will increase. Heavy crude oils yield lower proportions of light and middle distillate products and so additional processing of the heavier residues is needed.

Improvements in the quality of refined products will be required to facilitate the introduction of advanced vehicle-emission technologies designed to improve fuel efficiency and reduce tailpipe emissions. An increasing number of countries have set timetables for the introduction of such standards (Figure 4.13). In addition to government measures, a group of vehicle manufacturers has developed the *World-Wide Fuels Charter*[12] to promote improved understanding of fuel-quality needs and to harmonise fuel-quality standards. This charter provides a useful indication of the possible evolution of fuel-quality standards, which includes a goal of sulphur-free gasoline and diesel.

Figure 4.13: **Diesel Sulphur Standards in Selected Markets**

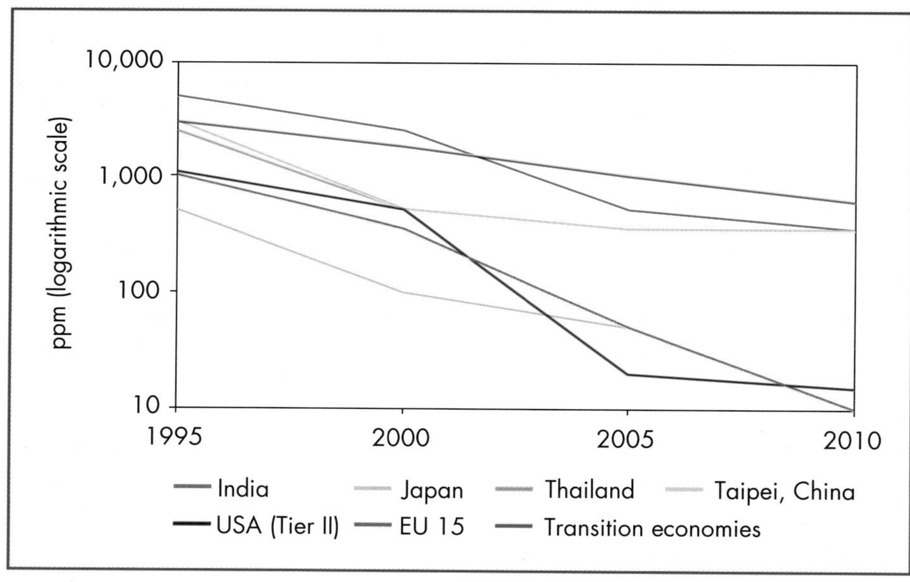

Investment requirements for product-quality improvements have been derived from published cost estimates for meeting worldwide the fuel-quality standards already adopted in various major markets, including Europe, North

12. World-Wide Fuel Charter 2002. http://www.autoalliance.org/fuel_charter.htm

America and Asia-Pacific, but not yet fully implemented even in the OECD. The assessment assumes that similar unit costs will be incurred globally. In practice, the exact cost of improving fuel quality in a region will depend on the interaction of many factors, including existing refinery configurations, crude slates and technological developments in refinery processing. Further fuel-quality improvements are likely through to 2030, but no allowance has been made for the associated investment because the nature of such future changes cannot be predicted. In addition to investments in improving the quality of road transport fuels, refiners are expected to incur considerable costs to meet environmental standards for other fuels, such as bunker oil, and to reduce refinery emissions.

Technology and Cost Developments

The current capital cost of building a new refinery in an OECD country is thought to be around $10,000 per barrel of daily capacity, though this cost is difficult to estimate as very few refineries have been built in recent years. Costs in non-OECD regions, at around $8,000 per barrel of daily capacity, are lower because of lower material and labour costs as well as lower costs associated with pollution abatement. Recent greenfield refinery projects in non-OECD countries have tended to be large and have, therefore, benefited from economies of scale. For example, the capital cost of Reliance Petroleum's 580-kb/d refinery in India – which is able to produce high-quality fuels – was less than $5,000 per barrel of daily capacity. In all regions, expansion at existing facilities remains the least costly way of increasing refining capacity.

The capital cost associated with conversion of the residual heavy fractions of the distillation process into more useful products varies both regionally and between refineries. Regions that already have a high conversion capability, such as North America, are increasingly investing in more expensive deep-conversion processes that are able to break down the heaviest residues. In contrast, there is greater scope to employ less sophisticated conversion technologies in the simple hydroskimming refineries that characterise the Asian industry. Capital costs for conversion processes are of the order of $10,000 per barrel of capacity per day in OECD countries and $8,000 in non-OECD countries. The cost of capacity creep of existing conversion capacity is significantly lower.

Investment associated with improving fuel-quality specifications is dependent on the extent to which standards are raised, on refinery configurations and on the quality of the crude oil processed. Reductions in unit costs have arisen from advances in refinery processing technologies, which are constantly evolving as experience with producing high-quality fuels grows. Typically, reducing sulphur levels involves the highest costs. It has been estimated that the investment in capital items required by the European refining industry to reduce gasoline and diesel sulphur levels from 50 parts per million (ppm) to 10 ppm, as required as part of their Auto-Oil II programme, will be $5.5 billion.[13] This equates to around $420 per barrel of refining capacity. Cost estimates for meeting reduced sulphur levels required by the Clean Air Act in the United States are around $650 per barrel of capacity.

Transportation[14]

Investment Outlook

Considerable investment to transport crude oil and refined products to market via tankers and pipelines will be necessary to accommodate the dramatic increase in international trade through to 2030. Because of lengthening supply chains resulting from increased production in the Middle East, far from the main consuming centres, an increasing share of trade will be waterborne. As a consequence, most of the investment will go towards expanding the capacity of the oil-tanker fleet. Whilst investment in pipeline capacity will be more subdued, it will nonetheless be important in terms of opening up landlocked reserves, particularly in the strategically important Central Asian region.

Cumulative global investment in the oil-tanker sector will amount to $192 billion over the period 2001-2030, or an average of $6.4 billion per annum (Figure 4.14). In comparison, investment in 2002 (excluding handy size tankers) totalled $4.7 billion.[15] This comprises investment to increase overall capacity and fleet renewal. Most oil tankers are built in Asian shipyards, mainly in South Korea and Japan.

Investment to replace older tankers and to meet environmental regulations will account for 52% of total spending on new oil tankers through to 2030, with expenditure particularly high in the first half of the *Outlook* period, after the entry into force of international pollution-prevention regulations. These regulations – which require single-hulled tankers to be progressively phased out of service – will result in the most dramatic modernisation ever experienced in the oil-tanker industry. Following the sinking of the *Prestige* off the Spanish coast in 2002, the European Parliament decided to bring forward the phase-out date to 2010 (2005 for vessels older than 23 years) and to ban the carriage of heavy oils in single-hulled tankers with immediate effect. In non-EU waters, the phase-out date for single-hulled tankers is currently 2015.

Most of the very large oil-pipeline projects that are currently under construction or being planned are in transition economies. The value of these

13. Concawe (2000).

14. Includes investment in international oil tanker and pipeline trade; spending on domestic and intra-regional trade is excluded.

15. Clarkson Research Studies (June 2003). Excludes tankers under 25,000 DWT.

Figure 4.14: **World Oil Tanker Cumulative Investment, 2001-2030**

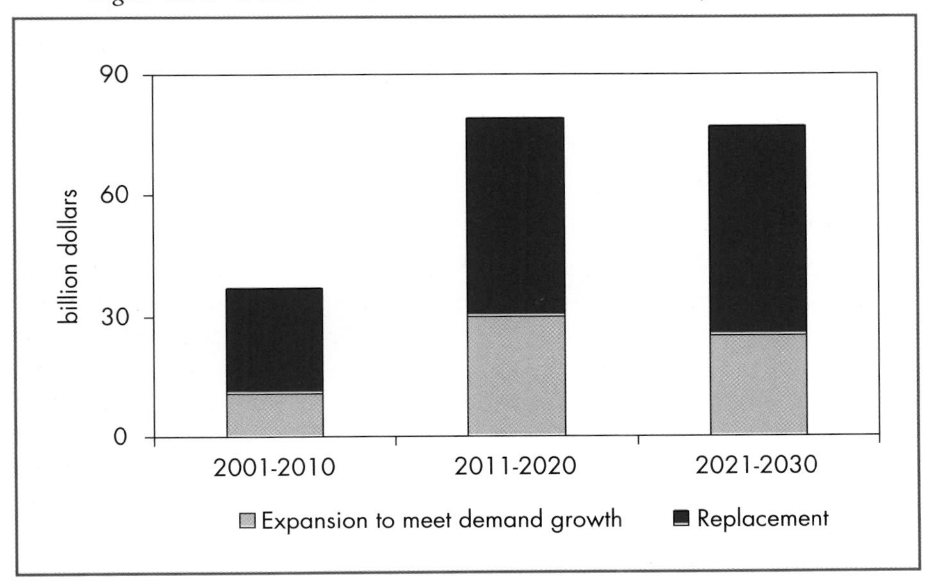

projects is around $22 billion. In total, global pipeline investment is projected to amount to $65 billion over the projection period.

Transport Capacity Requirements

International trade in oil and refined products will increase by around 80% between 2002 and 2030 (Figure 4.15). This will be driven by the widening regional mismatch between indigenous production of both crude oil and refined products and domestic demand. The Middle East is expected to see the biggest increase in net exports, the bulk of which will go to Asia. Oil exports from Africa, Latin America and the transition economies will also grow, but much less rapidly. The share of refined products in total oil trade will increase over the projection period. Oil tankers will remain the main means of transport, owing to lengthening supply chains.

In response to the growth in trade and changing trade patterns, the oil-tanker fleet[16] is projected to expand by over 90% to 522 million dead weight tonnes (DWT) by 2030 (Table 4.4). In addition to increasing capacity to meet demand growth, tanker-capacity additions will be required to compensate for scrappings. Around 60% of the existing oil-tanker fleet will have to be scrapped by 2015 in order to meet the requirements of international pollution-prevention regulations.

16. Tankers of over 25,000 DWT.

Figure 4.15: **Major Oil Trade Movements, 2002 and 2030** (mb/d)

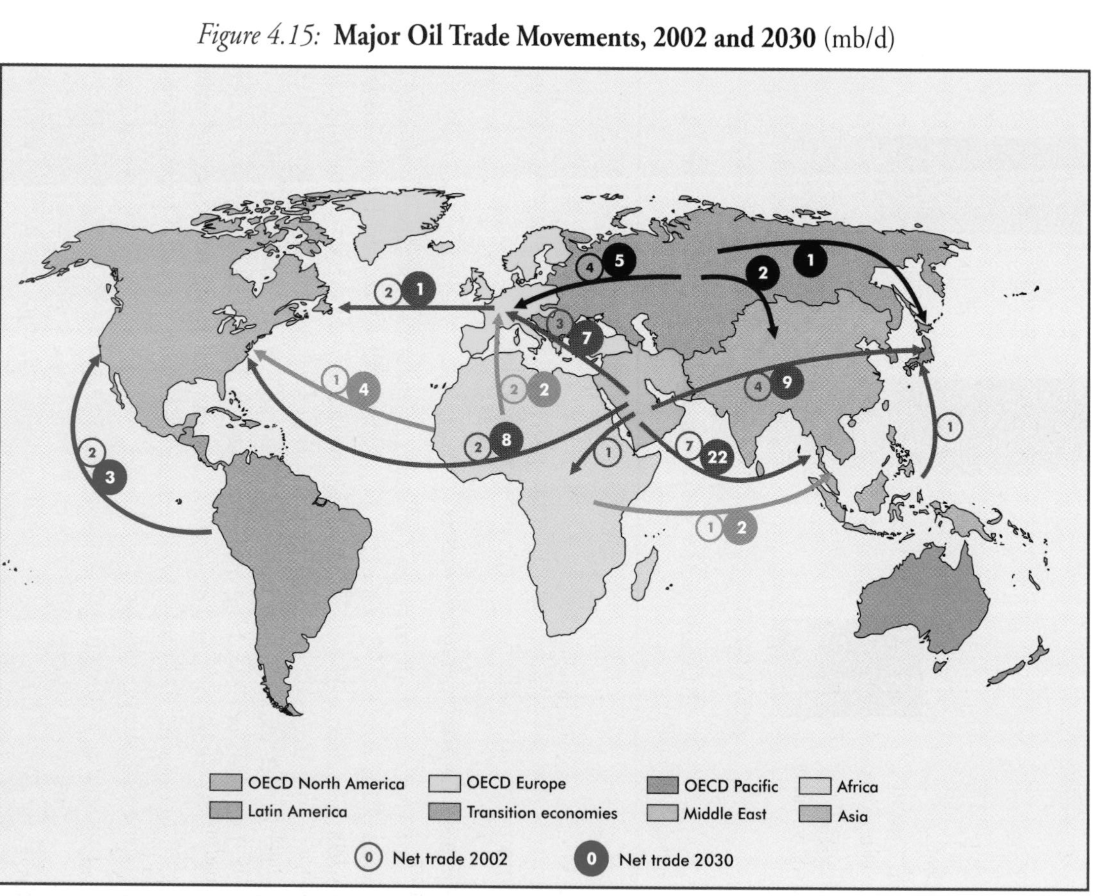

Source: 2002 data: BP (2003); 2030 projections: IEA analysis.

Table 4.4: **World Oil Tanker Fleet**

	2001	2010	2020	2030
Capacity (million DWT)	271	365	425	522

Source: Clarkson (2003); IEA analysis.

There will be a structural shift in the composition of the fleet towards larger vessels throughout the projection period. This will be driven by the lengthening of supply chains which will result from the Middle East being called upon to supply an increasing share of world crude oil and refined products. Larger vessels, particularly very large crude carriers (VLCCs), offer economies of scale on long-haul, high-volume voyages. Handy sized tankers (small vessels of less than 25,000 DWT) have been excluded from our analysis as these are typically used for domestic oil movements, not international trade. Analysis of the age profile of the handy tanker category suggests that much of it will need to be replaced soon.

Inter-regional crude oil and refined product trade via pipeline is projected to increase by 42% between now and 2030. Most of the additional capacity will be constructed in regions where oil reserves are landlocked, such as Azerbaijan and Kazakhstan, and in providing Russian oil with access to Asian markets (Table 4.5).

Table 4.5: **Major Pipeline Projects under Construction or Planned**

From	To	Length (km)	Status at end 2002	Expected completion	Cost ($ billion)
Azerbaijan	Turkey, Ceyhan	1,731	E	2005	2.7
Bulgaria	Albania, Vlora	906	P	2005	1.1
Canada, Athabasca	Canada, Edmonton	516	P	2005	1.0
Chad	Cameroon, Port of Kribi	1,047	C	2003	2.2
Ecuador, Lago Agrio	Ecuador, Esmeraldas	502	C	2003	1.2
Kazakhstan*	China	3,101	C	2007	2.4
Pakistan, Karachi	Pakistan, Mehmood Kot	816	P	2003	0.5
Russia	China, Beijing	2,500	E	2005	1.6
Russia	Coast of the Sea of Japan, Nakhodka	4,000	P	-	5.0
Russia	Kazakhstan, Atyrau	449	C	2003	0.8
Russia	Russia, Primorsk	270	P	-	1.2

C: Construction. P: Planned. E: Engineering.

* Assuming this pipeline runs from Kazakhstan's western Aktobe region, a further 4,000 km of pipeline would be needed to reach the peak demand of China's eastern coastal cities.

Source: *Oil and Gas Journal*, 25 November 2002; IEA database.

Technology and Cost Developments

Oil tankers fall into four categories on the basis of their size and the trade routes that they typically ply (Table 4.6). On long-haul voyages, larger vessels offer economies of scale, but physical limitations imposed by ports or channels prevent their use on certain routes. For example, the Suez Canal, the strategic waterway which connects the Red Sea with the Mediterranean, is currently too shallow and narrow for VLCCs. The Egyptian government, which generates much of its foreign currency income from Suez Canal transit fees, is currently undertaking a long-term project to widen and deepen the Canal to allow larger vessels to pass through.

Table 4.6: **Assumed Oil Tanker Construction Costs by Category**

Category	**Size** (thousand DWT)	**Cost per tanker** ($ million)
VLCC	> 200	73
Suezmax	120-200	49
Aframax	80-120	39
Panamax	60-80	36

Historically the cost of constructing oil tankers has declined owing to consistent surplus capacity and continued productivity gains, particularly in Japanese and Korean shipyards. The expansion of Chinese shipyard capacity and better productivity as the yards gain experience are expected to continue to exert downward pressure on prices.

Unit pipeline costs are highly variable, ranging typically from $0.7 million to $1.6 million per kilometre. The most important determinant of unit costs is capacity, because of the significant economies of scale involved. These savings need to be weighed against the advantages of multiple routes which offer risk insurance against temporary closures. The geography of the route also affects unit costs, particularly if there are major physical obstacles present. Annual operating costs are typically around 2% of capital costs.

Investment Uncertainties and Challenges

Opportunities and Incentives to Invest

Although the oil-investment flows projected for the next three decades will be large and will rise progressively, availability of capital is not expected to be an investment constraint. But that does not guarantee that all those investments will be made or that capital will flow to where it is needed. Several factors could discourage or prevent investment from occurring in particular regions or sectors. In other words, investment may be constrained by a lack of profitable business opportunities rather than any absolute shortage of capital. Table 4.7 summarises the necessary and sufficient conditions for an oil company to invest in a particular upstream project.[17]

Table 4.7: **Investment Conditions for Upstream Oil Investment**

Necessary conditions	Sufficient conditions
• Confidence in market demand • Resource base – quantity and quality • Access to reserves • Legal and institutional framework • Rule of law	• Profitability • Acceptable risk • Repeatability – opportunity to sustain profitable stream of investment • Fit with corporate strategy • Fit with portfolio of current and planned assets and projects

Where necessary investment conditions are fulfilled, profitability and risk are the key factors that determine whether the investment goes ahead. The required rate of return, or hurdle rate, on any investment varies according to the risks associated with it. Actual returns will depend on unit costs (including capital and operating costs), fiscal terms and oil prices. If oil companies expect actual returns to fall or if they raise their investment hurdle rates for a given region and type of project, capital flows decline. Equally, an expected improvement in returns relative to hurdle rates will normally lead to higher investment if the improvement is sustainable. As investment shifts to regions where country risk is likely to be higher, oil companies' average hurdle rates will rise.

The most volatile element in the investment equation is the oil price. Upstream global oil and gas investment in recent years has tended to fluctuate in line with oil prices. A price collapse, such as in 1985 and 1998, typically leads to a subsequent reduction in investment spending. Conversely, higher prices tend to encourage investment spending, as has been seen in recent years (Figure 4.16). The rate of investment, in turn, affects prices. There is also evidence that the increasing short-term volatility of oil prices, by increasing risk and therefore hurdle

17. For additional discussions on the investment decision-making process, see Chapter 3.

rates, is constraining investment.[18] Global upstream oil and gas investment is projected to fall significantly from recent record levels as prices fall, though average investment for the current decade, at $117 billion per year, is projected to be significantly higher than that in the 1990s ($93 billion per year).

Figure 4.16: **Global Upstream Oil and Gas Investment and Average IEA Oil Import Price**

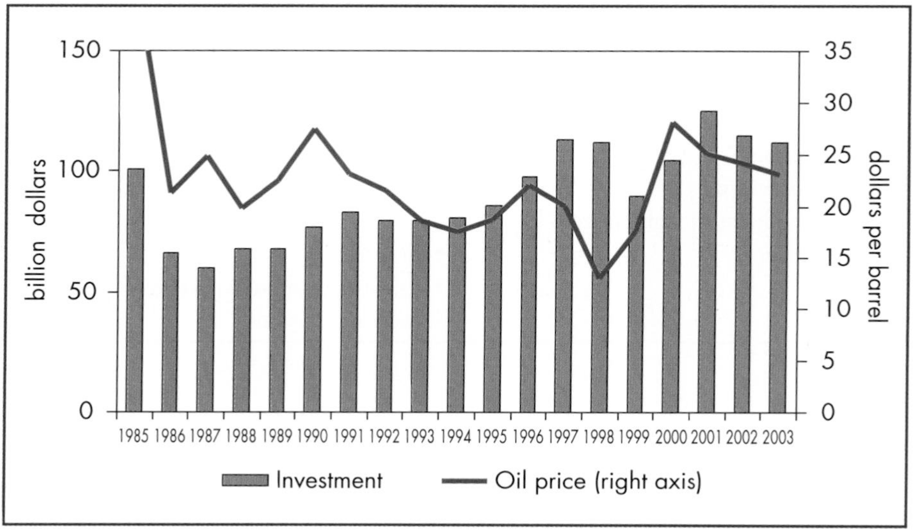

Source: IEA database; Lehman Brothers (2002).

Upstream investment decisions depend on oil company assumptions about future oil prices. Oil companies are continuing to factor oil prices of $15 to $18 per barrel into their investment decisions, even though prices have averaged over $25 per barrel since 2000. Such conservative price assumptions may reflect pressure from shareholders to maintain high investment returns and reluctance by management to risk criticism in the event that targeted returns are not achieved. Uncertainties surrounding the emergence of competition from low-cost regions also support the adoption of conservative oil price assumptions, leading to a decline in capital spending as a share of operating cash flow (Figure 4.17). As a result, some oil companies have seen their average returns on equity rise and their debt equity ratios decline (see Chapter 3).

18. IEA analysis points to a robust inverse relationship between upstream oil investment and price volatility; an increase in volatility results in a decline in investments, and *vice versa*. See IEA (2001).

Figure 4.17: **Oil Company* Capital Spending and Operating Cash Flow**

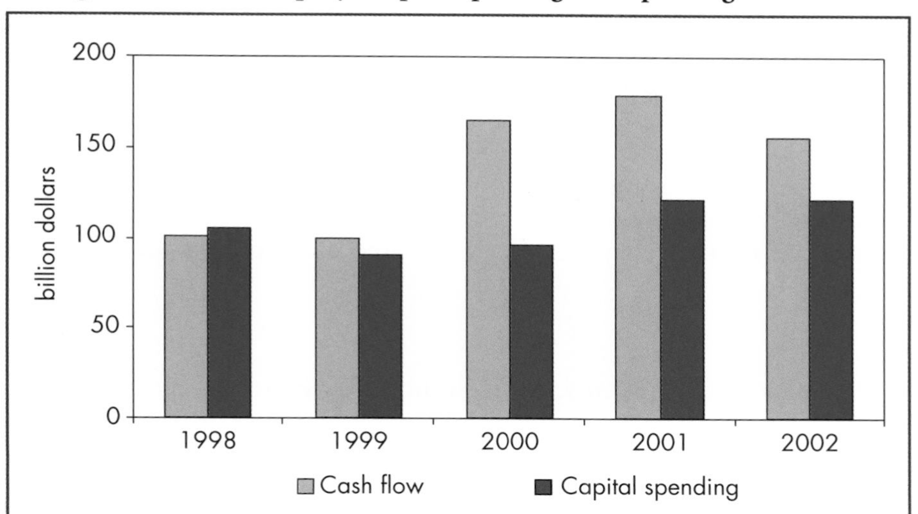

* Covers a sample of 36 integrated and independent exploration and development firms with a combined production of 36 mboe/d.
Source: Merrill Lynch (2003).

Access to Reserves

Opportunities to explore for and develop oil reserves depend on host country policies on foreign investment, depletion rates[19] and environmental protection (see below). The openness of those countries with large oil resources to foreign direct investment to develop their resources for export will be an important factor in determining how much upstream investment occurs and where. By the mid-1990s, most countries had at least partially opened their oil sector to foreign investment. However, in some of these countries, particularly Russia, China and Iran, foreign investment has proved difficult because of regulatory and administrative barriers and delays. Today, three major oil-producing countries – Kuwait, Mexico and Saudi Arabia – remain totally closed. However, plans are afoot in Kuwait to allow direct investment. Mexico has opened some areas to foreign oilfield services companies and may loosen its ban on direct investment in the future. Readiness to open up the Saudi upstream has so far been restricted to natural gas, but economic pressures could ultimately lead to some opening of the oil sector too. Internal political and socio-economic factors will determine the extent and pace of any such change in these countries.

19. Refers to the decrease in the amount of oil in a reservoir over a given period. It is a function of the remaining reserves at the start and end of the period and the rate of production. See IEA (2001) for details.

Several other countries which have opened up ownership restrict foreign investment to production-sharing and buyback deals, whereby control over reserves remains with the national oil company. These countries together with Saudi Arabia and Kuwait hold around 60% of the world's proven oil reserves (Figure 4.18). In those OPEC countries where foreign investment is possible, production may be subject to quotas adopted by that organisation – the so-called *OPEC risk*. Oil companies have to take this factor into consideration when evaluating possible investment in an OPEC country. Around 21% of world reserves are held by countries that offer private access through concession contracts.

Figure 4.18: **Access to Oil Reserves**

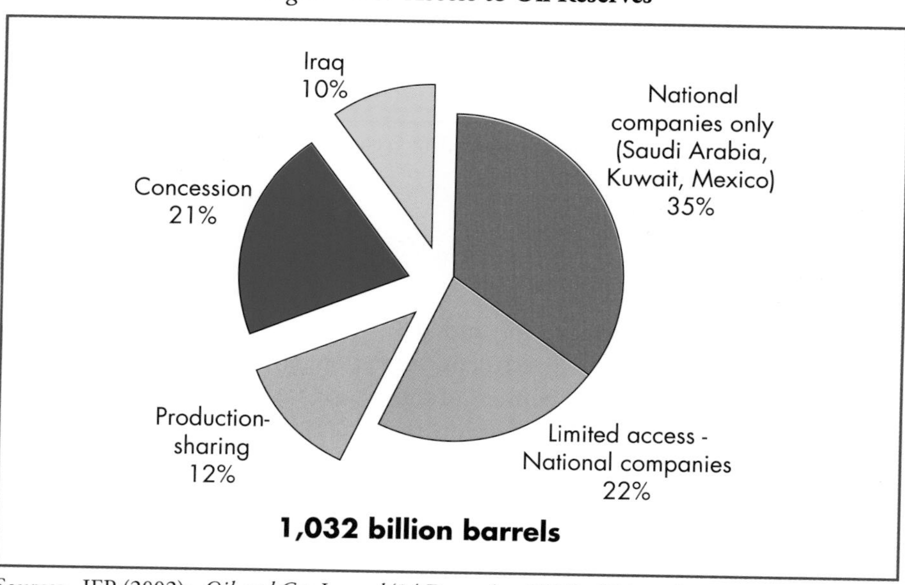

Sources: IFP (2002); *Oil and Gas Journal* (24 December 2001); IEA analysis.

Licensing and Fiscal Terms

The licensing and fiscal terms offered by host governments are a critical determinant of the attractiveness of an upstream investment. All governments have to strike a balance between maximising their share of the rent – the difference between the cost of production and the selling price of the oil produced – and encouraging investment. This is a matter of judgement, since the attractiveness of the investment conditions depends on perceptions of geological, economic and political risks relative to projected returns.

Governments must strike a balance between short- and long-term objectives: once investment has occurred, the host government may be inclined to raise taxes and royalty rates to increase revenues quickly, but at the risk of discouraging further investment. In practice, the rate of government take (taxes and royalties as a share of profits) varies considerably among countries and within countries according to the maturity of the upstream sector, short-term economic and political factors and investment risk. It also fluctuates over time. Take is typically lower in regions with a mature industry and relatively high extraction costs, such as the North Sea, and highest in those countries with the largest production potential and lowest development costs (Figure 4.19).

Figure 4.19: **Current Government Take and Proven Reserves**

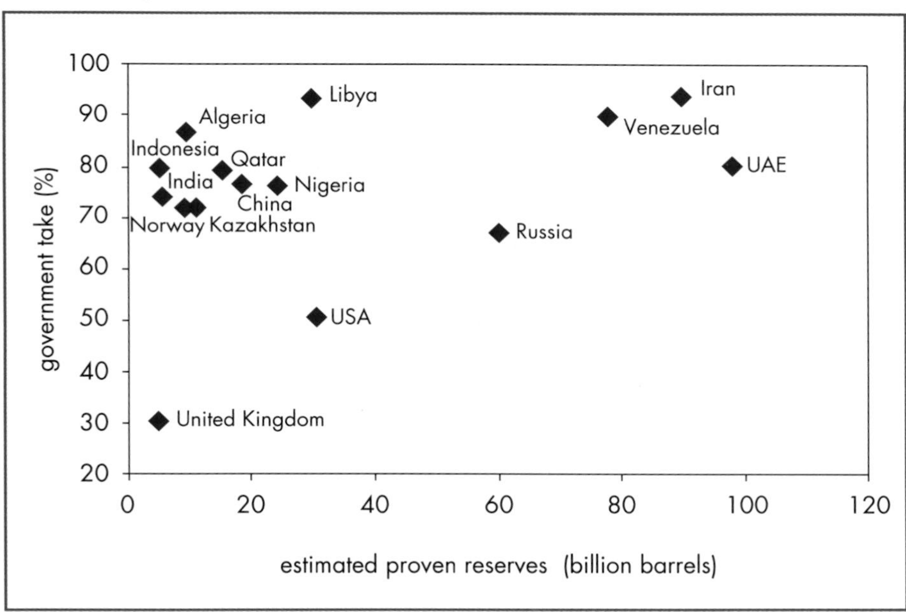

Source: IEA database.

The stability of the upstream regime is an important factor in oil companies' evaluation of investment opportunities. Frequent changes that retrospectively affect the taxation of sunk investments force investors to raise their hurdle rates for future investment decisions to accommodate the higher perceived risk. Production-sharing agreements (PSAs) – which effectively lock in fiscal terms for the life of a project and ring-fence it from future changes in the general upstream tax regime – are intended to get around this problem. PSAs have become a common form of upstream investment in some emerging markets, including those that have only recently opened up their oil sectors to foreign investment. Although PSAs were expected to be the main vehicle for foreign investment in Russia, there are signs that international oil companies are becoming more comfortable with investing under the general tax regime. However, this may change if prices drop (see Russia section below).

The detailed design of the fiscal regime, especially how sensitive government take is to movements in oil prices, shapes investment risk and the incentive to invest. A tax structure that protects oil companies' investment returns against a drop in prices shifts the price risk onto the host country and causes tax revenues to fluctuate in line with prices. This creates difficulties for short-term economic management – particularly for countries which rely heavily on oil revenues – and may be politically unacceptable. On the other hand, a regime that makes upstream returns vulnerable to a fall in oil prices will lead oil companies to impose higher hurdle rates to compensate for the fiscal risk. High royalty rates based on prices rather than profits that have been proposed in Algeria and Venezuela will deter investment in marginal fields with high development costs.[20]

Impact of Environmental Regulations

Environmental considerations increasingly affect opportunities for oil investment and the cost of new projects. Worries about the harmful effects of oil and gas drilling on the environment are raising the risk of, or simply holding back, investment in several countries. Even where drilling is allowed, environmental regulations and policies may impose restrictions or onerous investment and operating standards, driving up capital costs and causing delays. Public opposition to upstream and downstream projects on environmental and ethical grounds, even if ill-grounded, may block oil companies from investing in controversial projects (Box 4.3).

In the United States, moratoriums on drilling covering large swathes of federal onshore lands – including the potentially prolific Arctic National Wildlife Refuge (ANWR) – and offshore coastal zones have been in place for many years. Even if ANWR is opened to drilling, as the current administration hopes, severe restrictions aimed at limiting the environmental "footprint" are likely. In Norway, exploration and development activities can only proceed once detailed analyses of the risk of oil spills and of gaseous emissions have been completed. Gas flaring is banned and other emissions are regulated. Carbon dioxide emissions are taxed. In addition, drilling in the Norwegian and Barents Seas was halted recently to give time for an environmental impact assessment to be completed (see Northwest

20. Kellas and Castellani (2002).

Box 4.3: **Ethical Investment and the Oil Industry**

International oil companies are paying increasing attention to the ethical aspects of their investments in response to criticisms from non-governmental environmental and human-rights organisations about some of their activities. The need to pay attention to such issues is being increasingly emphasised by their shareholders. Poor publicity about controversial investments has had a detrimental effect on some companies' share prices, leading some of them to withdraw from particular projects and countries.

The mismanagement of oil revenues by some governments is a particularly sensitive issue. Several companies, including those operating in West Africa, have been criticised for not ensuring that revenues paid to host governments are used to benefit the local population. Nigeria's per capita GDP remains less than $1 a day despite more than $300 billion in oil revenues earned by successive governments in the past 25 years.[21] Some oil companies are responding to these challenges by seeking backing for sensitive projects from multilateral lending agencies. For example, an ExxonMobil-led consortium sought World Bank involvement in the $3.7 billion Chad-Cameroon petroleum development and pipeline project. The Bank imposed strict lending conditions to try to ensure that the Chad government spends the oil revenues on poverty alleviation and economic development rather than weapons.

The World Bank is also supporting a broader transparency initiative, *Publish What You Pay*, launched by two non-governmental organisations – the Open Society and Global Wisdom. The initiative aims to encourage international oil companies to disclose special payments to host governments

Europe Continental Shelf section below). The likelihood of further changes in environmental regulations is a major cause of uncertainty for investment in refineries and tankers. For instance, whilst shipping regulations have so far focused primarily on single-hull tankers, more stringent rules limiting the age of the entire fleet could also be introduced in the future.

Concerns over the loss of fisheries and potential physical and environmental hazards have prompted most countries to introduce regulations governing the decommissioning of obsolete offshore oil platforms and related infrastructure. Structures must usually be completely removed and brought to shore for dismantling once production has ceased, though partial removal is permitted in certain cases. Costs for decommissioning oil platforms in the North Sea

21. See *Petroleum Economist* (March 2003), "Becoming Model Citizens".

currently range from $50 million to $250 million depending on their size, type and location, but these costs are expected to fall as experience grows.[22] Decommissioning is becoming an important issue in this mature region, as more platforms reach the end of their production life.

Remaining Oil Resources and Technology

There is a reasonable degree of confidence about proven oil reserves – oil that has been discovered and can be recovered economically at current prices and using current technology. There is inevitably much less certainty about oil resources, a broader category that includes oil that is thought to exist and to be economically recoverable.[23] Nonetheless, most recent assessments indicate that there are sufficient proven reserves to meet expected increases in demand through the *Outlook* period and beyond. The *Oil and Gas Journal*,[24] a leading source of reserves data, reports that global *proven* oil reserves stood at 1,032 billion barrels at the end of 2001. Even without any new discoveries, which will obviously not be the case, this would be sufficient to ensure the cumulative projected production over the period 2001-2030 of about 960 billion barrels.

Uncertainty about oil reserves and resources is especially high in the Middle East. Data from this region are often criticised for being either under- or over-reported for a number of reasons. Estimates remain opaque because of a lack of credible external auditing of national claims. The degree of uncertainty is also high in the transition economies as independent evaluations of resources have not been possible until recently in many of them and are still not possible in others. Assessments of oil reserves and resources form a key input into the exploration decision-making process of oil companies. New reserve assessments could influence the location and timing of exploration activity, with knock-on effects on reserve additions and field development. As unit costs for each element of the supply chain vary regionally, such revisions would affect overall global investment requirements.

Regional Analysis

OECD North America

Upstream oil investment in North America will diminish as opportunities for profitable investment decline, especially if current restrictions on drilling on some federal lands remain in place. Cumulative investment will nonetheless be higher in the next three decades than in any other region. Non-conventional oil sands projects

22. UKOOA (2002). Decommissioning costs have not been included in our investment analysis.
23. See IEA (2001) for detailed definitions of categories of reserves.
24. *Oil and Gas Journal*, 24 December 2001. All reserves estimates cited in this chapter are from this source unless otherwise stated.

in Canada will account for a growing share of capital spending. Refining investment – small by comparison – will be largely focused on improving product quality at existing refineries.

United States and Canada

The investment that will be needed to support projected development in the oil sector in the United States and Canada over the next three decades is estimated at $545 billion – or $18 billion a year. Conventional oil exploration and development will account for 73% of this spending, although the share of non-conventional oil will grow over the projection period (Figure 4.20). Upstream capital spending will decline progressively, as opportunities for profitable investment diminish and drilling is focused more on regions outside North America and on gas. Upstream oil investment is projected to drop from an average of $16 billion per year in the current decade to $10 billion in the decade 2021-2030, which is significantly less than in the recent past. By that time, upstream gas investment will have risen to $17 billion. Total upstream investment (including gas) in the United States and Canada averaged almost $40 billion a year in the decade to 2002, peaking at $54 billion in 2001.[25] Almost all the projected $35 billion of cumulative investment in refining over the period 2001-2030 will go to secondary conversion and product treatment facilities, as only small amounts of additional distillation capacity are expected to be added.

Access to federal lands that are currently off-limits to drilling will be a decisive factor in the upstream investment outlook in North America. At present, some of the most promising regions for future production, including the Arctic National Wildlife Refuge (ANWR), the Rocky Mountains, offshore California and the Gulf of Mexico, are subject to strict access restrictions as a result of environmental concerns or multiple-use conflicts.[26]

Indigenous production of crude oil and NGLs is expected to remain broadly flat over the projection period. Onshore production in the 48 continental states, the most mature oil-producing region in the world, is likely to fall rapidly, as drilling investment and new discoveries decline and development costs rise. But production in Canada, Alaska and offshore Gulf of Mexico is projected to increase, at least over the next two decades. Production of bitumen and synthetic crude oil from the oil sands of Alberta in western Canada is projected to surge from 740,000 b/d in 2002 to around 5 mb/d by 2030 (see below). With primary oil consumption expected to maintain its upward trajectory, imports into the region will continue to grow from 11.5 mb/d in 2002 to 15.5 mb/d in 2030.

25. Lehman Brothers Equity Research databases.

26. An energy bill under consideration by Congress in the Summer of 2003 included a provision to open up drilling in some areas, including parts of ANWR.

Figure 4.20: **Annual Average Oil Investment in the United States and Canada**

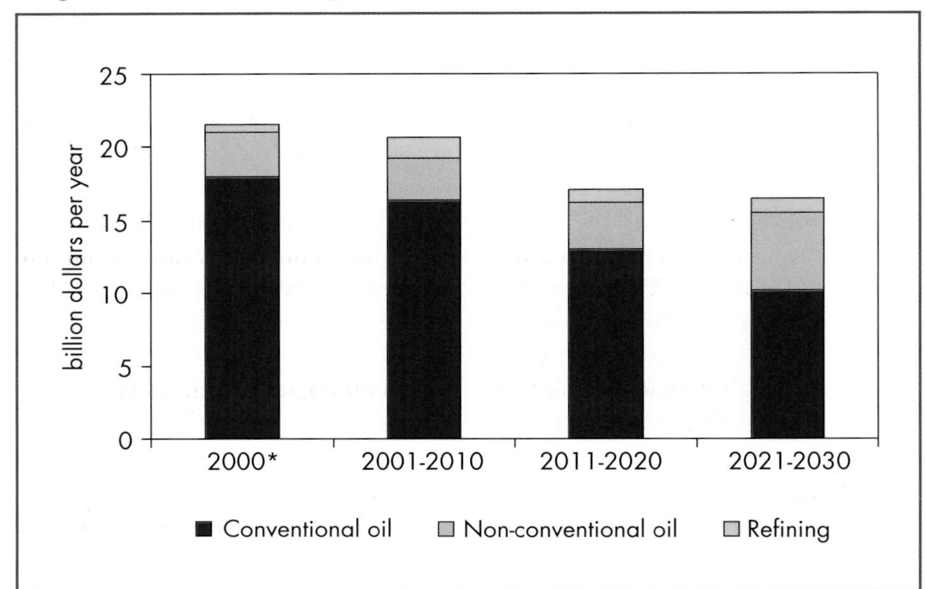

* Estimate.

Their share of total oil supply is projected to increase from around 46% in 2002 to 57% in 2030.

Canadian Oil Sands

Cumulative investment in Canada's oil-sands industry over the next three decades to 2030 is expected to reach $92 billion. This comprises investment for new project developments and for sustaining current production. In comparison, a total of $30 billion has been invested since commercial oil-sands production began in 1967. It is assumed that over 80% of projected spending will be on integrated projects that produce synthetic crude oil.

Canada's non-conventional oil production is centred on the province of Alberta. Alberta's oil sands contain approximately 50 billion cubic metres (bcm) of crude bitumen that is considered potentially recoverable under anticipated technology and economic conditions.[27] Of this total, 28 bcm is reported as "initial established reserves" or proven reserves that can be recovered under current technology and present and anticipated economic conditions. The primary hydrocarbon content, raw bitumen, is extracted from oil-sands deposits and then either upgraded into a high-quality synthetic crude oil (typically

27. Alberta Energy and Utilities Board (2002).

35° API, 0.09% sulphur) or diluted with a lighter hydrocarbon to facilitate transportation to refineries. Currently, crude condensate is the preferred diluent, though it is likely that alternatives, such as synthetic crude oil, will become more widely used as bitumen output increases, owing to expected condensate shortages. In 2002, Canadian production of raw bitumen totalled 829,000 b/d. Of this total, over 60% was upgraded to produce 435,000 b/d of synthetic crude oil and 305,000 b/d was sold as diluted bitumen. *WEO-2002* projects Canadian oil-sands production to reach 1.8 mb/d by 2010 and 5.1 mb/d by 2030.

Surface mining is employed where the bitumen deposit is close enough to the surface (less than 75 metres) to be recovered economically using standard opencast mining techniques. Proven reserves of such deposits amount to 5.6 bcm, about one-quarter of which is currently under active development. To develop deeper oil sands, in-situ recovery technologies are necessary. These involve the introduction of heat, normally via steam, into the oil sands to allow the bitumen to flow to well bores and then to the surface. In-situ reserves total 22.5 bcm. There are also primary recovery techniques that do not require the introduction of heat to produce the bitumen but rather use conventional drilling technology where the bitumen flows naturally without the use of steam.

Most of the output from Canada's oil-sands production is destined for markets in the United States. Investment in export capacity will depend on confidence that sufficient pipeline capacity from Alberta into northwestern and central US states will be brought on line soon enough. A related issue is whether the refining industry can install adequate coking and cracking capacity to handle the diluted bitumen and synthetic crude oil respectively. Other challenges include reducing greenhouse-gas emissions: producing a barrel of synthetic crude oil derived from thermally produced bitumen results in more than 100 kg of CO_2. Water usage is also becoming a serious challenge, as are the cumulative environmental effects of so many projects in one area and the availability of sufficient volumes of natural gas as the fuel for steam boilers and power generation. The Canadian oil-sands industry is addressing these challenges but it will take time to develop and implement the appropriate new technologies.

Oil-sands projects in Canada are capital-intensive, with costs varying on the basis of location, size, the extent of overburden and the process used. The indicative capital cost of an integrated mining upgrading project is currently around $25,000 per barrel of daily capacity. The initial capital cost for the recently completed Athabasca Oil Sands Project, a mining operation which will produce 155,000 b/d of bitumen (190,000 b/d of synthetic crude oil) at peak capacity, was $4.1 billion.[28] This is much more expensive than a typical North American conventional oil-development project. Capital costs for in-situ bitumen projects are considerably lower, at $6,500 to $11,000 per barrel of daily

28. Based on exchange rate of 1 Canadian dollar = 0.73 US dollar.

capacity. But refineries that use diluted bitumen as a feedstock incur significantly higher capital costs. Operating costs for oil-sands projects are extremely high, particularly for integrated mining/upgrading projects, because of the additional processing necessary to produce synthetic crude oil from the bitumen.

In addition to high capital and operating costs, development times for oil-sands projects tend to be very long – typically between five and seven years. A number of recent major projects have incurred significant cost overruns. These were due primarily to competition for skilled labour resulting from multiple projects being developed simultaneously. Another contributory factor was "teething problems" associated with new technologies and configurations that had not previously been deployed commercially.

The costs of oil-sands developments have fallen dramatically since the early 1980s thanks to process improvements and major innovations in truck-and-shovel mining and hydro-transport techniques. The cost savings have been most pronounced for mining operations: in recent years the cost reductions from improved in-situ techniques have been largely offset by rising natural gas prices (Figure 4.21).

Figure 4.21: **Canadian Oil Sands Supply Costs**

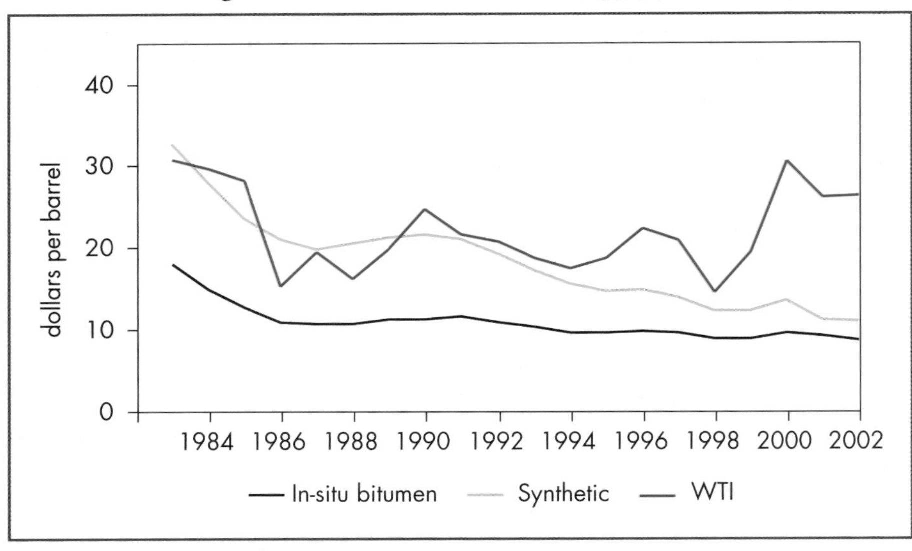

WTI: West Texas Intermediate price.
Source: Natural Resources Canada (2003) and IEA estimates for data after 1999.

According to a report by the Canadian National Energy Board (NEB), total supply costs range from C$ 15 to C$ 18 per barrel for integrated mining/upgrading projects and from C$ 8 to C$ 16 for in-situ operations.[29] These include all capital and operating costs, taxes and royalties associated with exploration, development and production, and a 10% real rate of return on investment. The NEB report states that an increase in the natural gas price of C$ 1 per gigajoule increases operating costs for an oil-sands operation by about C$ 1 per barrel. Considerable scope remains for further cost reductions – particularly as operating experience with much of the technology currently in use is still only two to three years. Over the longer term, further improvements will be partially offset by an expected decline in the quality of the resource base, because the best deposits are being developed first.

As large emitters of greenhouse gases, oil-sands projects will be affected by the Kyoto Protocol, which Canada ratified in 2002. The Canadian government has capped the cost of meeting emission reduction targets at C$ 15 (US$ 11) per tonne during the Protocol's initial implementation period. Suncor, the largest producer of Canadian oil sands, has estimated that this will increase their operating costs by only 20 to 27 Canadian cents per barrel by 2010. This is much less than the industry had previously estimated.

Mexico

Investment in the Mexican oil sector is projected to amount to $77 billion, or $2.6 billion per year, over the three decades to 2030. Most of this investment will go to the upstream sector, but the downstream sector will account for an increasing share after 2020 as crude oil production starts to decline. Total investment is expected to fall in the last decade. It may prove difficult to mobilise this investment, as the national oil company, Pemex, faces financing constraints and no foreign investment is yet allowed in the Mexican oil sector.

Mexican oil production is projected to peak at 4.1 mb/d around 2010 and to remain flat during the following decade. As a result, investment in exploration and development will increase up to 2020. After 2020, production is expected to decline sharply, reaching 2.7 mb/d in 2030, so upstream investment will fall, despite the rising cost of finding and developing new fields. More investment in refineries will be needed to upgrade existing capacity, to enhance product quality and to meet continuously rising domestic demand.

Uncertainties surrounding the development of Mexico's oil sector include the following:

- *Access to reserves:* With the exception of some foreign oil services companies, no exploration permits have been awarded to companies other than Pemex.
- *Tax policy*: Without much lower tax rates, it would be very hard to attract foreign investors even if the upstream sector were to be opened.

29. Canadian National Energy Board (2000).

- *Oil-market policy:* Mexico has at times lowered production for short periods in support of OPEC's efforts to keep prices up. The risk of such constraints in the future may discourage investors.
- *Resources:* Because of past under-investment in exploration, proven crude oil reserves, at 27 billion barrels, are relatively low, covering only 20 years of production at current rates. Reserves have fallen continuously over the past 17 years. The US Geological Survey (USGS) estimates that Mexico has some 23 billion barrels of undiscovered recoverable resources of oil and NGL.[30] Discovering and developing this oil would require massive investment. New enhanced oil recovery (EOR) techniques, including nitrogen injection, could help to offset declining production in major fields, such as Cantarell.[31]

Northwest Europe Continental Shelf[32]

Although production from the North Sea and the rest of the Northwest Europe Continental Shelf (NWECS) is expected to pass its peak in the next few years, investment will remain high as exploration and development shift to higher-cost locations and smaller fields. Upstream licensing and taxation policies, oil price developments and the impact of technology improvements on decline rates and recovery factors will be key determinants in sustaining capital flows in this mature producing area.

Investment and Production Outlook

Investment in the NWECS is expected to decline substantially over the next three decades as opportunities for profitable field developments diminish. Annual oil investment is expected to drop from an average of $8.3 billion in the current decade to $4.1 billion in the decade 2021-2030. In 2002, capital spending on both oil and gas amounted to about $5 billion in the United Kingdom and $8.2 billion in Norway. An increasing share of future investment will be channelled to projects aimed at enhancing recovery rates and stemming the natural decline in production from mature fields. For example, in the UK sector of the North Sea, 144 new projects in mature oilfields were being considered in 2002, compared with 96 in 2001.[33] The marginal cost of developing new fields will rise as drilling shifts to smaller fields in deeper water. Cumulative investment over the period is expected to amount to $182 billion, with the bulk of it coming in the first half (Figure 4.22).

30. USGS (2000).
31. IEA (2001).
32. The Northwest Europe Continental Shelf covers Denmark, the Netherlands, Norway and the United Kingdom, and comprises the North Sea, the Norwegian Sea and the Norwegian sector of the Barents Sea.
33. UKOOA/DTI (2002).

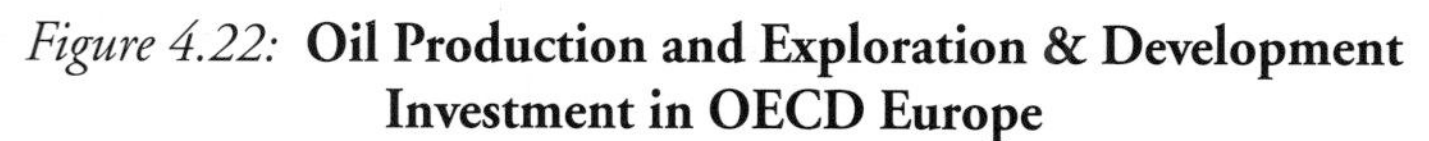

Figure 4.22: **Oil Production and Exploration & Development Investment in OECD Europe**

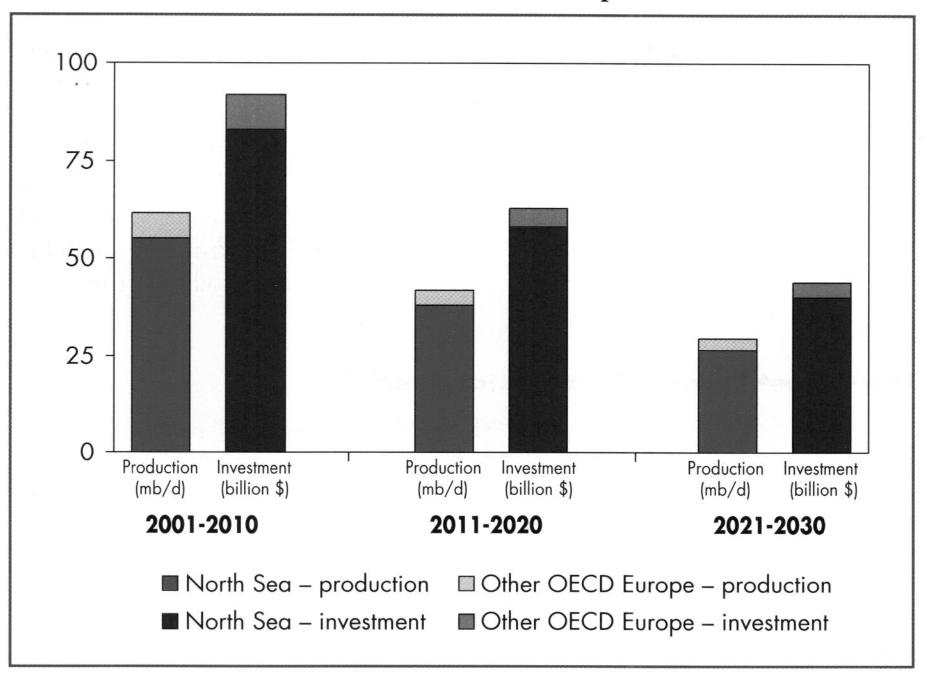

The North Sea is a mature oil region. UK production is already in decline, more than offsetting a continuing modest increase in Norwegian output. Remaining UK offshore reserves at the beginning of 2003 amounted to only 4.7 billion barrels, compared with 19 billion barrels already produced.[34] In Norway, where most of the remaining NWECS reserves are found, there is scope for increasing oil production from the Norwegian Sea and the Barents Sea. But higher output from these areas is not expected to be sufficient to offset declines in production elsewhere after the middle of the current decade. The Norwegian Petroleum Directorate's estimate of remaining proven oil reserves fell for the first time ever in 2003.[35] Remaining Norwegian reserves amount to 9.4 billion barrels compared with 16 billion barrels already produced. Overall NWECS oil production is projected to plateau at around 6 mb/d in the next few years and to decline thereafter to around 2.3 mb/d in 2030.

34. UK Department of Trade and Industry website (www.dti.gov.uk/energy).
35. Norwegian Ministry of Petroleum and Energy (2003).

Investment Uncertainties

Several uncertainties surround these investment and production projections, notably the impact of geology and technology on decline rates and recovery factor for each field, oil prices and their impact on profitability, the success of exploration drilling and government licensing policies. Advanced drilling technology, including horizontal wells, has increased the speed with which peak production is reached in each field and has also raised slightly the decline rate in the first two years after peak production is reached, though this is mostly due to the shift towards smaller fields, which have higher decline rates – 22% versus 9% for large ones (Figure 4.23). Overall, the decline rates of big and small fields have increased only slightly over the last two decades: 0.5% per year for small fields and 0.7% for big fields.

Figure 4.23: **Average North Sea Oilfield Decline Rates***

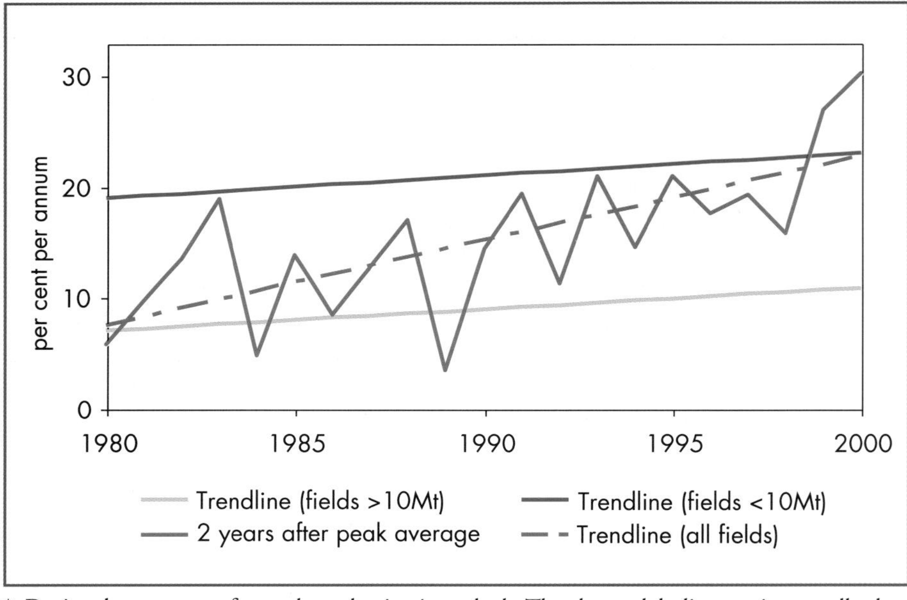

* During the two years after peak production is reached. The observed decline rate is normally close to the natural decline rate in the first two years, since additional investment to sustain production would normally occur later in field life.
Source: IEA analysis based on UK DTI (2002); Norwegian Petroleum Directorate (2003).

The *average* observed decline rate for all North Sea fields is much lower than that of new fields, because the decline rates of older fields, which still account for a significant proportion of total production, are somewhat lower (Figure 4.24).

Figure 4.24: **Average North Sea Oilfield Decline Rate* by Age of Field, 2003**

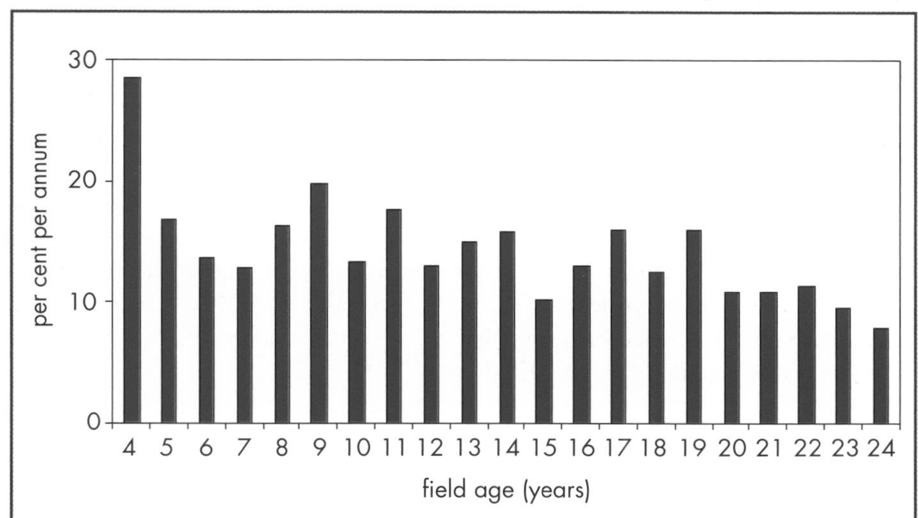

* Average annual field decline rate since peak.
Source: IEA Analysis based on UK DTI (2002); Norwegian Petroleum Directorate (2003).

The high cost of incremental capacity means that investment in exploration and development in the NWECS is highly sensitive to movements in oil prices and the terms of fiscal and licensing regimes. There has been a shift in recent years away from royalties towards profit taxes (Figure 4.25), which has made exploration and development less sensitive to short-term changes in oil prices. The 2002 Finance Act in the UK introduced a supplementary 10% profit tax, although this was partly offset by a 100% first-year depreciation allowance on investment and the abolition of royalties on the oldest oil and gas fields. According to an analysis by Wood Mackenzie, changes in UK taxes in 2002 led to a transfer of £5 billion of the remaining value of North Sea assets from investors to the government.[36] The UK government also introduced in February 2003 a new short-term licensing scheme aimed at encouraging smaller companies to explore for oil and gas. The major international oil companies have been reducing their interests in the UK sector of the North Sea in recent years.

Access to reserves will be critical to the rate of development of new fields in the Norwegian and Barents Seas. The Norwegian government has placed a moratorium on exploration in these areas until it completes an environmental impact assessment. In the longer term, the government aims to strike a balance

36. Wood Mackenzie (2002).

Figure 4.25: **Upstream Tax Take in the UK and Crude Oil Price**

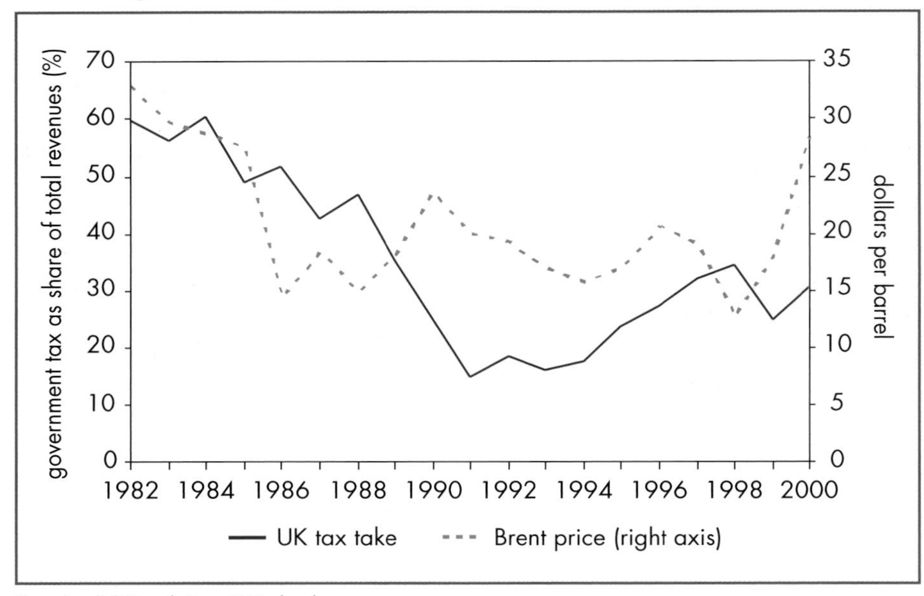

Source: DTI website; IEA databases.

between encouraging oil companies to extend the life of existing fields and gradually opening up access to new ones. The aim is to prolong offshore production and allow time for development of production technology that is less environmentally harmful.

Russia

Maintaining the momentum of the recent rebound in Russian production and exports will call for large and increasing investments in upstream and transportation facilities. A growing share of capital flows will be needed to maintain the production potential at current producing fields in western Siberia – the main producing region – and to develop new fields in frontier regions. Attracting investment in the upstream sector will depend on the expansion of crude oil export pipelines and sea terminals, as well as on continued government efforts to stabilise the legal and tax regime and improve the investment climate. Continued industry consolidation and foreign direct investment could give additional support to upstream development, especially for large-scale projects in frontier regions.

Investment Outlook

Total investment in the Russian oil industry is expected to amount to $328 billion, or about $11 billion a year on average, over the period 2001-2030

(Figure 4.26). Over 40% of this investment will be in projects to supply OECD markets. Exploration and development will account for over 90% of total investment. Spending has increased sharply in recent years, but will need to rise further to meet projected needs. Total upstream investment by Russian oil companies was around $6 billion in 2002, up from only $1.5 billion in 1998.[37] Most recent investment has been in boosting production from existing fields. Investment will be needed increasingly to explore for and develop new resources. Upstream investment needs will be highest in the last decade, when more drilling in less mature areas, such as Timan-Pechora, and frontier provinces, such as east Siberia, the Pechora Sea, or the Russian sector of the Caspian, will be needed to replace ageing fields in the mature western Siberian basins. Investment needs per barrel per day of capacity, estimated at around $13,000, are higher than in most other regions and well above those in the Middle East. Lifting (operating) costs are also relatively high.

Figure 4.26: **Investment in the Russian Oil Industry**

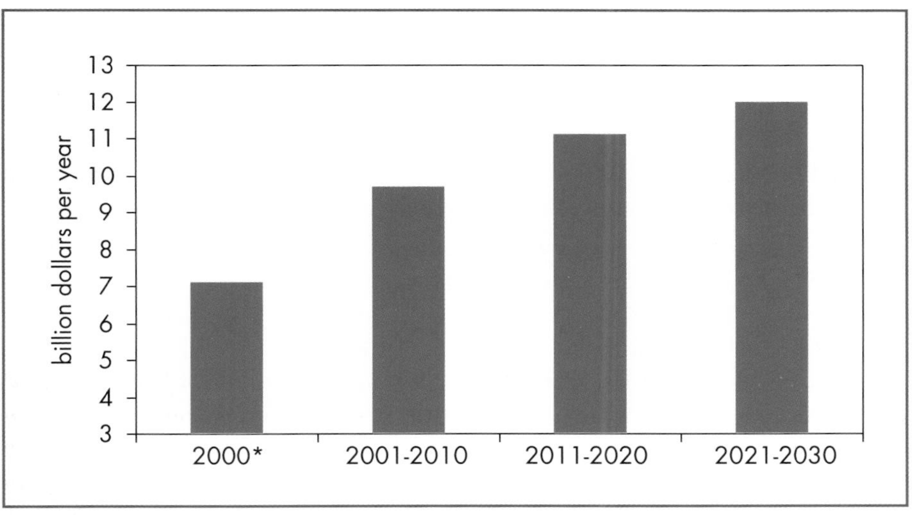

* Estimate.

Cumulative investment in the Russian refining industry over the period 2001-2030 will amount to $20 billion. Most of this investment will be related to refurbishment of outdated refining capacity.

The new *Russian Energy Strategy*, approved by the government in September 2003, estimates that cumulative investment of $230-240 billion will be required

37. IEA (2002b).

through to 2020. This equates to almost $14 billion a year – more than a third higher than our estimate. The difference may be explained by the more rapid shift to higher-cost producing areas assumed in the strategy, including the higher costs of associated pipeline construction.

Supply Prospects

Russia has large proven reserves of oil.[38] In 2002, Russian oil production climbed by just over 9% to an estimated 7.66 mb/d – an annual increase of 1.5 mb/d since 1999. A further surge in output of up to 11% is likely in 2003. The recent upturn in oil production has been largely a result of increased capital spending on enhanced oil recovery techniques, and on putting idle wells back on stream.[39] Higher spending was made possible by the high international oil prices prevailing since 1999 and by the 1998 rouble devaluation.

The *WEO-2002* projects that oil production will reach 8.6 mb/d by 2010 on the assumption that recent rapid rates of output growth are not sustainable in the near term. Production is expected to continue to rise thereafter, reaching 9.5 mb/d by 2030 (Figure 4.27). Western Siberia is expected to provide almost all of the increase in production to 2010, with eastern Siberia and the Far East contributing most of the growth thereafter. The official strategy projects faster growth in oil production in the near term, to 450 to 490 Mt (9.1 to 9.9 mb/d) in 2010, but slower growth in the following decade, with production reaching 450 to 520 Mt (9.1 to 10.5 mb/d) in 2020. There is considerable uncertainty about whether these production increases will be achieved.

Russian primary oil consumption stagnated for most of the 1990s in response to the economic collapse that followed the break-up of the former Soviet Union. Overall demand began to grow again at the start of the current decade but is still stagnant in many of the old Soviet era military and manufacturing cities. *WEO-2002* projects that demand will continue to grow throughout the projection period at an average annual rate of 1.7%, driven mainly by rising transport demand.

Russia is the world's second-largest oil exporter. In 2002, net exports of crude oil and refined products climbed to 4.3 mb/d as oil companies profited from high international oil prices. In the current decade, *WEO-2002* projects that Russia's domestic oil production is expected to grow faster than domestic demand. As a result, net exports will rise to 5.5 mb/d in 2010 assuming that sufficient new export capacity becomes available. Net exports will fall to about

38. There is a great deal of uncertainty about proven oil reserves in Russia. The *Oil and Gas Journal* estimates reserves at 48.6 billion barrels at end 2001, while IHS Energy estimates are far higher, at 140 billion barrels at end 2001.
39. IEA (2002b).

Figure 4.27: **Oil Balance in Russia**

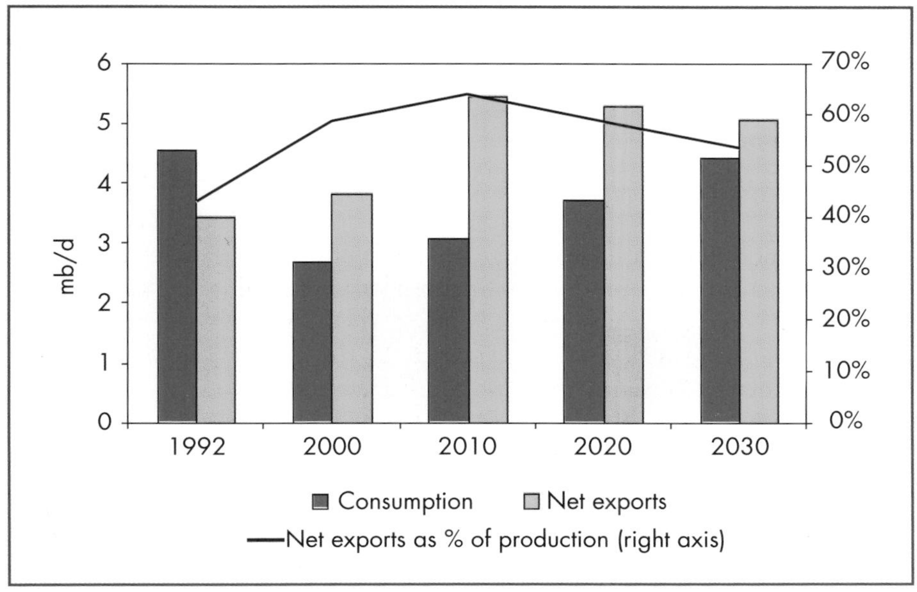

Source: IEA (2002a).

5.3 mb/d in 2020 as domestic demand growth outstrips production and will remain at about that level for the rest of the projection period.

A considerable amount of new oil-production capacity will need to be added over the next three decades to meet rising demand and to compensate for depletion of fields already in production. Average decline rates are expected to increase, especially in the short term, as more intensive production techniques are deployed and the focus of development shifts to smaller fields. As a result, cumulative capacity additions are projected to total 20 mb/d over 2001-2030 (Figure 4.28). More than 85% of these additions will be needed to replace existing wells that will be depleted and wells that will be completed and subsequently depleted over the projection period.

Investment Drivers and Uncertainties

It is uncertain whether the projected investment in upstream development will be forthcoming because of doubts about future investment returns and risks. Much of the recent surge in output has been achieved by rehabilitating existing fields. Three companies – Yukos and Sibneft (in the process of merging) and Surgutneftegas – have contributed around two-thirds of the increase since 1999. The first two companies have made extensive use of modern technologies, such as

Figure 4.28: **Oil Production Capacity Additions in Russia, 2001-2030**

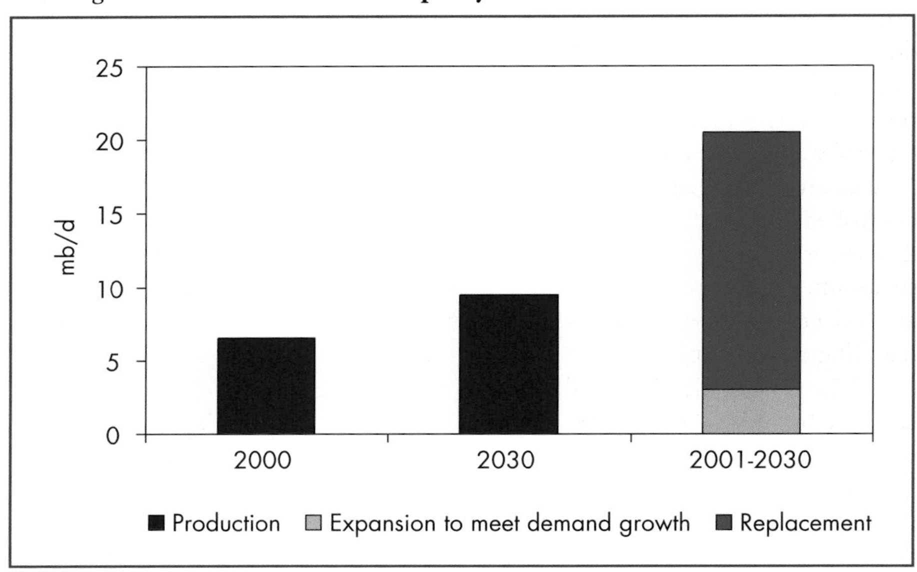

hydro-fracturing and horizontal drilling, by either buying the rights to technology for self-implementation or hiring services from leading western oilfield services companies. More efficient management practices have also contributed to productivity gains.

Despite these developments, there are concerns that fields have been damaged by long-standing practices such as quasi-systematic water injection to raise output as quickly as possible in order to meet aggressive government planning targets. Such techniques have resulted in an increasingly large share of water in the oil extracted. Only a very small amount of production results from re-injection of gas. The depletion of ageing fields and a dearth of exploration drilling in recent years has led to a deterioration in the structure of the Russian oil reserve base. An increasing portion of proven reserves falls into the "difficult-to-recover" category (55 to 60% currently). Over 70% of fields now in production yield such low flow rates that their operation is only marginally commercial: approximately 55% of total oil reserves now in development yield flow rates of 10 tonnes per well per day or less compared with average rates of 243 tonnes in the Middle East and 143 tonnes in the North Sea. Continuation of the present low level of exploration drilling would curb the growth of production below the projected level.

New export pipelines will be needed to provide outlets for new reserves and encourage upstream development. There is considerable spare capacity domestically throughout much of the existing Russian crude oil and refined products pipeline system, the result of the precipitous decline in production and domestic demand in the early 1990s. Current pipeline capacity is probably sufficient to accommodate the projected increase in domestic demand. But with production rebounding and export volumes growing, bottlenecks have emerged, especially at the country's few seaports. Congestion is particularly serious at Novorossiysk, Russia's major export harbour on the Black Sea. Transneft, the state-controlled monopoly oil pipeline company, has managed to increase capacity by debottlenecking, by eliminating the Siberian Light export grade and by bringing on stream the first phase of the Baltic Pipeline System. Several new pipeline projects are under development or planned (Table 4.8). In addition, Transneft is expanding handling capacity at Novorossiysk.

The projects detailed in Table 4.8 would eventually increase Russia's export capacity by more than 4 mb/d. However, there are doubts about whether the strategically important Murmansk pipeline, which accounts for around half that increase, will proceed in view of the high cost of the line. Continued *ad hoc* political intervention in operational and investment decisions and a lack of a clear government policy on the possible opening-up of the sector to private companies may hinder investment in new projects. The failure to establish a quality bank[40] may also undermine and distort upstream investment decisions.

Further efforts to clarify and strengthen the legal and regulatory framework and to reform the system of oil taxation would help to boost investment in Russia's oil industry. Attracting investment in exploring for oil and developing reserves in frontier regions is of particular importance to Russian oil production prospects. Recent legislation protecting the rights of reserve holders, reinforcing property rights and providing for international arbitration has helped to reduce uncertainty and to improve the investment climate. Other reforms, including the lifting of export controls aimed at boosting deliveries and lowering prices in the domestic market and the streamlining of licensing arrangements, have yet to be adopted. A more flexible tax system that takes account of the economics of oil production in different areas and of profitability would help to encourage investment.

Production-sharing agreements (PSAs) are no longer expected to play as important a role as previously envisaged. Until recently, investment in large-scale oil and gas development projects, particularly those being pursued by foreign companies, was generally believed to be contingent upon the establishment of a workable PSA framework. The original PSA law was passed in 1995 and other

40. A system that takes account of the different qualities of crude oil fed into the pipeline network and compensates producers of higher-value oil. At present, crude oil is mixed during transportation within the Transneft system producing a single export blend, benefiting producers of low-quality oil, such as Tatneft and Bashneft.

Table 4.8: **Oil Export Pipeline Projects**

Project/route	Capacity (kb/d)	Status (expected completion)	Comments
Caspian Pipeline Consortium (CPC)	160 (1st phase Russian allocation)	2001 (1st phase) 2015 (2nd phase)	Most capacity is for Tengiz production. Total export capacity is 600 kb/d, rising to 1 mb/d by 2009. 1st phase cost $2.4 billion and total cost could reach $4.2 billion.
Baltic Pipeline replacement and extension, and new terminal at Primorsk	600	Completed	Diversifies exports, but geography makes it costly.
Angarsk (eastern Siberia) to Daqing (China) pipeline	400 (initial) 600 (final)	2006 (initial) 2009 (final)	Go-ahead given, but construction delayed: now expected to begin in late 2004.
Adria pipeline reversal to flow to the Adriatic coast and Druzhba expansion	200	2006	Reversal of line between Hungarian border and Adriatic coast in Croatia, and extension to Schwechat refinery in Austria.
Sakhalin 1 and 2 pipelines and export terminal	440	2007	Pipelines across Sakhalin Island.
W.Siberia-Murmansk pipeline	Up to 2,400	2007	Proposed by 5 Russian majors. Transneft is carrying out feasibility study. Will be costly owing to difficult terrain and length (c. 1,800 km). Would allow export of oil from Timan-Pechora.

Source: IEA databases.

related laws amended in 1998, but normative laws and regulations to implement the 1995 law are still under discussion. As a result, no new PSA has been signed and implemented since then. In the meantime, domestic and foreign investors have responded favourably to moves to strengthen the general legal framework in Russia and improvements in fiscal terms in the oil industry in 2001 and 2002. As a result of these changes and higher oil prices, investment returns are now higher under the normal tax system than under PSAs in low-cost areas such as western Siberia. In addition, a shortening of the depreciation period from 15 to 5 years has made remote and offshore projects more attractive under the normal tax system. But profits and, therefore, the ability and incentive to invest could slump in the event of a fall in international oil prices.

Despite these concerns, the Russian government and most Russian companies are of the view that the legal and taxation systems are now sufficiently attractive and stable for PSAs to be no longer necessary for all but the largest oil investments. The government points to the creation of a joint venture between BP and TNK in early 2003 – the largest-ever foreign direct investment in Russia – as well as Shell's decision to develop the 730 million-barrel West Salym field in western Siberia under the general taxation regime as evidence of increased investor confidence. Moreover, in a reversal of its previous policy, ExxonMobil announced in September 2003 that it would be pushing ahead with its Sakhalin 3 project without a PSA. The Duma subsequently passed a law requiring new acreage to be offered by open tender. Only those plots that attract no initial interest would then be offered under PSA terms.[41]

The Russian oil industry continues to restructure and consolidate. Today, the industry is dominated by seven large vertically integrated Russian "majors", one of which (Rosneft) is wholly state-owned and three partially state-owned (Lukoil, Surgutneftegas and Tatneft). These companies accounted for 84% of total Russian crude oil production in 2001 (Table 4.9). Yukos-Sibneft will become the largest oil company in production terms, once their planned merger has been completed, overtaking Lukoil. Further consolidation is expected, although small independent producers specialising in maximising recovery from mature fields are likely to remain important.

Consolidation will reinforce the Russian majors' ability to pursue large-scale upstream investments in frontier regions and could provide a springboard for their expanding overseas. Other large investments by international oil companies are possible, though perhaps not on the same scale as BP's recent investment. Foreign direct investment would help Russian companies lower their cost of

41. PSAs signed after 1995 but not yet implemented may also be re-offered at open tender if the holders do not wish to proceed, though procedures have not yet been decided. Three PSAs signed before 1995 – Sakhalin 1 and 2, Kharyaga – and three others that have already been given the green light are not affected.

Table 4.9: **Main Oil Companies Operating in Russia**
(2001 data)

	Oil production (mb/d)	Oil exports (kb/d)	Capital expenditure ($ billion)
Yukos-Sibneft	1.6	600	0.7
Lukoil	1.3	440	1.1
TNK	1.0	349	0.8
Surgutneftegas	0.9	325	1.4
Tatneft	0.5	182	0.6
Rosneft	0.3	111	0.4
Slavneft	0.3	107	0.2
Others	1.1	414	1.3
Total	**7.0**	**2,528**	**6.6**

Source: International Petroleum Economist (2003).

capital and expand the range of projects that they will be able to pursue. Foreign investment will also improve their access to technology and project management expertise. Limited upstream equity opportunities in the Middle East and elsewhere will increase foreign companies' interest in taking big stakes in Russian firms, especially since Russian assets are still relatively cheap. However, lingering concerns about corporate governance and uncertainties about the stability and attractiveness of the tax regime – especially if oil prices were to fall – might discourage such developments, even if the Russian government was willing to permit international oil companies to take major new stakes in Russian firms.

Caspian Region[42]

The Caspian region, with its large oil (and gas) resources, is emerging as a major potential new supplier of oil to the world market. Production could approach 4 mb/d by 2010, assuming producers are able to access export pipelines. The sustainability of capital flows and production increases in the longer term will depend on a resolution of long-standing disputes over mineral rights in the Caspian Sea, on the commercial terms on offer to private investors and on the stability of the legal and political environment. Investment will remain sensitive to movements in oil prices as development costs in the region are relatively high.

42. Defined as Azerbaijan, Kazakhstan, Turkmenistan and Uzbekistan.

Investment and Supply Outlook

The *WEO-2002* oil-supply projections for the Caspian region imply a need for investment of around $112 billion over the next three decades – $23 billion of it in the current decade. Kazakhstan will need to attract the largest amounts of investment, most of it to complete development of the Tengiz and Kashagan fields.

These capital flows will underpin a continuing rise in Caspian oil production, most of it coming from Azerbaijan and Kazakhstan. By early 2003, oil production had already surged to almost 1 mb/d in Kazakhstan and to over 300,000 b/d in Azerbaijan. Proven reserves in these countries are large, especially in relation to current production levels. The lion's share of the increase in output in the next decade will come from the Tengiz and Kashagan fields in Kazakhstan. Tengiz, with an estimated 9 billion barrels of proven reserves, started to produce oil in 1993. The Kashagan field, the largest oilfield discovered anywhere in the world in the last 30 years, with 13 billion barrels of reserves, is due to start producing in 2006.

Investment Uncertainties

Access to export markets is a critical factor for the development of the region's oil and gas. Higher oil production will not be possible without the construction of new pipelines and agreements on transit. Russia will remain a crucial transit country. There are plans to raise the capacity of the CPC pipeline that runs from Kazakhstan through Russia to Ukraine to over 1 mb/d by the end of the current decade, allowing Kazakh production to rise (Figure 4.29). The first phase of the project was commissioned in 2001. There is concern about the resulting increase in oil-tanker traffic through the Bosporus Straits. An initial intergovernmental agreement has been reached on the 900,000 b/d Baku-Tbilisi-Ceyhan (BTC) pipeline, which will bring Azeri oil across Georgia to the deep water Turkish port of Ceyhan on the Mediterranean coast. Discussions are also under way with Kazakhstan over the possibility of exporting Kazakh crude oil through this line after shipment across the Caspian. Construction is expected to begin in 2004, once financing has been secured. First exports are expected in 2005. The pipeline alone is estimated to cost $2.7 billion, while investment in developing Azeri oil reserves is estimated to require $12 billion. Russia and Iran are strongly resisting proposals to build trans-Caspian pipelines.

Large amounts of capital have already been invested in major oil projects in the region and additional sums have been committed for the next few years. Foreign direct investment in the oil and gas sector in Kazakhstan, for example, amounted to $1.3 billion in 2001. But the outlook for investment in new projects further ahead will depend on several factors other than access to export lines, including the following:

- *Investment climate:* Concerns remain about the fragility and instability of the legal and regulatory framework governing foreign investment and

Figure 4.29: **Caspian Region Export Oil and Gas Pipeline Routes**

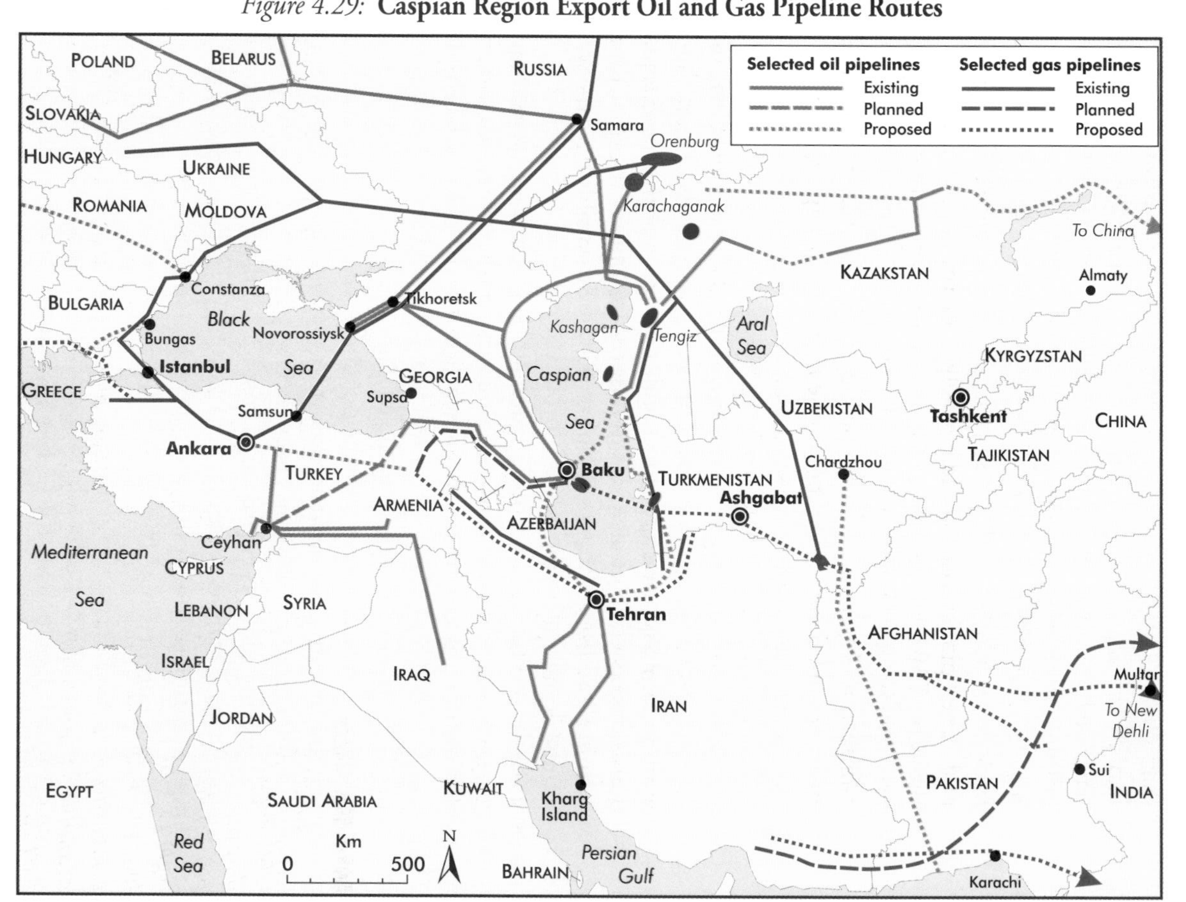

whether the general political climate favours foreign investment. Investors have also expressed concern that host governments are seeking to claw back incentives to encourage investment under PSAs. For example, a dispute has arisen in Kazakhstan over the Kashagan partners' liability to value-added tax, while recent changes in the investment law have reduced protection for investors. Weak judicial systems and arbitrary decisions by regulators are a major concern for investors in the region. Shifts in policies over government take and attempts to claw back shares of projects after foreign investors have carried the exploration and development risks could severely undermine investment in new projects. Hurdle internal rates of return for Caspian upstream projects are already very high, at around 25% in many cases. Respect for contracts and guarantees over pipeline tariffs will be crucial to bolstering investor confidence, lowering hurdle rates, sustaining capital flows and improving the return to the host countries.

- *Caspian Sea legal issues:* The five littoral states have made little progress in agreeing on the legal regime governing mineral rights in the Caspian Sea. In April 2002, a summit of the Caspian littoral heads of state failed to produce a multilateral agreement, although some of them subsequently signed bilateral agreements. Resolution of the remaining disputes would allow access to more resources and would possibly pave the way for trans-Caspian pipelines.
- *Oil prices:* The profitability of upstream oil investments in the Caspian region is particularly sensitive to movements in international oil prices, because of the structure of upstream taxation and high development costs.

Middle East

Investment in expanding oil-production capacity in the Middle East will be vital to global energy-market prospects in the medium to long term. Mobilising that investment will depend largely on the production and investment policies of the key producers, particularly Saudi Arabia. Although the costs of developing the region's vast reserves are lower than anywhere else in the world, restrictions on foreign involvement in many countries and the dependence on national oil companies for a large share of state revenues might constrain the amount of capital available for investment in increasing production. Investment prospects in Iraq are particularly uncertain.

Investment Outlook

The projected growth in the Middle East oil sector implies a need for around $523 billion of capital spending over the next three decades (Figure 4.30).

Investment flows will need to rise substantially from an average of $12 billion a year in the current decade to around $23 billion a year in the last decade of the projection period. Almost half of this investment will be to supply OECD countries.

Figure 4.30: **Middle East Oil Investment**

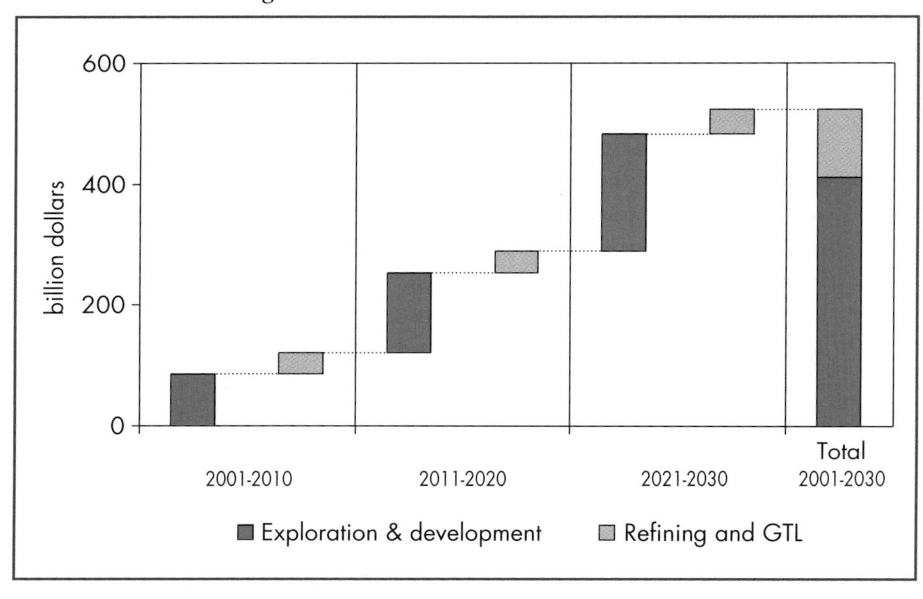

It is assumed that exploration and development costs for new supplies will be broadly constant for both low-cost onshore and higher-cost offshore fields over the projection period, but a gradual shift towards offshore drilling in some countries will raise average costs slightly in the region as a whole. Costs in the Middle East are nonetheless expected to remain the lowest in the world, and well below assumed price levels. Saudi Arabia, whose reserves are mostly onshore, has the lowest development costs, averaging around 60 cents per barrel. The average capital cost of new onshore capacity in Middle East OPEC countries is around $4,600 per b/d, compared with $10,200 worldwide and around $22,000 in the North Sea.

Downstream oil investment needs will also be substantial. Refining capacity is projected to surge from just over 6 mb/d in 2002 to 10 mb/d in 2010 and 15.6 mb/d in 2030 – equivalent to one large new refinery a year. This will entail capital outlays of around $99 billion, or $3.3 billion a year. Investment in GTL capacity, most of which will probably be built in Qatar and Iran, will amount to $12.7 billion, more than 30% of the world total.

Supply Trends and Prospects

The Middle East is the world's largest oil-producing region. It produced 20 mb/d of crude oil and NGLs in 2002, equal to over a quarter of global output. Around 16 mb/d of the region's production was exported. Production and exports have declined in three of the last four years, owing to supply cutbacks to meet OPEC production quotas in the face of weak global demand and rising non-OPEC production. Most Middle East oil producers are members of OPEC. Saudi Arabia remains the region's and the world's largest producer and has the largest reserves. At 686 billion barrels, the Middle East's remaining proven reserves of oil and NGLs are equal to 66% of the world total. The reserves of Saudi Arabia, Iran and Iraq alone amount to 464 billion barrels.

Low-cost Middle East producers are well placed to capture most of the expected growth in global oil demand over the next three decades. *WEO-2002* projects a sharp increase in Middle East crude oil production on the premise that the key OPEC producers in the region will be both willing and able to increase their long-term market share as oil prices rise.[43] Increasing marginal production costs are expected to hold back the growth in output in non-OPEC countries, allowing OPEC producers to expand steadily their market share.[44] Middle East production of crude oil and NGLs will surge to 28.3 mb/d in 2010 and 52.4 mb/d in 2030 – an average annual rate of increase of over 3%. On the basis of a projected average natural decline rate of around 5%, roughly 65 mb/d of incremental capacity will be needed over the projection period. Of this, around 60% will be needed to replace depleted capacity (Figure 4.31).

Investment Uncertainties

The principal uncertainties shrouding future Middle East supply and, therefore, investment needs is the rate of growth of global oil demand, the resulting call on OPEC supply and the supply policies producers choose to pursue. A slightly lower rate of increase in demand than projected in *WEO-2002* would have a particularly marked effect on investment needs in the Middle East as the main residual supplier of oil to the international market.[45]

43. The average IEA crude oil import price is assumed to rise linearly from $21/barrel in 2010 to $29/barrel in 2030.

44. The *WEO-2002* oil supply projections are derived from projections of global oil demand. OPEC conventional production is assumed to fill any gap remaining after taking into account production of conventional oil in non-OPEC countries and non-conventional oil worldwide.

45. The impact of much lower investment in the region on production and investment patterns in the rest of the world is analysed in the last section of this chapter.

Figure 4.31: **Middle East Oil Production and Capacity**

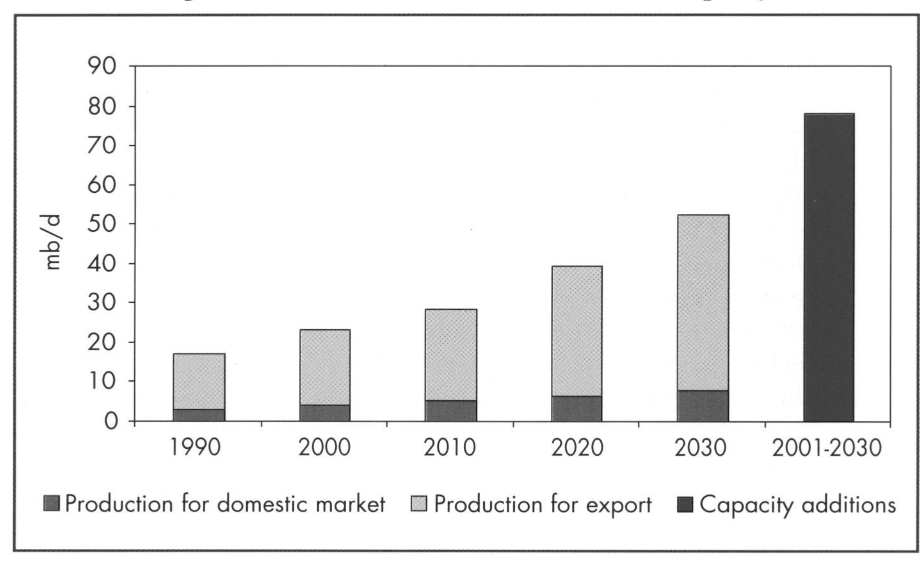

There are also uncertainties on the supply side. Host countries' policies on opening up their oil industries to private and foreign investment, the fiscal regime and investment conditions, and government revenue needs may constrain capital flows to the sector:

- *Access to reserves:* Saudi Arabia, Kuwait and Iraq remain the only countries that do not allow a company other than the national oil company to have any direct involvement in the development or production of oil or gas resources, although foreign companies operate there under technical services contracts. However, there are plans to seek foreign investment in the upstream oil sector in Kuwait and in gas field development in Saudi Arabia (see Chapter 5). Restrictions on where foreign companies are allowed to operate and the nature of their involvement remain in other countries. For example, foreign companies are allowed to invest in Iran only under a special form of fee-for-services deal known as a buy-back contract (see below). US sanctions impede US companies' involvement in Iran. The future role of foreign companies in Iraq remains uncertain.
- *Fiscal and commercial terms:* Even in those countries that have opened their oil sector to foreign investment, the commercial and fiscal terms on offer have not always been attractive enough to tempt investors. Stand-offs between host countries seeking the largest possible share of upstream rent and investors seeking returns commensurate with their perception of the risks involved could delay or block investment in the future. Negotiations

over the buy-back deals in Iran, for example, have proceeded slowly since the first deal was signed in 1996 because of political infighting and dissatisfaction on the part of foreign companies over terms. The major international oil companies have enormous cash resources and are keen to gain access to the region's low-cost oil reserves, but will make commitments only under appropriate commercial and fiscal terms. Oil services companies and smaller oil companies could play a more important role in assisting national companies to develop reserves without necessarily taking equity stakes.

- *Financing:* The bulk of investment by national and international oil companies is expected to be financed out of internal cash flows. How much of national oil company capital spending will be financed through debt will depend both on government revenue needs and on oil prices. Budgetary pressures could squeeze the amount of their earnings that the national companies are allowed to retain for investment purposes and, therefore, their need to borrow. Lower prices than assumed in our underlying projections would have a disproportionately pronounced impact on oil company cash flows and their need to rely on debt financing.

These factors must be seen against a backdrop of continuing political instability in the region. The US-led occupation of Iraq, endless and exhausting negotiations over a peace deal between Israel and Palestine, and social and political tensions throughout the Gulf compound the political and economic unpredictability of the region. Foreign oil companies cite threats to the personal security of their staff as a major deterrent to launching activities in certain countries. These factors raise the risks and, therefore, the costs of doing business in the region. Unsurprisingly, country risk is considered most acute in Iraq. It is currently rated lowest in Kuwait, Qatar and the UAE (Table 4.10).

Table 4.10: **Sovereign Ratings of Selected Middle East Countries*, July 2003**

	Moody's	Standard & Poor's	Fitch
Bahrain	Baa3	B–	A–
Iran	Not rated	Not rated	B+
Kuwait	A2	A+	AA–
Oman	Baa2	BBB	Not rated
Qatar	A3	A–	Not rated
Saudi Arabia	Baa2	Not rated	Not rated
UAE	A2	Not rated	Not rated

* Long-term foreign currency debt.
Source: Company reports.

Future trends in the natural decline rates of Middle East oilfields will have a major impact on how much investment will be needed. Much of the oil produced in the Middle East today comes from super-giant fields that have been in operation for several decades. For example, the Ghawar field in Saudi Arabia – the world's largest – started production as long ago as 1951. The ten largest fields in the Middle East, all of which were discovered before 1975, account for around half of total output.[46] Ghawar alone accounts for a quarter. The natural decline rates at some of these fields are thought to be rising.

The uncertainties described above have implications for the pricing and production policies of the key Middle East oil producers, the role of foreign investors and the timeliness of investment decisions. But as long as prices remain high, the governments of the major resource holders will have a very strong economic incentive to ensure that upstream oil investment occurs promptly to extract the very large rents available.

Saudi Arabia

With 262 billion barrels, Saudi Arabia has more remaining proven oil reserves than any other country, as well as the lowest production costs. Its reserves/production ratio currently stands at more than 85 years. Yet Saudi production of crude oil and NGLs has been declining in recent years albeit because of choice not geology. Having held steady at around 8.3 mb/d throughout the period 1991-1998, production was reduced to 7.9 mb/d in 2002 (not including the Saudi share of output from the Neutral Zone). It was increased to over 9 mb/d in the first half of 2003 but this was in response to fears of supply shortages in the wake of the US-led invasion of Iraq and disruptions in supply in Venezuela and Nigeria. Sustainable production capacity is currently estimated at 9.8 mb/d. Surge capacity could be 1 mb/d higher.[47] However, surge production is thought to exacerbate problems with water encroachment in oil wells.

Saudi Aramco is planning to raise production capacity by developing several onshore and offshore fields. Three projects already under way will expand capacity by 1.1 mb/d in the next few years, involving investment of $4-5 billion. The last significant addition to capacity occurred in 1998, when the 500,000-b/d Shaybah field came on stream after investment of $2.5 billion.[48] How much of this additional capacity is actually used will depend on the rate of growth in global oil demand and negotiations within OPEC on production quotas.

46. Simmons (2002).
47. Sustainable capacity is defined as production levels that can be reached within 30 days and sustained for 90 days (*IEA Monthly Oil Market Report*, June 2003).
48. Arab Petroleum Research Centre (2003).

Saudi Arabia is expected to maintain a margin of spare capacity over production to ensure that as a swing-producer it remains able to exert control over the market within the OPEC framework. The relatively low cost of installing new production capacity means that the case for holding spare capacity is strategic as well as economic since it can be brought into use at times of high prices. A shift in Saudi oil policy, aimed at raising market share and choking off non-OPEC conventional oil and oil-sands investment, is possible. This will hinge on the Saudi government's calculation of the short-term political and economic costs of lower prices and revenues and the longer-term benefits of higher production and revenues. The government's immediate need for tax revenues will be critical to its production and pricing policies. Under a price assumption of $20 per barrel, the budget deficit in 2003 will exceed $10 billion (Figure 4.32). An average price of about $25 per barrel over the year would be needed to balance the planned budget. Pressures to raise government spending will continue to grow in line with population, implying a need to raise exports if oil prices remain constant. A policy aimed at raising market share might entail larger budget deficits in the near term.

Saudi Arabia has no plans to open up the upstream oil sector to direct foreign investment. Public opposition to foreign investment in oil production is

Figure 4.32: **Saudi Arabia Budget Surplus/Deficit and Crude Oil Price**

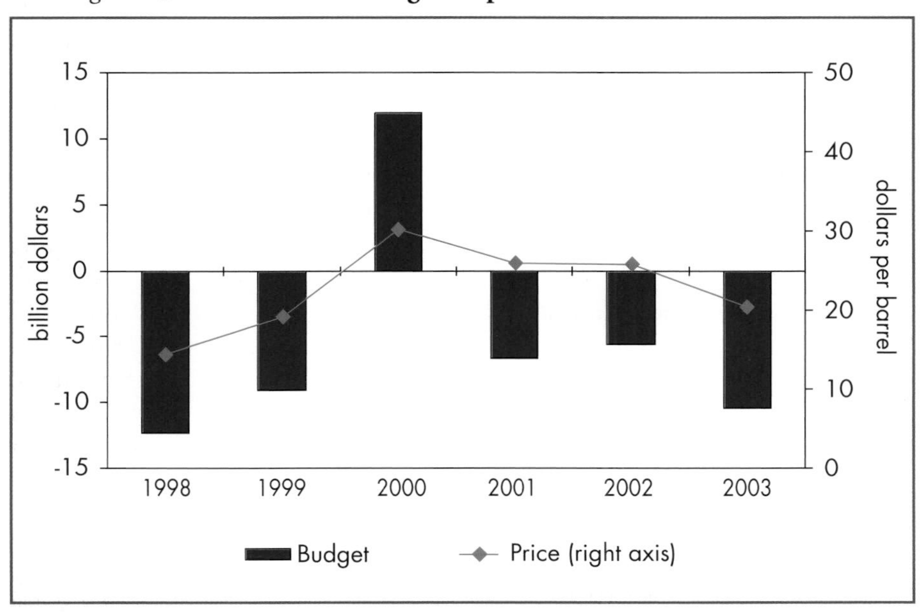

Note: The oil price is spot WTI. The price for 2003 of $20.5/barrel is that used in the budget.
Source: Merrill Lynch (2003).

strong across the Saudi population. Only persistently weaker oil prices and shortfalls in tax revenues might prompt a re-think of policies towards foreign investment in oil.

Iran

Remaining proven oil reserves in Iran amounted to 90 billion barrels at the end of 2001 – little changed from a decade before and equal to over 70 years of production at the current rate. Iranian crude oil production fluctuated between 3.5 and 3.7 mb/d from the mid-1990s through to 2001, dipping to a little under 3.5 mb/d in 2002. Production rebounded to 3.8 mb/d in early 2003 as OPEC quotas were eased in response to supply losses elsewhere. Much of the increased production stemmed from newly commissioned offshore fields.

Table 4.11: **Iranian Oil Buy-back Deals**

Year of award	Project	Contractor	Value ($ million)	Peak production (kb/d)	Status
1995	Sirri A & E	Total (70%), Petronas (30%)	760	102	Completed in 1999
1999	Dorood	Total (55%), Agip (45%)	998	85	EOR project. Production started 2002, completion due in 2004.
1999	Balal	Total (85%), Bow Valley (15%)	300	40	Production started January 2003.
1999	Soroosh & Nowruz	Shell	800	190	1st phase completed 2001; 2nd due in 2003.
2001	Masjid-e-Suleiman	Sheer Energy Cyprus	81	23	EOR project, completion in 2006.
2001	Dakhovin	ENI/Naftiran	1,000	180	Production due to start in 2003.

Source: Arab Petroleum Research Centre (2003).

The National Iranian Oil Company (NIOC) has targeted an increase in production capacity from an estimated 3.75 mb/d at present to 5 mb/d by 2005. Most of this additional capacity will come from the development of new fields and the deployment of enhanced oil recovery techniques in mature fields, for which the technology will be provided by international oil companies operating under so-called buy-back contracts. Under these contracts, the companies finance the entire cost of upstream development and recover their investment plus an agreed rate of return from the sale of the production that results. By early 2003, six buy-back contracts had been signed, production under four of which has been either fully or partially completed (Table 4.11). Three other deals, involving the Azadegan, Cheshmeh Khosh and Bangestans fields, are still under negotiation.

Buy-back contract negotiations have been hampered both by opposition from some political factions to foreign access to Iranian reserves and by foreign investors who complain about the rigidity of contractual terms, which penalise them for overspending or failing to achieve targeted production levels while not rewarding them for exceeding targets. A dispute over the first contract with Total and Petronas is an example. These problems, as well as the unstable political situation in the country, raise questions about future upstream investment. A change in, or reorganisation of, the Iranian leadership could lead to oil policies that are more receptive to foreign investment, including US company involvement. Higher oil prices than expected in recent years have boosted the government's cash reserves, which could have been used to finance new upstream developments but are more likely to have gone to other socio-economic priorities.

Iraq

Nowhere are short-term oil production prospects more uncertain than in Iraq, following the overthrow of the Saddam Hussein regime by US-led forces in early 2003.[49] The immediate objective of the administration now running the country is to restore production capacity as quickly as possible in order to generate the export earnings needed for reconstruction. Looting and sabotage have reduced sustainable capacity, which was estimated at 2.8 mb/d before the war. Restoring output to even that level may take many months as Iraqi oil infrastructure has suffered years of under-investment. Achieving the 1990 level of 3.5 mb/d is expected to take several years. The provisional administration announced in July 2003 that it had agreed a $1.1 billion investment plan to return crude output to the pre-war level.[50] Achieving this goal will depend on how quickly order can be restored to the country and the extent of the damage to oilfields caused by poor production engineering practices and maintenance during the 1990s.

In the longer term, there is considerable potential for expanding capacity. According to the *Oil and Gas Journal,* Iraq has 112 billion barrels of proven

49. The analysis in this section is based on information available at end September 2003.
50. Restoration of Iraqi Oil Infrastructure Final Work Plan (July 2003) (http://www.energyintel.com/resources/workplan.pdf).

reserves. Other estimates range from 95 to 120 billion barrels. Of 73 known fields, only 15 have been developed. But there are enormous uncertainties about the future development of the Iraqi oil industry, including:

- How quickly the transition to a stable government is achieved and how effective that government turns out to be.
- How extensive the damage caused by looting and attacks on facilities and pipelines is before law and order can be restored.
- Whether foreign oil companies will have any role, and if so, what commercial and fiscal terms they might be offered.
- The role and ownership of the Iraq National Oil Company (INOC) and the availability of oil revenues for re-investment in the industry.
- The availability of sufficient pipeline capacity for exports.
- The legal status of contracts signed or negotiated with foreign firms before the war.
- Iraqi policy *vis-à-vis* membership of OPEC and acceptance of production quotas.

An Iraqi government, once on its feet, will be keen to boost capacity as quickly as possible to increase revenues for reconstruction. The payback period on the investment would be very short given the productive resources available. It is estimated that raising capacity to around 3.7 mb/d by 2010[51] – the Reference Scenario projection – will require cumulative investment of close to $5 billion. But government revenues from production over the period 2003-2010 would be much higher, at over $20 billion. The economic return on upstream investment will, nonetheless, decline with rising investment. The marginal cost of an additional barrel of capacity will be particularly low in the short term, since investment will be mainly needed to restore existing infrastructure. But as production increases, the marginal cost of a barrel will increase substantially. Extending production beyond about 4 mb/d will require major new projects, which would call for investments in exploration, new production capacity and new export facilities.

To illustrate the uncertainties shrouding future Iraqi oil developments and their implications for investments, we have devised two alternative production profiles corresponding to faster and slower development (Figure 4.33). In a rapid growth case in which production reaches 9 mb/d in 2030, investment needs would be around $54 billion, about $12 billion more than in the Reference Scenario, where production reaches 8 mb/d. The rapid growth case assumes that capacity of 3.5 mb/d is reached within two years and that production rises in a linear fashion through to 2010. This trend implies that Iraq's share of total OPEC production would rise significantly. In the slow production growth case,

51. This assumes law and order is restored in a timely manner.

investment is about $12 billion less than in the Reference Scenario, with production reaching 3.5 mb/d only in 2007 and 6 mb/d in 2030. In this case, once production of 3.5 mb/d has been achieved, the subsequent production increases keep Iraq's share of total OPEC production more or less constant.

Figure 4.33: **Oil Upstream Investment Needs in Iraq for Different Production Profiles**

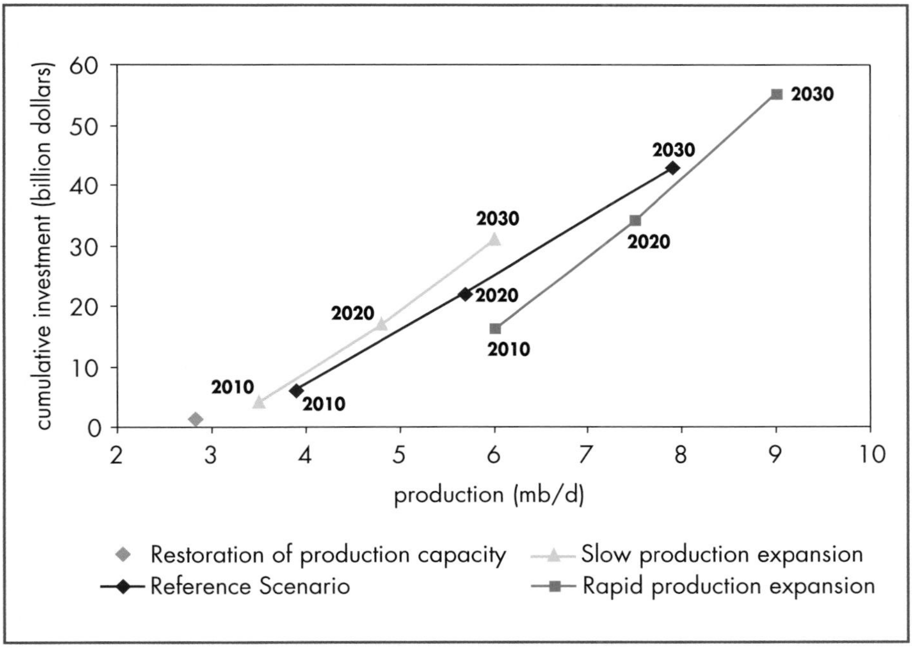

Source: IEA analysis; Restoration of Iraqi Oil Infrastructure Final Work Plan (July 2003).

Because of the many demands on Iraqi oil revenues and Iraq's inability to borrow from international financial markets, the new government will have a strong financial incentive to attract foreign investment. This must be balanced against a likely nationalist sentiment of the Iraqi population. An opening to foreign capital would also bring access to better technology and project management skills. The major international oil companies are keen to take up investment opportunities in Iraq, once law and order has been re-established. But the amount of capital that they will be prepared to invest in Iraq will be determined by the political and security risks. Negotiations over investment terms could be protracted. In any case, the international majors are unlikely to sign long-term deals before a new legitimate Iraqi government is installed and perhaps not until after the occupying forces are no longer required to maintain security. This situation is unlikely to be achieved before 2005.

At present, the most plausible investment scenario might involve a fairly rapid increase in capacity over the next two to three years through rehabilitation of existing fields by NIOC, involving relatively modest investments. Oil-services companies are expected to be central to this process, but they will initially be working with the unsophisticated technology and equipment in place today. This initial phase could be followed by further step-wise increases in capacity as new fields are developed, most likely with foreign assistance, and most likely through contracts with service companies or buy-back/production-sharing agreements with international oil companies. Investment flows will depend on a number of inter related factors, including how quickly political stability and civil order can be restored, the international oil price, the rate of growth in global oil demand and the investment terms offered to foreign investors. The pace of capacity additions may also be constrained by OPEC production quotas if Iraq, as expected, remains a member of that organisation.

Other Producers

Most new oil investment in the Middle East outside Saudi Arabia, Iran and Iraq is likely to be in Kuwait, Qatar and the United Arab Emirates (UAE). The government of Kuwait, with proven reserves of 96 billion barrels, has prepared a plan to attract foreign investment, under operating services agreements, to develop further five oilfields in the north. The aim is to double current production to 900,000 b/d. Initial investment is expected to come to at least $7 billion. The project has stalled in the face of strong opposition in parliament, but the government is hopeful that a law enabling investment to go ahead will be adopted soon. Around half of Kuwait's current production is from the Greater Burgan oilfield, which is the world's second-biggest after the Ghawar field in Saudi Arabia.

UAE's proven reserves, mostly in the super-giant Upper Zakum offshore field in Abu Dhabi, are estimated at 98 billion barrels. No new production capacity has been added since 1995, but work is under way to raise production at current producing fields. Capacity is targeted to grow from around 2.4 mb/d at present to 3 mb/d by 2005 and 4 mb/d by 2010.

GTL Projects

Much of the global growth in GTL production is likely to occur in the Middle East, centred initially at least on Qatar, where GTL plants are expected to provide the bulk of new hydrocarbon-liquids production. The country's remaining reserves of gas, at 86 billion barrels of oil equivalent, are far larger than those of oil, which are only 15 billion barrels. A joint venture between Qatar Petroleum and Sasol, is building the world's largest GTL complex. The 34,000 b/d plant, which will process 3.4 bcm a year of gas from the North Field, is due to begin commercial operation in late 2005. Investment in the plant will total $900 million, to be financed mostly on a project finance basis – a first for a GTL project. Upstream development costs are thought to amount to around $500 million. The project was officially launched in March 2003, dispelling concerns that large-scale investments in the Gulf would be delayed by the war in Iraq.

Other GTL projects are planned in Qatar. Shell is pursuing a two-train 140,000-b/d plant, using its proprietary middle distillate synthesis process. ExxonMobil has proposed a 100,000-b/d plant, while ConocoPhillips and Marathon have also proposed large-scale projects. Shell is trying to develop a plant in Iran, probably in partnership with NIOC, Sasol and Statoil. Gas from the 14th phase of the South Pars development is earmarked for the plant. The successful development of these and other GTL projects in the region will be particularly dependent on the sponsors' ability to negotiate attractive commercial terms for the supply of gas to the plant and on taxation polices, particularly if oil prices fall.

Africa

Investment in the African upstream oil sector has boomed in recent years, leading to rapid increases in proven reserves. Continuing restrictions on foreign involvement and political instability in the Middle East have encouraged oil companies to devote more of their upstream budgets to Africa, despite that region's even more volatile political environment. Attention will continue to focus on the western and northern states. Nigeria's position as the continent's largest producer is likely to remain unchallenged. Most major new discoveries have been found in deepwaters where development costs are high, but operations are insulated to some extent from civil unrest on the mainland.

Investment Outlook

Oil-sector investment in Africa is projected to total $360 billion over the period 2001-2030, or over $12 billion a year. In comparison reports indicate that oil sector investment in Africa in 2003 is expected to exceed $10 billion.[52] Around 85% of projected capital spending will be on oil exploration and development (Figure 4.34). This will contribute to an increase in production from the current level of 8 mb/d to 13 mb/d in 2030. Expansion of Africa's refining industry, where capacity is expected to more than double to 7.3 mb/d by 2030, will require investment of $42 billion. Of total oil investment in the region, almost half will be for exports to OECD countries.

52. US State Department (2002) (http://usinfo.state.gov/regional/af/a2012901.htm).

Figure 4.34: **African Oil Investment**

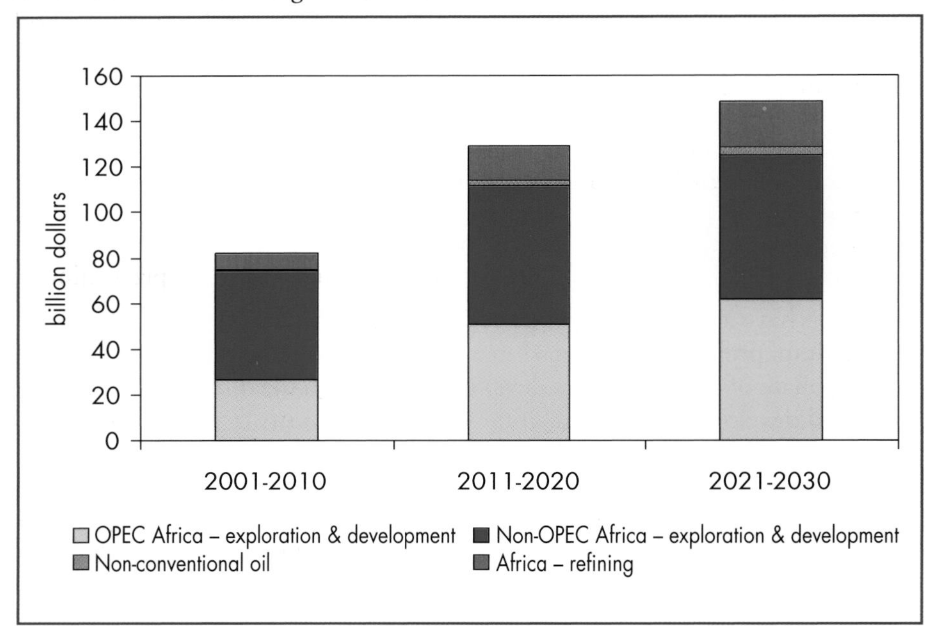

Exploration and development costs for new oil supplies in Africa are estimated to range from \$1.7 to \$4.5 per barrel. Costs are lowest for onshore discoveries in North Africa, most of which are in Algeria and Libya. The cost of developing deepwater offshore reserves, such as those in Nigeria and Angola, are towards the top of this range. Advances in exploration and development technology are leading to a steady increase in the share of offshore oil deposits in total reserves and are bridging the cost gap. These cost reductions, coupled with the threat of civil unrest to onshore production operations and facilities, are reflected in a gradual shift in favour of offshore production.

Although foreign investment in the African oil industry is dominated by major oil companies and large independents, there has been a growing influx of small, niche operators. This is partly due to the reluctance of some large companies, subject to legal obligations of transparency and corporate social accountability, to get involved in projects in troubled countries where corruption is rife and ethical concerns are to the fore (see Box 4.3 earlier in this chapter).

Supply Trends and Prospects

Africa produced around 8 mb/d of crude oil and NGLs during 2002 – or 11% of global supply – making it one of the major producing regions

(Table 4.12). Production is concentrated in western Africa, mainly Nigeria and Angola, and in North Africa, particularly Algeria, Libya and Egypt. Around 6 mb/d of the region's production was exported during 2002. Exports go mainly to the United States and Europe.

Table 4.12: **Oil Production and Proven Reserves in Africa**

	Production, 2002 (kb/d)	**Growth in reserves, 1992-2002** (%)	**Reserves/ production ratio** (years)
Algeria	1,521	0	17
Angola	897	260	16
Cameroon	69	0	16
Rep. of Congo (Brazzaville)	255	88	16
Egypt	754	–40	13
Gabon	248	257	28
Libya	1,381	29	59
Nigeria	2,116	34	31
Sudan	242	100	7
Equatorial Guinea	199	500	15
Other Africa	322	0	3
Total Africa	**8,004**	**25**	**27**

Source: Reserves: *Oil and Gas Journal* (24 Dec 2001); production: IEA databases.

African proven oil reserves were estimated at 77 billion barrels at the end of 2002, equal to more than 7% of the world total. They are dominated by reserves in Libya and Nigeria, which make up 70% of the total. Proven reserves have increased dramatically in the last decade thanks to sustained exploration success, particularly in deep water offshore acreage. The prospect of further reserve additions is good, since large offshore areas remain under-explored.

Nigeria is the region's largest producer with output of 2.1 mb/d in 2002 – the maximum level permitted under its production quota as a member of OPEC. Production capacity is well above this level, at close to 2.5 mb/d, and is expected to increase further in the near term as a series of recent high-volume deepwater discoveries come on stream. To date the Nigerian National Petroleum Corporation (NNPC) has not had to cap foreign-operated output, because operational difficulties have "naturally" constrained production. The prospect of further capacity expansions has led to calls from the Nigerian government to OPEC for their production quota to be raised by 25%. If a sufficient increase is not granted, limits might be placed on the pace of development of Nigeria's offshore reserves.

Algeria is the second-largest oil producer in Africa, with production of 1.5 mb/d in 2002, including around 0.5 mb/d of condensates and natural gas liquids. Over the last few years, some large oil and gas discoveries have been made, raising the prospect of a significant expansion in production capacity in the near term. These prospects are strengthened by the government's interest in further increasing the level of foreign involvement in the oil sector which is already high relative to most other OPEC countries. In anticipation of an expansion in oil production capacity, Algeria is pushing for an increase in its production quota.

With production in 2002 of 1.4 mb/d, Libya is the third-largest oil producer in Africa. The National Oil Corporation is keen to increase production to 2 mb/d by 2008. Although the country has abundant proven reserves – 29.5 billion barrels – it may prove difficult to achieve this target, because of sanctions and bureaucratic delays in awarding production licences.

A rapid expansion in production is also expected in Angola, where a number of large deepwater discoveries are to be brought on stream over the next five years. These have resulted from an active exploration programme that boosted the country's proven oil reserves from under 1.5 billion barrels in 1992 to 5.4 billion barrels in 2002, including the first giant oil discovery, Girassol, in 1,365 metres of water in 1996.

African oil-production expansion is not limited to offshore discoveries, as evidenced by the recent completion of the 1,050-km Chad-Cameroon pipeline, which will facilitate oil production in landlocked Chad for the first time. Chad's production is projected to build up to a peak flow of 225,000 b/d by mid-2004.

Financing and Investment Uncertainties

Given that the largest oil producers in Africa are members of OPEC, the organisation's production and pricing policies will have an important impact on the region's production prospects and need for investment. Uncertainty about future production quotas and the so-called "OPEC risk", that a part of capacity might be shut in to keep national output within quota, might discourage foreign oil companies from investing in some instances.

Sovereign risk associated with political and economic turbulence is high in most African countries relative to most other oil-producing regions. As a consequence, companies which invest in the region face less certain cash flows. In 2000, the Angolan state-owned oil company, Sonangol, stiffened contract terms and procedures, insisting on the "Angolanisation" of oilfield services. This included introduction of a new policy which seeks to extend the life of fields to ensure revenues for future generations. Foreign oil companies were dismayed, since optimal deepwater development requires fields to be developed rapidly in hubs and clusters in order to reduce costs.

African onshore oil production has been plagued by civil unrest and wars, which have at times led to sabotage of equipment, strikes and protests. Many disgruntled communities feel that they have been deprived of an equitable share of oil revenues. In March 2003, ChevronTexaco, Shell and Total were forced to suspend production in the Niger Delta region, at an estimated cost to the region's production of around 250,000 b/d. These political risks, which raise the cost of capital, may diminish in the future, as more investment is directed to offshore discoveries, with onshore opportunities deferred until quieter political times.

Increasing attention by international oil companies to the ethical aspects of their investments has a particular relevance to the future availability of capital for investment in Africa, where companies have attracted criticism from shareholders and non-governmental environmental and human rights organisations. Corruption is widespread in the oil industry in Africa. The International Monetary Fund has been reported as estimating that in Angola 20% of state revenue (most of which is generated by oil contracts) goes missing every year, whilst Nigeria regularly tops polls of the world's most corrupt countries. In an effort to combat such problems, the British government recently launched the Extractive Industries Transparency Initiative. The aim of this initiative, which has been endorsed by a coalition of institutional investors, is to increase the transparency of company payments to governments.

Uncertainty surrounds the ability of several African governments to raise the finance necessary to meet their plans to expand oil production capacity. Around 95% of Nigerian oil is produced by six joint ventures between the NNPC and major international oil companies. These ventures have often suffered from under-investment, because of a lack of state funding of NNPC, which usually holds a 55% or 60% stake. More recently, contracts have been structured as production-sharing agreements, which leave operational control and responsibility for raising capital to foreign investors.

Latin America

Projected oil supply trends call for growing investment in Latin America through to 2030. Investment will be dominated by conventional oil projects in Brazil and Venezuela, with heavy oil projects in Venezuela's Orinoco Belt region absorbing much of the rest. These countries will have to pay close attention to the fiscal and licensing terms and conditions on offer if they are to attract the foreign investment that will be necessary to meet their ambitious expansion targets.

Investment and Supply

Investment in Latin America's oil sector is expected to be dominated by projects in Brazil and Venezuela. Total investment will amount to $336 billion over the period 2001-2030. Because of strong growth in production, annual capital spending in the region will increase sharply, from an average of $9 billion in the current decade to over $13 billion in the decade 2021-2030.

Latin American production of crude oil and NGLs averaged 6.5 mb/d in 2002 and is expected to increase to almost 12 mb/d by 2030. Production is dominated at present by Venezuela and Brazil, with output in Argentina, Colombia and Ecuador accounting for most of the rest. The region's proven oil reserves stood at 99 billion barrels at the end of 2002, close to 10% of the world total.

Brazil

Brazil is projected to require the largest amount of investment of any country in Latin America. The latest strategic plan of Petrobras, Brazil's state oil company, envisages domestic capital spending of $29 billion for the five-year period to 2007. The country is aiming to become self-sufficient in oil by 2006. Achieving this target is expected to depend largely on capacity expansion from high-cost deep water or ultra-deep water fields in areas such as the Campos basin north of Rio de Janeiro. Brazil's oil sector has been open to foreign involvement since 1998 and Brazil has held five oil-lease licensing rounds since then. The latest round of bidding in August 2003 attracted little interest from foreign oil majors. This is believed to be because of the disappointing results of recent exploration drilling, which yielded relatively small deposits of heavy oil in deep water. New policies affecting oil producers, such as domestic participation and procurement conditions, and a proposed change to the tax system have also undermined Brazil's attractiveness to foreign oil companies.

Venezuela

With the world's largest oil reserves outside the Middle East, Venezuela will remain the largest oil producer in Latin America throughout the *Outlook* period. Production of crude oil, NGLs and extra-heavy oil, which was partially affected by the start of the nationwide strike, averaged 2.9 mb/d in 2002. PDVSA's investment plan for the period 2003-2008 targets an ambitious increase in output to 5.1 mb/d. Capital spending of $43 billion, covering expansion of production from existing and new oilfields and from the Orinoco extra-heavy oil deposits, is budgeted.

The Venezuelan government has acknowledged the need for foreign oil companies to become involved in order to meet this target. Venezuela's Hydrocarbon Law, which came into effect at the start of 2002, opened up the country's refining industry and exploration and development of light and medium crude oil to private investment, but limits private participation to 49%. It sets royalties at 20-30%. The Venezuelan Hydrocarbon Industry Association has claimed that these new fiscal arrangements need to be made more flexible in order to encourage foreign investment in mature and high-risk regions. This would help to compensate for the risk of output being capped by the government, as a result of OPEC production quotas. The last time this occurred was in April 2001.

Development of non-conventional extra-heavy oil, which the Orinoco Belt region contains in abundance, will account for a growing share of Venezuelan oil investment. Investment of $52 billion over the period 2001-2030 will be needed to meet projected production growth, the majority of which will go to projects based on partial or full upgrading of the region's extra heavy oil. Cumulative investment in such projects to date has totalled $13.3 billion.

Extra-heavy oil (8 to 9° API) can be produced like conventional crude oil without artificial stimulation. However, once at the surface, it must be diluted in order to pipe it to upgraders where it is converted into high-quality synthetic crude oil (typically 16° to 32° API, 0.07% sulphur). Venezuelan extra-heavy oil is also used to produce an emulsification with water known as Orimulsion which is marketed as a coal substitute for use in old coal- or oil-fired power plants. Orimulsion is excluded from Venezuela's OPEC production quota. Venezuelan production of non-conventional oil totalled 520 kb/d in 2002. Output is projected to increase to 950 kb/d in 2010 and 2.9 mb/d in 2030. Most of this additional output will be synthetic crude.

Venezuelan non-conventional oil projects will be more resilient to periods of low oil prices than most other non-conventional oil projects because their supply costs are lower. But there are other risks associated with the political and economic environment within the country. These risks became manifest during the strikes at PDVSA in 2003, which severely disrupted production at all the Orinoco heavy oil projects.

There are currently four heavy oil upgrading projects in the Orinoco Belt (Table 4.13). All of them are joint ventures between the state-oil company, PDVSA, and one or more foreign partners. Capital costs in each case were around $20,000 per barrel of daily capacity. PDVSA's business plan envisages a fifth project being undertaken before 2008.

Table 4.13: **Orinoco Belt Heavy Oil Upgrading Projects**

Project	**Partners (with PDVSA)**	**Peak capacity** (kb/d)	**Project life** (years)	**Start date**	**Capital investment** ($ billion)
Cerro Negro	ExxonMobil	120	35	2001	2.3
Petrozuata	ConocoPhillips	120	35	2001	2.4
Sincor	Total, Statoil	200	35	2002	4.2
Hamaca	ConocoPhillips, ChevronTexaco	190	34	2003	4.0

Source: IEA database.

The cost of upgraded Venezuelan extra-heavy oil is somewhat lower than in Canada, at an estimated average of around $8 per barrel. Upgrading accounts for more than half of the cost. Venezuela's lower costs arise largely from more favourable geological and climatic conditions. In particular, Orinoco oil is produced without thermal stimulation, because the oil-bearing formations are relatively hot (around 55°C compared to 8-13°C in Alberta). Higher ambient temperatures also make the oil easier to transport once extracted and result in less harsh working conditions.

Other Countries

Other countries in Latin America have potential to increase oil output. Expansion of capacity in Ecuador is currently being hampered by a lack of pipeline infrastructure. This will be rectified late in 2003, when the heavy crude oil pipeline, which will double current capacity, is due to open. Colombia has also been successful in attracting interest from foreign oil companies, despite the country's security problems.

China

Despite the potential to increase output in the Chinese Sea, capital flows to the upstream oil sector in China are expected to decline over the longer term in response to diminishing commercial investment opportunities. An increasing share of investment in China's oil sector will be needed to upgrade antiquated refineries and, after 2010, to boost capacity in response to surging demand.

Investment and Supply Prospects

Total domestic investment needs for the three decades to 2030 are estimated at around $119 billion. The upstream oil sector will account for about 60%, although annual spending in this sector will decline over the projection period as production falls, from around $2.7 billion in 2001-2010 to $2 billion in 2021-2030. A growing share will be needed in the refining industry (Figure 4.35).

Deteriorating upstream investment prospects will result in falling output over the projection period. *WEO-2002* projects that Chinese oil production will plateau at 3.4 mb/d through the middle of the current decade and then decline over the rest of the projection period, although more intensive efforts to promote exploration and development through industry restructuring and regulatory reform could delay the decline. The super-giant Daqing field, the world's fourth-largest producing field, which has been in production for forty years, and other mature fields in China's eastern region – the main sources of indigenous production – are in decline. Higher production in offshore areas and from fields in the west of the country is expected to compensate for these declines in the near term. Offshore exploration and development interest is focused on the Bohai Sea

Figure 4.35: **Oil Supply and Investment in China**

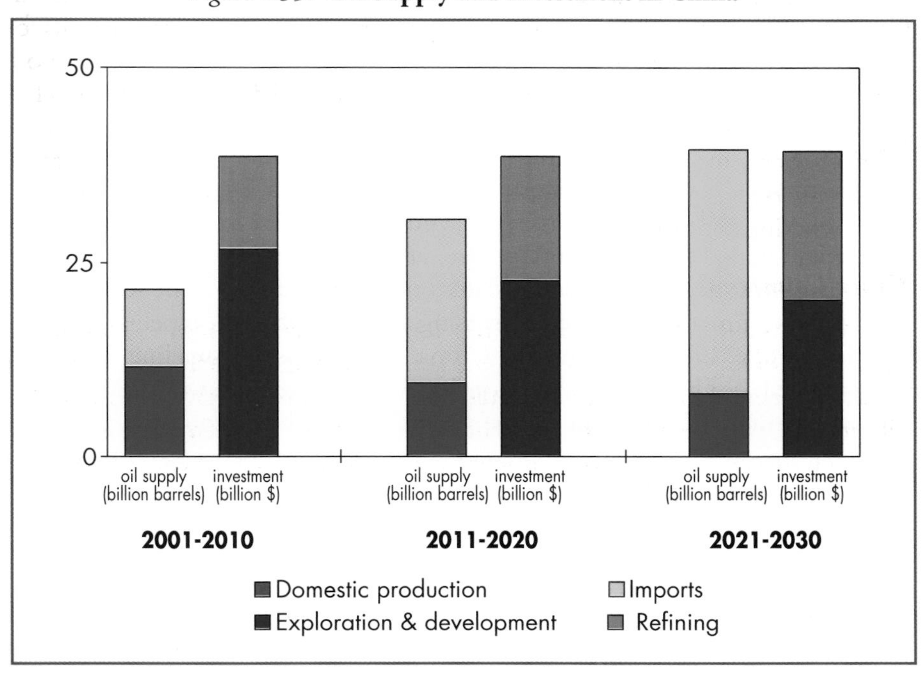

and Pearl River Mouth areas. China National Petroleum Corporation (CNPC), the largest state-owned oil company, is seeking foreign partners to help it enhance recovery rates and extend production at Daqing and other mature fields.

The government is seeking to secure direct control over foreign oil resources through CNPC, which has acquired interests in exploration and development in Azerbaijan, Kazakhstan, Sudan, Iran, Peru and Venezuela, but very little of this equity oil is actually shipped to China. Sinopec, another state firm, has also started to look into buying upstream assets overseas.

In contrast to the upstream production sector, spending on refineries will rise steadily over the projection period. In the first decade, most investment will be needed to upgrade existing refineries to handle heavier and more sour grades of crude oil from the Middle East, which are expected to meet most of China's growing import needs. The share of spending on new primary distillation capacity will grow after 2010, as current overcapacity is absorbed by rising demand. In the last decade of the projection period, capital flows to the refining industry are projected to reach $1.8 billion a year – equal to half of China's total oil-sector investment (not including pipelines and retail networks). The refining industry in China has experienced a degree of consolidation in recent years with the closure of small, unsophisticated refineries in producing regions.

Oil consumption is projected to surge from 5.2 mb/d in 2002 to 12 mb/d, driven largely by transport demand. All of this additional oil will have to be imported. Net oil imports are set to jump to almost 10 mb/d in 2030, up from around 1.8 mb/d in 2002. The share of imports in total oil demand is projected to reach 82% in 2030 compared with 35% in 2002.

Spending on cross-border pipelines and port facilities – not included in the above estimates – will also be substantial. Most imports, largely in the form of crude oil, will probably come from the Middle East. But imports from Russia could play an increasingly important role, which could mitigate or ease Chinese concerns about the country's dependence on Middle East oil. The Russian and Chinese governments have been discussing the feasibility of a pipeline to bring Siberian crude oil to north-east China. The Russian pipeline operator, Transneft, talks of building a $5 billion 1 mb/d pipeline from west Siberia to an export terminal at Nakhodka on the Pacific coast, which would open up the possibility of tanker exports to China and other Asian markets. The Russian major, Yukos-Sibneft, has proposed a 600,000 b/d spur line, costing $1.7 billion, which would link the Transneft line to the existing Chinese pipeline network at Daqing.

Industry Restructuring

Further restructuring of the oil industry and regulatory reform could have a considerable impact on both supply patterns and sources of capital. Removing the legal requirement that a Chinese firm must hold a controlling interest in joint ventures with foreign companies – a condition for China's entry into the World Trade Organization – could encourage investment by foreign companies in the upstream and downstream sectors. Oil import restrictions, which have long given the state-owned Chinese major a near-monopoly in the downstream market, are also being dismantled. Even so, most large-scale oil and gas projects in China will probably still involve one of the major Chinese firms.

Asia-Pacific[53]

Oil investment in the Asia-Pacific region will be characterised by declining spending on upstream activities, as the region's limited oil reserves are exhausted, and rising spending in the refining sector to keep pace with strong demand for transportation fuels. This reduction in indigenous oil production coupled with strong demand growth will result in a stark increase in import dependence. A large portion of projected refinery investments will go towards improving product quality, particularly in the region's OECD member countries.

53. This region includes East Asia, South Asia and OECD Pacific.

Investment and Supply Outlook

Oil-supply trends imply cumulative investment needs in the Asia-Pacific region of $207 billion over the period 2001-2030. Despite diminishing upstream commercial investment opportunities, spending on exploration and development activities will account for around 50% of total investment. Because of limited reserves, most oil producers within the region, with the exception of Indonesia, are expected to see their production decline through the period, resulting in a striking increase in the overall level of import dependence. The region's production in 2030 is projected to have dropped to 3.2 mb/d from 4.4 mb/d in 2002, whilst net imports are expected to have increased by 13 mb/d. Growth in import dependence will be particularly strong in India and has prompted the government's New Exploration and Licensing Policy which is attracting foreign involvement in exploration, an activity previously restricted to Indian state-owned firms.

The proportion of total investment in the Asia-Pacific region earmarked for the refining sector (45%) is high relative to most other regions. This reflects the strong growth in capacity expected within the region, from around 15 mb/d in 2002 to over 25 mb/d in 2030. A growing share of refinery investment will go towards increasing conversion capacity to process the heavy Middle East crude oils that are expected to fill much of the widening gap between the region's rising demand and declining indigenous production. Refining capacity in India is expected to double by 2030, the most rapid growth of any county within the region. As with the upstream sector, the Indian government is seeking to involve foreign companies in the expansion of its refining industry in acknowledgment of the magnitude of the investment challenge.

The balance of the projected oil investment in the Asia-Pacific region will be in GTL and other non-conventional oil projects. Over 20% of the projected global investment in GTL projects in the period 2001-2030 is expected to occur in the region, mostly in Indonesia and Malaysia. Australia's vast shale-oil resources are expected to absorb investment spending, although the extent will depend heavily on further technological developments to reduce supply costs and reduce emissions of greenhouse gases.

Oil-investment requirements in the OECD Pacific countries (Japan, South Korea, Australia and New Zealand) will total $44 billion, a mere 21% of total spending in the Asia-Pacific region. This investment will be fairly evenly split between upstream developments and the refining sector. Australia, the only producer of note in the OECD Pacific region, is expected to account for the vast majority of the region's projected upstream investment. Production declines are already occurring in many of Australia's mature basins. The possibility exists of new finds in the country's relatively under-explored deepwater and frontier areas, but these projects will be expensive to explore, develop and operate. Although a reasonable level of refinery capacity expansion is projected for South Korea, the bulk of refinery investment in the OECD Pacific region will go to improving product quality.

Restricted Middle East Oil Investment Scenario

This scenario assesses the implications of investments in Middle East OPEC countries occurring at a lower level than projected in the WEO-2002 *Reference Scenario.*

Background

The *WEO-2002* Reference Scenario projections of oil production, from which our investment projections are derived, are subject to several uncertainties. Expectations of the Middle East are particularly important in view of the region's central role in meeting global oil (and gas) demand in the coming decades, but there is a real risk that production capacities in the region will not be developed as quickly as expected in the Reference Scenario. The purpose of this section is to assess the consequences for global oil supply and demand, oil revenue and upstream investment of a scenario in which investment in the region is limited.

OPEC Middle East countries are expected in the *WEO-2002* Reference Scenario to account for two-thirds of the increase in global oil production between now and 2030, boosting the region's share in global oil supply by an additional 15%. This projection assumes that the requisite investment in the upstream oil sector will be forthcoming, enabling production in OPEC Middle East countries to grow from 21 mb/d in 2001 to 51 mb/d in 2030 (Figure 4.36). The depletion of existing fields and fields that will come on stream in the coming years will entail investment in another 44 million barrels per day of production capacity, in addition to the investment in the 30 mb/d increase in production that will be required over the next three decades to meet demand growth. The region's total upstream oil-investment needs are projected at $370 billion.

The investment needed to meet the projected growth in production may not be forthcoming for several reasons. These include the following:

- *Resource availability:* If reserves prove to be smaller than expected or difficult to recover because of operational difficulties[54], the need to find and develop new resources would be greater. Such new projects would not be developed as quickly as those projected in the Reference Scenario, since investors would have lower confidence in resource availability and the associated risks would make projects less attractive.

54. In some parts of the Middle East oil reservoirs suffer from high levels of water intrusion which increases the amount of water appearing in the oil produced. As this water must be treated, removed, and disposed of, overall production costs are raised, which impacts profitability.

Figure 4.36: **Investment Requirement and Production in Middle East OPEC Countries**

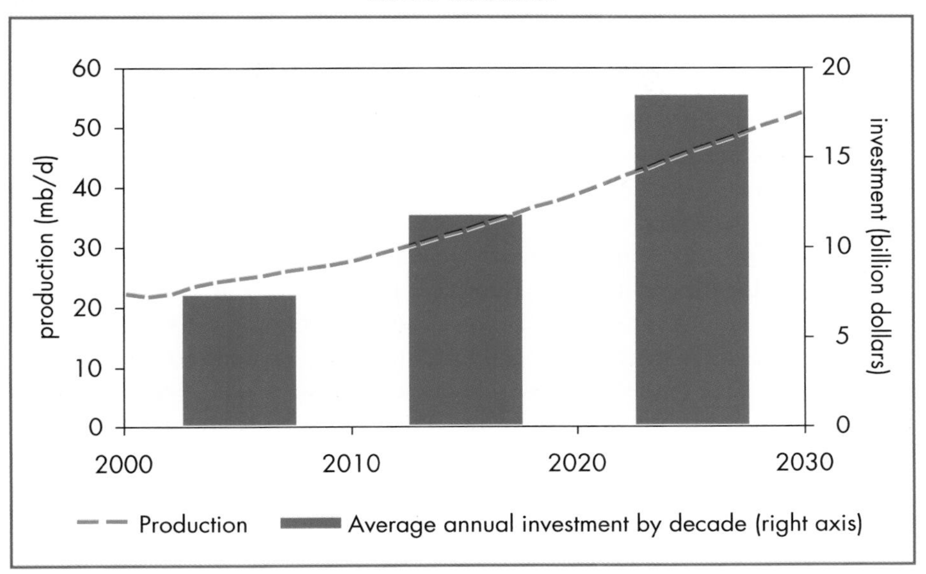

- *Infrastructure:* Inadequate infrastructure could constitute another barrier. If access to production sites is difficult, or roads, railways, pipelines or export facilities are not available, upstream projects could be delayed or cancelled.
- *Labour availability:* Operational and financial performance could be affected by a shortage of qualified labour.
- *National financing constraints:* In countries with national oil companies, financing new projects could become a problem where the national debt is already high and considerations of national sovereignty discourage reliance on foreign investment. The call of the national budget on future oil revenues could increase new financing costs. Strong guarantees could be difficult to find or involve a high insurance premium. Sovereign risk in many Middle Eastern countries is still high. A combination of these factors could delay or prevent investments.
- *Foreign investment policies:* The producing countries' policies on opening up their oil industries to private and foreign investment, the legal and commercial terms on offer and the fiscal regimes will have a major impact on how much capital Middle East producers will be able to secure.
- *Oil prices:* Prices will affect the ability of producing countries to finance investments from their own resources and, therefore, their need to turn to private and foreign investment.

- *Oil pricing and depletion strategy:* Governments could choose to delay development of production capacity in order to achieve higher profits by driving up international prices. Some may slow the development of their resources in order to preserve them for future generations.
- *Competition for financial resources:* In countries with state-owned companies and a rapidly growing population,[55] education, health and other sectors of the economy could command a growing share of government revenues and constrain capital flows to the oil sector. Even in countries open to foreign investment, and therefore less dependent on government revenues, the needs of an expanding population could lead governments to increase taxes, lowering the profitability of projects and so deterring investment.

In order to simulate the effect of restricted oil investment in OPEC Middle East countries, their share of global oil production is assumed to remain flat at 28%. This is close to the level observed in recent years. Under this assumption, production in the region keeps growing, but at a much slower pace than in the Reference Scenario (Figure 4.37). This is considered to be at the outer limit of what might be plausible. A scenario involving a falling share of Middle East OPEC in global supply is improbable, given the region's large reserves and low development costs. Producers in the region would be unlikely to accept a loss of collective market share.

Figure 4.37: **OPEC Middle East Share in Global Oil Supply**

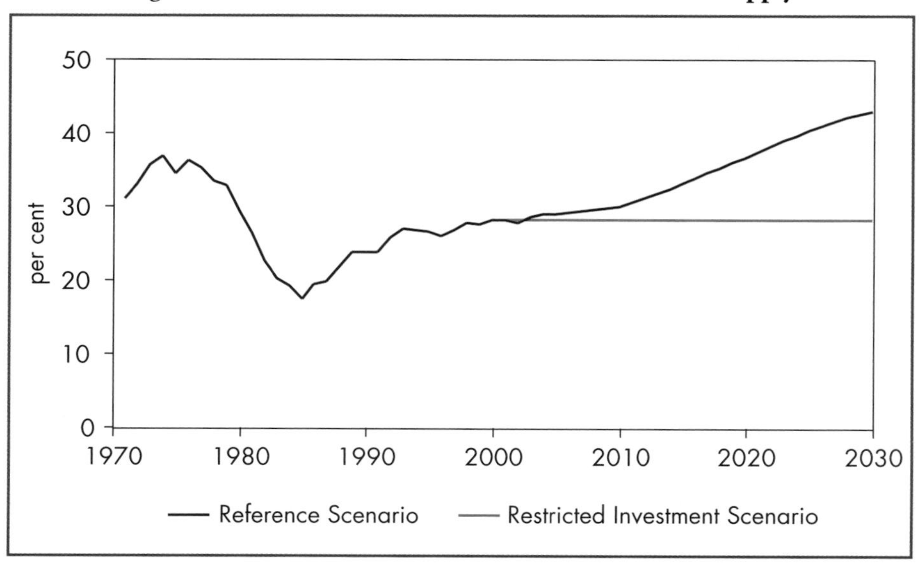

55. Middle East population is expected to grow by more than 2% per year between now and 2030 – the fastest growth of any WEO region.

Under the Restricted Investment Scenario, prices are assumed to be on average 20% higher than in the Reference Scenario.[56] As a result, the average price for imports into IEA countries rises to around $35/barrel in 2030 (in year 2000 dollars). Prices are also assumed to fluctuate around this rising trend, peaking in some years at around $40/barrel when global production approaches short-term capacity.

Results

Oil Demand

In the Restricted Investment Scenario, world oil demand grows by 1.3% per year, reaching 110 mb/d in 2030 (Table 4.14). This is 8% lower than in the Reference Scenario because of higher prices, which promote more efficient energy use and switching to other fuels. The share of oil in the global energy mix is 35% in 2030, compared with 38% in the Reference Scenario.

Table 4.14: **Global Oil Supply in the Reference and Restricted Investment Scenarios**

		2000	2010	2020	2030
Price *($ per barrel)*					
	Reference	28	21	25	29
	Restricted Investment		27	30	35
Demand *(mb/d)*					
OECD	Reference	47	52	56	60
	Restricted Investment		49	52	53
Non-OECD	Reference	28	37	48	60
	Restricted Investment		36	46	57
World oil balance *(mb/d)*					
	Reference	75	89	104	120
	Restricted Investment		85	97	110
Supply *(mb/d)*					
Middle East OPEC	Reference	21	26	38	51
	Restricted Investment		24	27	31
Rest of the world	Reference	54	62	66	69
	Restricted Investment		61	70	79

56. Although the higher oil prices in the Restricted Investment Scenario would lower economic growth rates (in oil-importing countries) which in turn would reduce oil demand, this effect has not been modelled in this analysis.

The impact on demand differs among regions. The reduction in demand is biggest in the OECD, where oil consumption is more than 10% lower in 2030 than in the Reference Scenario. Demand is 8% lower in the transition economies and 6% lower on average in developing countries, where there is less scope for energy savings and fuel switching. Demand in net oil-exporting regions, which are mostly in the developing world, is generally less affected.

Oil Supply

OPEC Middle East production in the Restricted Investment Scenario is 20 mb/d lower in 2030 than in the Reference Scenario, but it still increases from 21 mb/d in 2001 to 31 mb/d in 2030. The shares of non-conventional oil producers and other major oil-resource holders in OPEC are substantially higher, to make up for lower production in the Middle East (Figure 4.38). The share of non-conventional oil and GTL is six percentage points higher, at 16% of global supply, in 2030. Encouraged by higher oil prices, current non-conventional projects are developed more quickly. Rising production accelerates cost reductions through learning and economies of scale. Globally, non-conventional oil production reaches 18 mb/d in 2030, compared to 12 mb/d in the Reference Scenario. Production in Canada increases dramatically and GTL projects become more attractive economically.

For conventional oil in non-OPEC regions, the situation differs between regions with limited resources and those which still have large amounts of oil, such as the transition economies and Africa. For the first category, lower production in the OPEC Middle East countries affects the production profile more than the

Figure 4.38: **Global Oil Supply in the Reference and Restricted Investment Scenarios, 2030**

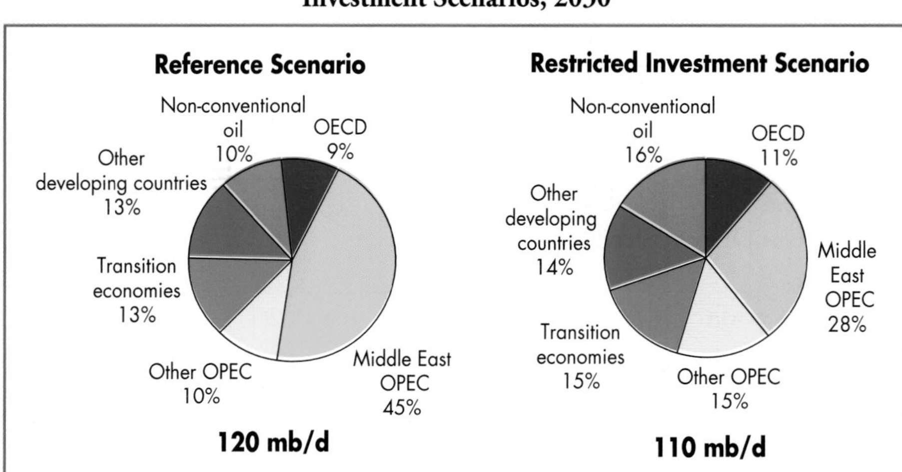

global amount produced. The more rapid increase in oil prices would result in faster depletion of resources than in the Reference Scenario. Higher oil prices would also increase the amount of recoverable resources as more efficient, but more costly techniques would be utilised. In the longer term, resource limitation concerns in non-OPEC regions would still arise and production in 2030 would be similar to that projected in the Reference Scenario.

The transition economies and non-OPEC African countries, with large resource bases, are in a better position to benefit from higher oil prices. They would be able to expand their capacities quickly and sustain the increase in output. In 2030, production in these regions is about 2 mb/d higher than in the Reference Scenario.

OPEC Revenues

Oil revenues in OPEC countries outside the Middle East grow continuously over the projection period in the Restricted Investment Scenario, thanks to higher oil prices and higher production. The situation is different in OPEC Middle East countries, where cumulative revenues are lower than in the Reference Scenario (Figure 4.39). For OPEC as a whole, higher oil prices do not compensate for lower production. Cumulative revenues over the projection period are more than $400 billion lower than in the Reference Scenario. Even on a discounted cash-flow basis, cumulative OPEC revenues over the projection period are lower.[57]

Figure 4.39: **OPEC and OPEC Middle East Cumulative Revenues in the Reference and Restricted Investment Scenarios, 2001-2030**

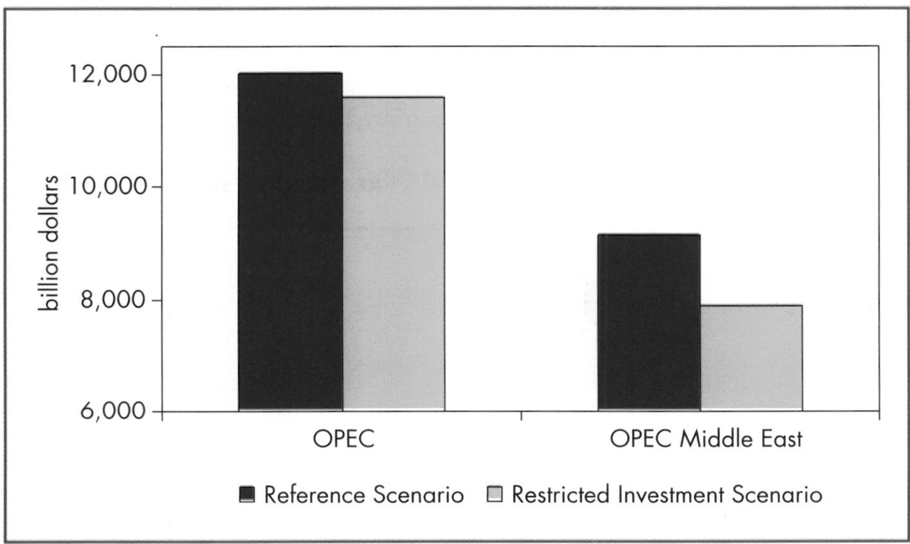

57. Estimated using a 10% discount rate.

Implications for Upstream Investment

In the Restricted Investment Scenario, despite lower global production, upstream oil-investment requirements are 3% higher over the projection period (Figure 4.40). Investment needs are higher in all regions, except OPEC Middle East, where they are $100 billion lower. The bigger investment requirements in Africa and the transition economies reflect faster production growth. But the bulk of additional investments goes to non-conventional oil, which becomes more economically attractive. Investments in non-conventional oil are more than four times higher in the Restricted Investment Scenario than in the Reference Scenario.

Figure 4.40: **Change in Investment Needs in the Restricted Investment Scenario Compared to the Reference Scenario**

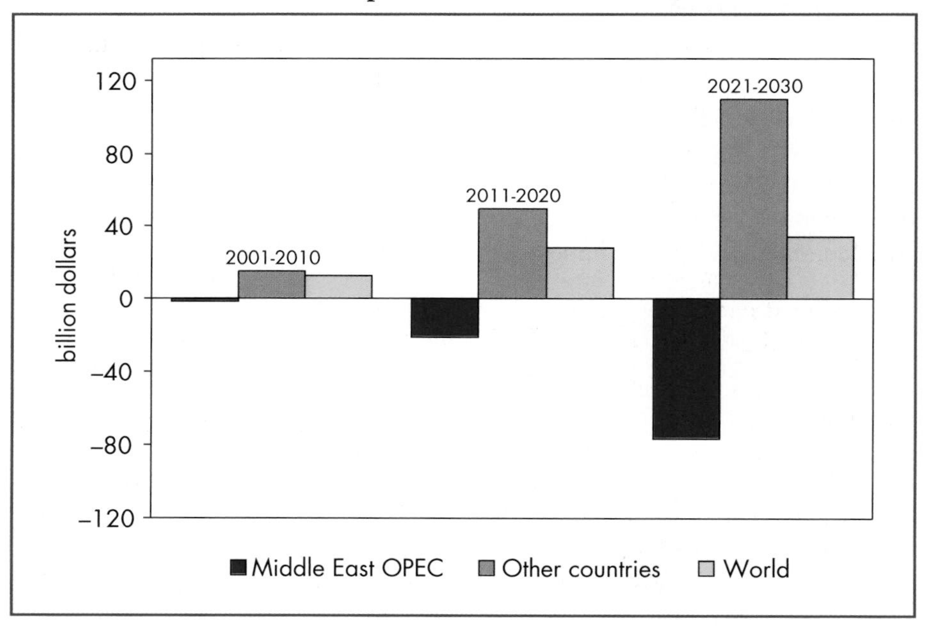

CHAPTER 5: NATURAL GAS

HIGHLIGHTS

- Cumulative investment in the global natural gas supply chain over the period 2001-2030 will amount to $3.1 trillion, or $105 billion per year on average. Much of this investment will be needed to expand capacity to meet a near-doubling of gas demand. Annual expenditures will increase over time from an average of less than $80 billion in the 1990s to $95 billion during the current decade and $116 billion in the third decade of the projection period.
- Exploration and development of gas fields, including bringing new fields on stream and sustaining output at existing fields, will absorb more than half of total gas investment. This will be needed to provide *each year* over the next three decades an average of around 300 bcm a year of new gas-production capacity worldwide — equivalent to the current capacity of OECD Europe. More than two-thirds of this new capacity will be needed to compensate for declining production at fields that are already in operation and others that will come on stream and decline during the projection period. The required rate of investment will increase rapidly, reaching $68 billion per year in the third decade — 55% higher than in the year 2000.
- Cumulative capital needs in the downstream will total $1.4 trillion. A large number of new cross-border high-pressure pipelines and LNG liquefaction plants, ships and regasification terminals will need to be built as trade expands and supply chains lengthen. Annual investment in the LNG chain will double from $4 billion over the past decade to around $9 billion in the period 2021-2030, supporting a sixfold increase in LNG trade. Investment in transmission and distribution networks will be much bigger, but will increase less rapidly.
- The United States and Canada, which have the world's biggest and most mature gas industry, will absorb well over a quarter of global gas investment over the projection period. Although demand will grow less rapidly than in most other parts of the world, high production decline rates will boost investment needs. The OECD as a whole will account for almost half of global gas investment. Most of the rest will go to the major current and emerging exporting regions — Russia, the Caspian region, the Middle East and Africa. LNG investment will be largest in the Middle East, while the transition economies — the former Soviet Union and Eastern Europe — will account for the largest share of investment in transmission pipelines and

storage facilities. Around 40% of non-OECD gas investments will be in projects for export to OECD countries.

- Sufficient capital is available globally, in principle, to support the large investment in gas-supply infrastructure needed to meet the projected increase in demand over the next three decades. However, energy-market reforms, more complex supply chains and the growing share of international trade in global gas supply, will give rise to profound shifts in gas-supply investment risks, required returns and financing costs.
- Greenfield projects throughout the world will involve very large initial investments. Obtaining financing for such projects — especially in developing countries where much of the investment will be needed — will be difficult, time-consuming and, therefore, uncertain. The private sector will have to provide a growing share of investment needs, because state companies will not be able to raise adequate public finance. In many cases, only the largest international oil and gas companies, with strong balance sheets, will be able to take on the required multi-billion dollar investments. Their share of global gas-industry equity is likely to rise.
- The lifting of restrictions on foreign investment and the design of fiscal policies will be crucial to capital flows and production prospects, especially in the Middle East and Africa, where much of the increase in production and exports is expected to occur.
- As a result of these factors, there is a danger that investment in some regions and parts of the supply chain might not always occur quickly enough. In that event, supply bottlenecks could emerge and persist due to the physical inflexibility of gas-supply infrastructure and the long lead times in developing gas projects. Such investment shortfalls would drive up prices and accentuate short-term price volatility.
- Governments need to tread very carefully in restructuring and reforming their gas markets, establishing long-term policy and a regulatory framework which sets clear and stable rules for gas and electricity markets in order to ensure that market structures do not impede investments that are economically viable. Long-term take-or-pay contracts in some form will remain necessary to underpin large-scale projects, at least until the transition to a truly competitive downstream gas market has been completed. Uncertainties about how quickly spot markets and market centres will develop, as well as more volatile prices and the possibility of lower wellhead prices in the future, are increasing investment risks now, tending to raise the cost of capital and skew investment towards smaller, closer-to-market projects.

The first part of this chapter summarises the results of the investment analysis, assesses the need for new capacity at each stage in the gas supply chain and reviews trends in technology and supply costs. An analysis of the factors driving gas investment, the uncertainties surrounding them and their implications for policy follows. The rest of the chapter provides an analysis of investment trends and issues by major region.

Global Investment Outlook

Over the period 2001-2030, cumulative investment in the global natural gas supply chain will amount to $3.1 trillion (Table 5.1), or $105 billion per year. Much of this investment will be needed to expand capacity to meet a near-doubling of gas demand (Box 5.1). Exploration and development of gas fields, including bringing new fields on stream and sustaining output at existing fields, will absorb more than half of total gas investment. Capital

Table 5.1: **Global Natural Gas Cumulative Investment by Region and Activity, 2001-2030** (billion dollars)

	Exploration & development	Transmission & storage	LNG*	Distribution	Total
OECD North America	553	145	44	189	931
OECD Europe	227	110	29	108	474
OECD Pacific	46	22	30	21	119
Total OECD	**826**	**277**	**102**	**318**	**1,524**
Russia	187	109	5	32	333
Other transition economies	85	56	0	19	160
Total transition economies	**272**	**165**	**5**	**51**	**493**
China	31	29	5	35	100
South and East Asia	168	51	18	31	270
Latin America	141	52	21	39	253
Middle East	140	65	64	12	280
Africa	153	34	37	3	226
Total developing countries	**633**	**230**	**145**	**120**	**1,129**
Total non-OECD	**905**	**395**	**150**	**171**	**1,621**
Total world	**1,731**	**673**	**252**	**489**	**3,145**

* Shipping is equally allocated among exporting and importing regions.

needs in the downstream[1] will total $1.4 trillion. A large number of new cross-border high-pressure pipelines and LNG processing and transportation facilities will need to be built as supply chains grow longer. Domestic transmission and distribution networks will also need to be developed in new gas markets and will need to be expanded in established markets. An increasing share of investment will go to LNG supply. Projected capital spending over the next three decades will total $252 billion dollars — more than twice the amount spent over the past 30 years.

Box 5.1: **Global Gas Market Outlook**

The projections of gas investment requirements presented in this report are based on the Reference Scenario of the *World Energy Outlook 2002 (WEO-2002).* Global natural gas demand (excluding feedstock to GTL plants) will almost double over the projection period, increasing by an average 2.4% per year from 2,569 bcm in 2001 to 5,047 bcm in 2030. The share of gas in the global primary energy mix will increase from 23% in 2001 to 28% in 2030. The power sector will account for a growing share of total primary gas consumption worldwide, reaching almost half by 2030. Demand is expected to grow most rapidly in the fledging markets of developing Asia, notably China, and in Latin America. Nonetheless, North America, Europe and Russia remain by far the largest markets in 2030.

Production will grow most in absolute terms in the transition economies and the Middle East (Table 5.2). Most of the incremental output will be exported to Europe and North America. Output will increase quickly in Latin America and Africa. The projected 2,685-bcm increase in production between 2001 and 2030 will require massive investment in production facilities and transport infrastructure.

The United States and Canada, which have the world's biggest and most mature gas industry, will absorb well over a quarter of total world investment over the projection period. Although demand will grow less rapidly than in most other parts of the world, relatively high production decline rates will boost investment needs. The OECD as a whole will account for almost half of global gas investment. Most of the rest will go to the major current and emerging

1. Transmission and distribution pipelines, underground gas storage and LNG liquefaction plants, regasification terminals and ships. Investments in gas-to-liquids plants are included in oil (see Chapter 4). Refurbishment costs are not included, because of the enormous difficulties in projecting future needs and because these costs are often classified as operating expenditures.

Table 5.2: **World Natural Gas Production** (bcm)

	1990	2001	2010	2030
OECD North America	643	783	886	990
OECD Europe	210	306	300	276
OECD Pacific	27	42	65	125
Total OECD	**880**	**1,131**	**1,251**	**1,391**
Russia	640	580	709	914
Other transition economies	196	161	205	308
Total transition economies	**835**	**742**	**914**	**1,222**
China	17	34	55	115
East Asia	77	150	213	409
South Asia	29	62	89	178
Latin America	61	101	217	516
Middle East	99	242	421	861
Africa	70	134	246	589
Total developing countries	**354**	**723**	**1,241**	**2,667**
Total non-OECD	**1,190**	**1,464**	**2,156**	**3,889**
World	**2,070**	**2,595**	**3,407**	**5,280**

Note: Natural gas production includes supply to GTL plants.
Source: IEA (2002a).

exporting regions — Russia, the Caspian region, the Middle East and Africa. The Middle East will have the largest requirement for LNG investment, while the transition economies will account for the largest share of transmission and storage investment. On average, 40% of total gas investment in non-OECD countries will be in projects for export to OECD countries (Table 5.3). In some key exporting regions, notably the Middle East and Africa, this share will be higher, at around two-thirds.

Annual capital expenditure will increase over time, due to the increasing need for new and replacement capacity. It is projected to grow from an estimated average of just under $80 billion in the 1990s to $95 billion per year during the current decade and $116 billion in the third decade of the projection period (Figure 5.1). Capital spending on exploration and development (E&D) will increase rapidly, reaching $68 billion per year in the third decade — 55% higher than in the year 2000. Although more drilling will occur in lower cost regions, a doubling of global production and a shift in

Table 5.3: **Energy Investment in Non-OECD Countries by Destination of Supply, 2001-2030** ($ billion in year 2000 dollars, % for share)

	For supply to OECD markets		For supply to domestic and other non-OECD markets		Total
	$ billion	%	$ billion	%	$ billion
Total non-OECD	**646**	**40**	**976**	**60**	**1,621**
Of which					
Russia	103	31	230	69	333
Middle East	196	70	84	30	280
Africa	147	65	79	35	226

Note: Shipping included.

drilling in offshore fields will push up overall upstream investment. Gas processing costs, included in E&D, may also rise, as the quality of reserves declines. As a result, investment in upstream gas will approach that in oil (Box 5.2). Development costs will account for more than 90% of total upstream investment. The share will be highest in the transition economies

Figure 5.1: **Global Gas Investment**

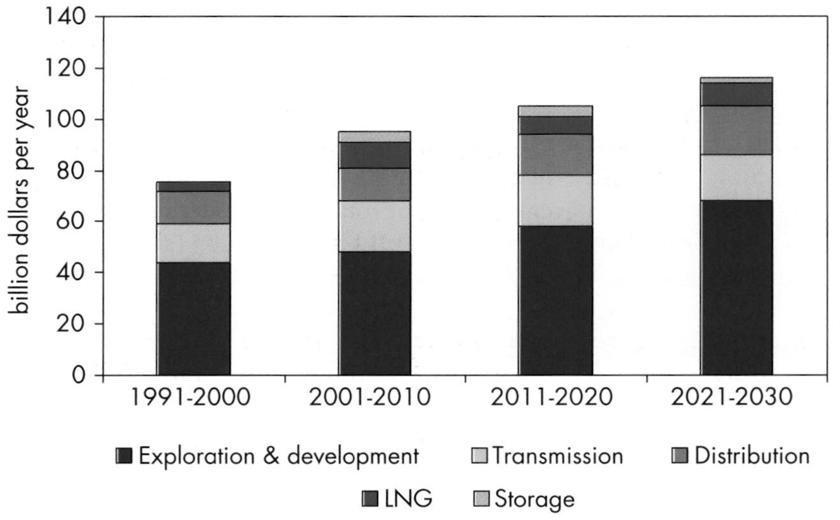

Note: Data for storage are not available for 1991-2000; the estimate for the same period for exploration and development is for the year 2000.

Box 5.2: **Exploration and Development Investments Shift to Natural Gas**

Interest in natural gas has steadily increased over the past decades and it will continue to do so over the next three decades. The clearest evidence of companies' heightened interest in natural gas is the shift in the composition of global hydrocarbon production towards natural gas. Combined global oil and gas production has soared from 90 million barrels of oil equivalent per day (mboe/d) in 1980 to 121 mboe/d in 2002, with the share of gas increasing from 29% to 39%. The higher profitability of gas production and rising demand explains this trend. Official US data show that upstream companies[2] have enjoyed higher profit margins for gas than oil for most of the 1980s and 1990s. The share of gas wells in total well completions worldwide has also been increasing, from 15% in the mid-1980s to above 60% in 2002.[3]

Combined global oil and gas production is expected to reach 207 mboe/d in 2030, with gas accounting for 42%. Investments in exploration and development are expected to follow a similar pattern (Figure 5.2). In the year 2000, oil and gas E&D investment equalled $108 billion, oil making up 60%. Annual average E&D investments are expected to increase over the next three decades, reaching $147 billion per year over the third decade. While oil investment will increase slightly, investment in natural gas will climb steadily. In the third decade, gas is expected to account for 46% of the annual average E&D of oil and gas spending, compared to 40% in the year 2000. The differential between annual oil and gas expenditure in the third decade will be $11 billion, compared to $20 billion in the year 2000. However, taking into account the additional cost of bulk transportation and oil refining, the capital requirement for one mboe/d of supply capacity over the next three decades will remain higher for gas than for oil: on average $17,000 for gas compared to around $14,900 for oil.

and in developing countries, where large proven reserves are awaiting development. Average annual investment in the LNG chain will double from $4 billion per year over the past decade to $9 billion per year in the period 2021-2030, driven by a sixfold increase in LNG trade. Investment in transmission and distribution networks will increase less rapidly.

2. Energy companies that report their financial and operating data to the Energy Information Administration Financial Reporting System.
3. See EIA (2001).

Figure 5.2: **Upstream Oil and Gas Investment and Indexed Production**

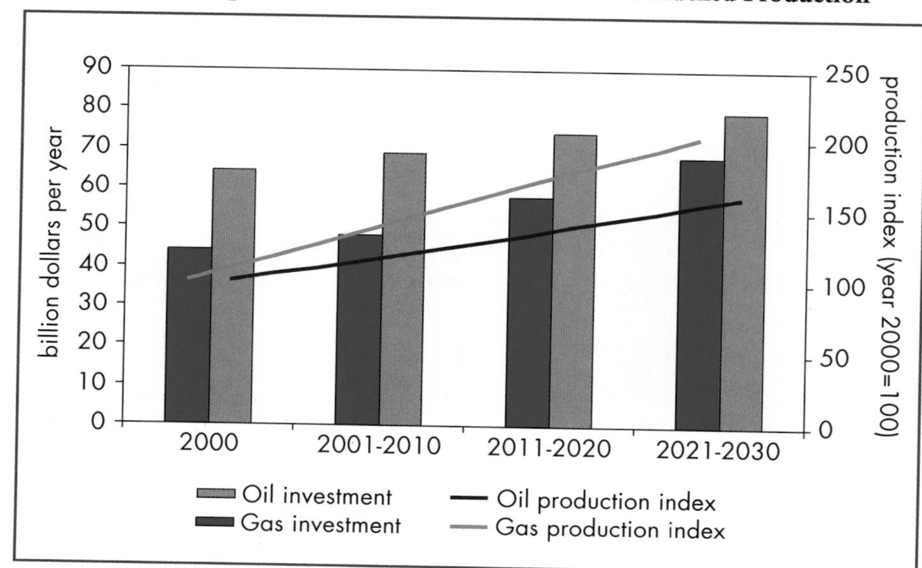

Capacity Requirements

The global capacity of gas supply infrastructure, including production facilities, high-pressure transmission and local distribution pipelines, LNG plants and ships, and underground gas storage facilities, will have to expand significantly to meet increasing demand over the next three decades.

Upstream

A total of 9 trillion cubic metres of new gas-production capacity will have to be installed worldwide over the next three decades, an average of around 300 bcm a year (which is the current capacity of OECD Europe). Only 30% of this new capacity will be needed to meet rising demand. The rest will compensate for declining production at fields that are already in operation and others that will come on stream and decline during the projection period (Figure 5.3). The rate of new capacity additions will reach around 360 bcm per year in the third decade.

The natural decline rate — the rate at which well production declines from one period to the next in the absence of any capital spending — is the crucial determinant of the rate at which new capacity will need to be added.[4] The decline rate globally is projected to average 6% per year, but it varies between 4%

4. See Chapter 4 for a discussion of decline rates.

Figure 5.3: **Average Annual Additions to Gas Production Capacity, 2001-2030**

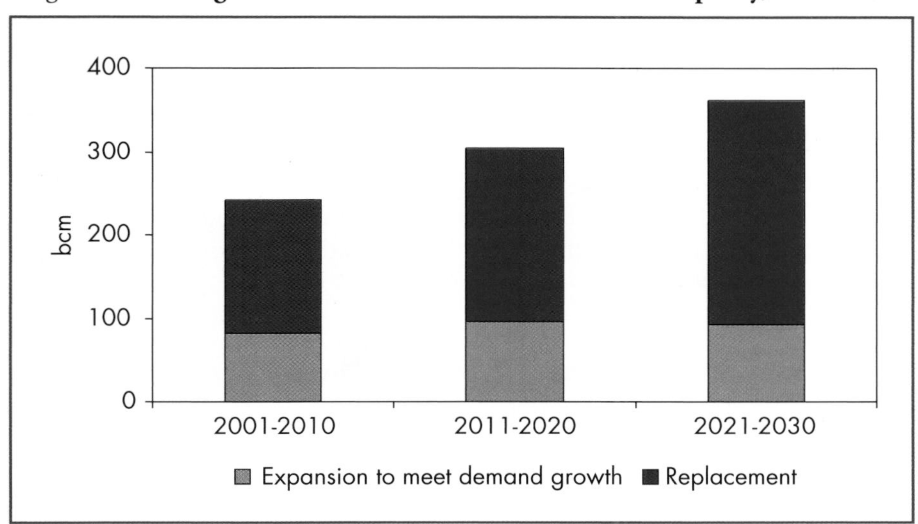

and 11% both over time and by region, according to the maturity of the producing area. Decline rates in the first one or two years of production can be much higher, but these declines are usually offset by slower year-on-year declines later in the life of a production well. The highest rates are projected for the end of the projection period in OECD North America and OECD Europe.

A quarter of global capacity additions will occur in OECD North America, where decline rates are high because of the age of fields, the falling size of new discoveries and extraction technologies that maximise initial production rates. Additions to production-well capacity will also be large in Russia (45 bcm per year) and the Middle East (43 bcm per year). Together, these two areas will account for 35% of the increase in gas production over the next 30 years.

In 2001, 71% of all the natural gas produced in the world came from onshore fields. This share is expected to drop to 64% in 2030, as drilling shifts to more lucrative offshore sites. OECD North America, the transition economies and the Middle East will account for two-thirds of the onshore capacity brought on stream over the next three decades. The Northwest Europe Continental Shelf and the Gulf of Mexico will together account for almost one-third of global offshore capacity additions. Asian countries will account for almost one-fourth.

Downstream

The expansion of the gas market and the mismatch between the location of demand and production will continue to drive rapid growth of inter-regional trade. Net trade between regions is projected to surge from 474 bcm in 2001 to almost 1,700 bcm in 2030 (Figure 5.4). The growing importance of the Middle East and Africa as exporters to Europe and North America will lengthen supply chains on average, boosting the attractiveness of transporting gas in liquefied form. LNG costs are expected to continue to fall relative to pipeline costs (see Technology and Cost Developments section). Inter-regional LNG trade is expected to increase sixfold over the next thirty years, becoming as important as pipeline trade by 2030. Most of the increase in LNG trade will be in the Atlantic basin, which will overtake the Pacific in volume. New cross-border transmission lines will be built, but the rate of growth in capacity will be slower than in the past.

Figure 5.4: **Net Inter-regional Trade and Production, 2001-2030**

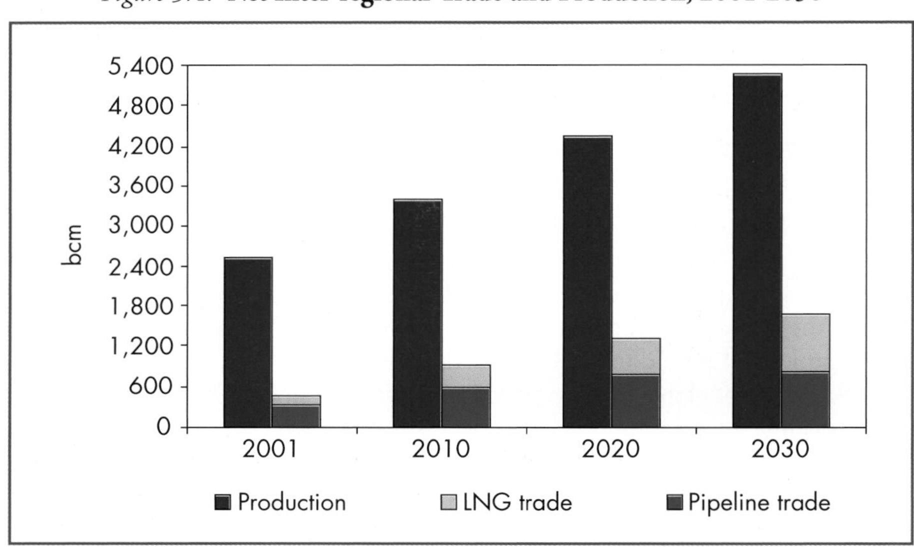

LNG liquefaction capacity will need to expand more than fivefold from 133 Mt per year in 2002 to 720 Mt in 2030 (Figure 5.5). This corresponds to about 100 new trains, assuming a steady increase in the average size of each train. There are currently 15 liquefaction plants with 69 trains in operation worldwide. The Middle East alone will account for 40% of the increase. Africa will account for another quarter, and Latin America and Asia for most of the rest.

Importing countries will need to add almost 900 bcm (660 Mt) of new regasification capacity. Capacity in 2002 stood at 388 bcm at 40 terminals, 24 of them in Japan. OECD North America will account for almost half of the additional capacity and OECD Europe for another 30%. Korea, China

Figure 5.5: **LNG Liquefaction and Regasification Capacity, 2001-2030**

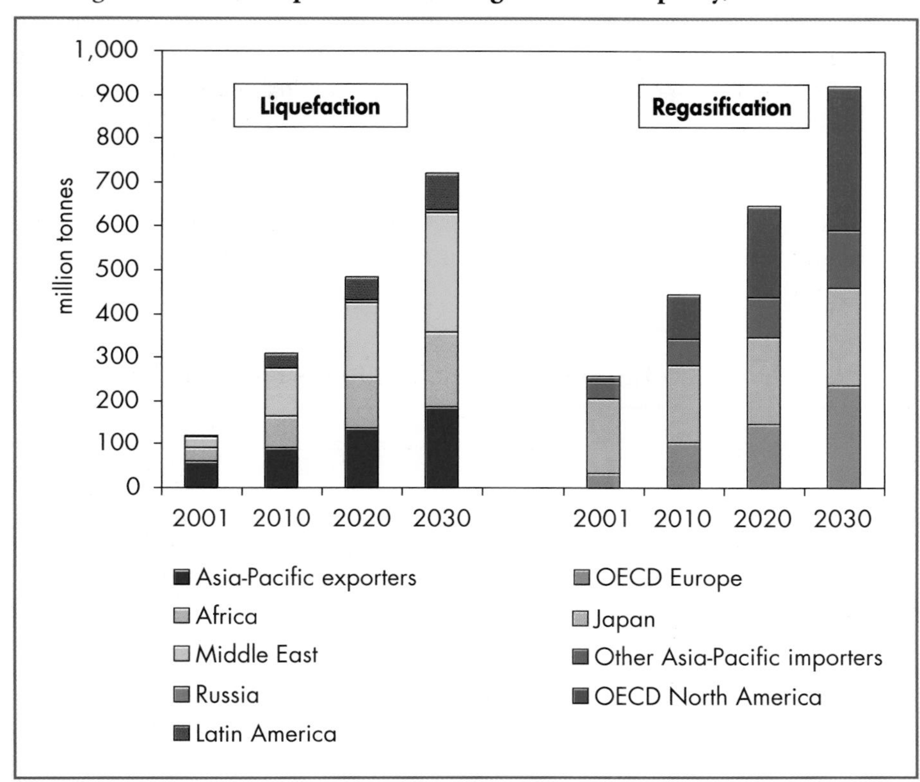

and India will also need to increase significantly the number of regasification facilities. Importing countries, particularly Japan, are expected to maintain some spare capacity for energy-security reasons. Global average utilisation rate is nonetheless expected to increase substantially (from only 40% in 2001[5]).

The world's **LNG shipping** fleet, which currently numbers 128 ships, will have to virtually quadruple to sustain the projected growth in trade (Figure 5.6). Some 58 new tankers have already been ordered for delivery between 2003 and 2006. Of these ships, 15 are not linked to long-term contracts. LNG liquefaction project developers control 60% of the existing fleet, but only 40% of the ships on order.[6] International oil and gas companies and LNG buyers account for over half of new orders. This change in control reflects the trend towards a more liquid and short-term LNG market, the financial difficulties of some of the leading gas merchants and the large amounts of capital involved.

5. The low utilisation rate is due mainly to overcapacity in Japan.
6. CERA (2003).

Figure 5.6: **LNG Shipping Fleet, 2001-2030**

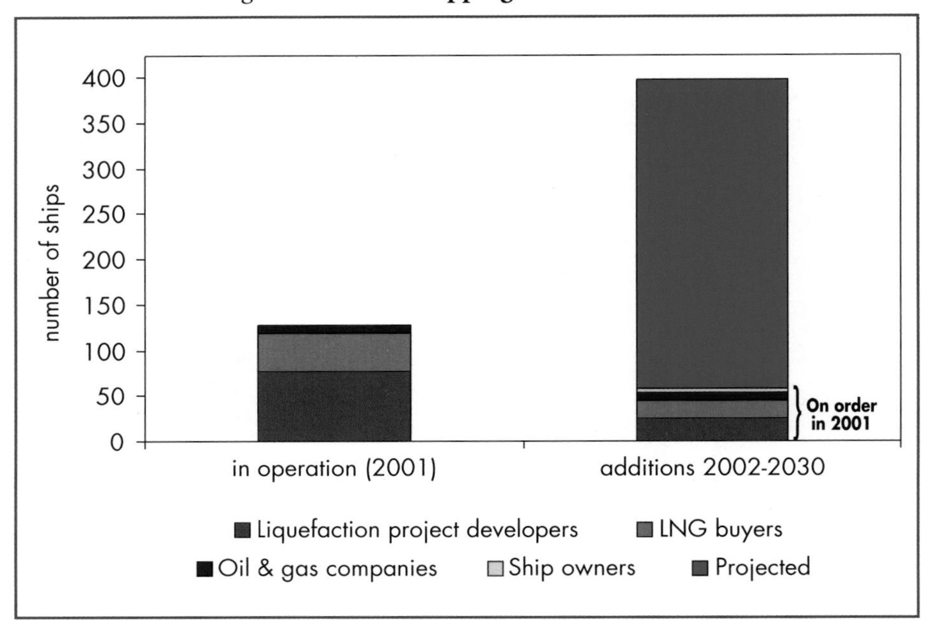

The length of the world's high-pressure gas **transmission pipelines**, which totalled 1.1 million km in 2000, will increase by around 80% over the next three decades. The diameter and capacity of new lines will increase during the projection period, so the length of the pipeline system will not grow as fast as throughput. The rate of growth in pipeline length will average 2% per year, with an expected growth in demand of 2.4% per year. Transmission capacity needs in each region are estimated on the basis of the historical evolution of the network and the growth in gas demand, exports and transit volumes.[7] Specific gas-pipeline projects under construction or planned have been taken into account.

In 2030, 75% of all gas transmission pipelines will be found in the mature gas markets of OECD North America, OECD Europe and the transition economies — down from 90% at present (Figure 5.7). The North American market will account for a quarter of global pipeline additions, driven by rising demand and a shift in supply sources as existing mature basins are depleted, new domestic sources are tapped and LNG terminals are built. The transition economies will need to expand their export pipeline systems, mainly

7. A more detailed explanation of the methodology can be found in Annex 2.

to Europe, and build new lines to the Far East. Producers in the Middle East and Africa will also build new transmission lines, mainly for export purposes but also to meet growing domestic demand. Offshore pipelines are expected to increase faster than onshore, since an increasing volume of production is expected to come from offshore fields and more export lines will be built offshore. Emerging Asian and Latin American markets will also need to expand rapidly their gas transportation infrastructure.

Figure 5.7: **Length of Transmission Pipelines, 2000-2030**

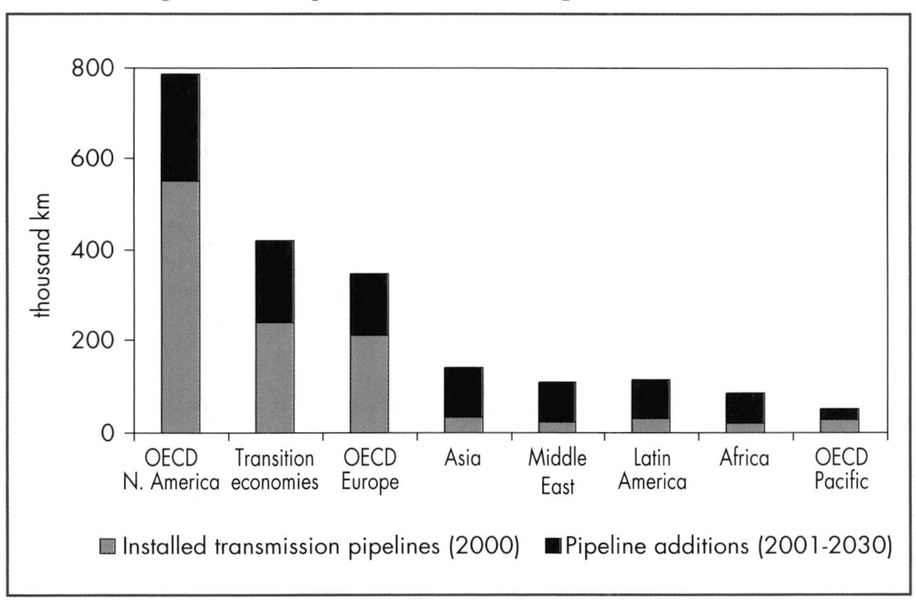

The next three decades will see the aggregate size of the world's local **distribution networks** grow from 5 million km in the year 2000 to 8.5 million km in 2030. Capacity will grow less rapidly than demand, because most of the increase in global demand will come from power stations, which are normally supplied directly off transmission pipelines. There is limited scope for increasing the use of gas in the residential and commercial sectors in many developing countries, because of low space-heating demand and the high capital cost of installing local distribution networks. OECD North America will require the biggest increase in distribution capacity in absolute terms (Figure 5.8). But Asia, Latin America and the Middle East will experience the highest growth rates, with capacity quadrupling between 2000 and 2030. As with transmission pipelines, the projections of distribution network length in each region are based largely on past trends in network development and projected residential and commercial gas demand, population and network densities.

Figure 5.8: **Length of Distribution Networks, 2000-2030**

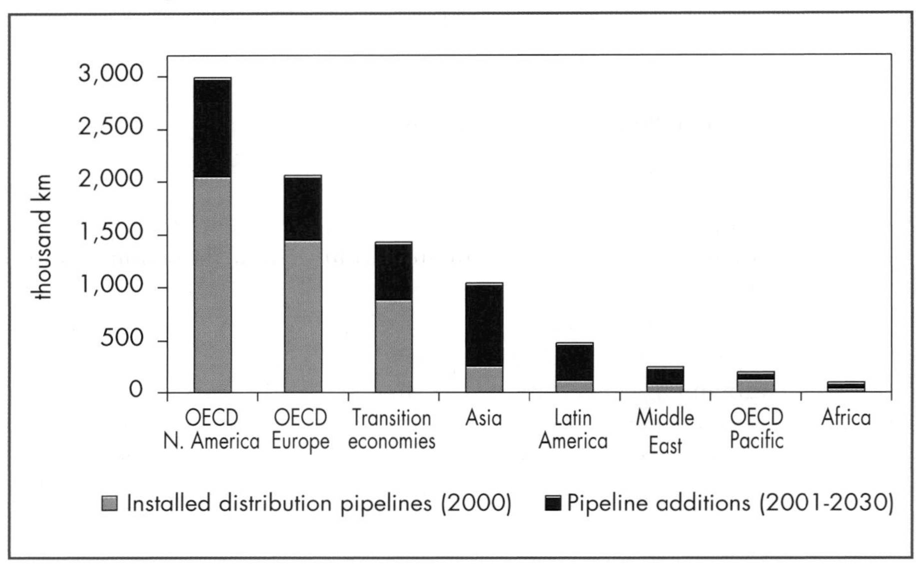

Global **underground gas storage** working volume is projected to grow from 328 bcm in 2000 to 685 bcm in 2030. More than 80% of this expansion will take place in OECD North America, OECD Europe and the transition economies, where there will be an increasing need to manage seasonal swings in demand in the residential sector (Table 5.4). Liberalisation

Table 5.4: **Underground Gas Storage Working Volumes, 2000-2030** (bcm)

	Working volume 2000	**Working volume 2030**	**Additional working volume 2001-2030**
OECD North America	129	215	86
OECD Europe	61	138	77
OECD Pacific	2	14	12
Transition economies	132	266	134
Developing countries	4	51	47
World	**328**	**685**	**356**

of gas markets, which will boost short-term trading and opportunities for arbitrage, will increase the demand for storage capacity. New markets with high demand growth rates, such as China, and exporting countries, such as Iran, are already planning to build their first storage facilities. This trend is expected to increase further over the projection period. Regional storage capacity requirements are projected on the basis of gas demand, exports, transit volumes and the degree of maturity of the gas market.

Technology and Cost Developments

Exploration and Development

Estimated gas upstream investment requirements depend on projected onshore and offshore production, estimated additional onshore and offshore discoveries and average exploration and development cost estimates. Figure 5.9 gives an overview of natural gas reserves locations and corresponding exploration and development costs. Regions with the largest resource endowment also enjoy the lowest E&D capital cost. The methodology used to evaluate gas upstream investment requirements is the same as that used for oil. A discussion on E&D costs is provided in Chapter 4, while a detailed description of the methodology can be found in Annex 2.

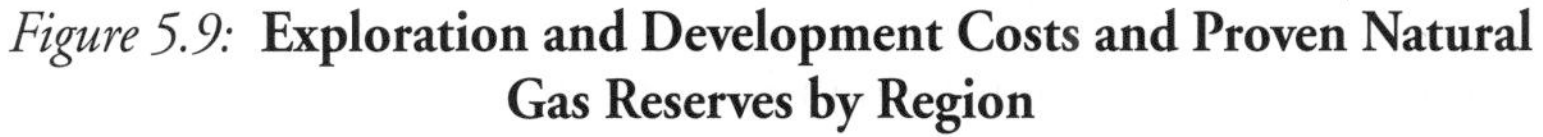

Figure 5.9: **Exploration and Development Costs and Proven Natural Gas Reserves by Region**

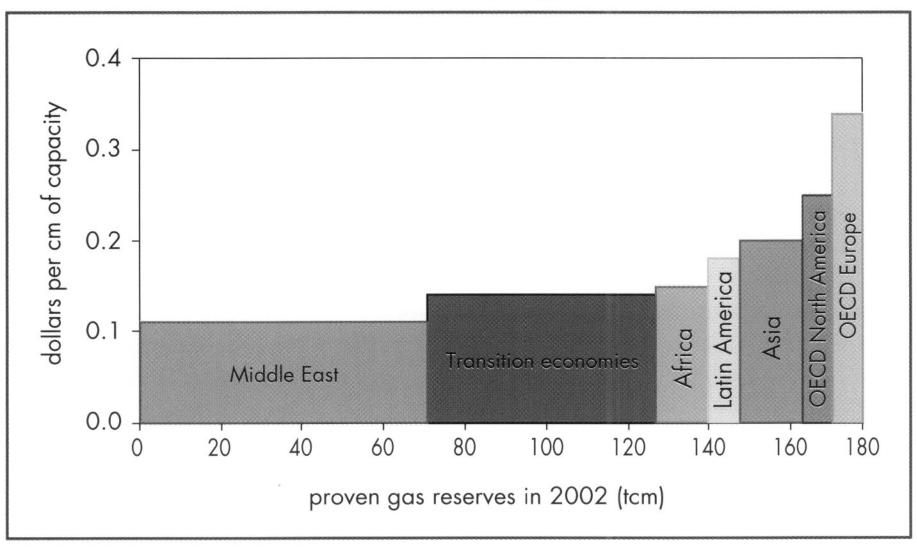

LNG

Technology developments and improvements in refrigeration and liquefaction techniques have lowered considerably the capital costs of liquefaction and shipping — the most expensive processes in the LNG supply chain. Falling investment costs are making gas reserves located far from markets economic to develop. Nominal liquefaction capital costs had fallen to about $240 per tonne of LNG capacity on average in 2002 from about $550 in the early 1990s, largely due to scale economies from larger trains. The latest designs being built today can process up to 7 Mt per year, up from about 2 Mt in the early 1990s. Economies of scale are achieved by sharing common facilities, such as utility infrastructure and storage tanks. As a result, the unit cost of a second train is typically much lower than that of the first train. Integrating terminals with power plants have also helped to reduce costs. Further economies through even larger trains and competition among liquefaction-technology providers will most likely continue to drive down capital costs. On average, liquefaction costs are projected to drop to $200 per tonne of LNG by 2010 and to $150 by 2030.[8]

The cost of building LNG carriers has fallen dramatically too, by more than 40% during the last decade. The average cost for a ship with four storage spheres and a capacity of 137,000 cubic metres (cm) has stabilised at around $160 to $175 million, down from $250 million (in money of the day) in the early 1990s. Larger capacities have lowered unit costs and this trend is expected to continue. The largest ship on order at present is 145,000 cm, which will cost $200 million, but a $230 million ship of 153,000 cm is planned for delivery in 2005.[9] When five spheres are introduced, capacity could reach 165,000 cm. By the end of the decade, carriers of 220,000 to 250,000 cm may be possible. ExxonMobil plans to order ships of over 200,000 cm to ship Qatari gas to Atlantic basin markets from 2008, but such large vessels are still at the design stage. The ability of receiving terminals to handle very large carriers may also limit the increase in carrier size. Our analysis projects an average decline of 20% in unit shipping costs over the projection period, although costs could rise at times of strong demand if a shortage of shipbuilding capacity occurs.

Offshore loading and receiving terminal concepts may provide a solution to difficulties in siting LNG liquefaction and regasification plants, caused by the absence of natural harbours or public opposition. Offshore facilities can be cheaper and quicker to build because they do not need harbour facilities. Floating LNG production and loading technology can also make the development of offshore reserves more economic and more secure, because it avoids the need for a pipeline to shore. Average unit regasification terminal capital costs are

8. See Annex 2 for an explanation of cost projections.

9. *World Gas Intelligence*, "France to Subsidise Biggest LNG Tanker" (9 July 2003).

currently around \$86 million per bcm per year of capacity, accounting for 20% of the total cost of an LNG chain, not including upstream development. On average, regasification unit costs are projected to fall, mainly thanks to economies of scale, from around \$86 million per bcm at present to \$77 million by 2010 and to \$65 million by 2030.

The overall capital costs of LNG supply chains are expected to continue to fall, but at a slightly lower rate than over the past decade as the scope for learning and exploiting economies of scale diminishes. Total capital requirements have fallen from around \$700 per tonne in the mid-1990s to around \$500 today. Costs are projected to fall to \$420 per tonne by 2010 and \$320 per tonne by 2030 assuming a shipping distance of around 4,000 km (Figure 5.10).

Figure 5.10: **Indicative Unit Capital Costs for LNG Projects**

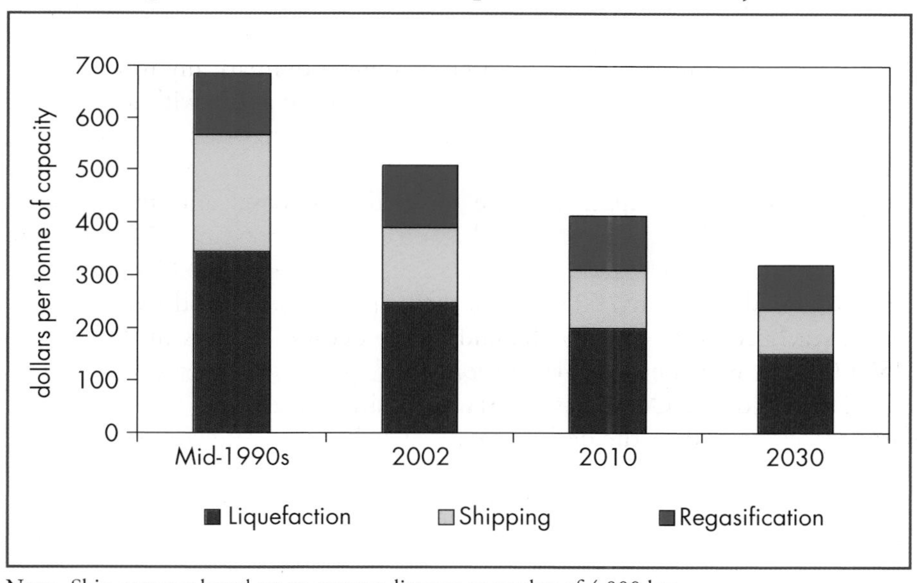

Note: Ship costs are based on an average distance to market of 4,000 km.

Transmission and Distribution

Diameter, operating pressure and length are the key technical parameters influencing pipeline construction costs. Geography and the nature of the terrain to be crossed also affect costs. Material, labour and rights of way usually account for most of the cost of building a pipeline. The degree of competition among contracting companies plays an important role in determining the final cost, as do safety and environmental regulations. In the United States, labour is the largest single cost component, followed by materials, for both onshore and offshore pipelines (Figure 5.11). The cost of building onshore pipelines there has risen over the last decade due to higher labour and rights of way costs. On the other hand, technological advances have pushed down the cost of materials and labour in offshore pipeline, reducing unit costs dramatically. This factor has been less important in the onshore pipeline construction, where technology is more mature.

Figure 5.11: **Pipeline Construction Costs in the United States**

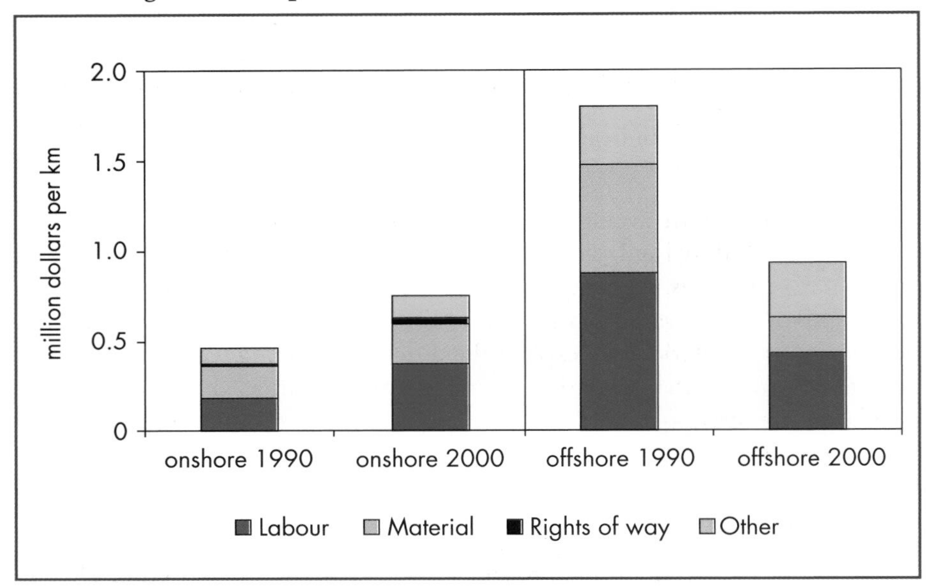

The relative importance of each cost component varies significantly among geographical regions. The share of labour is generally lower in developing countries. On the other hand, material costs can be higher in those countries, if they are imported. In China, for example, labour typically accounts for less than 10% of total capital costs, while material costs can be as high as 70%.[10] The breakdown of pipeline costs also depends on the size of the pipeline: small diameter pipes use less material, so labour takes a larger share of total construction costs.

The main advance in pipeline technology in recent years has been the introduction of high-strength steels that allow pipelines to be used at high

10. Royal Institute of International Affairs (2001).

pressures. This development enables a larger throughput for a given diameter and also makes possible larger diameter pipelines,[11] reducing the unit cost of large-scale projects. This is a critical factor, since materials account for 40% to 50% of the cost of large-diameter pipelines. Stronger steels and new high-pressure technologies, as well as the use of new fibre glass composite materials, will contribute to further declines in transmission unit costs in the future.

Onshore pipeline construction unit costs are expected to remain broadly constant over the next three decades. Technology advances will probably lead to lower materials and engineering costs, but these are assumed to be offset by higher costs for labour and for rights of way as permitting requirements become more and more onerous. Safety and environmental concerns, especially in populated areas, could limit the potential for cost reductions from the use of higher strength pipelines. On the other hand, the construction cost of offshore pipelines is, in general, expected to continue to decline, but at a lower rate than in the past.

A database with average regional transmission pipeline unit costs has been compiled, drawing from a number of sources, including gas companies and literature surveys. A similar database has been assembled for distribution costs. Labour costs, safety standards, environmental regulations, population density and the material used explain the differences among regional unit costs. Stricter environmental regulation, increasing urbanisation and tougher safety standards are expected to push up gas distribution unit costs over the projection period.

Underground Gas Storage

Most of the world's underground storage facilities are located in OECD countries and the transition economies. In 2001, 83% of working capacity was in depleted gas fields, 12% in aquifers and 5% in salt caverns. Depleted fields account for a slightly higher share — 85% — in OECD North America and the transition economies. European storage is more diverse, consisting of 63% depleted fields, 23% aquifers and 14% salt caverns. Investment costs vary according to storage type, working volume, maximum injection and withdrawal capacity, location and geology. Capital costs comprise the cushion gas, exploration expenditures, drilling of wells, leaching (for salt cavern storage), underground and surface equipment, and spending on connecting the facility to the transmission system. Average overall costs are lowest in the United States, partly because of less stringent health and safety regulations than in Europe (Table 5.5).

11. There are very important economies of scale in building pipelines.

Table 5.5: **Underground Gas Storage Investment Costs by Type and Region**
($ per cubic metre of working gas)

	Europe	United States	Transition economies
Aquifer	0.35 - 0.60	0.14	0.30
Depleted field	0.35 - 0.60	0.12	0.30
Salt cavern	0.70 - 1.00	0.30	n.a.

Source: IEA analysis based on UNECE (2000) and industry sources.

The development of new and emerging storage technologies has helped to reduce risks and costs of building new facilities, as well as to improve their efficiency and safety. Some recent cost-saving technologies such as horizontal drilling have allowed storage development to proceed with fewer injection/withdrawal wells and reduced cushion-gas requirement, lowering investment costs. New storage technologies and costs will be strongly influenced by technology developments in oil and gas exploration and production. Technology is expected to cut storage-investment costs further, but more restrictive security and environmental regulations are expected to offset this reduction. In our analysis, capital unit costs are assumed to remain constant between 2001 and 2030.

Investment Uncertainties and Challenges

Risks and Returns

Sufficient capital is expected, in principle, to be available globally to support the large volume of investment in gas-supply infrastructure needed to meet the projected increase in demand over the next three decades. There is considerably less certainty about cost developments, the outlook for prices and demand, and whether the required funding will always be forthcoming — especially in countries with little experience of major gas projects. A number of developments, including market reforms, longer supply chains and the growing share of international trade in global gas supply, will give rise to profound shifts in gas-supply investment risks, required returns and financing costs. As a result, there is a danger that investment in some regions and parts of the supply chain might not always occur in a timely fashion. In this case, supply bottlenecks could emerge and persist due to the physical inflexibility of gas-supply infrastructure and the long lead times in developing gas projects: investment decisions have to be taken well in advance of when demand is expected to materialise. Such investment shortfalls would drive up prices and accentuate short-term price volatility in competitive markets.

An increasing number of greenfield projects throughout the world will involve very large initial investments, often amounting to several hundred millions or even billions of dollars. For example, the cost of the proposed pipeline from Alaska to the lower 48 US states is estimated at $18 billion, which would make it the largest pipeline investment ever. Similarly, greater distances and larger capacities have tended to push up the total capital cost of LNG chains, despite significant reductions in unit costs in recent years. Greenfield projects are the most costly and challenging of all types of gas investment, since the infrastructure for the full supply chain — gas-field production facilities, high-pressure pipelines and/or LNG chains and local distribution networks — needs to be brought into operation simultaneously. The profitability of such projects depends heavily on how quickly all the supply capacity is put to use, because upfront capital expenditures dominate total supply costs. Project risks are particularly large where the market being supplied is immature and where there are doubts about the creditworthiness of the major consumers — typically power stations and large industrial plants — on which the project relies.

Obtaining financing for multi-billion dollar investments is difficult, time-consuming and, therefore, uncertain. The sheer scale of investment will preclude all but the largest international oil and gas companies, with strong balance sheets, from becoming involved. Even those companies will usually participate as part of consortia to spread investment risk. The growing importance of mega-projects may ultimately lead to further industry consolidation, through horizontal and vertical integration, strategic alliances and partnerships.

Country risk will become an increasingly important factor for a growing number of export pipeline and LNG projects as well as domestic downstream projects. Country risk, notably political and economic instability, can significantly increase overall project risk and the required return on investment, especially in the case of cross-border pipelines that transit third countries. Changes to tax system are another major risk. Geopolitical factors are especially important to the prospects for developing long-distance pipelines in the Middle East and Central Asia.

In many developing countries, where much of the investment will be needed, the private sector will have to account for a growing share of investment, because state companies will have difficulty raising sufficient funds. The sources of private finance will vary. In some cases, foreign direct investment by international oil and gas companies will grow in importance, particularly for export-oriented projects. In other cases, local and regional capital markets may provide the necessary funding, with national gas companies and/or foreign companies providing the technical know-how. Restrictions on foreign investment, typically motivated by political or nationalistic considerations, could undermine investment in some countries, notably in the Middle East.

Investment prospects are more secure for domestic downstream projects in OECD countries, particularly those that involve the extension or enhancement of existing pipeline networks, including the construction of new compressor or blending plants or looping of existing lines. This type of investment is usually considered to be relatively low-risk, particularly where demand trends are reasonably stable and predictable and where returns are protected by the regulator through explicit price controls. The returns that can be made on downstream investments generally depend to a large extent on the regulatory framework and on incentives to increase profit through efficiency improvements.

Where the investment is incremental and where the regulatory regime provides a high level of assurance to the investor that he will be able to recover his costs through regulated tariffs, the allowed rate of return is generally low relative to the average return on investment in the country, reflecting the lower level of risk. There is a danger that the regulator may fix the allowed rate of return too low, which can lead to under-investment. Nevertheless, the tariff-setting mechanism may provide the investor with scope to earn higher returns. Price-cap or revenue caps, for example, provide incentives to minimise costs and boost sales. Experience in the United Kingdom, for example, shows that this approach can provide effective incentives for new investment. However, the emergence of bottlenecks in parts of the transmission system in the late 1990s led the regulator to modify the regulatory regime, including the introduction of capacity auctions, in order to strengthen price signals to guide investment to where it is needed.

Impact of Market Reforms on Gas Investment

Market and regulatory reforms aimed at promoting competition in gas (and electricity) supply and reducing costs are having a profound impact on investment risk and financing arrangements. However, experience, notably in Canada, the United Kingdom and the United States, suggests that market reforms do *not* undermine long-term investment in gas infrastructure, at least not in mature markets. In North America, for example, investment in the upstream and downstream gas industry has increased since restructuring in response to strong demand growth and relatively attractive rates of return on investment (see the North America section below). In both Europe and North America, financial risk-management instruments have been developed to help producers deal with price volatility. But the collapse of Enron in 2001 and the financial difficulties of other gas-merchant companies have severely curtailed the use of such instruments and undermined liquidity in spot and futures markets. The major gas companies in both regions are becoming increasingly concerned about the impact of energy liberalisation on the development of very large-scale upstream projects, which will become increasingly vital to supply prospects in the coming decades. Most investments to date have been relatively small-scale and incremental, involving supplies from near-to-market and politically low-risk sources. Concerns are also growing about the effect on the cost of capital of uncertainty about future regulatory developments.

In Europe, large-scale cross-border investments have traditionally been made possible by stable relationships between national monopoly producers and marketing organisations and dominant downstream gas companies, based on long-term take-or-pay contracts. In North America, major pipeline projects are underpinned by long-term contracts with marketers or local distribution companies for fixed amounts of capacity at regulated rates. Securing external financing and equity investment for new pipeline and LNG projects is still generally impossible without such contracts. Nonetheless, as has already been observed in North America, there is likely to be a tendency for gas merchants to seek contracts of much shorter duration than the 20 to 25-year terms that are typical in Europe at present. They will also push for less onerous take-or-pay conditions and more flexible pricing terms, in recognition of the uncertainties about their future market share and the risk of being stranded with surplus gas that they might have to sell on at a loss. At the same time, upstream companies will look to integrate downstream with gas-merchant companies in order to spread risk and secure adequate financing at reasonable cost.

The combined result of these pressures as well as regulatory and structural developments in regional gas markets is uncertain. But long-term contracts in some form are likely to remain necessary to underpin large-scale projects in Europe and elsewhere, at least until the transition to a truly competitive downstream gas market has been completed. Once that stage has been reached, the need for those contracts to secure financing should, theoretically, disappear, since spot markets could then take any volumes that a gas merchant contracts for but is unable to sell directly. The ultimate guarantee of volume is the growth of demand in the market as a whole, together with liquid short-term markets. In Europe, the 1998 EU Gas Directive provides downstream gas companies with a degree of financial protection against the risk of being burdened with uneconomic long-term contractual liabilities, although most contracts provide for pricing and other terms to be renegotiated periodically in the event of significant changes in market conditions.

For now, the uncertainties relating to the evolution of the regulatory framework at national and, in the case of Europe, EU levels, together with the additional price volatility that is coming with the emergence of gas-to-gas competition, are leading to a perception of greater overall project risk on the part of investors and lenders. Uncertainties about how quickly spot markets and market centres will develop and how liquid they will be, as well as the possibility of significantly lower wellhead prices in the future, are also increasing risk. These factors are tending to raise the cost of capital and lead investors to favour smaller, closer-to-market projects. They may form a barrier to investment in technically riskier, multi-billion dollar projects.

Liberalisation of electricity markets is also contributing to the risks faced by developers of gas-supply projects, because of the importance of power-generation load. Major new gas-supply projects will need to be underpinned by firm long-term contracts between gas merchants and power companies, involving fairly rigid off-take commitments and pricing terms. But gas-fired power plants operating in competitive markets need prices that vary according to the marginal price of power to ensure that the plant is dispatched and to avoid a take-or-pay penalty. In many parts of the world, uncertainties about the future structure of the power-generation industry and changes in regulation, as well as the impact of government policies on electricity demand prospects, may make it harder for power generators to make long-term volume commitments (see Chapter 7).

Financing New LNG Chains

There is considerable uncertainty about the financing and contractual arrangements, and the structure of new LNG projects to supply North America, Europe and developing Asian countries. Market reforms and increased commercial and country risks are forcing players at different stages of the supply chain to change their ways of doing business. Deregulation, which is most advanced in Atlantic basin markets, is changing fundamentally the balance of investment risk. Traditional buyers no longer have a guaranteed market share and so will be less willing to commit themselves to very long-term contracts for large volumes of LNG, the basis for all projects to date, at least until a liquid spot market in LNG has developed (Box 5.3). But more flexible contractual arrangements will provide more opportunities to arbitrage between regional markets.

Until now, no liquefaction plant or receiving terminal has been built without long-term contracts covering the bulk of capacity.[12] But the expansion of capacity worldwide and growing competitive pressures are expected to encourage further growth in short-term trading. Spot sales accounted for 8% of total LNG trade in 2002.[13] Although long-term contracts will probably remain the backbone of the LNG industry, even in the Atlantic basin market, they will become shorter and take-or-pay commitments may

12. The Sakhalin-2 project in Russia is being developed without firm contracts in place, but most of the 9.6 Mt of capacity is expected to be covered by long-term contracts by the time the liquefaction plant is built.
13. Cedigaz (2003).

Box 5.3: **LNG Project Ownership and Financing**

Because of the large sums and risks involved, financing arrangements are crucial to the LNG project-development process. Almost all projects that have so far been developed have been on the basis of long-term contracts between the different parties along the supply chain: gas producers, the LNG liquefaction project sponsor, the LNG buyer and large final consumers. The upstream facilities and LNG liquefaction plant are often structurally separate, although the partners in both may be the same. The LNG buyer is normally responsible for financing the construction of the import terminal. Either the upstream/liquefaction project developer or the buyer is responsible for arranging financing for the ships, on a separate basis from both the liquefaction and receiving terminal projects.

Different players are involved in different parts of the LNG chain. In 2001, more than 60% of the equity in global LNG liquefaction capacity was owned by state companies, in some cases in a joint venture with a major oil and gas international company (see Table 5.6). Major international companies and big utilities, which often raise much of the substantial capital needs from their own cash flows, account for most of the rest of global LNG capacity. In many cases, a significant proportion of the capital has been raised from commercial banks and export credit agencies through project financing on a limited or no-recourse basis. The RasGas project, which started up in 2001, was financed partly through a $1.2 billion bond issue (see Middle East section).

become less onerous. Contract prices will probably be indexed to spot or futures gas prices, rather than oil prices, reflecting gas-to-gas rather than inter-fuel competition.

These developments will shift more of the integrated project risk onto upstream producers and liquefaction-project developers.[14] Raising debt finance in the traditional manner may, therefore, become more difficult and costly. To accommodate this added risk, suppliers will continue to pursue downstream integration. Buyers, on the other hand, will try to cover risks by securing a diversified portfolio of LNG supplies, carrier capacity and downstream off-take commitments. They are increasingly entering the shipping business. Because of the large capital expenditures needed for a complete LNG chain, vertical integration will be very costly: a single LNG chain involves investment of around $5 billion for a typical 6.6 million tonne two-train project involving a

14. Jensen, J. (2003).

shipping distance of around 4,000 km. The super-majors are best placed to achieve a high degree of diversification, but smaller companies might carve out niche positions in particular markets.

In developing countries, the creditworthiness of the ultimate customers for the gas and consumer prices will be critical factors in securing financing. Commercial banks are likely to show less interest in financing downstream projects there because of the greater risks involved, demonstrated by the enormous problems experienced with the Dabhol LNG project in India, even before its owner, Enron, went bankrupt. The involvement of export credit agencies and multilateral lending agencies will be essential to secure financing from banks in some cases. Where investment risks and difficulties in arranging project financing are particularly great, project sponsors may have to take larger equity stakes.

Table 5.6: **Equity Shares of LNG Capacity, 2001**

Rank	Company	Liquefaction capacity (Mt)	Equity share (%)
1	**Sonatrach**	**23.3**	**19.7**
2	**Pertamina**	**17.0**	**14.3**
3	**Petronas**	**10.0**	**8.4**
4	Shell	9.1	7.7
5	**Qatar Petroleum**	**8.9**	**7.5**
6	ExxonMobil	4.7	4.0
7	Mitsubishi	4.7	4.0
8	Jilco	4.6	3.9
9	TOTAL	4.4	3.7
11	**ADNOC**	**3.8**	**3.2**
12	**Omani State**	**3.4**	**2.9**
13	**Brunei government**	**3.4**	**2.8**
14	BP	3.0	2.5
15	**Nigerian National Petroleum Corporation**	**3.0**	**2.5**
16	**National Oil Corporation**	**2.6**	**2.2**
17	Vico	2.2	1.8
18	Unocal	2.2	1.8
20	Mitsui	2.0	1.7
21	BHP	1.4	1.2
22	Other (<1 Mt)	5.0	4.2
	Total state-owned companies	**75.4**	**63.5**
	Total	**118.7**	**100%**

Note: Companies with state participation are in bold. Equity refers to liquefaction projects only. The ownership of production assets can differ markedly.
Source: IEA databases.

Implications for Government Policies

The risks and uncertainties described above point to a need for governments to tread very carefully in restructuring and reforming their gas markets in order to ensure that the new rules and emerging market structures do not impede or delay investments that are economically viable. This is especially important with regard to cross-border pipelines and LNG terminals in importing countries. The management of the transition to competitive gas markets is especially critical to industry perceptions of uncertainty, the cost of capital and willingness to invest. Establishing a long-term policy and regulatory framework which sets clear, transparent and stable rules for gas and electricity markets would help to attract investment in gas-supply infrastructure and power plants. Cost-reflective pricing policies are particularly important in promoting investment in domestic downstream projects. Although it is impossible to remove completely uncertainty about future changes in the regulatory environment, investors will at least require assurance about the long-term evolution of market rules.

The public authorities may also need to provide special treatment for very large projects by exempting them from specific regulatory requirements, such as mandatory third-party access, that would otherwise jeopardise financing or increase costs. For example, in order to encourage investment, the US Federal Energy Regulatory Commission recently removed the requirement on LNG terminals to make available their capacity to third parties at regulated rates. The second EU Gas Directive permits national regulators to exempt LNG regasification capacity from third-party access requirements under certain conditions. And several national regulators, including Ofgem in Great Britain, have signalled that they will consider favourably applications for exemptions from LNG terminal developers. For their part, several companies considering investments in new terminals have indicated that they would be more likely to proceed if third-party access obligations were lifted.

Similarly, policy-makers will need to take account of the increased risks facing both upstream producers and merchant gas companies as a result of energy liberalisation in setting rules for long-term supply contracts and joint marketing arrangements. Downstream European gas companies are responding to the increased challenge of mobilising investment in large-scale gas-import projects by seeking a greater degree of co-operation and partnership with upstream operators within the constraints of competition law. This approach can help to mitigate risk and create a more reassuring environment for large investments. These partnerships involve joint investment in infrastructure projects and gas marketing ventures. For example, several European gas companies, notably Ruhrgas, are strengthening their commercial ties with Gazprom, while Gaz de France and Sonatrach have negotiated co-operation agreements.

Governments may also need to play a more proactive role in promoting investment in certain high-risk, large-scale gas projects — especially strategically important cross-border pipelines. Governments can help to lower country risk by intensifying political dialogue with the governments of supplier countries. This could contribute to a more stable investment climate and support closer collaboration between upstream and downstream companies. This is the goal of the EU-Russia energy dialogue, a formal process launched in 2000. The German government's support for the E.On-Ruhrgas merger was motivated partly by the strengthened financing capability that the merger would give to Ruhrgas for gas investments in Russia and elsewhere. And the UK government is giving strong political backing to the proposed Northern European Gas Pipeline from Russia to Germany and the Netherlands (see section on Russia, below).

The arguments for explicit subsidies to selected gas projects are less compelling, because of the market distortions that can result and the financial cost. The US Administration recently rejected calls for tax subsidies for the proposed Alaskan pipeline projects on these grounds. Nonetheless, there may be a case for some form of public subsidy for cross-border projects where there are significant strategic benefits to the country or region, such as diversity of supply or increased scope for competition between suppliers. A number of multilateral and regional institutions provide financial and other types of assistance to gas infrastructure projects in certain regions. The European Investment Bank and the European Bank for Reconstruction and Development are among those which have successfully provided financial and other types of assistance to such projects in the past.[15] Development banks, as well as national and multilateral export credit agencies, will continue to play an important role in backing major pipeline projects in the future. All forms of intervention will, nonetheless, need to be designed so as to minimise competitive distortions and promote the efficient operation of the industry.

Regional Analysis

Russia

Growing domestic and export sales, especially to Europe, will call for higher investment in all links in the gas supply chain in Russia over the next three decades. Most of the capital will be needed for upstream developments to replace the ageing western Siberian super-giant fields that have been the backbone of the Russian industry for decades. But a failure to implement much-needed market reforms,

15. A notable success was the Maghreb-Europe pipeline, which started operation in 1996, to which the European Investment Bank — the European Union's development bank — lent €1 billion.

including raising domestic prices to full-cost levels and giving independent producers access to Gazprom's monopoly national transmission system, could impede the financing of new projects and opportunities for the independents to develop their own reserves.

Investment Outlook

Cumulative investment needs in the Russian gas sector are projected to total just over $330 billion, or $11 billion per year, over the period 2001-2030. This compares with estimated investment of less than $9 billion in 2000. Exploration and development are expected to account for more than half of total investment needs; transmission pipelines, largely for export markets, and local distribution account for most of the rest (Figure 5.12). One-third of cumulative investment will be in projects for export to OECD countries. Average annual investment is expected to peak in the middle decade of the projection period and tail off slightly in the last decade. A sharp projected drop at that time in investment needs for transmission and storage is expected to more than offset a progressive increase in investment in the upstream and in distribution. Our investment projections are broadly in line with Russian government estimates for the period 2003-2020, although it is less optimistic about production prospects (Box 5.4).

Figure 5.12: **Gas Investment in Russia**

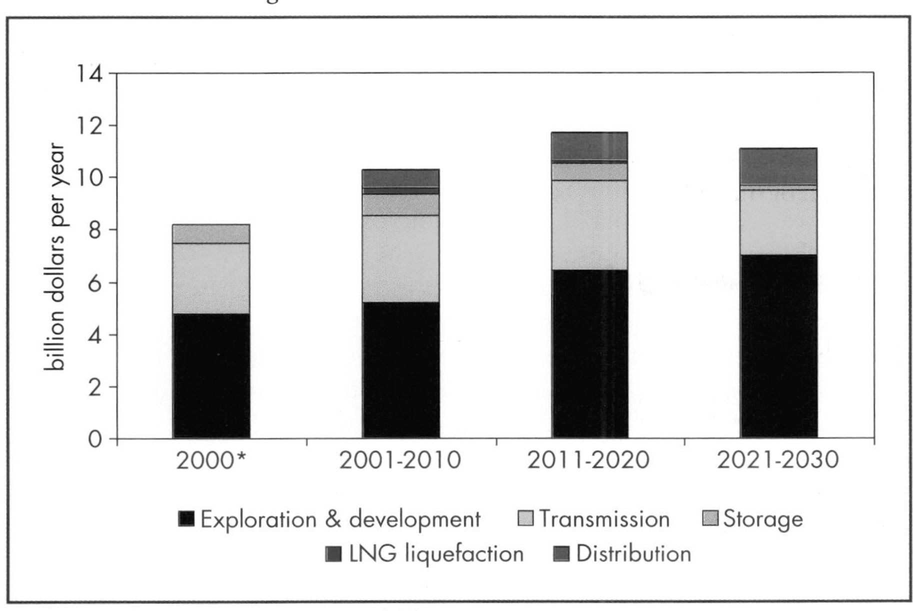

* Distribution and storage costs not available.

Box 5.4: **New Russian Energy Strategy**

The Russian government approved a new national energy strategy in mid-2003. The strategy projects that investment needs will amount to between \$170 and \$200 billion for the period 2003 to 2020, equal to average annual investment of \$9.4 billion to \$11.1 billion per year, depending on the rate of increase in production. Three scenarios for gas production are presented. In the base case, production rises to 680 bcm in 2020, with oil companies and other independent producers accounting for all of the increase. Gazprom is expected to see a small decline in its production, from an estimated 522 bcm in 2002 to 505 bcm in 2020, although Gazprom itself now expects to be able to raise production to 560 bcm. Output grows to 730 bcm in an optimistic scenario and stabilises at around 550-560 bcm a year after 2010 in a pessimistic scenario. Western Siberia is expected to remain the main producing area in all cases. The strategy expects domestic gas prices to rise from around \$24/kcm (\$0.70/MBtu) at present to \$37/kcm (\$1.10/MBtu) in 2006 and to full-cost levels of \$46-49/kcm (\$1.40/MBtu) in 2010. Prices are assumed to be fully decontrolled at that time.

Upstream investment needs are projected to rise steadily from an estimated \$4.8 billion in 2000 to \$5.2 billion per year in 2001-2010 and \$7 billion per year in 2021-2030. In general, the new fields targeted for development by Gazprom, the dominant national gas production and transportation company, and by other companies are smaller or located in more difficult operating environments. Production is expected to shift from western Siberia, the main producing area today, to new areas, including the Yamal peninsula, eastern Siberia, and the Barents Sea. As a result, per unit production costs are likely to be higher than in the past, boosting overall upstream investment needs.

Investment in transmission pipelines is expected to be heavier in the first two decades of the projection period, when exports are projected to increase most rapidly. Investment needs will average around \$3 billion a year. Around 50% of the estimated \$92 billion of investment in new pipelines over the period 2001-2030 will be needed to support exports. Most of the upstream development projects that are due to come on stream in the next few years will be able to link up with the existing pipeline system, much of which is underutilised. But some new large pipelines will be needed after 2010-2015, as fields in new producing areas are developed.

The figures presented here do not include refurbishment of pipelines, because of the acute uncertainties about how quickly existing pipelines will need to be replaced and how much of that investment would be classified as capital spending. Certainly, increasing amounts of capital spending will go to refurbishment of pipes and compressor stations. Gazprom is already spending over half a billion dollars a year in refurbishing its transmission pipeline system and storage facilities (see below). It is believed that the operating life of Russian pipelines is likely to be somewhat lower than those in Europe and North America, as the Russian pipelines were built to lower operating standards, using poorer materials. Of more than 150,000 km of high-pressure, large-diameter lines, around 70% was commissioned before 1985 and more than 19,000 km are beyond their design life span of 30 years.[16] A further 7,000 km are more than 40 years old. The average age of all lines currently in use is 22 years. In addition, 14% of compressor stations are past their amortisation period of 33 years and 64% of them are more than ten years old.[17] The amount of additional spending on refurbishment over the 30-year period is estimated to range from $30 billion to $56 billion, based on assumed average pipeline lives of 35 to 50 years (Figure 5.13).

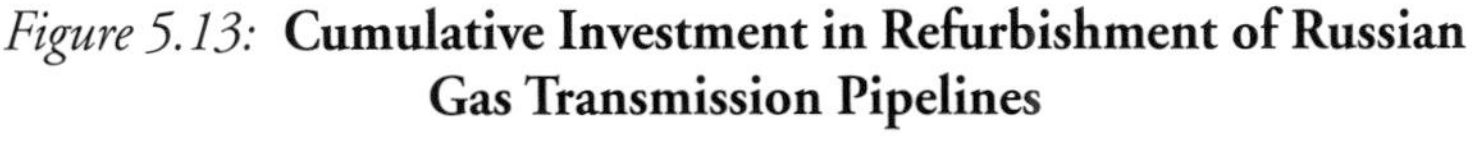

Figure 5.13: **Cumulative Investment in Refurbishment of Russian Gas Transmission Pipelines**

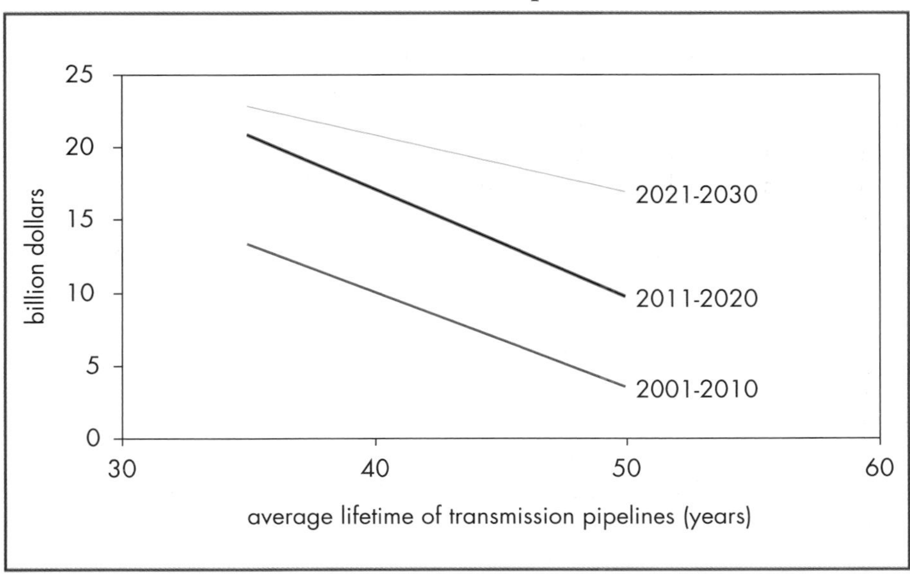

16. IEA (2002b).
17. *Gas Matters*, "Leaked 'Conception' Sets out Radical Agenda for Russia's Gas Future" (October 2002).

There is considerable spare capacity within most local distribution networks, so investment needs in that sector are expected to remain relatively modest. In addition, distribution capital costs are very low compared to other regions, because of lower-quality materials, engineering standards and labour costs. However, refurbishment costs, not included in the figures presented in this report, could also be substantial, in view of the poor state of networks.

At present, Gazprom is responsible for the bulk of investment in the Russian gas industry. Table 5.7 details the company's investments in 2002 and its budgeted outlays for 2003, including investments outside Russia. The high-pressure transmission system continues to attract the bulk of investment, with the most of the rest going to production. Of the 93 billion roubles ($3 billion) of investment budgeted for the high-pressure transmission system in 2003, 18 billion roubles have been set aside for refurbishment, mainly to replace pipes and compressors and to upgrade storage facilities. Most of the rest will go to expanding the export transportation network, mostly outside Russia. The development of the super-giant Zapolarnoye gas field, which started production in 2001, and the construction of two pipelines to link the field to existing transmission lines, account for over 40% of budgeted investment for 2003. The company expects most future gas investment to be focused on export pipelines and upstream development. Only limited information is available on upstream gas-only investments by other companies in Russia, but it is thought to have been running at around $1 billion a year in recent years. The government has called for increased investment by independents.

Table 5.7: **Gazprom Capital Expenditure, 2002 and 2003**

	2002		2003	
	Billion roubles	Billion dollars**	Billion roubles	Billion dollars**
Transmission	71.45	2.30	93.23	3.01
Production	65.57	2.12	77.91	2.51
Distribution	0.07	0.00	0.05	0.00
Other*	8.43	0.27	8.61	0.28
Total	**145.52**	**4.69**	**179.80**	**5.80**

* Includes gas processing and oil refining, but excludes expenditures for Sibur, Vostokgazprom and Purgaz.
** Based on the average exchange rate of 1 dollar = 31 roubles in 2002.
Source: Gazprom (2003).

Supply Trends and Prospects

WEO-2002 projects that natural gas, already the main fuel in Russia's energy mix, will become even more dominant over the next thirty years. The share of gas in total primary energy supply is projected to rise from 52% in 2000 to 56% in 2030. Its share of final energy consumption will increase from 27% to 32%. Most of the growth in primary demand for gas will come from the power sector. By 2030, gas will fuel almost 60% of total electricity generation, compared to 42% in 2000.

Indigenous resources are more than adequate to meet rising demand. Russia holds over 30% of the world's proven natural gas reserves. At the beginning of 2001, Russia's reserves stood at 46.5 trillion cubic metres, nearly three-quarters of them in western Siberia and most of these in the Nadym-Pur-Taz region. Reserves are still equivalent to about 80 years of production at current rates, even though they declined slightly during the 1990s, because of reduced exploration. The US Geological Survey estimates additional undiscovered resources at 33.1 tcm.[18]

Total Russian gas production is expected to increase continuously over the projection period, from around 580 bcm in 2001 to 709 bcm in 2010 and to 914 bcm by 2030. The bulk of this increase will meet rising domestic demand (Figure 5.14). Net exports will also increase, from 174 bcm in 2001 to 280 bcm

Figure 5.14: **Gas Balance in Russia**

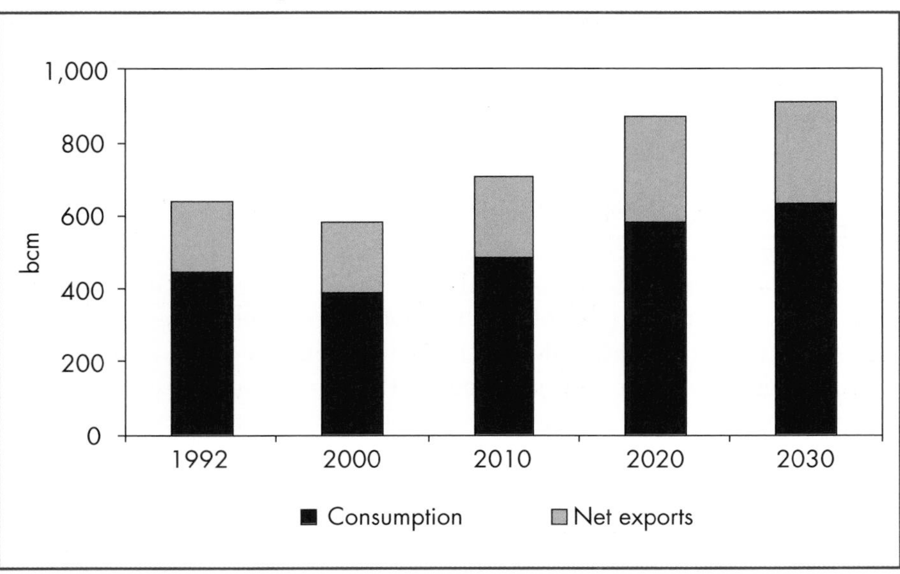

Source: IEA (2002a).

18. USGS (2000).

in 2030. As a result, Russia will remain the largest gas exporter in the world in 2030. OECD Europe will continue to attract the bulk of exports, but new markets, primarily in Asia, will also emerge. The Russian government is significantly less optimistic about the prospects for production and export availability (see Box 5.4 above). There is also considerable uncertainty about the impact of domestic price increases on demand.

The rate of decline of production from Russia's three super-giant fields, Medvezh'ye, Yamburg and Urengoy, in western Siberia is the main uncertainty for gas-production prospects in the near term and the primary reason why Gazprom and the Russian government expect gas production growth to slow over the next two decades. Output from these fields, which currently accounts for more than three-quarters of total Russian gas production, is expected to fall in the coming years. However, there are considerable doubts about the rate of production decline and the extent to which judicious investment could reduce it. Urengoy and Medvezh'ye have been in gradual decline for several years. Gazprom, which operates these fields, is projecting a sharp acceleration in decline rates at these fields.

Gazprom plans to offset partially these production declines in the next few years by raising output from a fourth super-giant gas field, Zapolyarnoye, which is also located in Nadym-Pur-Taz. Production from Zapolyarnoye, with 3.4 tcm of reserves, began in 2001. Gazprom plans to develop gas and liquids production from deeper strata of the field. It expects to complete the project by 2006, with production reaching 100 bcm a year by 2008. Gazprom is also planning to develop other deposits, mainly in western Siberia, over the next decade.[19] It plans to prioritise the development of new, smaller fields in the Nadym-Pur-Taz region so as to be able to make use of existing pipelines. These include the Kammennomysskoye fields, which lie just 150 km from the Yamburg field, with the potential to produce over 50 bcm by 2010. The company, nonetheless, also expects production of gas from new giant fields on the Yamal peninsula to begin soon, possibly as early as 2007, but this appears optimistic. Drilling conditions are more difficult and up to 1,000 km of new pipelines will be needed to connect these fields to the existing transmission system. If the Yamal fields are developed later, it may be possible to build shorter lines to connect with the Nadym-Pur-Taz system. Spare capacity in that system will emerge as production at Urengoy and the other super-giant fields declines. Outside Siberia, Gazprom is investigating with its partners when and how to bring the giant Shtokmanovskoye gas field in the Barents Sea into production.

New deals to import gas from Turkmenistan and other Central Asian republics will make it possible for Gazprom to postpone the costly Yamal and

19. Gazprom (2003).

Shtokmanovskoye developments until after 2010. In April 2003, Gazprom signed a landmark 25-year agreement with the state-owned firm Turkmeneftegas for imports of 5 to 6 bcm per year of Turkmen gas in 2004, rising to as much as 80 bcm from 2009. The price is fixed at $44 per thousand cubic metres from 2004 to 2006, to be paid for half in cash and half in bartered gas equipment and services, including refurbishment and upgrading of the existing transmission system in Turkmenistan and Kazakhstan to carry the gas to the Russian border. An associated intergovernmental agreement sets out the terms of the pipeline work, which will cost in total around $2 billion. Gazprom has also signed long-term co-operation agreements with Tajikistan and Kyrgyzstan.

A number of oil companies and independent gas producers are planning to raise output. Although the non-Gazprom producers account for only around 12% of production at present, they hold licences to develop around a third of the country's proven reserves. The largest independent is Itera, which emerged as a major player in gas supply to the Commonwealth of the Independent States (CIS) countries and the Russian market in the late 1990s. Its production was about 16 bcm a year in 2002, but will fall as the company's 49% stake in the Gazprom subsidiary, Purgaz, which accounts for 7 bcm, has been reacquired by Gazprom. Itera has invested $150 million in the Beregovoye field in Nadym-Pur-Taz, which is ready to produce over 2 bcm per year rising quickly to 11 bcm, but Gazprom has denied Itera access to the nearby 100-bcm/year Zapolyarnoye-Urengoye pipeline on the grounds that all the capacity is earmarked for Zapolyarnoye gas.

Among Russian oil companies, Surgutneftegas is the largest gas producer, with output of around 13 bcm in 2002, followed by Rosneft, with output of just under 6 bcm. Yukos, which plans to merge with Sibneft, and Lukoil, Russia's largest oil producers, also have important gas assets and plan to boost production. Rusia Petroleum, partly owned by BP, holds a controlling stake in the licence to develop the giant Kovykta field near Irkutsk in eastern Siberia. There are plans to develop the field to supply China and possibly Korea and Japan through a long-distance pipeline. The 25-bcm per year pipeline, which would most likely run parallel to a planned crude oil pipeline, would cost at least $7 billion. The Russian government recently appointed Gazprom to be the co-ordinator of all eastern Siberia gas export projects, although it has as yet no stake in any of the licences.

The biggest increase in non-Gazprom output in the next few years will come from the Sakhalin-2 project, which will export LNG. Shell, which is leading the entirely foreign consortium, gave the green light for the project to proceed in May 2003. The project involves upstream development of an offshore gas field and the construction of a two-train liquefaction plant with a capacity of 9.6 million tonnes a year. Total investment will amount to around

$9 billion. Gas supplies will also come from an adjacent oil and associated gas field. Total gas production from the project will amount to around 14 bcm. The less advanced Sakhalin-1 project, led by ExxonMobil, could involve a sub-sea pipeline for natural gas exports to either the North Island of Japan or Korea and China.

The *WEO-2002* gas-production projections for Russia imply a need for more than 1,350 bcm of cumulative additional production capacity over the period 2001-2030 (Figure 5.15). This is based on an average decline rate of 5%. The bulk of this capacity will be needed to replace current onshore producing capacity that is expected to be retired during the projection period. A growing share of capacity will be offshore. The biggest need for new capacity will be in the decade 2011-2020. This assumes a less rapid decline in production from existing fields than that assumed in the Russian Energy Strategy.

Figure 5.15: **Gas Production Capacity in Russia**

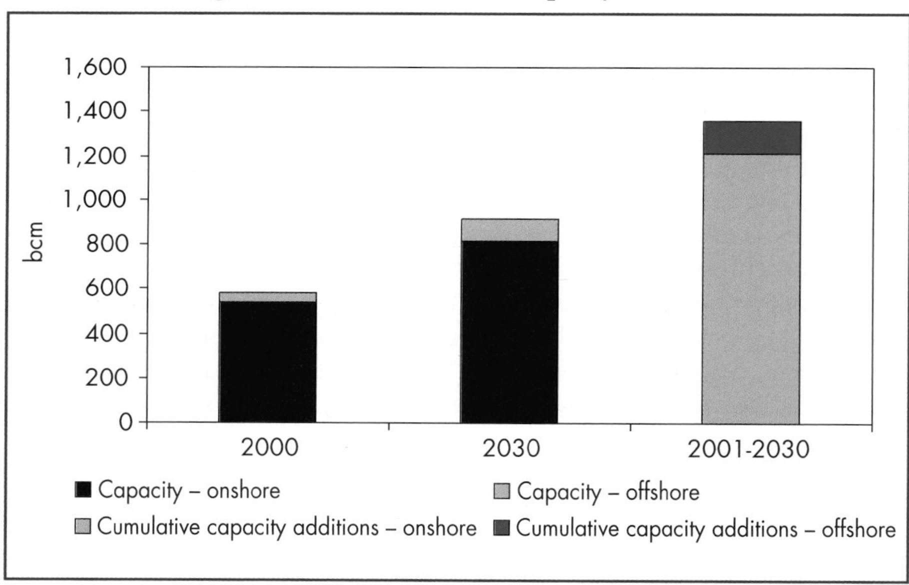

Financing Sources

Although Gazprom finances the bulk of its investments out of operating cash flows, the company rapidly increased its debt in the late 1990s to fund its investment programme and cover shortfalls in export earnings caused by lower oil prices. Total debt, much of which was secured against future convertible currency export earnings, reached $16 billion in 2001. Debt has fallen slightly since, with higher gas-export prices and operating cash flows. At the end of September 2002, the company's leverage ratio (net debt divided by net debt plus shareholder's equity), stood at 20%. Its net borrowings in 2001 — the last full year for which data are available — were equal to under 16% of capital spending. Gazprom management plans to reduce debt further in the near term through net repayments, mainly because most outstanding debt matures within the next two or three years. It also intends to shift its borrowings from short-term debt, which accounts for the majority of its total debt, to long-term debt. Much of Gazprom's short-term rouble borrowings are in the form of promissory notes, which are expensive. In the longer term, however, debt financing is expected to increase.

Independent gas producers and oil companies are expected to account for a growing share of total gas-industry investment in the coming decades. Lukoil, for example, intends to invest $1 billion over the period 2002-2006 in developing gas fields on the Yamal peninsula. Up to now, investments by non-Gazprom producers have been financed by a mixture of cash flow and short-term borrowing from domestic and foreign banks. Western investors have so far been reluctant to lend to most of them for more than two years. For these firms to increase investment as planned, they will need to gain access to credit over longer terms of five years or more. Unless they obtain higher prices for their output, they will have, in any case, little incentive to invest.

A significant share of future gas investment could come from foreign companies, typically through joint ventures. Foreign direct investment (FDI) in gas projects has so far been small, mainly because of the perceived high level of country risk (see below). The prospects for large export-driven projects may depend on attracting large amounts of FDI. The second phase of the Sakhalin-2 project will be by far the largest FDI ever in the gas sector. Shell owns 55% and is the operator of the project. Total investment, including oil production, processing and export facilities, is expected to run to about $9 billion. Financing will involve a mixture of equity and long-term non-recourse project finance, the bulk of which is expected to come from export credit agencies and multilateral lending agencies. Both the Sakhalin-2 and the Sakhalin-1 projects are covered by production-sharing agreements. Gazprom is seeking foreign partners for a planned $5.7 billion Baltic Sea pipeline, with a capacity of up to 30 bcm per year, that would run to Germany and overland to the Netherlands.

Improved access to local credit is an important factor in the prospects for investment in the Russian gas industry. Short-term rouble-denominated loans and bonds finance a significant proportion of current investment in the gas sector. The ability of Gazprom and independent domestic producers to continue to access domestic capital at reasonable cost will depend on the health of the banking and financial system, economic conditions in Russia generally and the extent of the exposure of individual Russian banks to Gazprom risk.

Investment Uncertainties

Future investment in the Russian gas industry is subject to a wide range of uncertainties, notably about underlying demand, price and cost factors. In particular, the ability of Gazprom and other gas suppliers to finance new supply projects and their incentive to do so are highly dependent on the prices that they are able to achieve on both domestic and export markets. The prices of gas exports, which provided 73% of Gazprom's turnover in 2001, are currently linked under long-term contracts to international prices for oil products. Even if there is a gradual de-coupling of oil and gas prices in supply contracts, oil prices will continue to exert a strong influence over gas prices through inter-fuel competition at the burner tip. Lower gas prices than assumed in the *WEO-2002* Reference Scenario and slower domestic price reform than expected would directly reduce cash flows and investment in upstream developments and in transmission capacity.

Beyond these underlying market factors, there are considerable uncertainties about the general investment climate in Russia, including developments in the legal, regulatory and institutional framework governing trade and investment generally and gas specifically. As economies in transition, Russia and transit countries are considered by investors to be subject to significantly higher economic, political and institutional risks than those in OECD countries. Although significant progress has been made since market reforms were launched more than a decade ago, significant concerns remain about the legal framework, particularly with respect to property rights, foreign investments and trade, corporate governance and transparency, and the enforcement of laws and regulations. Russia has signed the Energy Charter Treaty, which sets out common rules on energy trade, investment protection, transit and dispute resolution, but has yet to ratify it. Ratification of the treaty, as well as agreement on the Transit Protocol, currently being negotiated between member countries of the treaty, could play an important role in encouraging investment in long-distance gas-export pipelines to Europe and Asia.

The energy dialogue between the European Union and Russia, formally launched in 2000, could also help to lower country risk and foster investment in energy projects. The European Commission claims to have resolved a dispute over destination clauses in gas-supply contracts, which restrict resale rights. Gazprom continues to insist on the fundamental need for long-term take-or-pay contracts in order to secure financing for export projects. The commission accepts this, in principle, but argues that the contents of the contracts must be compatible with EU internal market and competition rules.[20]

20. CEC (2002).

As in many other emerging market economies, the initial legislative thrust for opening up the upstream oil and gas industry to private investment concerned production-sharing agreements (PSAs). As discussed in Chapter 4 (oil), the completion of the PSA regime was until recently regarded as a crucial step in providing the fiscal and legal certainty and long-term guarantees necessary for large-scale investments in the oil and gas industry. However, the State Duma adopted legislation in April 2003 effectively scrapping the PSA framework for new projects, beyond five agreements already approved under previous legislation (including the Sakhalin projects) and several other agreements signed over the past decade but not approved by the Duma. The government now believes that the legal and taxation systems are sufficiently attractive and stable for investors not to need PSAs, except for mega-projects. New reserves will be offered first at open tender and, if they are not allocated, then they will be re-offered under PSA terms. Indeed, many Russian oil companies now say they prefer to invest under the general regime, which is currently considered to be financially more attractive than PSA terms. This could change, however, if the tax regime were to change again or if oil prices were to fall sharply. A change in transfer-pricing rules could also make PSAs relatively more attractive in the future. As with oil, incentives to invest in gas will ultimately depend not just on the legal and taxation framework but also on the stability and predictability of that framework, whether it is PSA-based or not.

Uncertainties over the structure and level of taxes on gas sales and profits at the federal, regional and local levels and depreciation rules are a major source of uncertainty for future gas-sector investment. Gazprom is by far the largest source of tax revenues in Russia. In 2001, its profit taxes alone amounted to $6.5 billion (of which $3.6 billion were deferred) on total sales of $21.6 billion. Political pressures and difficulties in collecting taxes from enterprises in financial difficulty could lead to increases in the tax burden on Gazprom and independent producers and marketers, limiting their ability to invest. Adjustments to the taxation of oil and gas companies proposed by the government in summer 2003, which would come into effect in 2004, is not expected to have a significant impact on Gazprom's tax burden.

Government plans for the possible restructuring of the Russian gas industry are another major uncertainty affecting investment prospects. The pace and nature of any restructuring could have a profound impact on the dominance of Gazprom and the role of independent gas companies, and consequently on the opportunities for investment. The government has indicated that it intends to reform the gas sector, but has not yet decided how. A document drawn up in 2002 by the Ministry of Economic Development, called *The Conception for the Development of the Gas Market in Russia,* proposes a number of key reforms, including the structural separation of Gazprom's transportation business from its other activities and the establishment of a wholesale gas market to promote gas-to-gas competition through third-party access. The 2003 energy strategy provides no detail on how the industry might be restructured, although it does assume that prices are gradually raised to market levels and completely deregulated by 2010.

The establishment of an effective third-party access regime is likely to prove crucial to the outlook for investment by non-Gazprom companies. Although Gazprom is legally obliged to offer spare transportation capacity to third parties, few agreements have been reached, mainly because charges are considered prohibitive. A shortage of capacity at gas-processing facilities, all of which are owned by Gazprom, also hinders access. As a result, independents have no choice but to sell their gas, much of it associated, directly to Gazprom at low prices, or flare it. Selling directly to Gazprom, however, is considered to provide limited incentive to independents to invest in developing gas fields, because they are faced with a monopsony buyer and have little leverage over price. Gazprom currently pays independents only around $20-25 per thousand cubic metres and independents are worried that Gazprom may force them to accept lower prices in the future. Selling directly to end-users and negotiating separate transportation contracts with Gazprom would allow the independents to seek better pricing terms and give them stronger guarantees over future revenues. Concerns over gas flaring may increase pressure on Gazprom to improve access conditions.

Even if the government does not proceed with reforms that will improve the independents' access to Gazprom's transmission capacity, Gazprom may itself decide to improve access terms in exchange for financial help in upgrading its transmission system, even though such an approach would inevitably be discriminatory. Gazprom has been in discussion with oil companies about the possibility of co-operation along these lines. The government and Gazprom are likely to remain opposed to any move that would allow the independents to sell their gas directly on export markets, fearing that competition between Gazprom and other Russian gas producers could lead to lower export prices. Although higher export sales might offset this price fall, the government is concerned about the rate of depletion of Russia's gas resources.

Government regulation of domestic gas prices, which remain well below full cost, is a critical uncertainty for the financial health of the gas industry and its capacity to finance capital spending. Domestic prices also affect Russian gas demand and, therefore, the amount of gas that will be available for export. Domestic prices are significantly lower than the prices Gazprom obtains from its sales to Western Europe, even after netting back export and customs duties and transportation costs. Prices to residential customers are particularly heavily subsidised (Figure 5.16). The government plans to raise prices gradually and to remove cross-subsidies, but political factors may delay or impede those changes. Indeed, the government is not expected to raise domestic gas prices ahead of State Duma and Presidential elections in 2004. The energy strategy is premised on a gradual increase in prices to full cost levels of $45 to $50 per thousand cubic metres by 2010, but politics could slow or even stall that trend. A failure by Gazprom's customers to pay their bills or to do so in cash, a major problem in the late 1990s, is a further risk, as domestic prices rise.

Figure 5.16: **Average Gas Prices in Russia**

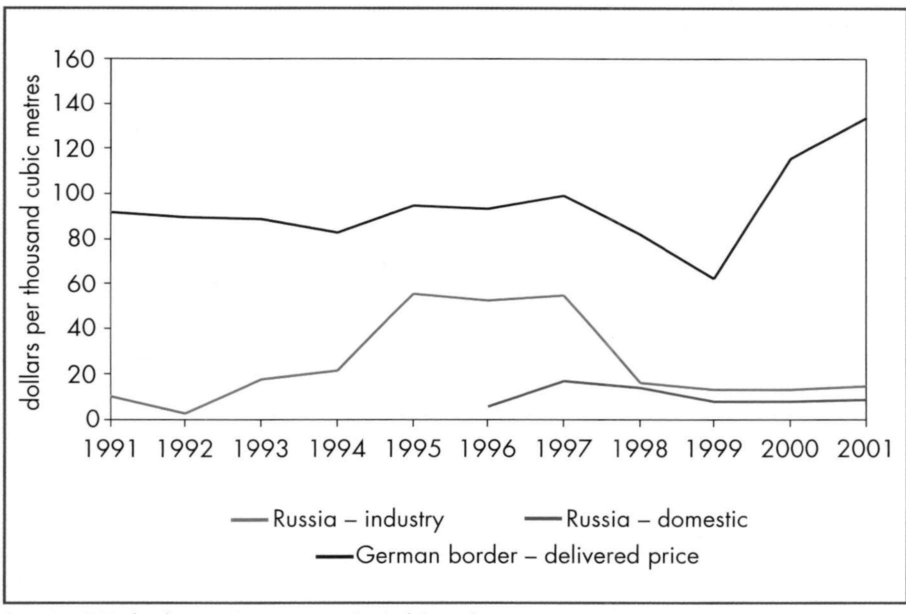

Source: IEA databases; Renaissance Capital (2002).

Moves to liberalise gas and electricity markets in Western Europe are expected to lead to increased gas-to-gas competition and could ultimately exert downward pressure on Russian export prices (see OECD Europe section, below). Liberalisation is also likely to affect the terms and conditions of future long-term contracts. European merchant gas companies will probably seek contracts of shorter duration, less onerous take-or-pay conditions and more flexible pricing terms in response to the increased market risks they face. These developments might increase the financial risk to investors in mega-projects such as the development of the Yamal gas fields or Shtokmanovskoye and raise the cost of capital.

Caspian Region[21]

The Caspian region, with important reserves, has the potential to become a major gas exporter, primarily to OECD Europe but also to Asia. Significant investment in export pipelines and in exploration and development will be needed, but geopolitical factors and worries about shifting government policies may hinder capital inflows.

Investment and Supply Outlook

The Caspian region will need to invest $107 billion over the next three decades in gas production and transportation infrastructure. Average annual capital needs will grow steadily, from $3 billion per year in the current decade to $4 billion per year in the decade 2021-2030. Upstream exploration and development and gas-export pipelines will account for three-quarters of this investment.

Caspian proven reserves stand at 7.3 tcm, larger than those of the United States and Canada combined. Production amounted to 127 bcm in 2002, 90% of which was in Turkmenistan and Uzbekistan. Over the past decade, gas production in the region has hardly increased and, in the case of Turkmenistan, fell sharply in the early 1990s, recovering only since 1999.

Output is expected to more than double by 2030, but this will depend on export demand. The existing gas transmission lines, built during the Soviet era, force most of the surplus gas produced in the region to flow north to Russia (Figure 4.29 in Chapter 4). It has proved impossible for the Caspian countries to negotiate transit agreements with Russia's Gazprom to allow them to access more lucrative markets in OECD Europe. But Gazprom is looking to the Caspian as a low-cost source of incremental gas supply. The Gas Alliance initiative launched in 2001 aims to optimise exports from Russia, Kazakhstan, Turkmenistan and Uzbekistan to Europe. Turkmenistan recently reached a long-term agreement with Russia on increasing exports to the latter (see previous section for details). These exports will allow Russia to export more gas to OECD Europe from its western Siberian fields and delay major new development projects. Uzbekistan continues to concentrate on domestic markets and on exports to neighbouring countries in the region, to avoid having to rely on sales to Gazprom.

Export Pipeline Prospects

Solving the problem of access to export markets as well as obtaining payment in hard currency is critical to the development of the region's gas reserves. Higher gas production will not be possible without the construction

21. Defined as Azerbaijan, Kazakhstan, Turkmenistan and Uzbekistan.

of new pipelines and agreements on the sale of gas to Gazprom and/or transit through its transmission network. As a result of these difficulties and the high cost of building new export lines bypassing Russia, oil and gas companies have so far shown greater interest in oil than in gas in the region.

The only new project to have been given the green light is the South Caucasus pipeline from Azerbaijan to Turkey. The line, which will cost an estimated $2.5 billion, will have an initial capacity of 16 bcm per year. It will run parallel to the Baku-Tiblisi-Ceyhan oil pipeline for most of its route. The first segment of the pipeline should be completed by 2006, connecting with the Turkish gas network at Erzerum. Gas will come from the offshore Shah Deniz field, one of the world's largest gas discoveries in the last 20 years, containing recoverable resources of roughly 400 bcm. The field is being developed by an international consortium led by BP at a cost of over $3 billion. There were plans to build a trans-Caspian pipeline from Turkmenistan to link up with the South Caucasus pipeline, but they have stalled because of the cost of the project, resistance from Russia and doubts about the amount of gas available to fill the line, taking into account the recent deal to sell Turkmen gas to Russia.

In the longer term, China and India could also be potential export markets. A pipeline from Turkmenistan to China that could pick up volumes from Kazakhstan and Uzbekistan has been under consideration for several years. Our projections assume that this project will go ahead in the last decade of the projection period. Another line from Turkmenistan to India has also been proposed, but geopolitical and economic factors are expected to prevent the project from going ahead before 2030.

Middle East

The Middle East, with its vast reserves of natural gas, will play a pivotal role in meeting the gas needs of other regions as well as its own rapidly growing markets over the coming decades. The region has recently emerged as a major exporter of LNG, and new projects are being developed. Long-distance pipelines to Asia and Europe are also likely to be built in the longer term. Capital requirements will grow rapidly, reaching $10.6 billion a year in the last decade of the projection period. As with oil, government budget constraints and limitations on national company borrowing mean that a growing share of this investment will have to come from private sources, including foreign oil and gas companies given host. Pricing terms, including the amount of government-tax take, are likely to be a stumbling block for many gas projects and will be a major source of uncertainty for investment in the region.

Investment Outlook

Expanding gas-supply infrastructure in the Middle East for domestic and export markets will cost an estimated $263 billion over the next three decades.

This is equal to an average \$8.8 billion per year — twice the estimated capital spending in 2000. More than half of projected investment, or \$4.7 billion a year, will be needed in the upstream sector. High-pressure transmission pipelines and LNG liquefaction plants account for most of the rest (Figure 5.17). Over two-thirds of projected investment will be needed for exports to OECD countries.

Figure 5.17: **Gas Investment in the Middle East**

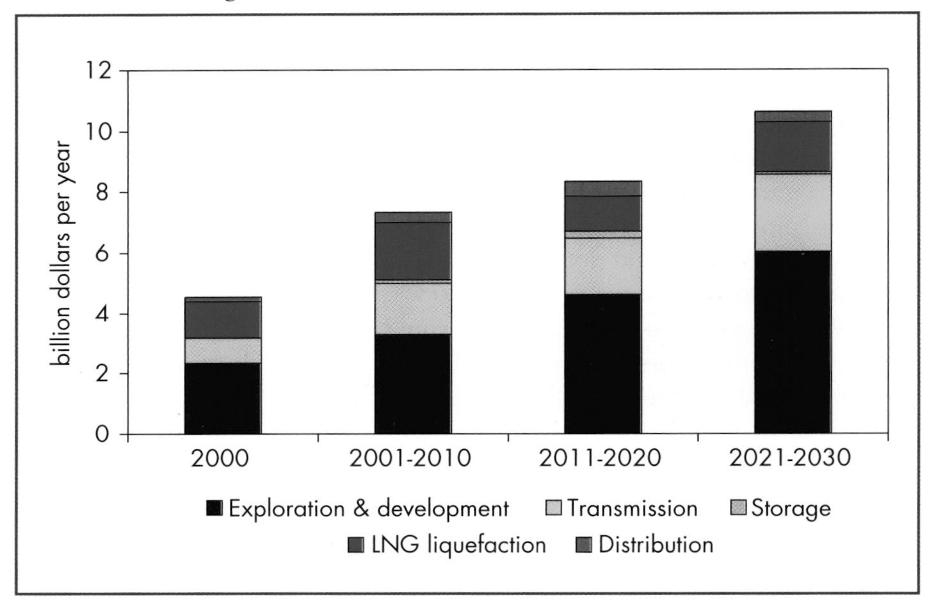

Although upstream investment needs will grow rapidly during the projection period, they will remain relatively low per unit of capacity. The Middle East has the lowest exploration and development costs for oil and gas of any major world region, with capital costs estimated at around \$0.2 per MBtu or \$7.5 per thousand cubic metres of gas produced. Spending will be almost twice as high in the last decade compared to the first decade of the projection period, driven by rising capacity needs.

Liquefaction plants will account for 18% of total gas investment in the region over the period 2001-2030. Annual investment needs are projected to drop from around \$1.9 billion in the current decade to \$1.2 billion in the second decade, due to falling construction costs. An even faster increase in new LNG capacity pushes up annual investment to \$1.6 billion in the period 2021-2030. Investment in transmission capacity, at just over \$2 billion per year, will mostly support domestic consumption. New transmission pipelines to connect gas fields to new LNG liquefaction plants will, in most cases, be short. Capital spending on distribution networks will account for a very small share of total spending. Most of the increase in domestic consumption in the region will be in the power generation, water desalination, heavy industrial and fertilizer sectors, which will be supplied mostly directly off the high-pressure transmission system.

Supply Trends and Prospects

The use of natural gas in the Middle East is growing rapidly. At 211 bcm in 2001, gas already accounts for 45% of primary energy needs. Oil accounts for almost all the rest. Gas is used mainly in industry, mostly as a petrochemical feedstock and for water desalination. Power generation takes almost a third of gas consumption and this share is growing. Iran and Saudi Arabia have the largest markets. Total gas demand in the region is expected to more than double between 2000 and 2030, driven largely by the power sector.

Gas production is expected to grow even faster over the next three decades, underpinning a huge increase in exports. *WEO-2002* projects Middle East gas production to surge from 242 bcm in 2001 to 861 bcm in 2030. Iran, with reserves of 26 tcm, is expected to see the biggest increase in output. Gas resources in the whole Middle East are very large, both in absolute terms and in relation to current production. Cedigaz estimates proven gas reserves were 71 tcm at the beginning of 2002, equal to almost 40% of global reserves.[22] Six countries — Iran, Iraq, Kuwait, Qatar, Saudi Arabia and the United Arab Emirates (UAE) — hold 97% of Middle East reserves. The USGS reckons that undiscovered resources amount to a further 36 tcm[23], while Cedigaz puts ultimate gas resources at between 115 and 136 tcm. Around 40% of Middle East gas reserves are associated with oil.

Much of the increase in Middle East production will go to exports outside the region. These are projected to soar from 30 bcm in 2001 to above 360 bcm over the next three decades. As a result, the share of exports in total production will jump from only 13% in 2001 to 42% in 2030. Incremental exports will come largely from Iran, Qatar, Oman, the UAE and Yemen in the medium term. Iraq could emerge as a significant exporter towards the end of the projection period. Europe and North America are expected to take the bulk of new gas exports (Figure 5.18). At present, almost all the region's exports go to Asia as LNG. The bulk of the increased exports will also be in the form of LNG, although export pipelines to Europe and Asia — from Iran and possibly Iraq — are expected to be built towards the end of the projection period. The only current export pipeline, which was commissioned in 2001, runs from Iran to Turkey.

22. Cedigaz (2003). Middle East reserves jumped by 20% between 2001 and 2002.
23. Mean estimate, USGS (2000).

Figure 5.18: **Middle East Net Gas Exports by Region**

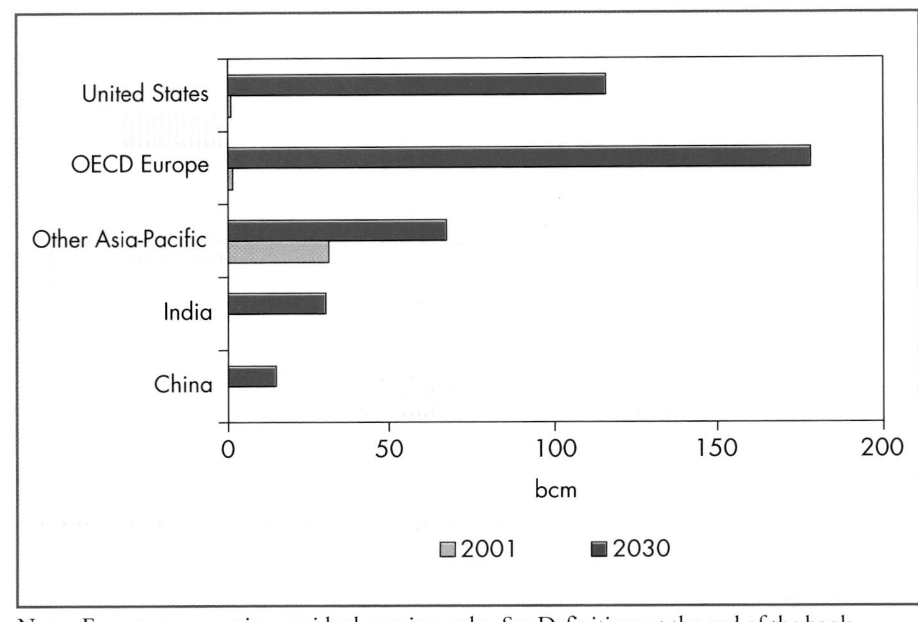

Note: Exports to countries outside the region only. See Definitions at the end of the book.
Source: IEA (2002a).

New gas-production capacity requirements to meet the projected jump in domestic and export demand and to replace current production will be enormous (Figure 5.19). The decline rate for gas fields is assumed to remain constant at around 4%. New capacity needs will be most pressing in the last decade of the projection period, when demand increases most in absolute terms and a significant proportion of existing capacity needs to be replaced.

Investment Risk and Uncertainty

There is no doubt that the region has sufficient gas reserves to underpin the projected increase in production. But whether the required investment can be mobilised is highly uncertain. The nature of the investment challenge varies considerably among countries, partly because of differences in country and project risks. The greater those risks, the more limited the access to international financial markets, the longer the delays in implementing projects and the higher the required return on investment.

Despite political tensions and conflicts in the region, most Gulf countries enjoy reasonably good risk ratings (see Chapter 4, Middle East section). Project risks in the region vary according to geopolitical and technical factors:

Figure 5.19: **Gas Production Capacity in the Middle East**

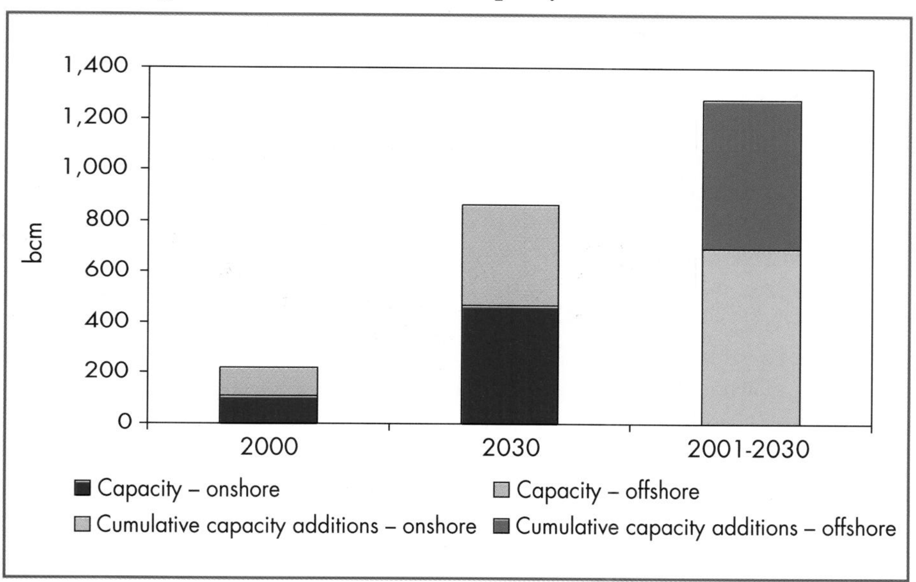

cross-border pipeline projects are considered highly risky in view of regional political tensions, while LNG projects in those countries such as Qatar that have successfully implemented such projects in the past enjoy strong credit ratings (see below). Still, the leading rating agencies have recently put the debt ratings of Qatar's RasGas LNG project and Oman LNG under review for a possible downgrading — mainly because of concerns about insurance cover in the event of terrorist attacks or war. Continuing instability in the Middle East would jeopardise credit ratings for new projects and raise the cost of capital. LNG will certainly remain easier to finance than pipeline projects in the medium term.

To date, access to capital has not been a major problem for new gas development projects in the region. Most projects have been funded out of a mixture of retained earnings, state budget allocations and, in the case of most export-oriented projects, project finance and/or international bond issues. National oil companies still dominate the gas industry in most of the major producing countries. However, there are signs that financing may become more of a challenge in the future. In many countries, the ability of the state to finance growing capital needs for new projects will undoubtedly be constrained by budget deficits and competing demands for state financial resources.

Increased recourse to project finance and international capital markets may not make up the difference, given the limited lending capacity of regional banks, the high cost of capital and growing competition for credit from the power sector. Although project financing is well established in the region, international banks have shown less interest of late in extending credits to large energy projects (see Chapter 3). This is mainly because the banks believe they can earn better returns in other regions and sectors, commensurate with the risks involved. Some international banks have pulled out of the region completely. Regional banks have limited funds and cross-country lending remains limited.[24] In Saudi Arabia, for example, the most that can normally be raised from Saudi banks for a large energy project is around $500 to $600 million. Non-Saudi Middle Eastern banks have much less lending capacity. And international bond issues are also proving difficult, despite the success of the RasGas LNG project bond issue in Qatar in 1996 (see below). The cost of capital may have to rise to attract the necessary capital inflows from international banks in the future. This, in turn, could undermine the viability of some projects whose economics are already marginal.

In addition, an increasing proportion of gas investment, as for oil, will have to come directly from private sources. Most countries in the region are now seeking to allow foreign companies to play some role in new gas projects as a means of ensuring investment as well as benefiting from the technological and project-management expertise of those companies. The approach to market opening and the pace of negotiations vary according to political and cultural factors. How successful each country is in attracting foreign participation will depend on the terms offered. Foreign direct investment will be key to the long-term development of gas-supply projects in the region.

Pricing terms including the amount of government-tax take are likely to be a stumbling block for many gas projects and will be a major source of uncertainty for investment in the region. The rent available on gas projects may be small in many cases, and certainly smaller than for most upstream oil investments. This factor will make gas investments in the region highly sensitive to oil and gas prices. Where capital is scarce, gas projects are likely to struggle to compete against oil projects, which typically yield higher tax revenues to the host countries. New gas projects, especially cross-border pipelines and ventures involving foreign investors, may also be delayed by protracted negotiations over investment terms and intergovernmental agreements. Interventionist government approaches to upstream oil and gas projects may hinder new investment in the region. The prospects for and uncertainties about investment in Middle East gas-supply projects are discussed below according to the type of project.

24. According to APICORP, local banks contributed in total around $6 billion a year to total expenditure of $16-18 billion on oil, gas and petrochemical projects in the Middle East and North Africa in 2002. See *Middle East Economic Survey* (17 March 2003), "Financing Capacity Becoming an Issue for Arab Energy Projects", (Energy Finance, B1).

Domestic and Regional Projects

The most important domestic and regional gas-development projects in the Middle East are centred on the South Pars field in Iran, the expansion of the Master Gas System in Saudi Arabia and the Dolphin project in Qatar. The initial phases of Iran's multi-billion dollar South Pars development are aimed at boosting gas supplies to the growing domestic market to free up oil for export and to reinject into ageing onshore oil fields.[25] The Pars Oil and Gas Company (POGC), a subsidiary of the National Iranian Oil Company (NIOC), has jurisdiction over all South Pars-related projects. It has entered into buy-back or production-sharing contracts with foreign companies for most phases. The exceptions are phases nine and ten, which are to be project-financed. Under the buy-back deals, the contractors provide upfront financing and are reimbursed in the form of a share of output over a period. The rate of return to the foreign investors, partly indexed to the prevailing cost of capital, is guaranteed, as long as the project is completed to budget. Ten project phases have so far been awarded, with a total sustainable peak gas-production capacity of just over 100 bcm a year (Table 5.8), as well as 684,000 b/d of condensates. There are plans to award at least another four phases aimed at LNG and GTL projects, and other phases may follow later. Phases two and three, involving gas production of almost 21 bcm a year, were completed on-time and on-budget in 2002 at a cost of $2 billion.

Table 5.8: **Iran South Pars Gas Development Plan**

Phase	**On stream**	**Sustainable peak production**	**Gas production** (bcm/year)	**Foreign participants**
1	July 2003	November 2003	10	–
2 & 3	March 2002	October 2002	21	Total, Petronas, Gazprom
4 & 5	August 2005	December 2005	21	ENI
6, 7 & 8	June 2006	Late 2006	31	Statoil
9 & 10	November 2006	Early 2007	21	LG
Total awarded			**104**	

Source: FACTS Inc (2003a).

Investment prospects for the development phases that have not yet been completed are uncertain. Phase one, which was scheduled for completion in 2001, has encountered major problems. The $1 billion project, led by a NIOC

25. Iran is also negotiating to export 3 bcm a year of gas to Kuwait, with supplies beginning some time after 2005.

subsidiary, is at least two years behind schedule, because of project-management and financing problems. The other phases involving foreign companies seem to be moving ahead more smoothly. Uncertainties are most acute for phases nine and ten, given the Iranian participants' lack of project-financing experience.

Saudi Arabia reached a break-through agreement with Shell and Total in July 2003 on joint gas exploration with the national oil company, Saudi Aramco, under the country's *Natural Gas Initiative.* This is the first deal since the initiative was launched by the government in March 2000 and the first time the Saudi government has authorised foreign investment since the 1970s. The deal is a more limited form of the third of three core ventures to explore for and develop gas reserves that the government had been negotiating with international oil companies. The other ventures have been abandoned following disagreement on a number of issues, notably the internal rates of return (IRR) and access to reserves. The initiative aims to expand domestic gas use in power generation, petrochemicals and desalination. The decision to bring in foreign companies was motivated by financing, technical and political considerations. New foreign investment legislation was introduced in 2001 to stimulate capital flows and streamline paperwork, and tax laws have been relaxed.

The Shell/Total deal covers 200,000 square kilometres of the Empty Quarter. Shell has a 40% stake, while Total and Aramco hold 30% each. The new deal is expected to pave the way for others involving international oil companies. The Saudi government has entered into discussions with more than 40 companies on smaller upstream projects extracted from the original core ventures. Bid packages were to be sent out to interested parties in August and September 2003. Saudi companies, including Aramco and SABIC, the state-owned industrial conglomerate, are expected to take on the downstream parts of the initiative, which are of less interest to foreign investors.

The Dolphin Project entails the construction of an 800-kilometre sub-sea pipeline to bring gas from Qatar's North Field to Abu Dhabi in the UAE. The line would connect with the Jebel Ali line from Abu Dhabi to Dubai commissioned in 2001, which could ultimately be extended to Oman. The Dolphin Project, which will have an initial capacity of 20 bcm a year, is expected to cost $3.5 billion. Dolphin Energy is considering raising up to $2.5 billion in project finance, but this sum may be reduced if the three shareholders (Total, Occidental and the UAE Offsets Group) decide to put up more equity. The tax-paid price of the gas delivered to end-users in Abu Dhabi has been fixed at $1.30/MBtu, but the final price delivered to Dubai has not yet been agreed. Final agreement on price may form part of a broader deal on social and economic matters being negotiated by the two emirates. Construction work is due to begin in 2004, with first gas deliveries scheduled for 2006.

Qatar is also developing a project to export gas to Kuwait via a sub-sea line across Saudi waters. Kuwait and Qatar signed a protocol in 2002 for the supply over 25 years of 7.75 bcm a year of dry gas from end-2005, but a final inter-governmental agreement has not yet been reached. The Saudi government gave its approval for the pipeline at the beginning of 2003. In the longer term, Iraq would be well placed to export gas by pipeline to East Mediterranean countries, Kuwait and Saudi Arabia. Iraq exported small volumes of gas to Kuwait before the Iraqi invasion in 1990. Saudi Arabia could eventually become an important regional market if its gas demand outstrips associated gas output and its capacity to invest in non-associated sources.

LNG Export Projects

The Middle East has recently emerged as a major LNG exporting region, with plants now operating in Oman, Qatar and the UAE. The first plant was built by Adgas in Abu Dhabi in 1977. The second, Qatargas, started operating commercially in 1997. By 2001, the three countries accounted for 21% of world LNG trade. Several new LNG projects and expansions of existing projects are under construction or planned (Table 5.9).

If all the currently planned projects proceed, export capacity would expand nearly fourfold from 35.7 Mt/year at present to almost 170 Mt/year by 2010. But their near-term prospects are uncertain in view of recent problems in lining up long-term contracts with major buyers in Asia and Europe. Our projections suggest that more than 80% of this capacity will be commissioned by 2010. Pricing has been the main problem facing LNG-project developers. LNG must be priced lower than in existing contracts with buyers in Japan, Korea and Chinese Taipei if it is to be competitive in new markets, such as India and China, where gas competes with low-cost coal. Government support for gas or environmental restrictions on coal use will also be necessary in those markets. RasGas-2, which is planning to build a second train in Qatar, signed an agreement with Petronet of India in 2001 for the supply of 7.5 Mt/year over 25 years from early 2004 at a FOB price of $2.20/MBtu based on an oil price of $18/barrel. This implies a delivered gas price after transportation and regasification of around $3.20/MBtu. The Korean gas company, Kogas, the sole buyer of gas under long-term contract from RasGas-1, is thought to be paying a significantly higher base price. The Yemen LNG project for the supply of LNG to India, which had been in advanced negotiations, has stalled over pricing terms.

Prospects for the three planned LNG projects in Iran will also depend critically on pricing terms and competition from other LNG suppliers. Several consortia have made bids for projects based on phases 11 to 15 of the South Pars development: Iran LNG (BP and Reliance), Pars LNG (Total and

Table 5.9: **Existing and Planned LNG Projects in the Middle East**

Country	Project	Capacity (million tonnes)	Number of trains	Start-up	Status
Abu Dhabi	Adgas	5.8	3	1977	In operation.
Iran	Iran LNG/Pars LNG/ENI (Pars phases 11&12)	17	4	Not known	Planned.
	Persian LNG (Pars phase 13)	10	2	Not known	Planned.
	NIOC LNG (Pars phase 15)	10	2	Not known	Planned.
Oman	Oman LNG	6.6 +3.3	2 + 1	2000	2 trains in operation. A 3rd train under construction is due on stream in 2005.
Qatar	Qatargas -1	6 + 3.2	3	1997	In operation. Debottlenecking will boost capacity of existing 3 trains to 9.2 Mt/year by mid-2005.
	Qatargas - 2	15	2	2006/7	Planned. ExxonMobil to take all output to supply UK/Europe.
	Qatargas - 3	18.4	2 + 2	2008/9	Planned. ConoccoPhillips has agreed to take 9.2 Mt; Total is negotiating a similar volume.
	RasGas - 1	6.6	2	1999	In operation.
	RasGas - 2	9.4	2	2004	1st train under construction, due on stream in 2004. 2nd planned for 2005.
	RasGas - 3	15	2	c.2010	Planned. ExxonMobil negotiating for export to US.
Yemen	Yemen LNG	6.2	2	Not known	Planned. Negotiations with buyers have stalled.
Total		**132.5**	**31**		

Source: IEA database.

Petronas) and ENI for phases 11 and 12; Persian LNG (Shell and Repsol) for phase 13; and NIOC LNG for phase 15. For the time being, only one project is likely to move ahead, possibly involving gas supplied from different phases.[26] NIOC LNG looks to be the best placed at present, although it may be based on phase 13 rather than 15, which has in any case not yet been officially sanctioned. BG and ENEL, who are primarily interested in securing LNG at competitive prices, are negotiating to join the NIOC LNG project as downstream partners. The government has indicated that it may provide as much as 50% of the financing for the project out of its reserve funds. BG is looking to secure gas for a planned receiving terminal in Pipavav in India. It is seeking a pricing formula that keeps gas competitive with coal, by indexing the base price only to inflation and not oil prices. If it succeeds, other buyers in India and China would undoubtedly push for similar terms. Whichever project goes ahead first, the earliest that the first shipments of LNG are likely to occur is 2008.

Financing new LNG export projects in the Middle East should not pose major problems if acceptable long-term sales and purchase agreements can be negotiated. As with recent LNG projects, they would most likely be funded by a mixture of equity and project finance, with guarantees provided by future revenue streams. It is uncertain whether international bonds could part-finance new projects, as was the case for RasGas-1. Some $1.2 billion of the total cost of $3.4 billion for that project was raised through a bond issue, with equity providing $800 million and bank and export credit agency lending providing the rest. The increasing emphasis on LNG sales to poorer developing countries, where country and currency risks are higher, may limit opportunities for project-bond financing. Bond issues have the advantage of offering longer repayment periods than bank loans.

Export Pipeline Projects

Several pipeline projects for the export of gas to countries outside the region have been proposed in recent years. At present, the only pipeline exports from the Middle East are from Iran to Turkey through a 10-bcm a year line commissioned in late 2001. Volumes delivered so far have been small. The most ambitious new project is a large-diameter pipeline running from the South Pars field in Iran to Pakistan and on to India. Iran has signed Memoranda of Understanding with both countries. The delivered cost of piped Iranian gas to India, assuming sales of at least 10 bcm a year, would almost certainly be competitive with LNG. But the project will not proceed until there has been a significant improvement in political relations between India and Pakistan. Sales to Pakistan alone will not be sufficient to make

26. FACTS (2003b).

the project viable. Even in the event of a thawing of relations and an intergovernmental agreement to build the line, financing would be extremely costly and difficult to arrange. Running the line offshore through international waters, in order to bypass Pakistan, would probably make the cost of the delivered gas in India uncompetitive.

Qatar is also considering exporting North Field gas to Pakistan, through an extension of the planned line that will link Dubai with Oman. Initial exports are expected to amount to 1-2 bcm a year. The Pakistan part of the project is expected to cost in the region of $2 to $3 billion. The project could be completed before 2010.

In the longer term, major pipelines from Iran and Iraq to Europe are expected to be developed. Capacity is projected to reach 40 bcm by 2030. The fate of these projects depends very much on geopolitical developments as well as trends in pipeline costs relative to those of LNG. Iraq is geographically well positioned to supply Europe compared with other Middle East countries.

Africa

Investment in Africa's gas-supply infrastructure will surge in the coming decades, driven mainly by exports to Europe and the United States. Upstream investments will represent the largest component. New LNG projects and cross-border transmission pipelines will account for an increasing share of investment. Access to foreign capital for upstream development will be crucial to timely expansion of production. Financing might be a hurdle to investment in new large-scale cross-border pipeline and LNG projects depending on cost developments and geopolitical factors.

Investment Outlook

Cumulative investment needs in the African gas sector are projected to total $216 billion, 7% of global gas investment, over the period 2001-2030. Annual capital needs will rise steadily, from an average of just over $3 billion over the past ten years to $4.8 billion in the current decade, to almost $10 billion in the last decade of the projection period (Figure 5.20). Upstream development will absorb the bulk of these requirements, as production rises spectacularly from 134 bcm in 2001 to 589 bcm in 2030. Investment in exploration and development, which was running at about $2 billion in the year 2000, is expected to rise briskly from an average of $2.8 billion per year over the current decade to $7.5 billion per year in the third decade. Capital spending will increase both because of rising production and because an increasing share of output will come from offshore fields, which cost more to develop.

Figure 5.20: **Gas Investment in Africa**

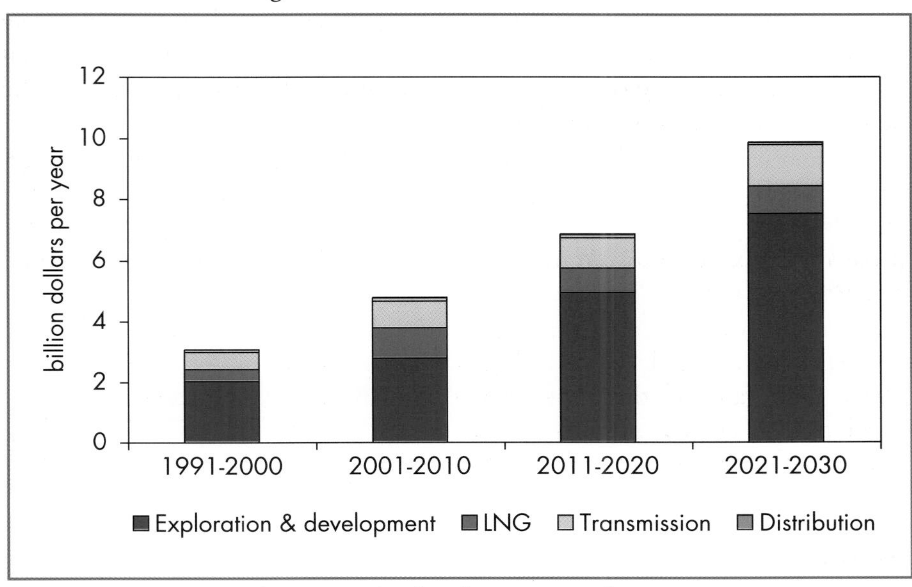

Note: The E&D estimate for 1991-2000 is based on year 2000 data.

More than $26 billion will be needed over the next three decades to build and expand LNG liquefaction facilities. This compares with slightly more than $10 billion invested to date in LNG plants in Africa. New investments will boost LNG export capacity from 47 bcm in 2002 to around 230 bcm in 2030. The expansion and construction of international and domestic transmission lines will require an additional $33 billion. Around two-thirds of projected investment will be needed for exports to OECD countries.

Supply Trends and Prospects

African proven natural gas reserves amounted to 13.1 tcm in 2002, 7% of the world total. The continent is still relatively under-explored, so its gas potential has not been fully appraised. Undiscovered gas resources are thought to be evenly distributed among onshore and offshore fields and half are associated with oil. More than two-thirds of the proven reserves are equally distributed between two countries, Algeria and Nigeria. In 2001, almost the entire production of the continent was supplied by four countries: Algeria (84 bcm), Egypt (23 bcm), Nigeria (16 bcm) and Libya (6 bcm). Most of the gas produced was flared or re-injected to enhance oil recovery. It is estimated that Nigeria alone accounts for 12.5% of all the gas flared in the world.

Production is expected to increase from 134 bcm in 2001 to 589 bcm in 2030, as associated gas is increasingly utilised and non-associated production increases. Offshore production, especially in Nigeria and Angola, is expected to rise significantly. Total African offshore production is expected to grow from 51 bcm in 2001 to more than 200 bcm in 2030. African gas demand is expected to increase at 5.2% per year over the next 30 years, faster than any other primary fuel. Nonetheless, more than half of production will be devoted to export, mainly to Europe and the United States. Given the long distance from the markets, most of the additional volume of gas traded will be transported as LNG.

Table 5.10: **Africa Gas Supply** (bcm)

	2001	2010	2020	2030
Production	134	246	389	589
Demand	67	95	155	239
GTL	3	6	22	50
Net export	64	145	212	299

Current liquefaction capacity in the region amounts to 34.4 Mt, 23 Mt in Algeria, 8.8 Mt in Nigeria and 2.6 Mt in Libya. Sonatrach, the Algerian national oil company, has the largest equity share in LNG plants in the world, and is expected to remain a leading LNG exporter. To meet the expected increase in LNG trade, 140 Mt of additional liquefaction capacity will be needed over the next thirty years, more than four times existing capacity. Capacity will need to double in the current decade alone. New countries, including Egypt and Angola, are planning to enter the LNG business for export to Europe and the United States, where import needs are expected to grow rapidly. At least nine projects are currently under construction or planned, five of them in Egypt (Table 5.11).

Exports to Europe by pipeline are also expected to expand rapidly. The 30 bcm of piped exports in 2001 are expected to triple over the next three decades. Currently two pipelines connect Algeria to Europe: the Transmed pipeline to Italy via Tunisia, and the Maghreb-Europe pipeline to Spain via Morocco. The Transmed line has a capacity of 26.5 bcm per year and involved total investment up to the Italian border of almost $3.5 billion, for a total length slightly more than 2,600 km. Flows in the Maghreb-Europe line started in 1996. This line has a capacity of 8 bcm per year and involved total investment (including the Spanish and Portuguese sections) of $2.3 billion.

Table 5.11: **Major New LNG Projects in Africa**

Location	Scheduled start-up	Cost ($ billion)	Capacity (Mt/year)	Current status	Company
Egypt					
Idku	2005	1.35	train 1: 3.6	under construction	BG, Petronas, EGPC, EGAS, GDF
	2006	0.55	train 2: 3.6	under construction	BG, Petronas, EGPC, EGAS
	2007	1.5	train 3: 4.0	planned	EGPC, BP, ENI
Damietta	2004	1.0	train 1: 5.0	under construction	EGPC, EGAS, Union Fenosa, ENI
	2006	1.0	train 1: 4.0	planned	EGPC/Shell
Nigeria					
Bonny Island	2005	2.1	trains 4/5: 8	expansion	NNPC, Shell, Total, ENI
	2007	n.a.	train 6: n.a.	planned	
Angola					
Luanda	2005	2.0	train 1: 4	planned	NOC, Chevron Texaco
Guinea					
Bioko Island	2007	n.a.	train 1: 3.3	planned	Marathon

Source: IEA databases.

An additional compressor station will increase the capacity by the end of 2003 to 11 bcm per year. A pipeline from Egypt to Jordan was recently commissioned at a cost of $200 million. A number of new export lines are planned, including sub-sea lines from Algeria to Spain and Libya to Italy, and a land line from Nigeria to Ghana and Algeria (Table 5.12). These projects and others that will follow will require large investments.

Many African countries are trying to implement local gas distribution schemes to promote domestic consumption. Existing distribution networks are a tiny 40,000 kilometres, with 60% of them located in just two countries, Algeria and Egypt. More than $3 billion will be necessary to more than double the existing network. Nonetheless, residential, commercial and industrial gas consumption is expected to remain very low, reaching only 52 bcm in 2030.

Table 5.12: **Main Planned African Pipeline Projects**

Pipeline	**Origin – destination**	**Capacity** (bcm)	**Length** (km)	**Year of operation**	**Cost** ($ billion)
GME	Algeria – Spain (via Morocco)	expansion (+3)	1,620	2004	0.2
Medgaz	Algeria – Spain	8	1,100	2006	1.4
Galsi	Algeria – Italy	8	1,470	2008	2.0
Arab Mashreq	Egypt – Jordan	n.a.	248	2003	0.2
Green Stream	Libya – Italy	8	540	2005	1.0
WAGP	Nigeria – Ghana	3	990	2005	0.5
Trans-Saharan	Nigeria – Algeria	10	4,000	After 2010	7.0

Source: IEA databases.

Financing and Investment Uncertainties

Until recently, the development and utilisation of gas in Africa has been hampered by several factors. Among these are hydrocarbon laws which have tended to concentrate on oil and uncompetitive tax and incentive schemes. Full exploitation of gas reserves has also been discouraged by the lack of local or regional markets and the long distances to markets outside the region. Political risks in countries with exploitable gas reserves and mistrust and conflicts between neighbouring states make the creation of regional markets difficult. Other impediments include the weak financial position of potential local gas customers, leading to payment risks, corruption and inefficiency that pushes up costs, and problems with convertibility and repatriation of profits. Improving the investment climate, including establishing transit protocols, will be vital to attracting new project financing, most of which will have to come from abroad given the limited financial resources in the region (see Chapter 3). The fiscal regime, the stability of the country and the degree of openness and transparency of the upstream sector will determine which African countries attract investment and its timing.

There are opportunities for foreign investors to participate in upstream gas projects in all North African countries. Foreign participation is essential to ensure the level of investment needed to expand production and to provide the technology to develop offshore fields and handle associated gas. The two main arrangements are concessions and partnerships (joint ventures by domestic and foreign companies). Egypt and Libya have implemented production-sharing agreements, in which foreign investors independently operate the production field and remunerate the country with a part of the production. Algeria has opted for partnerships between Sonatrach and foreign investors. Foreign investment has helped Algeria and Egypt to boost their gas production considerably over the past decade. Libya has had less success, barely managing to sustain the 6 bcm per year reached at the beginning of the 1990s. However, the government is becoming aware of the need to make its gas industry more attractive to investors. The Nigerian government has increased incentives for companies to utilise associated gas production in order to reduce flaring.

Government policies will need to set clear rules for gas companies. The EU has recently stressed the importance of the development of a Euro-Mediterranean energy policy, with the aim of enhancing energy security. Among the priorities that have emerged is the creation of a regional energy market, including Morocco, Algeria, Tunisia, and possibly Libya (depending on its progress in taking part in the Barcelona Process).[27] This will involve an agreement setting rules for gas markets that reflect the principles of reciprocity, competition, safety and security. The European Investment Bank (EIB) has declared its intention to invest at least $300 million over the next four years in North Africa in order to build energy infrastructure within the framework of the Euro-Mediterranean Investment and Partnership.

Figure 5.21: **Example of Gas Financing Arrangements in Africa**

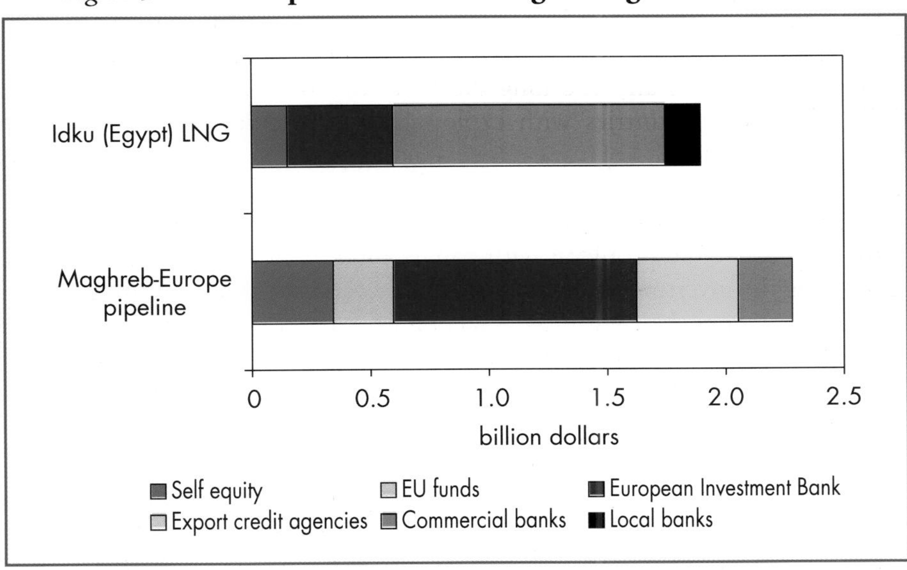

Financing arrangements and the cost of capital for gas projects will vary. Country risk is a key factor, but has not proved to be insurmountable in Africa.

27. The objective of the Barcelona Process is to turn the Euro-Mediterranean basin into an area of dialogue, exchange and co-operation guaranteeing peace, stability and prosperity. One of the main objectives of the process is the creation of a free trade zone by the year 2010.

Nigeria, for example, though considered one of the riskiest countries in the world in which to invest, was able to secure at the beginning of 2003 an international loan of more than $1 billion to finance the $2.1 billion two-train expansion of the Bonny Island facility. The key to successful financing is risk-sharing among a number of entities. The participation of multilateral banks, such as the EIB, can help gas projects raise funds from commercial and local banks (Figure 5.21).

China

China's natural gas industry, which is at an early stage of development, is poised for rapid expansion. The government is committed to a rapid increase in the share of natural gas in the country's energy mix. This policy is driven by concerns about the environmental impact of heavy dependence on coal and the energy-security implications of rapidly rising oil imports. Investment needs will be large, but capital availability is good. Nonetheless, the pace of development of gas infrastructure will ultimately depend on policy reforms to clarify the investment and operating environment and proactive government measures to boost the competitiveness of gas against cheap local coal.

Investment and Supply Outlook

Cumulative investment of just under $100 billion in supply infrastructure will be needed over the period 2001-2030 to meet projected increases in demand. Annual capital investment would need to average $3.3 billion (Figure 5.22). Distribution networks will attract the largest amounts of capital — more than a third of total investment. Gas investment is relatively modest

Figure 5.22: **Gas Investment in China**

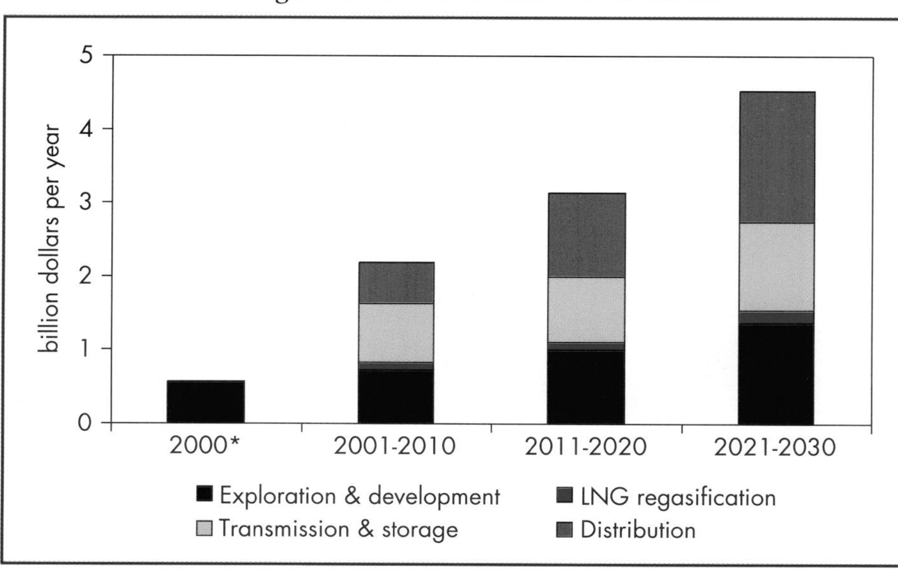

* Only E&D costs are available for 2000.

compared to overall energy investment, which is projected to average $75 billion per year. This is largely because gas will remain, in relative terms, a marginal fuel in China's primary energy mix.

Investment needs are projected to increase progressively over the projection period, in line with rising demand and the need for new transportation capacity. Rising upstream investment will be driven mainly by the need to replace depleted capacity at mature fields and by rising development costs.

These investment projections are underpinned by a projected surge in gas demand, from a modest 32 bcm in 2000 to 61 bcm in 2010 and 162 bcm in 2030 (Figure 5.23). The role of gas in the country's primary energy supply will, nonetheless, remain small, at 7% in 2030, unless major new discoveries are made. Official projections suggest even more rapid growth: the State Development Planning Commission, for example, forecasts gas demand of 120 bcm in 2010 and 200 bcm in 2020.[28] It is expected that most of the increase in gas demand will be met by indigenous production, but imports in the form of LNG and by pipeline from Russia and possibly Central Asia will make a growing contribution over the next three decades.

The 3,900-km West-East pipeline to transport gas to Shanghai from the Tarim basin in the west and the Ordos basin in central China is the centrepiece of the government's plan to establish a national gas market (Figure 5.24).

Figure 5.23: **Gas Supply in China**

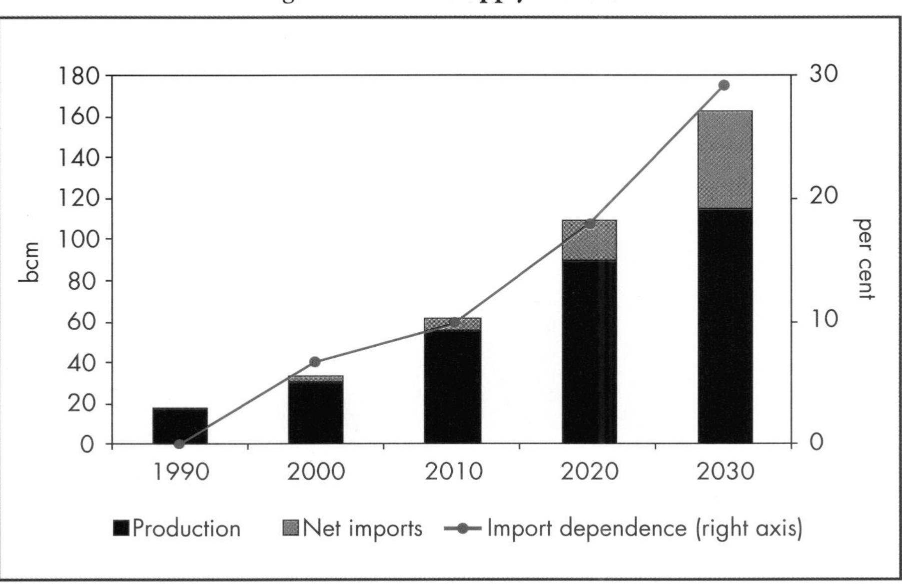

Source: IEA (2002a).

28. ERI/SDPC (2001).

Figure 5.24: **Future Gas Supply Infrastructure in China**

This is one of several major infrastructure projects to promote economic development in the poor central and western provinces. The line, which will have an initial capacity of 12 bcm per year, is expected to cost around $6 billion, accounting for almost a third of the total gas investment projected for the current decade. Investments in upstream and downstream facilities connected with the West-East pipeline network could bring the total cost close to $18 billion. A consortium made up of PetroChina (50%), SINOPEC (5%) — China's largest downstream oil and petrochemicals company — Shell and Hong Kong China Gas (15%), Gazprom and Stroytransgaz (15%), and ExxonMobil and Hong Kong Light and Power (15%), is building the pipeline and upstream facilities. First gas from the Changqing field in the Ordos basin is due to reach Shanghai in 2004.

The LNG terminal at Guandong, being built by a consortium led by CNOOC and BP, will have an initial capacity of 3 Mt/year. In 2002, China and Australia signed a contract under which gas will be supplied from Australia's North West Shelf Projects for 25 years. First gas is expected in 2005. CNOOC is planning a second terminal in Fujian. In the longer term, several other gas import terminals will be needed to meet projected demand growth. China is also expected to import gas by pipeline from Russia, most likely from the Kovykta field in east Siberia, but China now plans to pursue this project at a later stage, as priority has switched to the West-East pipeline. Imports from Kovykta are, nonetheless, expected to begin during the decade after 2010. A pipeline from Sakhalin-1 in Russia has also been mooted. Our investment calculations assume that it will be built in the decade 2011-2020. Pipeline imports from more distant foreign sources, such as Turkmenistan or Kazakhstan, have also been proposed, and it is assumed that these will begin in the decade 2021-2030. Over the next three decades, an interconnected national gas-pipeline network that would link the West-East pipeline, LNG terminals in the coastal areas and import pipelines is expected to take shape gradually.

Financing Infrastructure Projects

Although the amount of investment needed for natural gas projects is large, finance is available from a variety of sources. How projects are financed will affect the cost of capital; but lack of capital is not expected to be a major obstacle. Until now, most upstream gas and pipeline investments have been carried out by state-owned oil and gas companies, while distribution networks have been built by municipal gas companies.

Domestic banks, which have considerable lending capacity, are likely to meet much of the gas sector's funding needs. Chinese citizens have high savings rates, with most of those savings going into state bank accounts and government bonds. Most of the equipment and materials needed for gas-supply infrastructure is expected to be sourced from within China, so borrowing in the local currency, the yuan, would reduce any foreign currency risk. Interest rates on commercial loans by state banks, set by the Central Bank, are attractive. Current rules allow foreign joint-venture investors to raise large yuan loans fairly easily, although repayment periods are restricted to five years and need to be supported by hard currency deposits in China or by financial pledges.

Other potential domestic sources of finance include treasury bonds issued by the central government, bonds issued by state-owned policy banks, such as the China State Development Bank, and corporate bonds issued by state-owned enterprises themselves. Under current rules, provincial and municipal authorities are not allowed to borrow on their own account. As a result, they have to rely on the central government to issue bonds for local infrastructure projects. They can, however, provide direct funding from local tax revenues and can raise funds in other, less formal ways. In the longer term, domestic equity markets are expected to play an increasingly important role in funding gas projects. In 2001, Chinese companies — including state-owned and private firms — raised around CNY 100 billion ($12 billion) on the Shanghai and Shenzhen stock exchanges.

Foreign investment will most likely provide an important source of finance for the development of gas-supply infrastructure. China began to open up its upstream oil and gas sector to foreign investment in the early 1980s. In 2002, restrictions on foreign investment in the gas industry were significantly reduced with the release of revised *Guidelines on the Direction of Foreign Investment* and a new *Catalogue Guiding Foreign Investment in Chinese Industries.*[29] The catalogue widened the range of gas-related activities that are to be encouraged, and reduced activities that are subject to some investment restrictions to local distribution only. Foreign firms can now provide direct equity participation of up to 49% in local gas-distribution projects through joint ventures with local firms and, in theory, can own up to 100% of any high-pressure pipeline. A number of pipelines have been built in recent years, with varying degrees of foreign involvement and the West-East pipeline and Guandong LNG projects will involve significant foreign equity stakes. By opening up the gas sector to foreign investment, the government hopes to gain access to foreign technology and managerial expertise, as well as capital, and to spread investment risks.

The sources of capital and the financial structure of new projects will vary according to the type of project. Equity financing will remain the dominant form of investment in most projects, but project financing will probably play a key role in LNG import projects. Foreign direct investment will probably

29. See IEA (2002c) for details.

remain concentrated in large-scale upstream developments, LNG import terminals and long-distance pipeline projects, which are commercially and technically complex. Foreign investors are showing little interest in local distribution as yet, but this may change as networks expand and the regulatory regime evolves. Gas distribution was only recently removed from the list of prohibited areas for foreign investment, though restrictions remain. Most existing local distribution companies, many of which distribute gas manufactured from coal, are inefficient and unprofitable divisions of municipal governments.

Factors Driving Investment

In spite of the uncertainties surrounding policy reforms and how quickly demand will grow, large amounts of domestic and foreign capital are already being invested in gas-supply projects in China. Contrary to standard practice in the gas industry elsewhere, large projects are moving ahead even though long-term gas sales contracts have yet to be signed with major end-users. None of the 45 letters of intent from potential buyers of gas to be delivered through the West-East pipeline have so far been converted into firm sales. Similarly, no firm sales contracts for gas from the Guandong LNG terminal have yet been signed. Investors have been given assurances from the government about pricing and demand on a project-by-project basis, but they are becoming very concerned about the slow progress in developing consistent policies and regulations that would facilitate the signing of sales contracts.

One factor underpinning investor confidence is the enormous long-term potential for gas-demand growth. Investors are also counting on the government putting pressure on large energy users, including state-owned enterprises, to buy gas at high official prices. Nevertheless, maintaining the momentum of foreign investment will depend on major policy initiatives aimed at creating a more stable and predictable operating environment, reducing market risks and lowering the cost of capital. The type of legal, regulatory and policy framework for gas that evolves and, more specifically, the types of measures that are adopted to promote the use of gas will have a crucial impact on the attractiveness of investing in China's gas industry and the cost of financing.

This is especially important for the downstream sector. Beyond its stated goal of increasing gas use, the government has not yet adopted a national law covering gas transmission, distribution, importation and supply. Nor has it announced any concrete measures to encourage the long-term development of the gas market. The project-by-project approach so far favoured by the Chinese authorities, while providing flexibility in dealing with the specific characteristics of each undertaking, has led to confusion and inefficiency. Each project is subject to special rules and exceptions, while provincial and local authorities apply different bureaucratic procedures to authorising, licensing and regulating downstream activities.

A recent IEA report on China's gas market identified the need to clarify policy goals and establish effective laws and regulations as an urgent priority. This factor is a major source of uncertainty for future investment.[30] A new law, or set of laws, is urgently needed to codify the roles, rights and responsibilities of different players, set out regulatory principles and provide for regulations on technical norms and standards to be applied across the industry and throughout the country.

Active measures will be required to promote switching to gas and investment in the different links in the supply chain. How effective policies and measures are in making gas more competitive against other fuels will have a particularly important impact on market development and investment flows. The Chinese government has been pursuing a "supply-push" market-development strategy based on large-scale infrastructure projects. It has attempted to create demand for new gas supplies largely through regulatory and administrative means. For example, it has banned or placed restrictions on the use of coal within specified areas in large cities where pollution is a serious problem and has mandated the conversion of district-heating systems to gas-fired co-generation plants. It is also requiring provincial authorities to take minimum volumes of gas from the West-East pipeline for local distribution. These measures have proved necessary at this early stage of gas-market development, because gas has generally been unable to compete against other fuels, especially coal. To sustain the long-term growth of gas demand, more market-based measures, aimed at encouraging rather than forcing end-users to switch to gas, will be needed.

Gas-pricing and energy-taxation policies will, therefore, be of crucial importance to gas-market development. At present, gas prices are fixed and volumes are allocated to wholesale buyers and end-users by the Central government. Prices do not reflect costs or willingness to pay: high-volume industrial and commercial consumers pay higher wellhead prices (to which transportation fees must be added) than residential consumers, in inverse relation to the underlying costs of supply (Figure 5.25). Nor do wellhead prices take account of differences in production costs from individual fields or regions. On average, wellhead prices are high by international standards, making it particularly hard for gas to compete against coal and oil in industry and in power generation. Transmission and distribution mark-ups are also high, due to inefficiencies and additional charges levied by local authorities.

Ultimately, gas suppliers will need to price their gas according to its market value in relation to competing fuels. For this to happen, the government will either need to decontrol wellhead and wholesale prices or at least sanction the

30. IEA (2002c).

Figure 5.25: **Wellhead Gas Prices in China by Consumer Type, 2001**

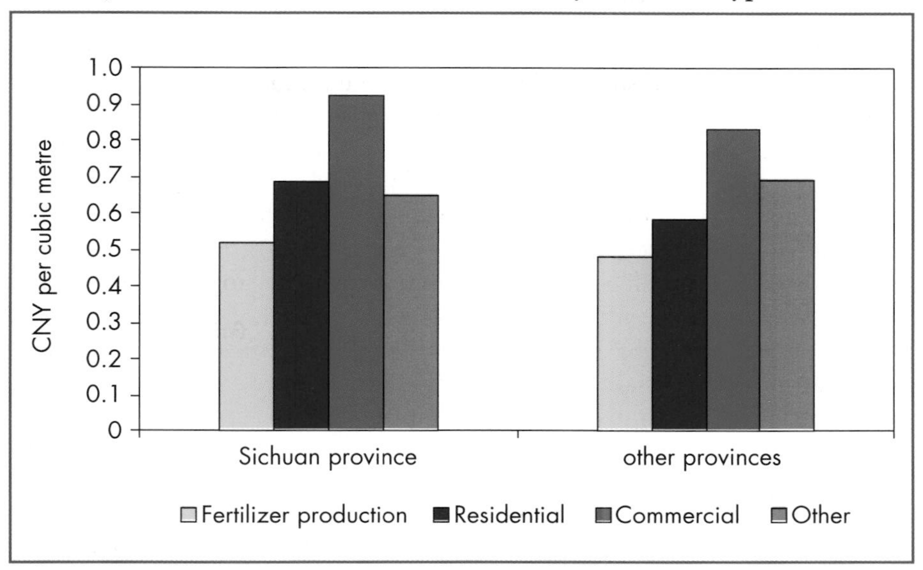

Source: IEA (2002c).

adoption by the appropriate regulatory body of market-based pricing principles for setting the prices for new gas projects.

The extent of the challenge to make gas competitive against coal in power generation is demonstrated in Figure 5.26. The curve indicates the ratio of gas to coal prices that would produce the same levelised long-run marginal cost of electricity, assuming gas is used in a standard combined-cycle gas-turbine plant and coal is used in a simple plant without flue-gas desulphurisation (FGD) equipment. For example, at current coal prices in Shanghai of around 280 CNY/tonne, the gas price would have to be below 0.85 CNY per cubic metre ($2.86/MBtu) to be competitive. This is well below the Shanghai city-gate price of 1.35 CNY/cm ($4.55/MBtu) that the government has set for gas to be delivered through the West-East pipeline, which would require a coal price of 500 CNY/tonne to be competitive.

The competitiveness of gas-fired plants will also depend on the number of hours these plants operate. For example, at a gas price of 1.35 CNY/cm, the cost of electricity generated would vary between 4.3 US cents/kWh and 5.4 cents/kWh depending on whether the plant is running in baseload mode of 8,000 hours per year or just 3,000 hours per year. The introduction of regulations requiring the use of FGD or limiting SO_2 and NO_x emissions would boost the competitiveness of gas by raising the cost of coal-fired generation. Similarly, the widespread introduction of time-of-day electricity pricing at

Figure 5.26: **Relative Competitiveness of Natural Gas and Coal in Power Generation in China**

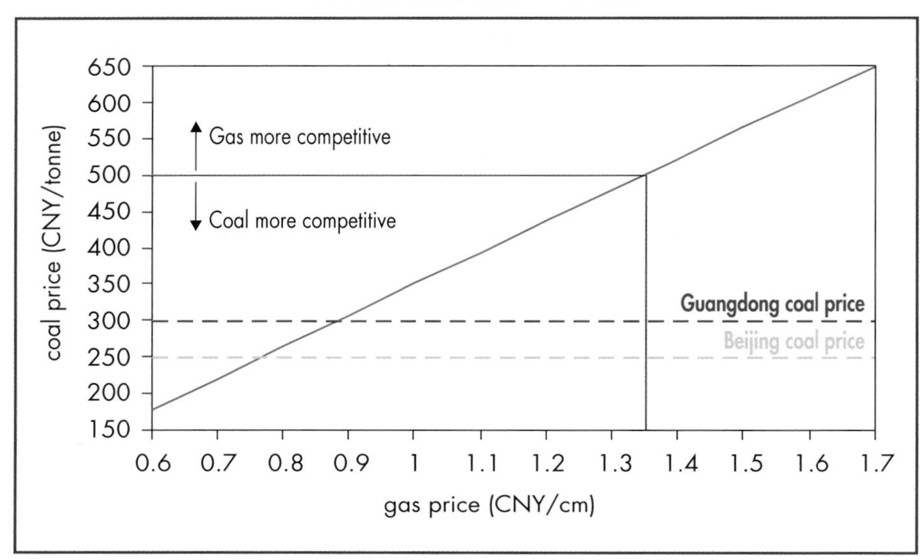

Source: IEA (2002c).

wholesale level that fully reflects the higher cost of generating power at peak would improve the competitiveness of gas against coal. Wholesale electricity tariffs are currently mostly set on a cost-plus basis, removing incentives for generators to reduce marginal generation costs and improve efficiency.

Tax policies will have an important impact on the competitiveness of gas *vis-à-vis* other fuels and on the profitability of potential investments along the supply chain. Taxing other fuels more than gas to give a price advantage to gas could be a powerful tool to boost the market for gas, but levying taxes on coal may be politically difficult. Improving the competitiveness of gas through taxation would be justified by the environmental benefits that higher gas use would bring about. Profits taxes, royalties, concession fees and value-added taxes (VAT) also affect the profitability of gas-industry investment and incentives to re-invest. The current system of VAT, which does not allow for the recovery of taxes on capital goods, and high duties on imports of LNG and pipeline materials and equipment, undermine the economics of gas projects and hinder investment. Whether the central and provincial governments are willing to reform the system of energy taxes sufficiently to support gas-market growth and stimulate investment is a major uncertainty.

Asia-Pacific[31]

Gas-supply prospects and investment needs in other Asian and Pacific countries vary considerably according to the maturity of their gas markets and access to reserves. Upstream investment — the largest chunk of total gas investment in the region — will remain concentrated in a small number of countries, notably Australia, Indonesia and Malaysia, which have large proven reserves. Japan and Korea will remain the main export markets for gas produced in the region, shipped entirely as LNG, although India and some other countries will become significant importers too. Financing new supply chains to supply emerging markets will require investment terms that take full account of financial risk and legal and regulatory frameworks that protect large-scale long-term investments.

Investment and Supply Outlook

Projected gas-supply trends imply cumulative investment needs in the Asia-Pacific region of around $380 billion over the period 2001-2030. Exploration and development account for around 60% of total investment, and transmission and distribution networks for most of the rest. Investment in LNG liquefaction and regasification plants is projected to amount to $35 billion, equal to less than 10% of total capital requirements (Figure 5.27).

Figure 5.27: **Cumulative Gas Investment in Asia-Pacific, 2001-2030**

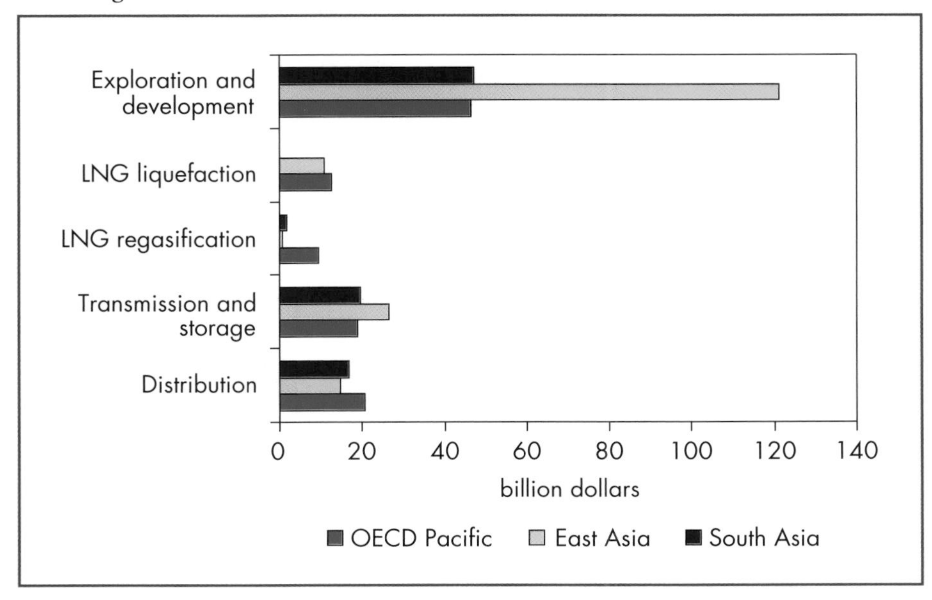

31. This section covers East Asia, South Asia and OECD Pacific.

Investment needs will be largest in East Asia, mainly for upstream developments in Indonesia and Malaysia.

Gas resources are scarce in most countries in the Asia-Pacific region. Consequently, most countries have either very small gas markets or rely heavily on imports. Consumption totalled 296 bcm in 2001. Japan has the region's largest gas market by far, consuming over 80 bcm. Indonesia, Malaysia, India, Pakistan, Australia and Korea account for most of the rest. Demand is projected to grow rapidly, reaching 695 bcm in 2030.

Asia-Pacific production amounted to 224 bcm in 2001. Indonesia, Malaysia, Australia, India, Pakistan and Thailand, the largest producers in descending order, account for the bulk of the region's output. Imports make up the shortfall in gas supply. Japan has virtually no gas reserves and imports almost all its needs as LNG. Three-quarters of its imports come from producers within the region and most of the rest from Qatar and the United Arab Emirates. *WEO-2002* projects that production within the region will grow more strongly than demand over the projection period, so that net imports from outside the region will decline after 2010 (Figure 5.28). Asia-Pacific countries will increase their imports from the Middle East, and Russia will soon begin exporting LNG to Japan and possibly later to Korea. But exports from South-East Asia to North America are expected to more than offset these increases in imports.

Figure 5.28: **Gas Production and Net Imports in Asia-Pacific**

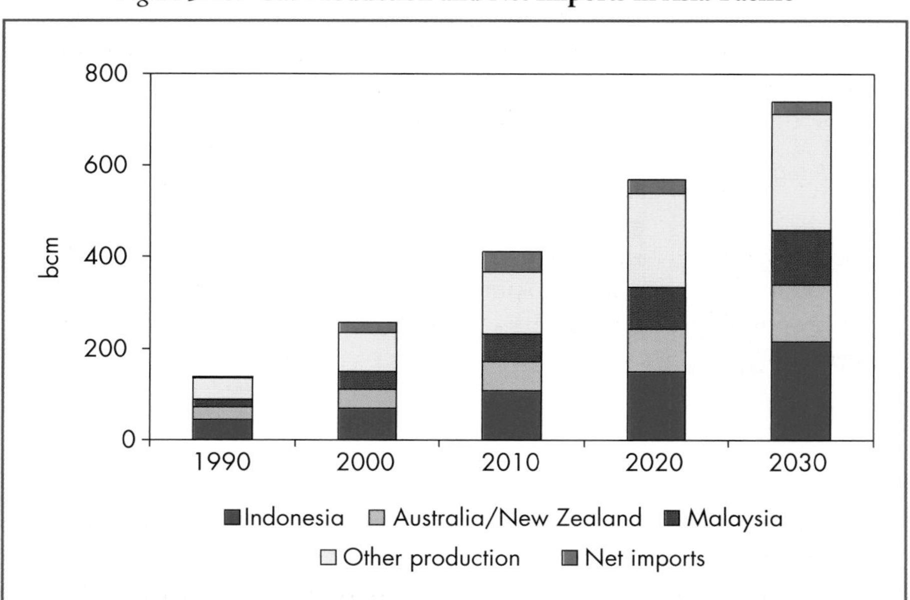

Box 5.5: **Gas Demand and Investment Uncertainties in Japan**

Natural gas currently represents 13% of the energy mix in Japan and accounts for 22% of electricity generation. Two key factors are likely to influence strongly the future evolution of gas demand in Japan and therefore the investment requirement in the gas sector. First, gas could play a bigger role than expected in the power sector, if the government's nuclear plans are not realised on schedule. Nuclear energy is a strategic fuel in Japanese energy policy, both ensuring energy security and limiting CO_2 emissions. In 2000, more than 30% of electricity generated in Japan was nuclear. However, in 2003, TEPCO, the country's biggest utility, was forced to shut down all its nuclear reactors for emergency inspections. As a result, it had to buy power from other producers while increasing electricity production from thermal plants. TEPCO's gas-fired generation surged in response to the nuclear shortfall, reflected in the purchase of 3 Mt of LNG on the spot market, equivalent to 5% of annual Japanese LNG imports. Increasing local government opposition to nuclear power is likely to threaten the government's longer-term plans to build more reactors. This would favour gas-fired generation, particularly if coupled with faster liberalisation of the electricity market. New entrants are, in fact, very likely to opt for gas-fired plants.

The key obstacle to higher penetration of gas in Japan is the lack of a domestic gas transmission network. The 24 LNG import terminals are not interconnected by gas pipelines, so gas demand is concentrated around the terminals and the market is fragmented into isolated areas of consumption. The length of high-pressure pipelines in Japan is only 3,000 km, compared to the UK network of 18,400 km. Expanding domestic infrastructure would facilitate an increase in consumption and competition in both electricity and gas markets. Low gas demand and the prohibitive cost of building pipelines in Japan has so far made it more economic to build new LNG terminals rather than extend pipelines from existing terminals. Improvements in price competitiveness with other fuels will also be essential to spur gas demand. Competition in the gas market would push prices down, raising demand and infrastructure needs.

Investment Uncertainties

A wide range of uncertainties surrounds these projections. Upstream investment prospects will depend on whether investors can secure investment terms that take full account of financial and political risks, as well as on proving up reserves in the key producing countries. For example, security concerns and uncertainties about the impact of a new oil and gas law in Indonesia have recently put a brake on new investment by international oil companies. Because of economies of scale, new LNG-production projects in Indonesia and other countries will proceed only where dedicated reserves are large enough and production-sharing or general taxation rules are attractive. Downstream LNG investments in emerging markets such as India will hinge on the establishment of effective legal and regulatory frameworks and appropriate pricing terms.

The outlook for investment in the region will also depend on whether governments adopt policies and measures to address growing concerns about the impact of rising gas imports on energy security and the implications for greenhouse-gas emissions of rising fossil-fuel use generally. The outlook for nuclear power — especially in Japan — is also critical to gas demand and, therefore, the need for investment in gas infrastructure (Box 5.5).

India

Cumulative investment needs in the Indian gas sector are projected to total $44 billion, over the period 2001-2030. Average annual investment is expected to increase from $1.1 billion per year in this decade to $1.7 billion per year in the third decade. The increase in production from the current 22 bcm to 58 bcm in 2030 and the related exploration activities will require half of total gas investment. In late 2002, a new discovery in the offshore basin of Krishna-Godavari was announced, with 140 bcm of reserves. India holds significant reserves of coal-bed methane, amounting to around 400 bcm according to official government estimates. Nonetheless, India's domestic production is not expected to keep pace with demand, so India will need to start importing gas soon. Gas-import requirements are expected to reach 38 bcm in 2030. So far, tenders for exploration, development and production under the New Exploration Licensing Policy have not succeeded in attracting significant amounts of foreign capital, vital to raise production.

Several private companies are pursuing a number of new LNG projects, despite problems with financing, due largely to pricing of the gas in India and worries about consumer creditworthiness. The 5-Mt per year Dabhol terminal was almost completed when the construction was halted in 2001, due to a dispute over pricing. It is likely to be completed once a buyer is found for Enron's share of the project. RasGas is expected to supply LNG to Petronet when a 2.5-Mt terminal at Dahej is completed in early 2004. Its capacity could be doubled later.

The prospects for a rapid increase in LNG imports will depend critically on power- and fertilizer-sector reforms. These sectors will be the main consumers of gas, but they both sell their output at subsidised prices. The

Supply Trends and Prospects

WEO-2002 projects that the region's primary supply of gas will grow by an average of 1.5% per year from 2000 to 2030. As in most other regions, the biggest increase in gas use is expected to come from power generation. Gas demand will also increase in the residential, services and industrial sectors, but at more pedestrian rates. Production in the region is expected to continue to meet the bulk of demand, but imports into the United States — mostly in the form of LNG — will play a growing role (Figure 5.30). Aggregate gas production in the United States and Canada is projected to grow from 736 bcm in 2001 to a peak of around 860 bcm by around 2020 before beginning to decline to 842 bcm in 2030. Net imports of gas into the United States and Canada, which amounted to a mere 7 bcm in 2001, are projected to reach 109 bcm in 2010 and 371 bcm in 2030. Canada will continue to export large amounts of gas to the United States.

Figure 5.30: **Gas Supply in North America**

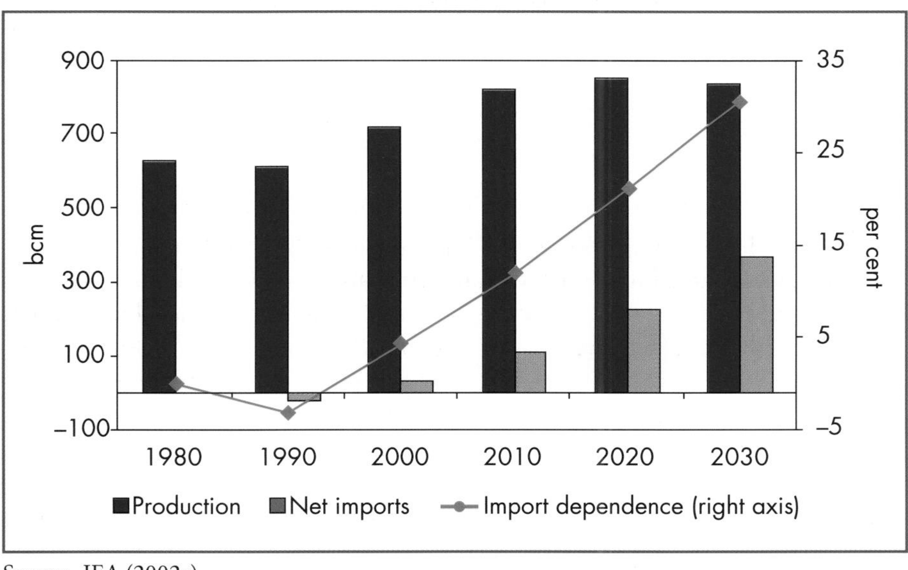

Source: IEA (2002a).

Investment Uncertainties

The prospects for North American gas production, demand and, therefore, the need for imports, remain very uncertain (Box 5.6). The costs of developing gas fields in existing and new basins and importing LNG will have a critical impact on the need for investment all along the gas-supply chain. Investment

Box 5.6: **North America Gas Supply and Investment Uncertainties**

WEO-2002 projects that North America will become a significant net importer of LNG over the next three decades. This projection differs substantially from the latest official projections of the US Department of Energy, which show much faster production growth and slightly faster demand increases.[33] *WEO-2002* projects gas imports of 109 bcm by the end of this decade, while the US DOE projects that imports will barely reach half this level by 2020 (Table 5.13).[34]

The DOE gas-production projections imply much greater reliance on non-conventional resources, such as tight gas, shales and coal-bed methane, and more costly conventional resources, both onshore and offshore. The differences in the projections are largely explained by the assumed rate of technology developments in the area of deep-water drilling and non-conventional sources, as well as the cost of developing conventional resources in the Rocky Mountains, in central and west Gulf of Mexico and Alberta. To reach the production levels projected by the DOE, investment requirements in domestic exploration and development and in transmission lines would be much higher than we project. On the other hand, the investment in new LNG import facilities would be substantially reduced.

Table 5.13: **North American Natural Gas Import Dependence**

	WEO		DOE	
	bcm*	%**	bcm*	%**
2000	24	3.3	26	3.5
2010	109	13.2	29	3.5
2020	228	26.7	53	5.2

* Net imports in bcm.

** Net imports as a percentage of primary gas supply.

Sources: DOE/EIA(2003c); DOE/EIA (2003e); IEA (2002a).

33. *Annual Energy Outlook* and *International Energy Outlook*: US Department of Energy/EIA (2003a and b).

34. A National Petroleum Council study, released on their website in September 2003, highlights the crucial importance of LNG imports to the US market. In their Balanced Future Case, LNG import capacity reaches 160 bcm, or 17% of US natural gas supply by 2025. The steep declines rates experienced recently in natural gas domestic production are another key highlight of the study. See NPC (2003).

risks might rise as focus shifts to larger-scale, more challenging upstream and transmission projects, which could raise the cost of capital and deter investment in some marginal projects.

Gas reserves, at 7 tcm or 3.9% of the world total,[35] remain modest relative to other regions. Reserves increased only by 4.5% in 2001, despite a leap of 58% in the number of exploration wells drilled to 961 — the highest annual total since 1985. In addition, the unit costs of gas production at existing basins, mainly in the southern and central United States and in western Canada, are rising in line with declines in the productivity of new wells as mature fields are depleted. The average size of new fields being developed is falling and the rate of decline in production from new wells in the first year after they start producing has accelerated from 17% in 1990 to 27% in 2002.[36] The average production of gas per well in the United States fell from 80,000 cubic feet in 1990-1995, to 67,000 cubic feet in 2001.[37]

As a result of these factors, US gas production has hardly increased in the last five years, despite a 64% increase in the completion of gas-development wells in the period 2000-2002 compared to 1997-1999. Wells less than three years old now account for more than half of total US production, compared with under 40% at the beginning of the 1990s. Average exploration and development costs are estimated to have risen from around $1/MBtu in 1996 to over $1.85/MBtu in 2002, while total exploration and production costs have increased from $1.95/MBtu to $2.55/MBtu.[38] Production from recently drilled wells in the Western Canadian Sedimentary Basin is also declining at much higher rates than from older wells and overall output has begun to stagnate.[39] The average rate of production decline from newly drilled wells reached just over 50% in 2000, compared with 18% in 1990.

Wild swings in gas prices have caused sharp fluctuations in exploration and development drilling activity in recent years, but there is growing evidence that the declining size and productivity of new wells and rising development costs are deterring drilling in mature areas. Small independent producers, who account for a large proportion of drilling, also find it difficult to gain access to credit, due to falling equity values, which may also be constraining upstream

35. Cedigaz (2003).
36. Data from John S. Herold cited in *World Gas Intelligence* (14 May 2003). A 2003 study by IHS Energy and Petroconsultants of oil and gas production in shallow waters in the Gulf of Mexico found that one-year decline rates for gas wells ranged from 25% to 100%, averaging 83%. Simmons (2002) also provides evidence of rising gas-well decline rates in Texas.
37. Almost 95% of all gas wells drilled in the United States in the ten years to 2002 were development wells. See EIA (2003a).
38. Data from John S. Herold cited in *World Gas Intelligence* (14 May 2003).
39. NEB (2002).

investment.[40] These factors suggest that production from these areas will, at best, hold steady in the coming years and could even fall significantly. Production from relatively undeveloped and new basins is expected to offset most of this decline. These basins include deeper-water locations in the Gulf of Mexico, tight-gas and shale-gas formations and coal-bed methane reserves, mostly found in the Rocky Mountains,[41] and new Canadian sources, such as offshore Atlantic basins in Labrador, Newfoundland and Nova Scotia and the Mackenzie River Delta/Beaufort Sea region in northern Canada. The Alaskan North Slope is expected to provide another major new source of gas, although developing these reserves, as well as those in the Mackenzie River Delta, will require the construction of a large-diameter, long-distance pipeline. Both projects are expected to proceed before 2020. In the longer term, gas reserves in frontier regions such as the Arctic Islands and the Northwest Territories are expected to be developed too. US production prospects would be boosted significantly if current restrictions on drilling on federal lands were lifted.

Capacity expansions at the four existing LNG terminals on the Gulf coast and East coast and potential investment in *new* terminals are expected to provide an increasingly important source of gas supply to the region in the long term. LNG could be imported directly, or via Mexico or the Bahamas if public opposition impedes the siting of terminals on US coasts. The Mexican government has already approved three LNG terminals, and others are planned.

Higher gas prices and technology-driven cost reductions in recent years have revived interest in LNG projects to supply the US market. More than 20 new terminals in the United States or Mexico are currently under consideration — mostly located in Louisiana, Texas or California — and a mothballed terminal at Cove Point in Maryland has been reactivated.[42] The combined capacity of these projects is almost 240 bcm per year. Three other terminals are already operating. At the beginning of 2003, the Federal Energy Regulatory Commission (FERC) gave preliminary approval for the first new LNG terminal in 25 years to be built by Sempra Energy in Louisiana. Many planned projects involve floating facilities sited a short distance offshore, connected by sub-sea pipelines. Several terminal projects have been proposed for the Bahamas, with the gas being transported via pipeline to markets on the US mainland. How much LNG-import capacity is ultimately needed is, however, very much dependent on indigenous production, demand and costs. The current cost of LNG imports from the closest Atlantic basin sources is well

40. *Petroleum Economist*, "Living with Market Volatility" (May 2003).

41. Coal-bed methane produced in the Rocky Mountains has accounted for over half of the overall increase in US gas production since 1990.

42. *Petroleum Intelligence Weekly*, "LNG Gets Ready for Second US Coming", 9 September 2003.

Figure 5.31: **Indicative Levelised Cost of LNG Imports into US Gulf Coast**

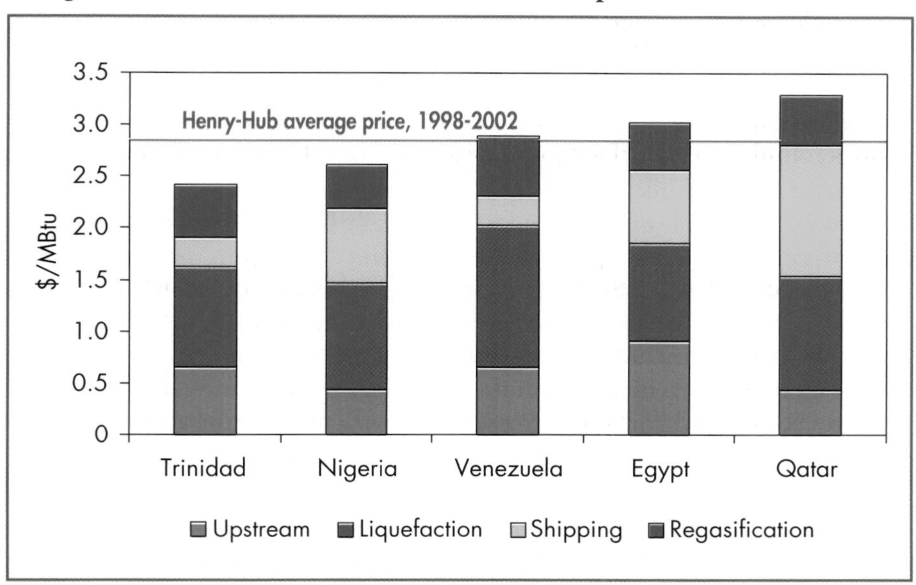

Source: IEA analysis based on data for planned projects from industry sources.

under $3/MBtu — below the average spot Henry Hub gas price for the last five years and much lower than recent prices of more than $5/MBtu (Figure 5.31).

Much of the new investment in transmission pipelines over the next three decades will be in looping and adding compression on existing systems and in replacing and refurbishing obsolete facilities. But more costly greenfield projects to bring gas from new supply sources to the main markets in the United States and eastern Canada will account for a growing share. For example, a 5,800-km, 35-bcm/year pipeline from Alaska proposed by BP and ConocoPhillips could cost as much as $18 billion. A separate line connecting fields in the Mackenzie Delta in northern Canada with existing pipeline systems in Alberta would cost around $4 billion. Major new investments in expanding transmission capacity from the Rocky Mountains are already being made to relieve system bottlenecks in moving the gas to California and Nevada. A project to double the capacity of the Kern River Transmission System to 1,800 mcf/d was completed in May 2003. Future transmission investment needs will nonetheless be mitigated to some degree by an expected increase in the average system load factor. Much of the increase in gas demand in the next three decades will come from the power sector, which has a relatively high load factor. An increase in the average load factor means that less pipeline capacity is needed relative to overall gas demand.

Until now, most individual upstream and transmission investments in North America have been relatively small-scale. For example, only 11 out of 54 US natural gas pipeline construction projects completed in 2002 cost in excess of $100 million and only one cost more than $1 billion (although several long-distance pipelines completed in 1999 and 2000 involved investments of over $1 billion).[43] Projects costing more than $100 million nonetheless accounted for the bulk of total investment. Most of these investments were carried out by a single operator and were typically financed out of cash earnings and corporate debt. In the case of most pipeline expansions, investment risks have been reduced by FERC and the National Energy Board in Canada adopting, in the 1990s, "rolled-in" tariff-setting principles, whereby the additional cost is included in the pipeline company's overall cost base for setting minimum revenue needs and tariffs. This practice lowers the tariffs that need to be charged for the incremental capacity and reduces the risk that throughput does not increase quickly enough in the early years of operation.

A growing share of capacity additions in the future will involve single investments of hundreds of millions or billions of dollars. Financing these investments will be less straightforward. Only the biggest oil and gas companies and gas pipeline companies with large cash reserves and access to cheap capital will be able to undertake these projects. Small and medium-sized exploration and production firms will be unable to carry the price risk inherent in large-scale upstream projects with much longer payback periods than are typical at present. Financing major pipeline projects requires most of the capacity to be reserved under long-term contracts. The most costly recent pipeline project was the Alliance pipeline from western Canada to the northeast United States, which was commissioned in 2000 and involved investment of around $2.9 billion. The trend towards larger-scale projects could encourage a further round of consolidation in the pipeline industry.

Earlier fears that the demise of long-term take-or-pay contracts, a result of deregulation of the inter-state gas industry, would discourage investment in large-scale grass-roots projects, such as the Alliance pipeline, appear to have proved unfounded. The degree of competition between pipeline companies to build major new lines suggests that the companies consider the returns that can be made from such investments to be attractive in relation to the risk. That risk is reduced by cost-of-service tariff regulation. Although rates of return on pipeline investments in North America have typically been lower than for industry generally, they have nonetheless been higher than for the pipeline companies' unregulated activities. According to a 1999 US Department of Energy study, the return on investment (operating income divided by assets) for the 14 largest US

43. EIA/DOE (2003d).

pipeline companies over the period 1990-1997 averaged 7.6% for transportation activities, compared with 4.2% for all other lines of business.[44]

Nevertheless, changes in pipeline capacity, contracting practices and the development of alternatives to simply reserving pipeline capacity are creating uncertainty about future demand for firm capacity and the ability of pipeline companies to recover their costs in regulated tariffs. Since the mid-1990s, a growing number of local distribution companies and marketers have been turning back part of the firm capacity they previously held under long-term contracts, when those contracts expired, and are relying more on secondary capacity markets. This poses a problem for pipeline companies in recovering pipeline investment costs, since pipelines usually have to discount the price of turned-back capacity to be able to market it to other shippers. Raising their tariffs to remaining customers to protect overall revenues may not be an option. If those customers are able to use other pipelines or storage facilities, it would lead to yet further reductions in capacity reservations and revenue losses. In some cases, pipeline investors have had to accept a reduction in their rate of return or retire early some assets. FERC has attempted to address this issue by offering pipelines more flexibility in tariff-setting.

The US and Canadian governments may decide to adopt measures to support future mega-projects, such as the Alaska-Lower 48 project, through targeted measures like accelerated depreciation and loan guarantees. The Canadian government has already deployed such techniques to promote frontier developments, including the Hibernia project off the coast of Newfoundland. Both governments have, however, rejected a proposal to introduce a special wellhead tax credit for when Alaskan gas prices fall below a pre-set floor, because of the market distortions that it would cause. Regulatory changes may also be used to promote new projects. In approving Dynergy's planned LNG terminal, FERC eased its requirement for open access. This ruling is expected to encourage investment in so-called "proprietary" terminals where the facility's owner controls LNG shipments. Major oil and gas companies have indicated that they would be much more likely to invest in new LNG terminals on this basis.

Although the investment in local distribution networks needed over the period 2001-2030, at $182 billion, is very large, financing that investment is not expected to be a major concern. Most of this investment will involve the extension or enhancement of existing networks. These are low-risk investments, given that demand trends are relatively stable and predictable and the state regulatory commissions protect investment returns by allowing the distribution company to pass through gas costs and recover operating costs, including depreciation.

44. DOE/EIA (1999).

OECD Europe

Investment in gas-supply infrastructure within Europe will decline over the next three decades, due to stagnating production and slower demand growth. Upstream investments will remain the largest component as costs rise with the depletion of reserves in the North Sea. Distribution networks will absorb a growing proportion of the region's investment. Investment in regasification terminals will be high in the current decade, but is expected to dip thereafter as costs fall. Financing will not be a major concern for domestic downstream investments, but might be a hurdle to investment in new large-scale cross-border pipeline and LNG projects, depending on cost developments, geopolitical factors and regulatory uncertainties.

Investment Outlook

Total gas-sector investment needs in Europe are projected to amount to $465 billion over the period 2001-2030, or almost $16 billion per year. Although most of the increase in supplies will come from imports, exploration and development are expected to account for close to half of total capital expenditure. This is mainly because development costs will rise as UK, Norwegian, Danish and Dutch gas reserves on the Northwest Europe Continental Shelf are depleted. This investment will, nonetheless, not be sufficient to prevent a gradual decline in production over the projection period. Decline rates are assumed to average 8% over the projection period.

Annual investment needs are expected to decline gradually after peaking in the current decade, as production stagnates and demand growth slows, reducing the need for new transmission capacity (Figure 5.32). Only distribution will see a rising trend in capital spending, due to increasing unit costs. Transmission and storage investment will increase in the current decade, but will decline thereafter as the increase in import capacity needs slows and technology drives down unit costs.

Supply Outlook

Natural gas will remain the fastest growing primary fuel in absolute terms in OECD Europe over the next three decades. Consumption is projected to grow by an average 2.1% per year from 2000 to 2030, driven mainly by rising power-generation demand. Demand growth will be highest in the decade to 2010, when gas will be most competitive against coal. Output from the North Sea — a mature producing region — and the rest of the Northwest Europe Continental Shelf is expected to peak in the middle of the current decade and slowly decline over the rest of the projection period. UK output is expected to decline steadily over the next three decades, and the country will become a major net importer of gas before the end of the current decade. Production in

Figure 5.32: **Gas Investment in OECD Europe**

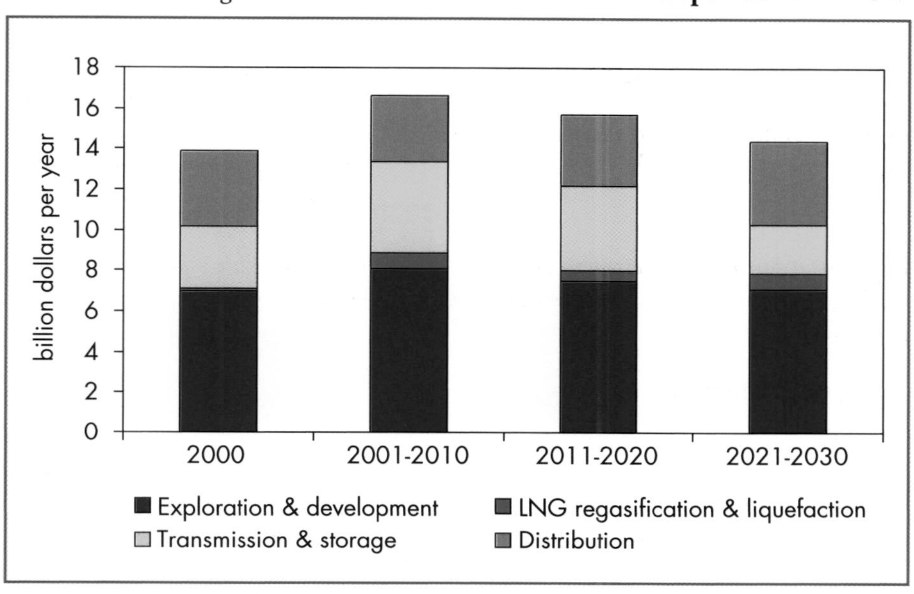

the Netherlands is also expected to begin to decline gradually in the near future. These declines will more than offset a continued rise in Norwegian output, net of reinjections. A new 22-bcm per year sub-sea line from the Ormen Lange field in Norway to the United Kingdom is due to be completed in 2007. Due to the decline in production and the projected increase in demand, net imports will have to grow dramatically. Gas imports are projected to rise from 180 bcm in 2001 to 625 bcm in 2030. Their share of total primary gas supply will increase from 37% to 69%.

Most of the region's gas imports are likely to come from its two main current suppliers, Russia and Algeria (Figure 5.33). The rest will probably come from a mixture of piped gas and LNG from elsewhere, including Libya (via pipeline), Nigeria, Trinidad and Tobago, Egypt, Qatar and possibly Iran (LNG). Venezuela may also emerge in the longer term as a bulk supplier of LNG, while spot shipments of LNG from other Middle East producers may also increase. Pipelines from Iraq, Iran and the Caspian region could play an increasingly important role in the longer term. Gas reserves in these countries are more than adequate to meet Europe's needs, but the unit costs of getting that gas to market will probably rise as more remote sources are tapped. Piped gas from North Africa and the Nadym-Pur-Taz region in Russia are the lowest-cost options, but supplies from these sources will not be sufficient to meet

Figure 5.33: **Gas Supply in OECD Europe**

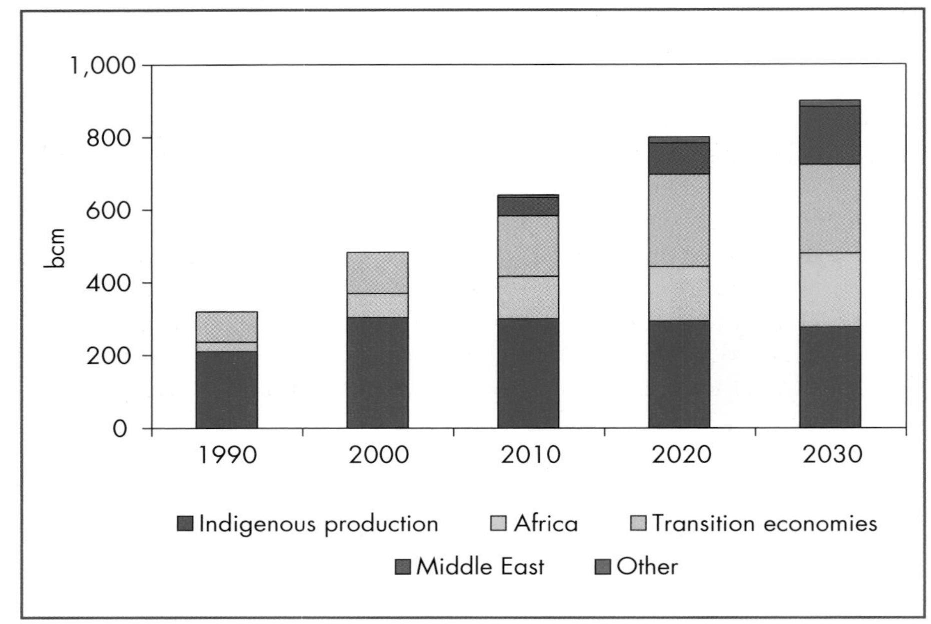

Source: IEA (2002a).

projected demand after 2010.[45] LNG is expected to account for a rising share of total imports. The region will also become a small exporter of LNG, when a liquefaction plant supplied with gas from the Snovhit field in Norway is completed. The plant, which will supply annually 5.7 bcm of gas to US and EU markets, is due to begin operating in 2006.

Investment Uncertainties

Financing expansions of local distribution, national transmission networks and interconnectors is not expected to be a major hurdle to investment in most European countries. Existing operators will most likely continue to finance most new investment, mainly out of operating cash flows. Tariffs for the use of transmission and distribution systems are generally regulated on a cost-of-service basis, effectively providing a high level of assurance to the investor that he will be able to recover his capital and operating costs for as long as demand for capacity is high enough. In addition, the European Union provides financial and other forms of assistance for new gas-infrastructure projects in member

45. See CEC (2003) for a detailed analysis of the cost of supplying gas to Europe.

states, as well as in transit countries in Central and Eastern Europe under the Trans-European Networks programme.

There is more uncertainty about investment in projects to bring gas to Europe beyond the current decade. This relates essentially to supply costs, geopolitical factors and the impact of liberalisation on financing large-scale projects. Over 95% of Europe's projected gas needs in 2010 are covered by existing gas-import contracts and projected indigenous production. There are sufficient undeveloped reserves located close to European borders or to existing pipeline infrastructure to make up the supply shortfall at relatively low cost. Indeed, there is a danger that some countries, notably Turkey, Italy and Poland, may be oversupplied with gas in the near term, as demand has risen less rapidly than forecast. But beyond 2010, the cost of gas is expected to rise significantly, as the region has to turn to new, more distant sources to replace existing supplies and meet rising demand. New sources in Russia and the Caspian region, for example, will be much more expensive to develop and transport to Europe.

While technological advances are expected to lower the unit cost of high-pressure offshore transmission pipelines and of LNG liquefaction, shipping and regasification, the pace of cost reductions and their impact on the relative economics of pipeline and LNG projects are highly uncertain. The most promising pipeline projects include the Baltic pipeline, which would bring Russian gas to Northern Europe (see Russia section), and new direct sub-sea lines from Algeria to Spain, and from Libya to Italy. There has recently been a surge of interest in building LNG regasification terminals in Europe (Table 5.14). Plants under construction and planned would raise import capacity from 51 bcm per year at present to at least 118 bcm by the end of this decade.

Geopolitical factors will become increasingly important as attention shifts to potential projects in the Middle East and the Caspian region. A more stable political environment in those regions and enhanced relations with Europe would lower investment risk and the cost of capital, making it more likely that investments will be made and reducing the upward pressure on gas prices. Pipeline projects are particularly vulnerable to perceptions of political risk. Long-distance pipelines need to have high capacities to be economic because of large economies of scale. As a result, any single project involving the piping of gas from the Caspian region or the Middle East to Western Europe will involve investment of several billions of dollars.

Uncertainties about how the regulatory framework will change as market reforms progress and about the impact of the development of competitive markets may impede the financing of large-scale import projects. Gas-to-gas competition has developed much less in continental Europe than in Great Britain. A second EU Gas Directive, agreed in 2003, is intended to accelerate the

Table 5.14: **New LNG Import Projects in OECD Europe**

Country (location)	Start-up	Capacity (bcm/year)	Investment ($ millions)	Operator
Under construction				
Portugal (Sines)	2003	Phase 1: 2.4	290	Transgas
	-	Phase 2: 4.8	-	
Spain (Bilbao)	2003	2.7	238	Bahia de Bizkaia Gas
Italy (Adriatic Sea)	2005	4.6	600	Edison, ExxonMobil
Spain (Valencia)	2005	5.5	-	Union Fenosa
Turkey (Aliaga)	Stalled	4.0	-	-
Planned				
Spain (El-Ferrol)	2004	2.8	200	Reganosa Group
France (Fos-sur-Mer)	2006	8.2	350-502	GDF
Italy (Brindisi)	2006	Phase 1: 4.1	390	BG Group
		Phase 2: 8.2	-	
UK (Milford Haven)	2006	-	-	Petroplus
UK (Milford Haven)	2007	18	1,600	ExxonMobil
UK (Isle of Grain)	-	4.5	-	Transco

Source: IEA databases.

process. The directive extends choice of supplier to all customers by 2007, requires each member state to set up a regulator with well-defined functions and obliges gas companies to unbundle legally their accounts and publish network tariffs. The directive also allows member states to apply to the Commission for a temporary derogation from certain provisions, where their implementation would undermine investment in a geographically limited area.

For major cross-border pipelines and new LNG terminals in continental European markets to obtain financing, long-term contracts will remain necessary in most cases, at least until such time as liquid spot and futures markets are well established, as in Great Britain. Gas merchants will nonetheless push for shorter terms, less onerous take-or-pay conditions and indexation based on gas rather than oil prices. The second Gas Directive permits national regulators to exempt LNG regasification capacity from third-party access requirements under certain conditions. Several national regulators, including Ofgem in Great Britain, have signalled that they will consider favourably applications for exemptions from LNG terminal developers.

The prospect of a significant increase in Europe's dependence on imports from Africa, the former Soviet Union and the Middle East and uncertainties about the impact of liberalisation of EU energy markets on investment are driving a number of initiatives to strengthen political and commercial ties. These include the formal energy dialogue between the European Union and Russia and European support for the Energy Charter Treaty (see the discussion on Russia earlier in this chapter). The European Union also helps to promote and finance gas-supply projects in non-EU Mediterranean countries that export to Europe or could do so in the future. Several European gas companies are strengthening their commercial ties with Gazprom. The German government's support for the E.On-Ruhrgas merger was motivated partly by the strengthened financing capability that the merger would give to Ruhrgas for investments in Russian and other gas-supply projects.

Figure 5.34: **Cumulative Gas Investment in EU15 in the Reference and Alternative Policy Scenarios, 2001-2030**

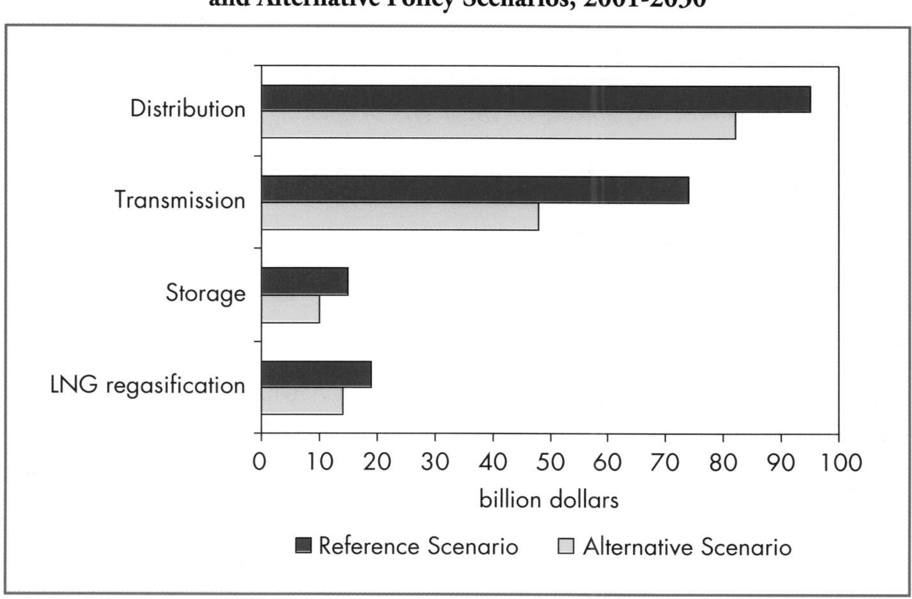

Note: The analysis assumes no change in indigenous production, such that imports fall in line with demand.

In the *WEO-2002* Alternative Policy Scenario, which takes into account more rigorous market intervention, cumulative investment needs for the downstream gas sector over the period 2001-2030 are significantly lower. This is mainly due to lower demand in power generation and consequently a reduced requirement for new transmission capacity (Figure 5.34). Investment in local distribution networks is also lower due to slower demand growth. This scenario takes into account the impact of policies that EU countries[46] are currently considering in order to address concerns about the impact of rising gas imports on energy security and the implications of rising fossil fuel use generally for greenhouse-gas emissions. These include the full implementation of the EU Renewable Energy Directive, which sets a target of 22% for the share of electricity to be generated from this source by 2010, compared with under 14% in 1997. Policies to promote combined heat and power and the penetration of highly efficient combustion and transformation technologies are also taken into account. Gas-import needs in EU15 would be 512 bcm in 2030, compared with 632 bcm in the Reference Scenario, cutting the region's import dependence from 81 to 77%.

Latin America

Investment in Latin American gas-supply infrastructure will grow steadily to meet rising demand and export volumes. Upstream investments will remain the largest component, but the development of transmission networks will absorb an increasing share of investment. Domestic demand and exports are set to rise significantly. Upstream financing will require large inflows of foreign capital. This might be problematic, depending on the degree of opening of the sector and on the political stability of the region. Cross-border pipeline and LNG projects will come on stream if a clear and stable regulatory framework is set and co-operation among countries develops fruitfully.

Investment Outlook

Cumulative investment needs in the Latin American gas sector are projected to total $247 billion, or more than $8 billion per year, over the period 2001-2030, amounting to 8% of global gas investment. Upstream development will absorb more than half of total capital flows, transmission and distribution pipelines accounting for another 36% (Figure 5.35). Investment in exploration and development was $2.1 billion in the year 2000. It is expected to rise steadily to $2.8 billion per year in the current decade and $6.8 billion per year in the third decade. Rising spending on E&D will be needed to sustain increasing production. Higher unit production costs will also add to capital needs, as a growing share of output will come from offshore fields.

More than $49 billion, or $1.6 billion a year, will be needed over the next three decades to build and expand cross-border pipelines and national transmission lines. The average annual capital expenditure is expected to increase over time. Spending will average about $1 billion per year during the current

46. EU15 only.

Figure 5.35: **Cumulative Gas Investment in Latin America, 2001-2030**

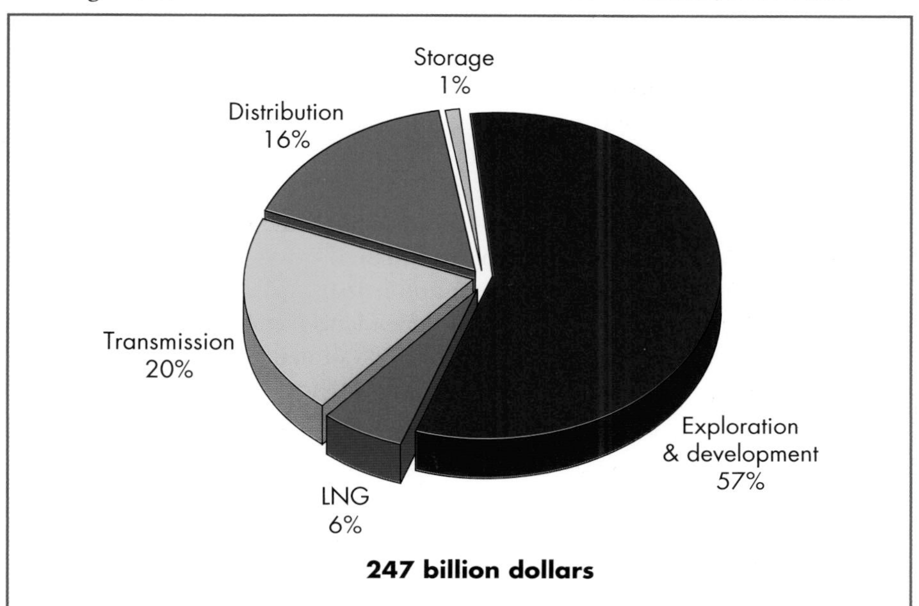

decade, compared with $700 million during the past decade. Investment will reach an average of $2.3 billion per year in the third decade, due to more technically challenging projects, with high per unit cost, and the faster expansion of the transmission network. The development of the domestic distribution network and underground gas storage will call for an additional $41 billion, or $1.4 billion per year. Some Latin American countries will invest heavily in LNG export facilities. Most of the LNG will go to the North American market and to importing countries in the region. Investment in liquefaction plants is expected to total $15 billion over the period 2001-2030.

Supply Trends and Prospects

Latin America's proven natural gas reserves amounted to 7.2 tcm in 2002[47], 4.5% of the world total. Probable and possible reserves could add another 6 tcm. Brazil's deep offshore fields are thought to have large potential and recent large discoveries in Bolivia and Trinidad and Tobago suggest that intensified exploration could lead to a substantial increase in Latin America's proven reserves.

47. Cedigaz 2003.

Venezuela holds 58% of the proven reserves, followed by Bolivia (11%), Argentina (11%) and Trinidad and Tobago (8%). Latin American natural gas production in 2001 was 102 bcm. However, gross production was much higher since large quantities of gas were reinjected and flared.

Production is expected to expand significantly over the next three decades, reaching 516 bcm in 2030 (Table 5.15). The share of offshore production is expected to climb from the current 20% to 32% in 2030. The growth in domestic and international demand will drive this expansion. Domestic demand is expected to grow fast, from 101 bcm in 2001 to more than 370 bcm in 2030. Demand in the power generation sector will account for more than half of this increase, spurred by a need in many countries to reduce dependence on hydropower. Brazil is expected to lead the growth, with a spectacular 7% annual average demand growth over the next thirty years. Brazil will account for 20% of the region's gas demand in 2030, and will play a pivotal role in the region's gas infrastructure evolution.

Table 5.15: **Production and Net Exports in Latin America** (bcm)

	2001	2010	2020	2030
Production onshore	81	168	248	352
Production offshore	20	49	91	163
Demand	101	167	251	373
Net exports *	0.9	50	88	143

* Net exports include GTL.
Source: IEA (2002a).

The development of gas pipeline interconnections is most advanced in the Southern Cone.[48] Both Argentina and Bolivia have abundant non-associated gas reserves that they are eager to export to neighbouring countries. More than $7 billion have been invested in transmission pipelines over the past 10 years, including the $2.1 billion Bolivia-to-Brazil pipeline and the first stage of the $250 million Argentina-to-Brazil pipeline. Several new pipelines are planned or under construction (see Table 5.16), providing the basis for a sub-regional gas transportation network.

Large gas reserves in the North offer potential for LNG projects. Trinidad and Tobago operates a three-train liquefaction plant (9.6 Mt per year) and exported 5.4 bcm of LNG in 2002, mainly to the United States. A fourth train is being built and a fifth is planned. Venezuela has enough gas reserves to become a major LNG exporter, but projects for liquefaction plants have been stalled by a combination of poor economics and lack of political support.

48. The Southern Cone encompasses Brazil, Argentina, Chile, Bolivia, Paraguay and Uruguay.

Table 5.16: **Main Pipelines under Construction and Planned in Latin America**

	Length (km)	**Diameter** (inches)	**Capacity** (bcm)	**Investment** ($ million)
Under construction				
Argentina-Uruguaiana Porto Alegre	615	20	4.4	260
Argentina-Uruguay (*Cruz del Sur*)	208	24	2.4	120
Planned				
Argentina-Brazil (*Mercosur*)	3,100	36	9.1	1,800
Argentina-Brazil (*Trans-Iguacu*)	n.a.	n.a.	12.0	n.a.
Bolivia-Chile (*Mercosur*)	850	20	2.2	285
Bolivia-Argentina-Paraguay-Brazil (*Gasin*)	5,250	n.a.	n.a.	5,000
Peru-Bolivia	900	36	14.6	900
Peru-Brazil	3,550	32	11.0	3,215
Venezuela-Colombia	200	n.a.	2.1	120

Source: IEA databases.

Bolivia is also investigating using some of its reserves for an LNG plant on the coast in Chile or in Peru, which would allow exports to the United States or Mexico. Peruvian gas from the giant Camisea field might also be exported as LNG. The region's LNG exports are expected to grow rapidly over the next three decades, potentially reaching 90 bcm in 2030. This would require more than $15 billion in LNG liquefaction plants alone.

Financing and Investment Uncertainties

One of the main uncertainties surrounding the pace of development of the transport infrastructure and, therefore, investment levels is the rate of growth of gas demand in the region, especially in Brazil. Regulatory uncertainties, gas pricing issues and the difficulties of introducing gas-fired plants have delayed thermal power projects over the past few years and might continue to do so in the future. Similarly, investment in LNG liquefaction plants will depend critically on the outlook for prices and demand in the key

US market. Another crucial factor will be the relative economics of LNG shipments and cross-border pipelines within Latin America: LNG regasification terminals in the region may be a viable option in areas where it would be too costly to extend the transmission lines, such as Suape in Brazil. The Dominican Republic has recently started a 3 bcm per year regasification terminal. Honduras and Jamaica are considering building regasification facilities too. LNG projects in Latin America, however, will have to be competitive with GTL projects and LNG projects outside the region.

Latin America has succeeded in attracting relatively large amounts of foreign direct investment in recent years, with most of it going to Brazil (see Chapter 3). Brazil succeeded in securing project financing for several large investments, including the $850-million Cabiunas gas-processing plant, the $1.2-billion EVM deep-water development and the $2.5-billion Barracuda and Caratinga oil and gas fields. Other countries, notably Bolivia and Argentina, had been successful in bringing in foreign investment in the gas industry, prior to the recent financial crisis in Argentina. But even before that, poorer countries struggled to attract investment in their upstream industry, because of political uncertainties, devaluations, banking crises, high unemployment and social unrest. Macroeconomic and political stability, together with moves to establish transparent, efficient and stable legal, regulatory and institutional frameworks governing the gas and other energy industries will be vital to reviving domestic and foreign capital flows. Gas-sector policies will need to be integrated with electricity policies, as gas-to-power projects are the key to ensuring the financial viability of the gas chain.

CHAPTER 6: COAL

HIGHLIGHTS

- Some $400 billion needs to be invested in the coal industry, globally, over the period 2001-2030: 88% in coal mining, 9% in shipping and 3% in ports. This will fund growth in global coal supply from 4,595 Mt in 2000 to 6,954 Mt in 2030. International trade in coal will increase faster than coal production, from 637 Mt in 2000 to 1,051 Mt in 2030.
- Investment is split evenly between developing countries and the rest of the world. Investment in OECD countries is expected to be $131 billion.
- The key uncertainty facing future coal demand and investment is environmental policy. This uncertainty is discouraging investment. In the OECD Alternative Policy Scenario, new environmental policies cut global investment in coal by $25 billion, as demand relative to the Reference Scenario falls by 524 Mt (7.5%). Investment in the OECD is around 11% lower in 2030.
- Production of one unit of energy of coal is about one-fifth as capital-intensive as producing the same unit of energy in oil and one-sixth as capital-intensive as gas.
- Robust inter-fuel competition and the highly competitive nature of the international coal market will keep coal investment and production costs down. Productivity tends to be lower and production costs higher in countries where the domestic industry is subsidised or otherwise shielded from competition.
- China accounts for 34% ($123 billion) of the global investment needs (excluding shipping) over the period 2001-2030. China needs to raise mechanisation and productivity levels, improve safety standards, reduce distorting social requirements and subsidies, and continue investing in infrastructure if it is to meet demand growth.
- The United States and Canada will remain the second-largest coal market in the world and account for 19% ($70 billion) of world investment (excluding shipping). Nearly all this will be required to maintain production capacity, which will grow in the west of the United States as production in the east declines.
- India will require investment of $25 billion over the *Outlook* period. Its financing will require continuing efforts to improve the profitability of electricity generators, the railways and the coal industry.

- Power generation will account for about 90% of global coal demand growth. Investment in the coal industry and in coal-fired electricity generation plants combined is expected to be around $1.9 trillion. Including the fuel chain in the capital requirements for electricity generation reduces the capital cost advantage of a gas-fired power station over coal by about half.
- Most coal investment is likely to be funded from retained earnings or on the strength of the balance sheet. Project financing is likely to play a much smaller role than in the oil, gas and electricity sectors. Foreign investment will be required in developing countries to meet demand growth, but will only occur if the investment environment is attractive to foreign capital. It will facilitate technology transfer and the opportunity to close the productivity gap and lower costs.
- The risk that the needed investment may not be forthcoming is greatest in countries where ownership remains in government hands (notably India and China).
- Advanced coal-fired power generation and carbon sequestration technology could greatly enhance the prospects for coal investment. Clean coal technologies in conjunction with carbon capture and sequestration, if their costs fall, could allow coal to continue to provide low-cost electricity in a carbon-constrained environment. However, the impact of these new technologies is limited in the scenarios considered here.

The first section of this chapter summarises the results of the investment analysis and assesses the need for new capacity in coal production, ports, and shipping. The second section reviews developments in costs and technologies and assesses the risks and uncertainties facing investment in the coal industry. The following section provides an analysis of investment needs and issues by region. Finally, the impact on investment of the World Energy Outlook 2002 *OECD Alternative Policy Scenario is presented.*

Global Investment Outlook[1]

Overview

Cumulative global investment needs in the coal industry are expected to be just under $400 billion over the period 2001-2030, including mining, shipping and ports. This investment is needed to replace production

1. The preparation of this chapter benefited from the assistance of the IEA Coal Industry Advisory Board.

capacity that will close during the projection period and to meet rising demand and trade. World coal demand grows from 4,595 Mt (2,356 Mtoe) in 2000 to 6,954 Mt (3,606 Mtoe) in 2030 (Table 6.1).

Mining investment over the period 2001-2030, at around $351 billion, represents 88% of projected investment in the coal industry (Figure 6.1). Cumulative coal-related global investment required in the bulk-dry cargo fleet is $34 billion (9%) and $13 billion (3%) in ports. OECD countries will account for 36% of the global investment in mining and ports. Around 8% of the investment in non-OECD countries is needed to supply the exports of coal from non-OECD countries to OECD countries.

Figure 6.1: **Cumulative World Coal Investment, 2001-2030**

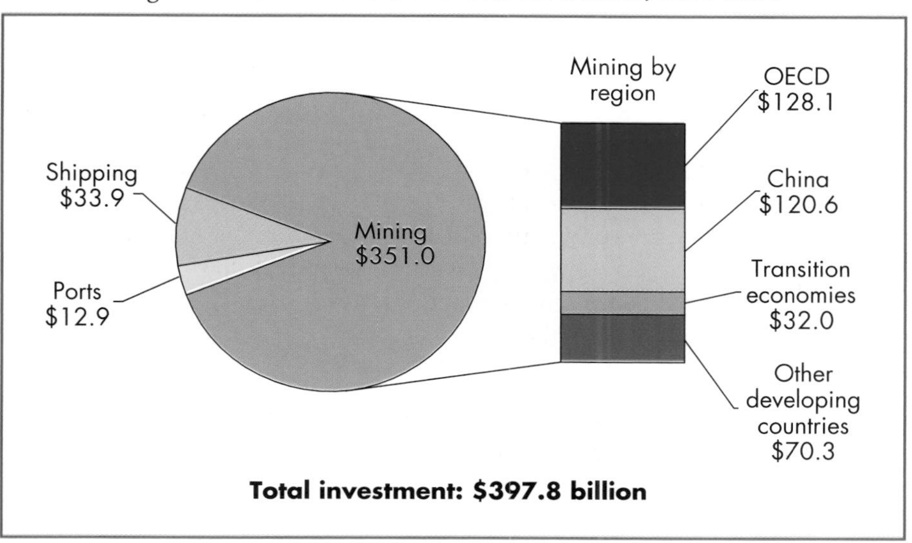

The world's coal resources are much more widely dispersed than those of oil or gas, allowing a much closer geographic fit with demand. Nonetheless, a number of factors have provided a stimulus to international coal trade, including highly competitive international markets, growing demand for high-quality coal, a narrower distribution of coking coal reserves, growth in demand for coal in high-cost producing areas and electricity-market liberalisation. Coal trade is projected to grow by 65% between 2000 and 2030, faster than total production.[2] This is driven by dramatic import growth in East Asia[3] and Korea

2. Unlike oil and gas, coal trade includes all cross-border coal trade between nations.
3. See Annex 3 for the regional definitions.

Table 6.1: **Summary of World Coal Production, Trade and Investment**

	Volumes			Investment	
	2000	**2030**	**Additions needed**	**Average investment cost** ($/unit)	**2001-2030** ($ billion)
Production (Mt)	4,595	6,954			
Expansion (Mt)			*2,588*		*91.3*
Replacement (Mt)			*3,603*		*117.2*
Total new capacity			6,191	*34 $/t. capacity*	208.5
Sustaining investment				*0.8 $/t. produced*	142.5
Total mining					**351.0**
Total trade (Mt)	637	1,051			
Port capacity					
exports (Mt)			457		
imports (Mt)			402		
Total port			859	*15 $/t. capacity*	**12.9**
Shipping capacity (M DWT)	67	116			
Expansion (M DWT)			49		
Replacement (M DWT)			59		
Total shipping			108	*314 $/DWT capacity*	**33.9**
World total					**397.8**

Box 6.1: **Mining Investment Categories**[4]

To project investment needs for coal mining, two types of investment were identified:

- **New production capacity investment** — this is investment needed to make available new production capacity, either in new mines (greenfield developments) or at existing mines (brownfield/expansion developments). It involves a one-off cost that is incurred in order to

4. See Annex 2 for a description of the model framework. Although the mining investment assumptions include investment in new railway infrastructure to connect to the existing rail network, no estimate was able to be made of investment in the existing rail network.

make production capacity available. This category includes both surface and underground facilities, machinery, and infrastructure. The amount of new capacity needed each year for a given country or region is determined by the annual change in production and the annual rate at which productive capacity from existing mines is closed. This also allows new capacity investments to be split into those needed to replace exhausted capacity and those to meet any increase in demand.

- **Sustaining investment** — this is the investment that is required on a regular basis in order to allow production at existing mines to continue until the mine's economic reserves are depleted and production stops. This category includes replacement of machinery and equipment, accessing new reserves within a mine, extending infrastructure, etc. It may result in an improvement in labour productivity. The amount of investment is a function of production, rather than capacity.

and the continued removal of producer subsidies and protection which will shift production to the most competitive sources of new supply. This trend has an important impact on the location and size of investment requirements.

Coal Investment by Region and Category

China is expected to account for 34%[5] (around $123 billion) of the total global investment required in the coal industry through to 2030 (Table 6.2). This is driven by significant growth in coal-fired power generation to help meet strong electricity demand growth and also by China's relatively high new mine development costs. Significant investment will be required in port facilities in China to handle increased coal exports, growth in internal trade and imports.

The North American domestic coal market is expected to remain the second-largest coal market in the world during the next three decades, although the influence and share of US and Canadian coal producers in export markets will continue to decline. Investment in the US and Canadian coal industry is projected to account for 19% of the global total (around $70 billion). Most investment will be needed in order to maintain production, with a relatively small amount required to expand production to meet demand growth. Investment in the eastern United States will remain significant, despite an anticipated decline in production, as virtually all mines currently in production in this area will close within the next 20 to 25 years.

5. All regional shares of the world total are calculated using only investment in mining and ports ($364 billion) and exclude shipping, as shipping investment is not attributed to individual regions.

Table 6.2: **World Cumulative Coal Investment by Region, 2001-2030**

	Mining	Ports	Total
	Cumulative investment 2001-2030 ($ billion)		
OECD North America	70.4	0.2	70.6
US and Canada	*70.0*	*0.2*	*70.1*
OECD Europe	24.9	0.2	25.1
EU15	*10.1*	*0.1*	*10.3*
OECD Pacific	32.8	3.0	35.7
Japan, Australia & NZ	*32.7*	*2.2*	*34.9*
OECD total	**128.1**	**3.4**	**131.5**
Transition economies	**32.0**	**0.3**	**32.4**
Russia	*13.1*	*0.3*	*13.4*
China	120.6	2.1	122.7
East Asia	15.4	4.4	19.9
Indonesia	*9.8*	*1.8*	*11.6*
South Asia	24.4	0.8	25.2
India	*24.1*	*0.8*	*24.9*
Latin America	8.6	1.2	9.8
Brazil	*0.5*	*0.1*	*0.6*
Middle East	0.1	0.1	0.2
Africa	21.8	0.4	22.2
Developing countries	**190.9**	**9.2**	**200.1**
World total	**351.0**	**12.9**	**363.9**
World total with shipping			**397.8**

Investment in the OECD Pacific region is dominated by Australia, with little investment required in Japan, Korea and New Zealand. Australia is currently the largest hard coal exporter in the world and has a highly competitive export coal industry. Australia's exports are expected to grow strongly as the country increases exports to all its current export markets, except Japan,[6] and will supply 55% of the increase in the imports of East Asia

6. This is due to the expected decline in Japan's imports and a slight decline in Australia's market share of Japan's imports.

and Korea. Investment in the OECD Pacific region is expected to be around $36 billion, or 10% of the coal industry's global investment between 2001 and 2030. Port investment in this region is expected to be $3 billion, with $2.2 billion in Australia. The OECD Pacific's share of global port investment will be relatively high, at around 23%, because much of the growth in production is for export and Korea will require significant new import facilities.

India is expected to account for 7% (around $25 billion) of global investment in coal mining and ports over the period 2001-2030. Despite its large coal reserve base, India faces many challenges in meeting future demand growth from domestic sources. These include low productivity, poor coal quality and the poor financial state of the coal industry and power sector. Growth in imports and in the internal shipment of coal is modest, requiring an additional $0.8 billion of investment in ports over the next three decades.

Investment in Africa will account for 6% (around $22 billion) of the global cumulative investment in the coal industry over the three decades to 2030. South Africa will account for virtually all of this investment. South African export growth is modest, reflecting the fact that export-quality coal in the most established coal-producing area is being rapidly depleted. There will be a need for significant investment to meet growth in domestic coal consumption for electricity generation and to maintain and expand exports. The expansion of export facilities at ports, most likely at Richards Bay, will be modest because of the relatively small growth in exports expected.

The modest investment requirements in OECD Europe, at 7% of the world total ($25.1 billion), reflect the reduction in production that will result from the phasing-out of subsidies. Hard coal production is likely to decline significantly from 2000 levels in the United Kingdom, Germany, Poland and Spain, and will soon be phased out entirely in France. However, some brown coal production will remain competitive, even without subsidies, and this will require investment in order to replace depleted mine capacity and to boost productivity to ensure the industry remains competitive with natural gas and hard coal imports. There will be minimal additional investment in import facilities at ports.

Investment requirements in Indonesia (around $12 billion) and Latin America (around $10 billion) together amount to 6% of the global total. These investments are modest relative to the world total, but are very large relative to the current size of their industries. They are expected to significantly increase their share of world trade (from a low base) over the projection period.

In the OECD, where primary coal demand growth is generally low and there is a high emphasis on productivity, sustaining capital investment exceeds that needed to replace depleted mines and meet demand growth (Figure 6.2).

Figure 6.2: **World Cumulative Coal Mining Investment by Region, 2001-2030**

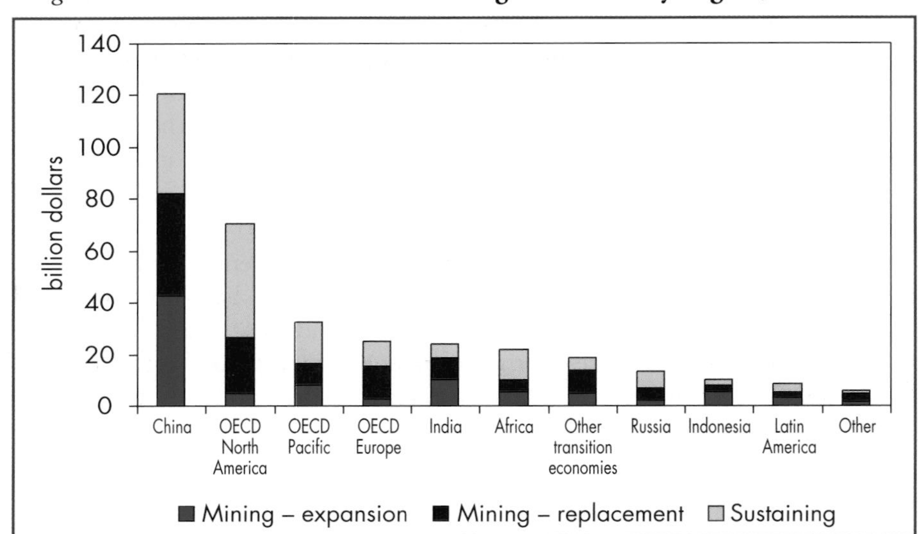

By contrast, in China, 68% of the mining investment is needed in order to add new capacity (Figure 6.3). This is accounted for by the dramatically higher production increase than in the OECD (around 2.8 times larger in

Figure 6.3: **World Cumulative Coal Mining Investment by Type, 2001-2030**

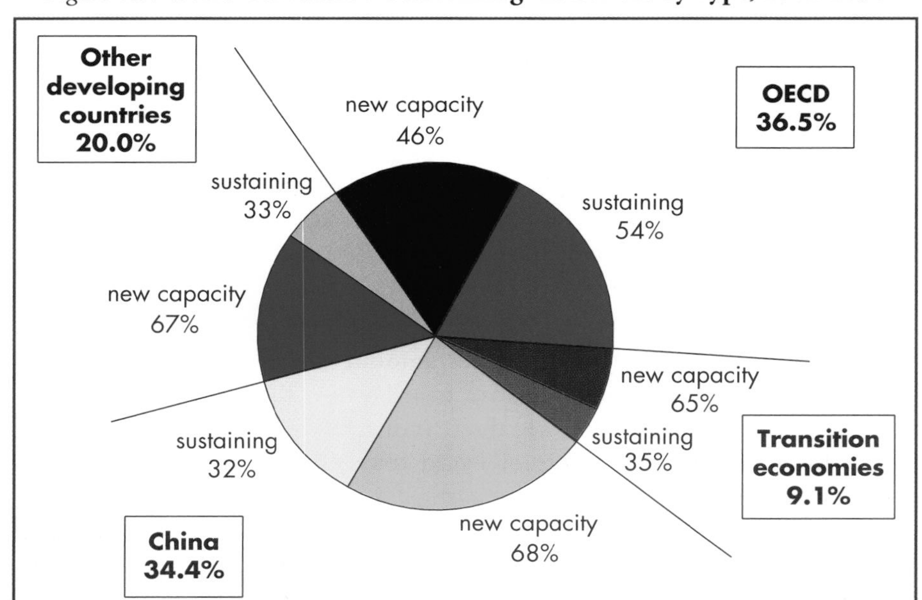

tonnes) and the lower emphasis on raising productivity. In the remaining developing countries and the transition economies, the share of mining investment for new capacity requirements is slightly lower than in China. This reflects lower production increases over the *Outlook* period and, in cases such as Colombia and South Africa, a higher share of sustaining investment to remain competitive in the export market.

Investment in Shipping and Ports

Global port investment needs will be modest at around only $13 billion over the *Outlook* period. Just over half of the new capacity required is for coal-handling facilities for exports. The importing regions of East Asia (excluding Indonesia) and Korea will require investments totalling $3.4 billion (27% of the world total) in import-handling facilities (Figure 6.4).

Figure 6.4: **Coal Port Investment by Region, 2001-2030**

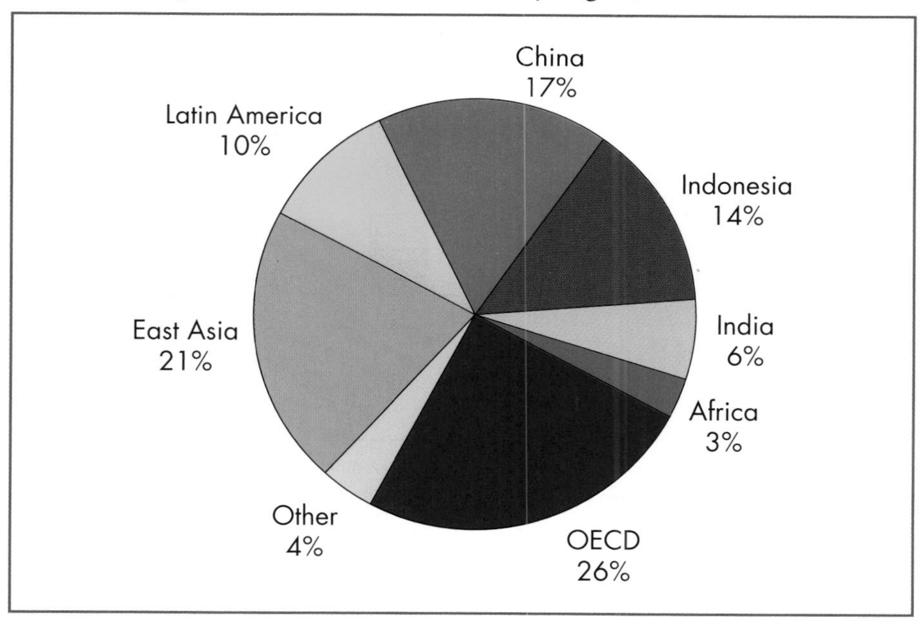

Total cumulative investment in the bulk-dry cargo fleet required for coal trade is expected to be around $34 billion. This will be driven by growth in the world's international seaborne coal trade of 419 Mt between 2000 and 2030 and by growth in the internal shipping requirements of India, China and Indonesia of 105 Mt. World seaborne coal trade is projected to reach

977 Mt in 2030.[7] The combined internal trade of India, Indonesia and China will reach 246 Mt in 2030.

Coal Investment and Electricity Generation

Most of the growth in demand for coal over the projection period will come from the electricity sector, where coal is in keen competition with gas. Cumulative investment in coal-fired electricity generation plants over the *Outlook* period is expected to amount to almost $1,480 billion (see Chapter 7). This is almost four times higher than the total investment in the coal industry over the *Outlook* period.

Figure 6.5: **Indicative Capital Requirements of Electricity Generation from Gas and Coal, 2010**

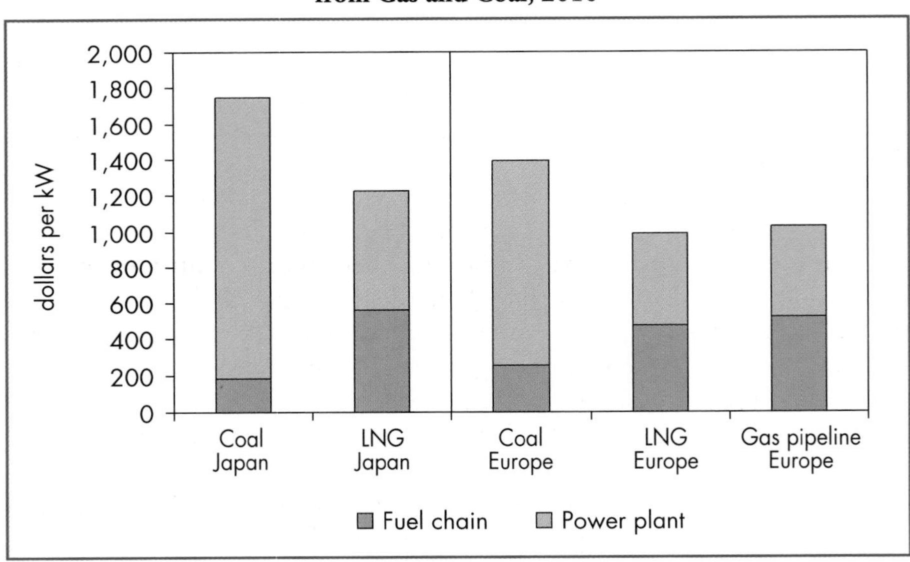

Note: Fuel chain capital costs are a weighted average of the capital requirements of imports (production through to delivery at the power station) in 2010. It therefore includes ports, shipping and pipelines where appropriate. Power plants are CCGT and conventional coal-fired plant.

Figure 6.5 compares the capital cost of adding an extra kW of electricity capacity in 2010, using coal or gas, including the capital requirement of the fuel chain of coal and gas. Including the capital requirement of the fuel chain in this way reduces the advantage of gas over coal in terms of the total capital required; but the total capital requirement for gas is still around 30% less than

7. World coal trade covers all coal trade, including secondary products such as coke oven coke.

coal. The levelised cost of producing a unit of electricity from coal in 2010 is expected to be around 18% higher than from gas in the EU15 and 8% higher in Japan.

The fuel chain needs of gas account for around half of the total capital requirement for gas-fired electricity generation in both Japan and the EU15, while for coal it varies between around one-tenth and one-fifth in Japan and the EU15 respectively. In Japan, the capital requirement of the fuel chain for gas in generation is around three times higher than for coal, whereas it is only twice as high in the EU15.

Coal Demand, Production and Trade

WEO-2002 projects global coal demand to grow by 1.4% per annum between 2000 and 2030. The fastest growth is experienced in developing countries, notably in South Asia, East Asia and Latin America, while demand declines in the OECD Pacific and OECD Europe regions. In absolute terms the primary energy demand for coal grows by the largest amount in China, India, East Asia, and OECD North America. Around 88% of the growth in the primary demand for coal will come from electricity generation.

Higher rates of electricity demand growth in the developing countries, coupled with the competitiveness of coal-fired electricity generation and the abundant availability of coal, either in their own country or on the international market, ensures that global demand growth for coal remains robust, if not spectacular. The lower growth in coal demand in the iron and steel sector, compared to electricity generation, is expected to result in steam coal increasing its share of world trade.

Global annual coal production is projected to grow by around 51% between 2000 and 2030 or by 2,359 Mt, reaching 6,954 Mt (Figure 6.6). This growth is roughly equivalent to today's combined production by China, Canada and the United States. The growth in coal production in China is expected to account for 1,072 Mt, or 45%, of this increase, while India, Australia, the United States and Canada, Indonesia and Africa will together account for virtually all the rest, at 1,161 Mt. The EU15 is the only region to experience a significant decline in production, by 106 Mt between 2000 and 2030.

China's 87% increase in production from 2000 levels means its share of world coal production will increase between 2000 and 2030 from 27% to 33%. OECD Pacific, India, Indonesia and Africa also see their shares of world production grow over the *Outlook* period. The most dramatic increase will be in Indonesia, albeit from a low base, where it will more than double to 4% in 2030.

The continued deregulation of electricity markets and the removal of subsidies, import barriers and other market distortions will continue to drive

Figure 6.6: **World Coal Production by Region**

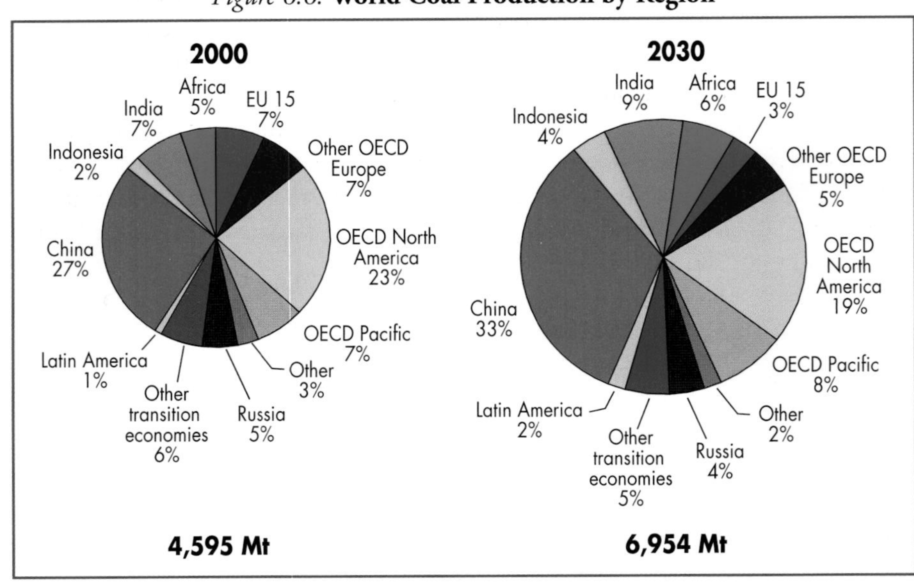

a shift in coal production to the lowest-cost production regions. The international trade in coal is projected to grow at around 1.7% per annum, from around 406 Mtoe (637 Mt) in 2000 to 672 Mtoe (1,051 Mt) in 2030, somewhat faster than demand at 1.4% per annum.

East Asia and Korea will drive the projected growth in import demand, together accounting for around 60% of the growth in coal trade. OECD Pacific will be the only region to experience a significant (12 Mt) decline in imports, as a result of the decline in Japanese imports. The increase in Asian import demand will primarily benefit exporters in Asia-Pacific, particularly Australia, Indonesia and China.

Mining, Port and Shipping Capacity Additions

The need for new production capacity over the *Outlook* period will greatly exceed projected demand growth, because the closure of existing productive capacity will require new mines to replace this capacity. In some producing regions, such as the eastern United States and Indonesia, a large proportion of the existing mine capacity will close over the *Outlook* period and if production is to be maintained, this capacity will have to be replaced. This high rate of mine closure is driven by a number of factors, including reserves, geology, regulations and the faster rate at which today's high-productivity mines can exhaust their economic reserves.

To maintain existing production levels and add capacity to meet projected demand growth between 2000 and 2030 will require 6,191 Mt of additional production capacity (Figure 6.7). This is equivalent to more than three times the current production of OECD countries combined. Capacity additions will have to average around 206 Mt per annum. It is expected that around 58% of the new mine capacity required will be needed to replace mine capacity that will be retired over the *Outlook* period. The balance will be required to meet demand growth.

Figure 6.7: **Cumulative Additional Coal Mining Capacity, 2001-2030**

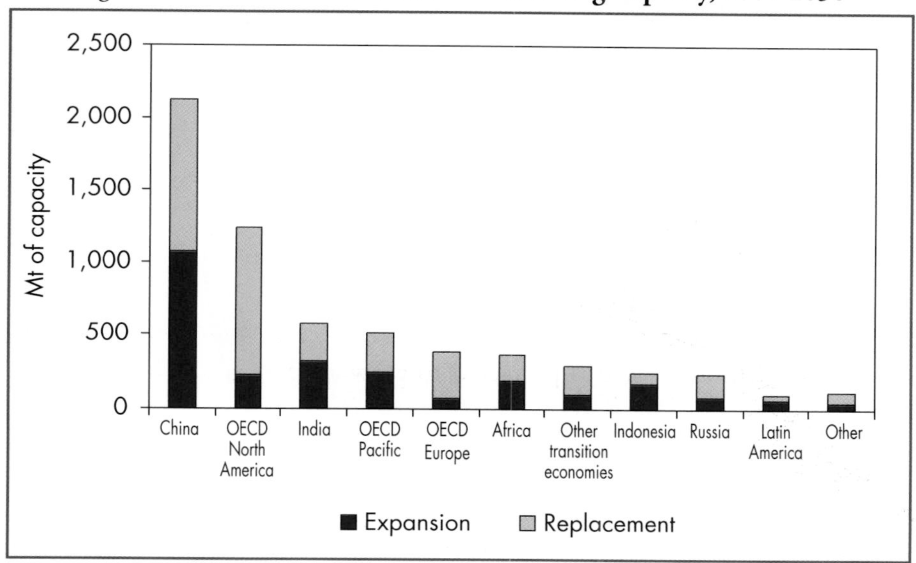

China alone will need to add around 2,130 Mt of production capacity, around half of which will be to meet demand growth.[8] This is equal to around 34% of the global new mine capacity additions and is some 70% more than the next largest region, the United States and Canada, where around 1,230 Mt of new capacity will be needed.

Ports and Shipping

Although, some regions currently have surplus import or export coal-handling capacity at ports, the growth in global seaborne coal trade (see Figures 6.8 and 6.9) will result in significant additional infrastructure

8. The high proportion of total new capacity needed just to replace capacity that will close over the projection period is, in part, driven by the Chinese government's desire to consolidate more production in the large state-owned mining companies.

Figure 6.8: **Major Inter-regional Coal Export Flows, 2000** (Mtce)

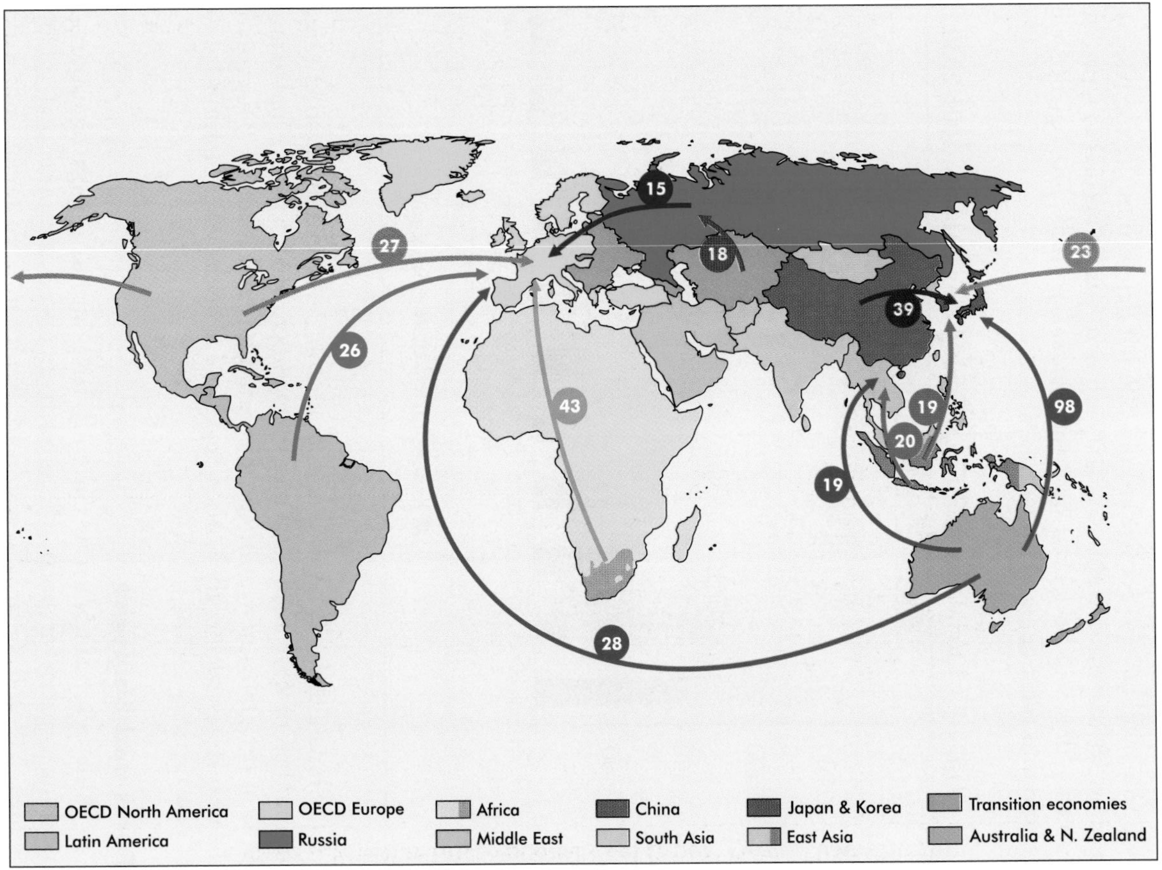

Figure 6.9: **Major Inter-regional Coal Export Flows, 2030** (Mtce)

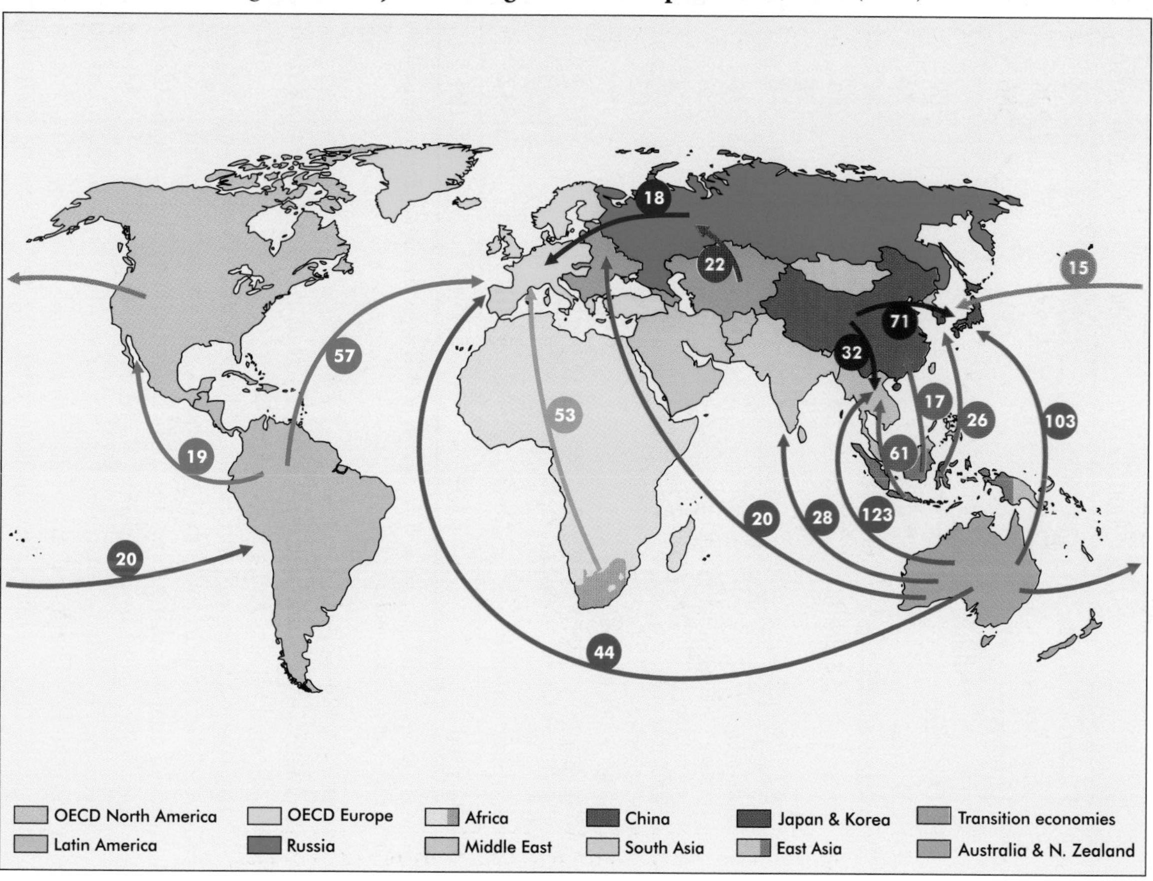

Table 6.3: **World Coal Production and Capacity Additions**

	Production (Mt)		**Additional production capacity** (Mt)
	2000	2030	2001-2030
OECD North America	1,056	1,299	1,234
US and Canada	*1,045*	*1,289*	*1,226*
OECD Europe	646	556	386
EU15	*340*	*233*	*162*
OECD Pacific	318	553	509
OECD total	**2,019**	**2,408**	**2,128**
Transition economies	**528**	**645**	**531**
Russia	*242*	*290*	*236*
China	1,231	2,304	2,126
East Asia	198	406	350
Indonesia	*77*	*248*	*244*
South Asia	332	660	586
India	*329*	*652*	*578*
Latin America	54	115	103
Brazil	*7*	*12*	*10*
Middle East	2	2	2
Africa	230	415	366
Developing countries	**2,047**	**3,901**	**3,532**
World total	**4,595**	**6,954**	**6,191**

Note: Additional production capacity includes capacity to meet demand growth and replace capacity at mines that have depleted their economic reserves.

requirements at new and existing ports. A large increase in the bulk-dry cargo fleet dedicated to coal will also be required.

International import and export requirements, as well as the internal coal trade needs of China, India and Indonesia, will mean that around 859 Mt of additional coal-handling capacity will be needed over the period 2001-2030. Around 53% of this additional capacity is expected to be for export requirements, mostly in Australia, China, Colombia, Indonesia, South Africa, and Venezuela. The locations of additional coal-handling port facilities for imports are more widely distributed, reflecting the diversity of importers. Around 60% of the additional import-handling facilities at ports will be needed in Korea, East Asia, Brazil and the Middle East.

With the growth in seaborne coal trade, tonne-kilometres for coal will increase by around two-thirds between 2000 and 2030. Although larger quantities of coal will be shipped over longer distances, the growth in coal trade in Asia-Pacific, with its relatively short freight distances, means that the average distance travelled by coal cargos will hardly change over the *Outlook* period. Also offsetting the growth in tonne-kilometres will be innovations in the coal market, such as electronic trading and coal derivatives. Electronic trading is already having an impact on physical coal trade in the Atlantic market.

It is expected that the deadweight-tonnage of the bulk-dry cargo fleet necessary for coal trade will increase by around 70%, or 49 million deadweight tonnes, to a total of 116 Mt in 2030. With losses and retirements to this fleet, this corresponds to the building of an additional 108 million deadweight tonnes of bulk-dry cargo capacity over the period 2001-2030.

Cost and Technology Developments

The coal industry's efforts to improve productivity and lower costs, in conjunction with periods of excess capacity in the coal market forced down coal prices during the 1990s. The industry responded to increased competition, falling real prices, excess capacity and often poor profitability in a number of ways that included:

- Higher productivity; this lowered the labour cost of coal produced.
- Production from the most economic reserves; with a trend to larger mines, with lower capital costs per tonne and greater potential for high rates of productivity.
- Lowering non-labour costs; outsourcing of services and improved production planning and inventory control.
- Industry consolidation; which lowered corporate and administrative costs and improved negotiating power.
- Lowering transport costs; efforts were made by transport providers and port operators to improve efficiency.
- Lowering transaction costs; by innovations in trading, notably electronic trading.

The coal industry operates in a highly competitive global market and pressures to improve technology, increase productivity and lower production and capital costs are unlikely to abate in the foreseeable future. Figure 6.10 shows the trends (weighted by production) in productivity, investment costs and coal prices for Australia, Canada, the United States, Colombia and South Africa. All of these countries have a significant presence in the coal export market or, in the case of the United States, a highly competitive domestic coal market. The steady decline in the coal price over the last twenty years has been matched by an increase in productivity and a decline in investment costs per tonne of capacity.

Figure 6.10: **Weighted Average Coal Productivity, Price and Investment Costs**

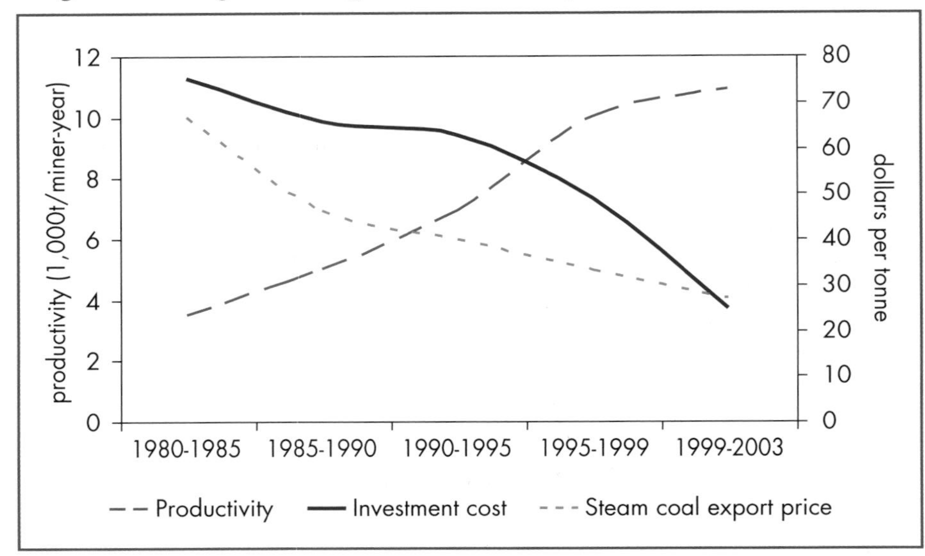

Note: Based on data covering Australia, Canada, the United States, Colombia and South Africa. The average steam coal price is the steam coal export price for each country weighted by exports.

Prior to the Asian economic crisis in 1997, producers responded to price volatility and declines by lifting output to maintain cash flow. Since then, individual producers have tried to match output more closely with demand in order to reduce price volatility.

Table 6.4 presents the capital cost and capacity closure rate assumptions used for the calculation of investment needs. For new capacity to meet demand growth or replace capacity that is closed, a one-off investment in dollars per tonne of annual capacity is used to derive the investment needed for new capacity. This unit cost is generally fixed over the *Outlook* period, but is assumed to increase or fall in some regions, depending on local circumstances. Investment in sustaining production at existing and new mines is calculated by applying a charge in dollars per tonne of coal produced each year.

The new capacity investment cost assumptions are based on recent data for new greenfield and brownfield/expansion developments. Importantly, these data include the capital cost of connecting to existing transport infrastructure (conveyor systems, road, rail, etc.). They have been adjusted where recent developments are not regarded as representative of future investment costs. Averages of greenfield and brownfield/expansion costs have been used as there is no evidence that the share of new capacity met from each type will change in the future. In any event, in most regions there are insufficient data available to

Table 6.4: **Coal Investment Cost and Capacity Closure Assumptions**

	Capacity cost ($/tonne)	**Capacity closures** (average annual retirement rate as % of 2000 capacity)	**Sustaining cost** ($/tonne /year)
United States			
Western steam coal	15	2.6	1.25
Metallurgical coal	32	3.7	1.25
Eastern steam coal	29	3.7	1.25
Brown coal	15	3.7	1.25
Canada			
Metallurgical and steam coal	49	2.6	1.25
Brown coal	32	2.6	1.25
EU15*	20	n.a.	0.45
Other OECD Europe	40	2.2	0.60
Australia and NZ	32	2.9	1.25
Russia	29	2.6	0.75
Other transition economies	47	2.6	0.50
China			
Metallurgical coal and large-scale mines	44	2.6	0.70
Province and municipality mines	35	3.0	0.70
Indonesia**	32	3.2	0.40
India***	32	2.6	0.40
Colombia and Venezuela****	56	2.6	1.25
South Africa	27	2.6	1.25

* Capacity and sustaining costs are for brown coal. Total hard coal investment is based on an assumption of $4.5/tonne of production.
** Average for 2001-2030. The capacity cost is assumed to increase from $22/tonne in 2001 to $35/tonne in 2010, remaining flat thereafter.
*** Average for 2001-2030. The capacity cost is assumed to increase from $28/tonne in 2001 to $35/tonne in 2030.
**** Average for 2001-2030. The capacity cost is assumed to increase from $44/tonne in 2001 to $60/tonne by 2015, remaining flat thereafter.

assess separate cost figures. The cost assumptions for sustaining investment vary by region depending on the degree to which advanced mining equipment is used, on productivity levels and on the intensity of competition. In some regions, sustaining costs are also influenced by the financial health of the region's coal industry.

Coal Production Costs, Technology and Productivity

Substantial advances in coal mining technology have occurred in the last 30 years. These have contributed to major improvements in the areas of health and safety, environmental performance, labour productivity and extraction costs. This application of advanced mining techniques has tended to occur more in developed countries, because high labour costs favour more capital-intensive production.[9] It is noticeable that in countries where government subsidies or other policies protect coal producers from market forces, growth in labour productivity has lagged (Figure 6.11).

Coal supply costs consist of a capital cost component, the costs of coal extraction and preparation, and then transportation to the end-user. The capital cost includes the costs of reserve acquisition and control, exploration and engineering, connection to existing transport infrastructure, mining equipment, coal preparation plant, site preparation and land reclamation at the end of the project. Additional financial costs can be incurred if the approvals

Figure 6.11: **Trends in Coal Industry Labour Costs and Productivity**

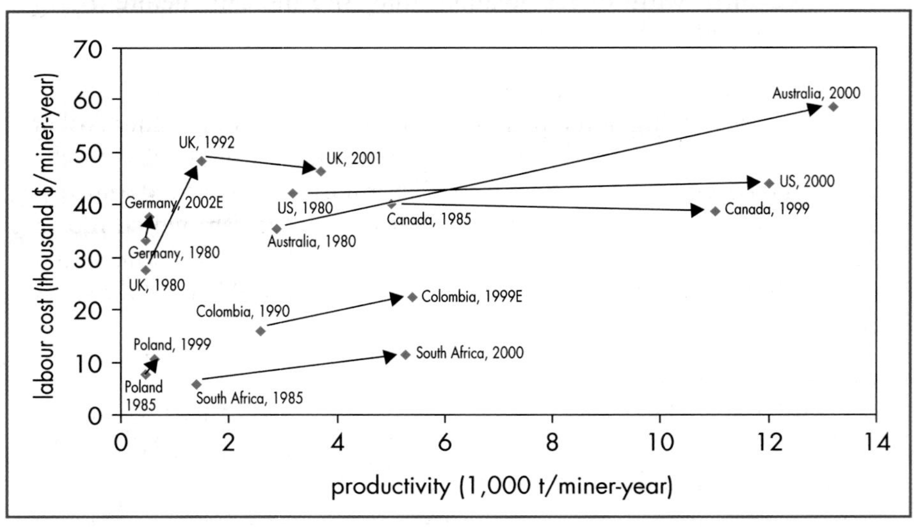

Sources: *IEA Coal Information*. The data for Colombia (1999) and Germany (2002) are IEA Secretariat estimates.

9. This has also been influenced by a lack of access to reasonably priced capital in developing countries.

Box 6.2: Mining Techniques and Technology

Coal is produced by a variety of methods ranging from highly efficient mechanical means underground, such as longwall mines, and large-scale drag-lines on the surface in the industries of the developed countries, to manual pick-and-shovel methods in some cases in developing countries. Coal preparation can be an important component of mine operating and capital costs.

Underground Mining – Room-and-pillar

In its most basic form, room-and-pillar extraction can be accomplished with manual labour for coal cutting, loading and haulage to the surface. There are two distinct approaches to mechanised room-and-pillar mining:

- Conventional mining – where coal is "undercut" by a cutting machine, drilled and blasted with explosives, and then loaded into shuttle cars for transport to a belt conveyor or underground rail haulage system which transports the coal to a shaft or slope for final delivery to the surface.
- Continuous mining – where coal is cut with a drum-type mining machine with direct loading into shuttle cars before being conveyed as above for conventional mining.

Underground Mining – Longwall

Longwall mining developed in Europe as the need to produce coal at very great depths required leaving large coal pillars which substantially lowered recovery rates in room-and-pillar mines. An advanced longwall mine employs a large "shear" or coal-cutting mechanism which moves back and forth on a panel of the coal-face that can be 300 or more metres wide. Hydraulic supporting devices are used above and behind the shear to hold up the roof.

Opencast Mining

Coal production by opencast excavation requires removal of soil and rock (overburden) from above the coal. Overburden removal is often the main activity, and the most expensive, in the operation of an opencast mine. Opencast mines range from small-scale operations removing coal from exposed outcroppings to huge surface mines using several drag-lines and shovels and a fleet of transport trucks. The reclamation and renewal of the mine site during mining can constitute a substantial portion of the mine operating costs.

process for new capacity is long. The variable costs of coal supply comprise the costs of extraction and preparation, including the costs of coal mining, crushing, washing and other treatment, and the transportation costs, including loading, haulage by truck, rail, barge or ship, handling when transferring from one mode of transport to another and storage.

Coal-extraction costs depend on many factors, including; the geology and location of the coal reserve, the mining technique, labour productivity, power and fuel costs, capital costs, the level of coal preparation, government policies on royalties, severance costs, health and safety regulations and environmental regulations. Coal preparation can be an important component of mine operating and capital costs. Coal preparation can take the form of simple breaking, crushing and screening to ensure adequate flow and uniform size, to the addition to that process of intense washing and drying to meet stringent ash, moisture and sulphur standards.

There are two broad categories of coal mine — underground and opencast. Globally, around two-thirds of hard coal is extracted from underground mines, though the proportion of opencast mines is higher in countries such as Australia, Canada, Colombia, Indonesia and the United States.

Opencast mining has expanded rapidly in the past twenty years, as more advanced excavation and materials-handling systems have been developed that have lowered extraction costs. The primary advantage of opencast mining is its scale and productivity. Many opencast mines are multi-million tonne-per-year operations, with very competitive productivity levels and low per-tonne costs.

Most modern underground coal extraction methods are a version of either "room-and-pillar" or "longwall" mining. Room-and-pillar mining has lower capital costs and causes less subsidence at the surface, but its disadvantage is that coal recovery seldom exceeds 60%. While initial capital investment and ongoing capital costs in longwall extraction can be five or six times higher, labour productivity is often four to five times higher and 80% to 90% of the coal seam can be recovered.[10]

Trends in Labour Productivity

Although interdependent, the improvement in labour productivity in the main coal-producing countries has contributed to the decline in real coal prices since the early 1980s. Between 1980 and 2000, average labour productivity in eight major coal-producing countries, Australia, Canada, Colombia, South Africa, the United Kingdom and the United States, increased on average by

10. IEA Coal Research (2001).

7.6% per year. The rate of productivity improvement has slowed in recent years to 4.9% between 1998 and 2000 in these countries.[11]

Figure 6.11 shows the relationship between labour costs and productivity for selected coal-producing countries. Although care needs to be taken in comparing different countries, owing to differences in the data, a key point emerges. In general, producers exposed to competitive domestic or international markets have achieved large increases in productivity with modest increases in labour costs, while those coal producers supported by subsidies and/or other protectionist policies have made little progress in improving productivity and have been dramatically left behind by countries such as Australia, Canada and the United States.

Care must be taken in interpreting the impact of productivity growth on coal production costs. Although productivity improvements can be achieved in some cases at little or no investment cost, most of these improvements have been achieved by the substitution of machinery for labour, resulting in additional capital costs which need to be recovered.

Improvements in labour productivity have been driven by a number of factors, including:

- A shift to opencast mining, which generally has higher productivity and lower labour costs than underground mining. In addition to this there has been a shift to:
    1. Even larger-scale opencast mining operations that have economies of scale and higher productivity.
    2. Exploitation of thicker coal seams with lower overburden ratios.
- Strong price competition, which causes less efficient producers to leave the industry.
- Increased use of longwall mining equipment in underground operations.
- Continued application of technological advances across the mining process in conjunction with a management and workforce focus on productivity and safety.

Three factors will continue to stimulate improvements in labour productivity:

- Increasing size of individual mines, especially in developing countries and in countries active in the international coal market. This will be driven by a continued shift to opencast mines and the more intensive use of mining equipment.

11. IEA (2003). Much of this decline in the growth rate is driven by a flat productivity in the United States over 1998-2000. However, the available data for 2001 support the slowing in productivity growth in recent years.

- Continued exploitation of only the most favourable reserves by a better trained and more flexible workforce.
- Improvements in coal extraction, preparation and transport technology will permit even larger-scale mining units and more efficient utilisation of labour.

Thus there remains scope for further gains in productivity in the international coal industry through economies of scale, exploitation of contiguous resources and further improvements in working practices. But these improvements come at a cost and require time, management skills and capital. Further improvements will occur, but it is likely that growth in labour productivity in the major coal-producing countries will continue at a more modest pace than the average 10% to 15% per annum growth sustained in the 1990s.

Investment Uncertainties and Challenges[12]

Coal mining is a high-volume low-margin activity. Further efforts to boost productivity and minimise costs are essential to achieving an adequate rate of return. The line between profit and loss is narrow and a clear understanding of the commercial and non-commercial risks a producer faces is critical to making informed investment decisions.

Project financing is likely to play a much smaller role in the coal industry than in the gas, oil and electricity sectors. Since there are very few investment projects that will have long-term off-take contracts for even a part of their production, most of the capital requirements for investment will be funded from retained earnings or on the strength of a company's balance sheet. Project finance may play a small role in developing countries where development aid and international financial organisations are involved.

A project will only go ahead if the expected rate of return exceeds the company's target rate of return for projects. Many coal producers are part of large integrated mining companies and therefore must compete for capital with a portfolio of projects across mining sectors. A potential barrier to coal investment is permitting and siting requirements. These regulations need to balance local and national environmental and economic costs and benefits.

Excess capacity, low demand growth and declining real prices in recent years have greatly increased risks for new investment projects that expand output. In this climate, for projects to be approved, they will need to show an adequate rate of return even under "worst case" scenarios. Difficult commercial conditions currently favour investments in consolidating production and improving productivity.

12. See IEA (2002a) for a discussion of the methodology used to generate the energy demand projections of the Reference Scenario and the uncertainties surrounding them.

In countries where the coal industry is closely controlled by the government, such as China and India, pricing and investment decisions are often influenced by political or social goals. This can lead to inefficient investment, which can act as a drag on the economy, and have an adverse impact on the financial performance of the industry and its ability to fund future investment.

Economic Risk

The primary goal of coal producers who operate in a normal commercial environment is to achieve a profit that results in an adequate return on the capital invested. Coal producers' main concern is thus to manage the economic risk they face in order to maximise the return on their shareholders' investment. In Table 3.1 this economic risk is broken down into four components: market, construction, operation and macroeconomic. The coal industry's main risks are related to the market and operational risks. Macroeconomic risks can be significant for exporters, as most contracts are denominated in US dollars.

Market risk is essentially a question of price. Coal producers must ensure that they manage their price and volume risks to ensure that they receive sufficient revenue to allow them to cover their cost of capital. While weak prices and uncertain demand prospects remain, investment will continue to be focused on lowering costs and replacing depleted capacity. However, producers will respond quickly to expand capacity in order to meet new demand if required. Given the abundant options for supply and the high level of competition, it is unlikely that prices will need to rise significantly to accommodate this investment.

Operational risk plays a significant role in the coal industry, larger than in many other sectors of the economy, because the production process, although well understood, is subject to significant geological uncertainty. For a small or even medium-size coal producer, serious mechanical failure or unexpected geological problems can convert a small profit into serious loss. This has raised industry productivity by focusing production in the largest, most efficient operations. Consolidation has also helped to limit operational risks, as larger producers, with production spread across a number of mines, are better able to manage this risk.

Political Risk

Coal producers in most developed countries face little risk of direct government interference or regulatory changes that dramatically alter their situation without compensation. However, some low-level risk remains. Examples include changes to land reclamation requirements, local community obligations, environmental standards at mine sites, pension requirements and royalty regimes.

In developing countries, the perceived level of political risk is often higher. This reflects investors' experiences in the past and their fear that governments might act in the future in a way that would adversely affect investment. This kind of risk can be offset to some extent by involving local partners, but political risk cannot be entirely eliminated, in particular the risk remains that changes to the initial conditions, laws and regulations, under which a decision was made to invest, might be detrimental to the projects' profitability.

In general, political risk is not perceived as a major obstacle to coal investment, but there are some regions where the perceived political risk remains high, for instance in China and India. Although this is not necessarily an obstacle to investment in the high risk and return gas and oil industries, it is for the coal industry, as it is a low-margin activity that is unable to support the additional political risk.

Environmental Policy and Technology

The main uncertainty surrounding the demand outlook for coal, and therefore the need for investment in the coal industry, is the impact of future policies and measures to address environmental concerns. This uncertainty creates a barrier to investment, given the risk that investments in coal might be stranded owing to environmental policy developments.

Changes in environmental requirements can have a large impact on coal industry activity and profitability. For instance, the recent growth in production of low-sulphur coal in the western United States has been driven, in part, by the stringent sulphur-emission standards that were introduced for coal users.

In recent years, little or no new coal-fired capacity has been built in developed countries outside Asia, partly because of uncertainty over environmental policy, but mainly because the cost of gas-fired generation, particularly in Europe and North America, has been lower than coal. Significant new coal-fired capacity is expected to be built in these regions after 2010 as tighter gas markets push up gas prices to a point which makes coal-fired plant competitive. The additional coal-fired generating capacity will boost global coal demand and trade. But there exists a risk that new climate change policies, including demand-side policies, could alter this picture and mean that much of this growth in coal demand might not materialise.

The CO_2 emitted from coal-fired power plant generally fall with higher capital costs (Figure 6.12). Although the most efficient coal-fired technology, integrated gasification combined cycle (IGCC), emits twice as much CO_2 per kWh generated as a combined cycle gas turbine (CCGT)

Figure 6.12: **CO_2 Emissions and Capital Costs by Electricity Generating Technology**

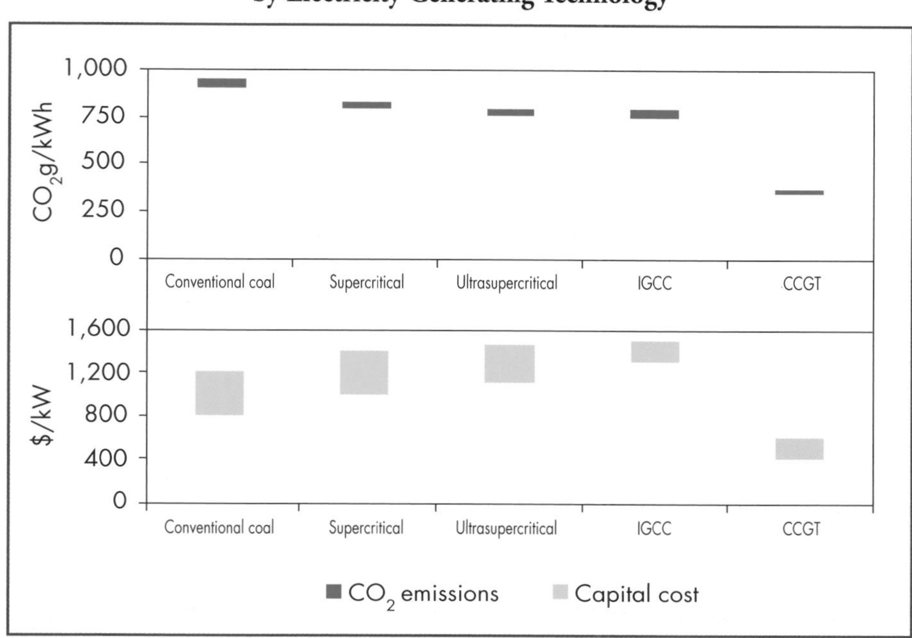

Source: IEA. Based on current and near-term technology and costs.

plant, it produces a concentrated stream of CO_2 which has advantages when it comes to sequestration. IGCC plants are currently in the demonstration phase of their development and are not commercially viable at this stage. However, continued research into clean coal technologies (CCT) offers the potential for further improvements in the performance of coal-fired power plants.

There remain a number of barriers to the adoption of clean coal technologies, but the most important of these is their high cost. However, government policies, such as increased research, could help to reduce these costs. The pace at which clean coal technologies penetrate the market will be crucial to future coal trends.

Clean coal technology and carbon sequestration research (see Chapter 8) could, if it leads to cost falls and commercially competitive applications, result in coal remaining a low-cost source of electricity generation in a carbon-constrained environment. If fuel cells were also to overcome the significant hurdles faced in their development, then coal could also play an important role in their future.

Regional Analysis

United States and Canada

Cumulative coal investment of around $70 billion will be required in the United States and Canada over the *Outlook* period, or around 19% of the world total. Production is expected to grow from 1,045 Mt in 2000 to 1,289 Mt in 2030. Primary energy demand for coal in the United States and Canada is projected to rise by around 0.6% per annum, from around 572 Mtoe (1,035 Mt) in 2000 to 675 Mtoe (1,264 Mt) in 2030. Exports will continue to decline to just 55 Mt in 2015, recovering to around 72 Mt in 2030. Imports will rise steadily, but from a low base, reaching around 48 Mt in 2030.

Coal Sector Profile

The United States is the second-largest coal producer in the world, exceeded only by China. The United States recoverable coal reserves are estimated at around 250 billion tonnes of hard coal and lignite, or around 25% of the world total, compared to Canada's 8 billion tonnes.[13] The United States and Canada have two very different coal markets. In the United States, the domestic market for coal is the most important, which is dominated by the demand of the power sector (over 90% of demand), with exports being marginal (5% of production). In Canada, production is much more focused on export markets, with around 46% of production exported in 2000. Canada also imports a considerable quantity of steam coal from the United States.

Over the past decade the deregulation of energy markets, increased competition both domestically and internationally, and periods of excess capacity resulted in declining real domestic and export prices. This has put a strain on the financial performance of an industry that has not only sustained, but expanded, production during this period.

Lower mine-mouth prices in the United States have been possible thanks to higher productivity, a better trained and more flexible workforce, increased consolidation of productive capacity within the industry and larger average mine sizes. Productivity has increased significantly in the United States and Canada, but the rate of growth has slowed in recent years (Figure 6.13).

Mines with more than 500,000 short tons of coal production increased their share of production from around 69% in 1986 to 85% in 1997, while their ratio of recoverable reserves to production dropped from around 34 years to 19 years. Larger coal mines, while allowing the recovery of fixed costs over a larger output, also tend to have higher levels of productivity, part of which,

13. World Energy Council (2001). All references to coal reserves are sourced from this report, and refer to reserves at the end of 1999.

Figure 6.13: **United States and Canadian Coal Production, Prices and Productivity**

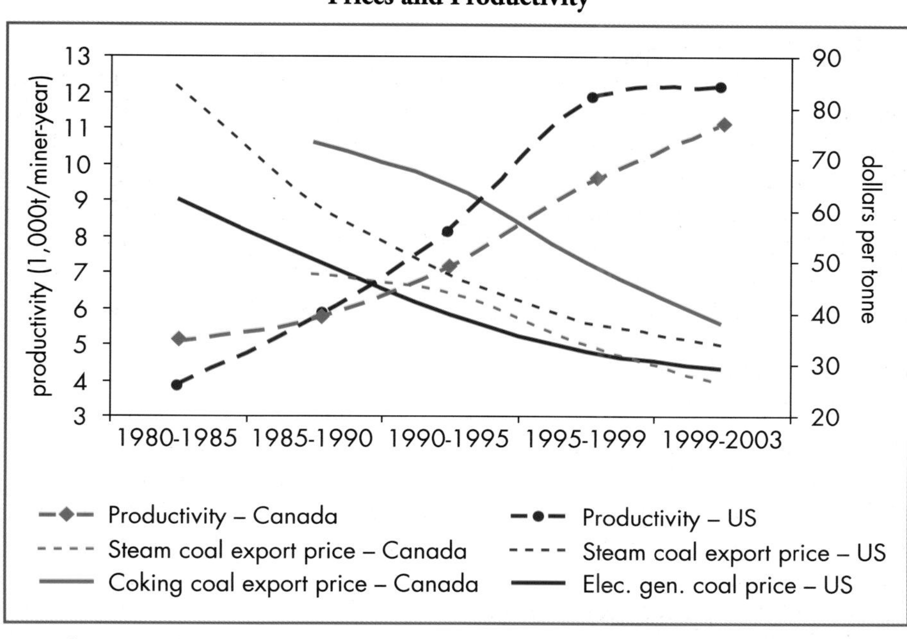

Note: The steam coal and metallurgical coal prices are export fob prices and the electricity generation price is delivered to the power station.

is explained by the higher productivity of opencast mines which make up the lion's share of the largest mines. Concentration of production in the largest companies has also increased sharply over the last decade (Figure 6.14).

Investment in the United States coal industry declined between 1977 and 1992 owing to a reduction in the development of new reserves and a focus on investments that would improve productivity from existing reserves. Investment needs were also reduced by acquiring reserves through mergers and acquisitions at cost levels below those of new developments. However, there is a limit to the extent that companies can postpone investment expenditure while boosting production; thus, investment increased between 1992 and 2000.

The United States has always been a marginal player in export markets, often selling into export markets only at marginal cost in order to avoid building stocks, while Canada's export competitiveness is hampered by high inland transport costs. Given the continuing poor returns exports offer United States and Canadian producers, United States coal export volumes have dropped sharply, from around 96 Mt in 1990 to 44 Mt in 2001 and to an estimated 35 Mt in 2002. Canadian exports (predominantly of coking coal)

Figure 6.14: **Concentration of United States Coal Production by Producer**

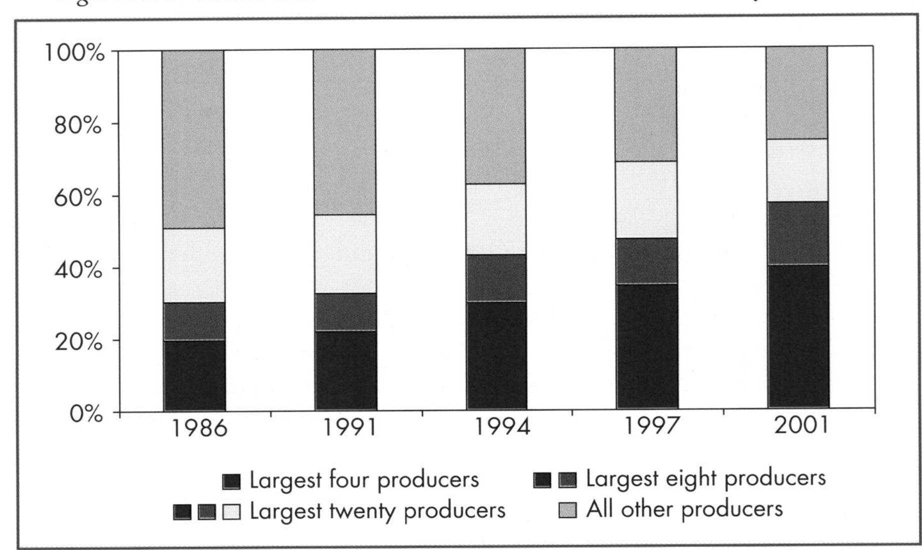

Source: EIA, *Annual Coal Report*, various years.

have declined from a peak of 37 Mt in 1997 to around 30 Mt in 2000 and to an estimated 27 Mt in 2002.

Mining and Port Investment

The cumulative coal investment of $70.1 billion over the *Outlook* period implies average investment of around $2.3 billion per annum. This is somewhat higher than the industry's investment in 2000 of around $2.0 billion. Of the total coal mining investment needs in the United States and Canada of $70.0 billion, around $43.7 billion or 62% of the total, will be required in equipment and works to sustain the productive capacity of mines. Investment in new mine capacity to replace depleted capacity will require $21.4 billion, or 31% of total mining investment. The investment in mining capacity to cater for demand growth is only $4.8 billion, or around 7% of total mining investment.

Investment to expand capacity represents a small share of investment needs, partly because this expansion is expected to occur in the western United States, where capital cost requirements per tonne of capacity, at around $15 per tonne, are around half the equivalent level in the East (excluding opencast brown coal).[14]

14. The western producing area is defined as the states of Alaska, Arizona, Colorado, Montana, New Mexico, North Dakota, Utah, Washington and Wyoming.

Mining investment requirements in Canada are estimated at $5.8 billion, around half of which will be to sustain production (Figure 6.15). After 2020, production growth is almost negligible, and virtually all investment will be needed in sustaining and mine replacement.

In the United States, the high depletion rates and mine development costs in the eastern states mean that, although production declines in the east, investment needs are still $31.2 billion or 49% of the United States total. This represents a significant shift in the regional investment pattern, as the west's share of total investment over the *Outlook* period, at around half the total, is around double its 1997 share.[15]

Investment in coal-handling facilities at ports for imports and exports will be small, as exports are expected to remain below 2000 levels throughout the projection period. Imports into the United States and Canada are expected to grow from around 38 Mt per annum in 2000 to 48 Mt in 2030, but much of this represents cross-border trade between the two countries: US exports to Canada will account for around 23 Mt of total imports in 2030. This means the increase in coal-handling capacity required at ports is only around 8 Mt per annum in total over the *Outlook* period, costing around $200 million.[16]

Figure 6.15: **Coal Mining Investment by Region in the United States and Canada**

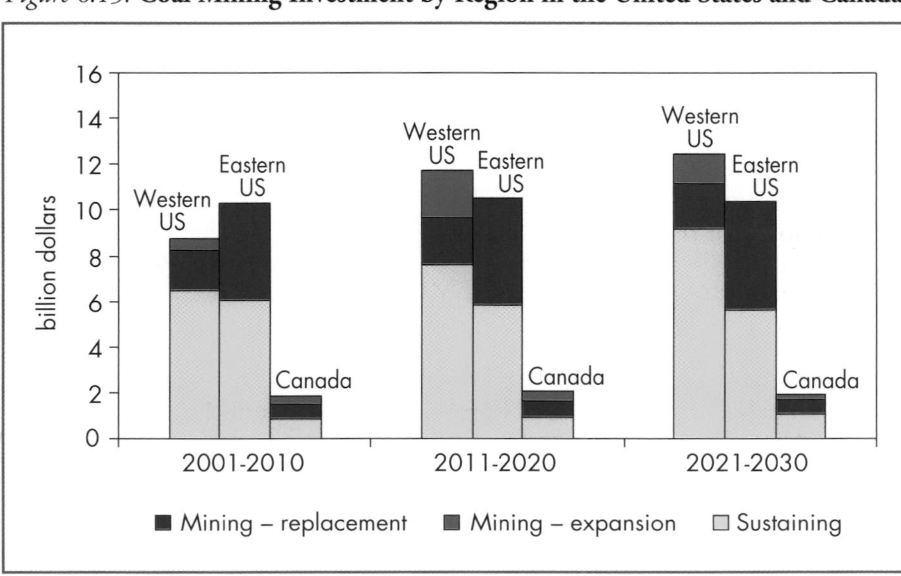

15. US Census Bureau (2001).
16. Some low-cost conversions of export-handling facilities to take imports are also assumed.

Mining investment will be needed to increase coal production in the United States from 976 Mt in 2000 to 1,204 Mt in 2030, and from 69 Mt to 85 Mt in Canada. Production growth in the United States is projected to occur in the western producing area, where production is expected to rise at 1.6% per annum over the *Outlook* period, from around 464 Mt in 2000 to 758 Mt in 2030. Production in the western producing area will grow faster than demand, because production in the eastern regions declines from around 512 Mt to 446 Mt.[17] This reflects the continuing cost competitiveness of western coals, even with the substantial transport costs incurred to move them to consumers, and the fact their low sulphur content is preferred by electricity generators without emission-control technologies.

Additional mining capacity of 1,226 Mt will be required in the United States and Canada in order to replace depleted productive capacity and to meet demand growth. The United States dominates these figures, with less than 80 Mt needed in Canada. The additional capacity required is mainly for thermal coal (1,024 Mt), with the balance split between brown coal (133 Mt) and metallurgical coal (68 Mt). Around 997 Mt of the additional capacity required is needed to replace productive capacity that will close over the next three decades.

OECD Europe

Cumulative coal investment of around $25 billion will be required in OECD Europe over the *Outlook* period. Coal production in OECD Europe is expected to decline from 646 Mt in 2000 to 556 Mt in 2030. Primary energy demand for coal is projected to decline by around 0.4% per annum from around 319 Mtoe (816 Mt) in 2000 to 283 Mtoe (771 Mt) in 2030. Exports are expected to continue to decline from 53 Mt in 2000 to around 22 Mt in 2030. Imports are expected to rise steadily, as production declines faster than demand, from around 204 Mt to 237 Mt in 2030.

Coal Sector Profile

Coal production in OECD Europe countries has declined significantly since 1990, from around 1,032 Mt in 1990 to 646 Mt in 2000. The largest declines in production between 1990 and 2000 have occurred in Germany

17. This is somewhat more pessimistic than the EIA's *Annual Energy Outlook* 2003, which projects an increase in production in the east of 0.1% per annum between 2000 and 2025. Interestingly, investment in the United States and Canada is not particularly sensitive to the projected split in production between the west and east, as long as all production growth is sourced in the west. For example, if eastern production were to decline to around 300 Mt, and imports and exports were unchanged, mining investment is only around 3% lower than that presented here.

(229 Mt), the United Kingdom (62 Mt), Poland (53 Mt) and the Czech Republic (36 Mt).

Currently, much of the brown coal production is commercially competitive, and unsubsidised, although there remain some exceptions. Much of the hard coal production remains uneconomic and is dependant on subsidies and/or policies that protect its domestic market.

Irrespective of the costs of extraction, OECD Europe has large recoverable coal reserves estimated at around 105 billion tonnes or around 11% of the world total. Around half of these reserves are brown coal and the hard coal reserves are relatively expensive to mine.

Estimates of the producer subsidy equivalent (PSE) per tonne of coal equivalent (tce) of production are given in Figure 6.16. Agreements for the reduction, or phasing-out, of aid to the coal industries in France, Spain and Germany were negotiated between 1995 and 1997. French coal production is to cease by 2004. In Germany, subsidies will decrease from €4.7 billion in 2000 to €2.8 billion in 2005, with subsidised hard coal production falling to a target of around 30 Mtce in 2005.

Subsidised production in Spain is protected by minimum local coal purchasing requirement levels for 15 power stations, but these are set to

Figure 6.16: **Estimates of Hard Coal Producer Subsidy Equivalent**

Note: tce is tonne of coal equivalent.

decline by 28% over the period 1998 to 2005. The United Kingdom once had a very high level of subsidised production, but at a very low rate per tonne. Subsidies ended in 1995, but were brought back between April 2000 and July 2002. Since then an investment aid scheme has been approved under EU laws that will allow up to 30% of investment needs to be met by grants.[18]

Mining and Port Investment

OECD Europe is unique in that it is the only major coal-producing region projected to experience a decline in production. This complicates assessments of additional capacity required and the related investments. The investment requirements presented here for OECD Europe, therefore, need to be treated with caution.

Hard coal production is expected to decline significantly in the EU15, from around 84 Mt in 2000 to only 17 Mt in 2030, while brown coal production stabilises at around 200 to 210 Mt after 2015.

Total investment needs of around $10.3 billion in the EU15 over the *Outlook* period reflect declining hard coal production, but also the high level of investment required to maintain what hard coal production remains. Investment in the hard coal industry in the EU15 is expected to be $4.4 billion over the next three decades. This represents an average of around $148 million per annum, or around half that of the 2000 level of $304 million. Investment in the brown coal industry is expected to be $5.7 billion, given its relatively lower capital requirements.

Hard coal production in OECD Europe outside the EU15 declines more slowly, from 121 Mt in 2000 to 73 Mt in 2030, reflecting the slower pace of restructuring, for social reasons. Brown coal production is expected to increase until 2020, before reaching a plateau of around 250 Mt per annum.

Investment in the OECD Europe countries outside the EU15 is higher than in the EU15, reflecting the expected increase in brown coal production and the additional, relatively more expensive, investment required in the hard coal sector. Total cumulative investment is expected to be around $14.9 billion over the projection period.

At an aggregate level, OECD Europe currently has a significant surplus of coal import capacity at ports. However, the pattern of imports will mean that some $200 million of new investment is required over the period to 2030. This investment could be higher, or conversely unnecessary, depending on the actual pattern of imports that eventuates.

18. UK Department of Trade and Industry (2003).

Japan, Australia and New Zealand

Cumulative coal investment of around $35 billion will be required in Japan, Australia and New Zealand over the *Outlook* period. Coal production is expected to increase from 313 Mt in 2000 to 553 Mt in 2030. Primary energy demand for coal is projected to decline by around 0.2% per annum, from around 142 Mtoe (275 Mt) in 2000 to 135 Mtoe (297 Mt[19]) in 2030. Exports grow strongly, from 181 Mt in 2000 to around 389 Mt in 2030, while imports decline, from around 145 Mt to 133 Mt in 2030.

Coal Sector Profile

Japan, Australia and New Zealand have very different coal sectors. After the EU15, Japan is the second-largest importer.[20] In contrast, Australia has one of the world's most competitive coal industries (located in Queensland and New South Wales) and is the world's largest hard coal exporter. High-quality metallurgical and steam coal reserves are abundant, while the domestic electricity generation industry also uses brown coal. Australia has proven recoverable coal reserves of around 82 billion tonnes, some 8% of the world total, of which around 54% is hard coal.

The declining real coal prices of the 1990s resulted in a decline in the profitability of the industry in Australia, with some producers making losses. International oversupply and the reduced premium over spot prices paid by Japanese consumers contributed to this decline. The industry response was to focus on reducing both variable and fixed costs and improving productivity. Productivity increased by around 9% per annum over the period 1990 to 2000, and in 2000 Australia became the most productive hard coal producer in the world, with average productivity of 13,200 tonnes per annum per miner, which was 2.4 times higher than 1990 levels. This has helped Australia lower its cash costs for export steam coal significantly, from around AUD 42 per tonne fob in 1995 to AUD 30.5 per tonne in 2000.[21]

Figure 6.17 presents the trends in coal price, productivity and investment costs over time. The trends in capital investment costs for new mine capacity

19. Primary energy demand increases in tonnes, despite the fall in tonnes of oil equivalent, owing to the increasing use of brown coal in Australia, with a calorific value of less than half that of steam coal.
20. For simplicity, because Japan accounts for virtually all imports in this region and Australia virtually all exports, references to these two countries will be used (rather than to the region as a whole) when discussing imports or exports. As an approximation, New Zealand accounts for around 1% of exports over the period.
21. McCloskey (2002a).

Figure 6.17: **Australian Investment Costs, Productivity and Coal Prices**

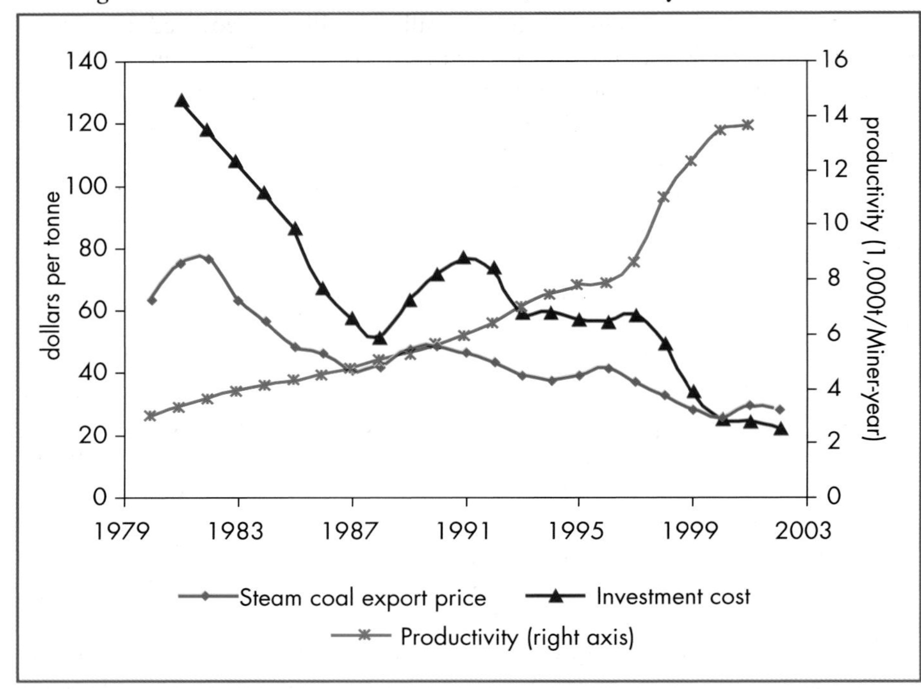

have followed those of the coal price, reflecting commercial pressures on the industry. The trend in productivity appears to be less closely linked to price.[22]

Consolidation of the industry in Australia has resulted in a larger share of steam and metallurgical coal export capacity concentrated in the hands of global mining companies. This has helped to limit excess capacity in Australia, with the margin between metallurgical coal capacity and exports extremely tight in 2000 and only a small excess capacity in the steam coal market.

Rail freight rates for coal in Australia were historically quite high, but since 1994, restructuring of the rail industries in each state and changes in policy have resulted in more commercial pricing policies, improved efficiency and lower charges. Australia has nine major coal-exporting facilities at seven ports, with a total capacity in 2002 of 256 Mt per annum. The port of Newcastle's total capacity of 89 Mt per annum, at its two coal-handling facilities, makes it the largest coal port in the world.

22. Although care must be taken in this conclusion, as productivity is the average industry level, and not matched solely to the new mine projects.

Mining and Port Investment

Cumulative investment of $34.9 billion will be required in Japan, Australia and New Zealand over the period 2001-2030. This is around 10% of the world total and is almost exclusively required in Australia. Coal mining will call for around $32.7 billion, with investment of another $2.2 billion required for coal-export facilities at Australia's ports. Coal mining investment, at an average of around $1.1 billion per year over the period 2001-2030, is somewhat higher than the average for 1990-2001 in Australia, of $800 million.

The mining investment needs are driven by a projected increase in coal production of 239 Mt between 2000 and 2030. Metallurgical coal production will grow at around 0.9% per annum, from 106 Mt in 2000 to 140 Mt in 2030. Brown coal production will increase by 1.0% per annum, from 68 Mt to 91 Mt. Steam coal production will grow fastest at 2.8% per annum, from 140 Mt in 2000 to 322 Mt in 2030. Steam coal production grows at 2.0% per annum until 2015, after which the rate of demand growth increases significantly to 3.6% per annum.

The recent rise in the Australian dollar, from an average of 54 US cents in 2002 to an average of around 63 US cents in the first ten months of 2003, has eroded margins,[23] leading a number of producers to announce production cuts.[24] However, the cash-cost reductions of the 1990s have helped put Australian exporters on a firm footing to exploit growing import demand around the world over the next three decades.

Nonetheless, to 2010 at least, they will continue to face strong competition from Indonesian and Chinese exporters in terms of price and market share. After 2010, slower export growth from China and Indonesia, combined with sharply increasing import growth in East Asia and Korea, means that Australia's exports will grow much more rapidly. Australian exports, which include exports to Japan and other regions, are therefore expected to increase from around 181 Mt in 2000 to 254 Mt in 2015, then to experience more rapid growth to 389 Mt in 2030.

Australia's exports increase in absolute terms to all regions it currently exports to, except Japan (Figure 6.18). Exports to East Asia will increase most, rising from around 11% of Australia's exports in 2000 to 34% in 2030. Despite competition from China and Indonesia, Australia's exports to Korea will grow significantly, but their share of total Australian exports will remain

23. This has a significant impact on profitability, as the difference between an exchange rate of 55 and 65 US cents for Australian steam coal exports is equivalent to an average $4 per tonne decline in the price received.

24. Production cuts of around 4.5 Mt had been announced for 2003. McCloskey (2003).

Figure 6.18: **Australia's and New Zealand's Coal Exports**

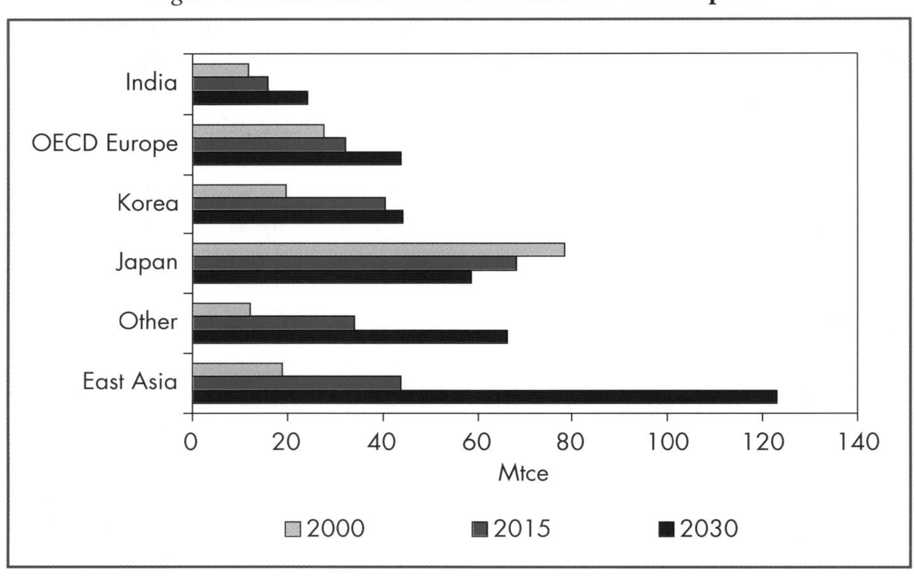

broadly flat at 12%. Most of the growth in demand for imports in East Asia and Korea will be driven by power-sector demand for steam coal. This will reduce the share of metallurgical coal in Australia's exports from around 58% in 2000 to around 34% in 2030.

Of the total cumulative mining investment, around 50%, or $16.4 billion, is needed in order to add new capacity of around 509 Mt to meet production growth and replace capacity at depleted mines. Australia accounts for around 500 Mt of this total. Around 263 Mt of the total new capacity needed in the region will be to replace capacity at mines that close over the next three decades, accounting for $8.5 billion, or around 26%, of the total mining investment needs. New mine capacity required to meet the regions production growth, both domestically and for exports, is expected to be around 246 Mt between 2000 and 2030. The cumulative investment required for this expansion is around $7.9 billion, or around 24% of the total mining investment requirements over the *Outlook* period.

New steam coal production-capacity needs are expected to be split evenly between NSW and Queensland, at least over the next 10 to 15 years, while Queensland is expected to continue to dominate metallurgical coal production. As a result, overall production will grow much more significantly in Queensland than in NSW, requiring proportionately more of the additional investment in rail and port infrastructure.

Sustaining investment needs will be $16.3 billion over the projection period and will be crucial if Australia is to compete against increasingly stiff competition in Asian markets from Indonesia and China (at least until 2010) and in Atlantic markets from Colombia and Venezuela.

Australia's ports currently have coal-handling capacity in excess of that required to service exports. However, the projected 208 Mt increase in exports between 2000 and 2030 will require an additional 185 Mt[25] of coal-handling capacity at ports. This is expected to require investment of around $2.2 billion over the *Outlook* period, or around 17% of the world total.

China

China is the world's largest coal producer and consumer and will require cumulative coal investment of around $123 billion over the *Outlook* period. Production is expected to increase from 1,231 Mt in 2000 to 2,304 Mt in 2030. Primary energy demand for coal in China is projected to grow by around 2.2% per annum, from around 659 Mtoe (1,208 Mt) in 2000 to 1,278 Mtoe (2,220 Mt) in 2030. Exports are expected to grow from 70 Mt[26] in 2000 to around 123 Mt in 2030. Imports grow from around 8 Mt to 40 Mt in 2030.

Coal Sector Profile

Coal is vital to the Chinese energy sector, supplying 69% of primary energy demand in 2000, and 87% of all fuel consumed in the power sector. China has very large proven recoverable coal reserves, at around 115 billion tonnes, or around 12% of the world total. Around 84% are estimated to be bituminous or sub-bituminous coal, with the balance being brown coal.

China's coal reserves and production are concentrated in the north and north-west of the country, with 22% produced in the Shanxi province, 10% in Inner Mongolia, 9% in Shandong, 7% in Hebei, and 5% in Anhui. The long distances between producing regions and many consumers mean that coal transportation is a significant logistical and cost problem in China. In the southern consuming regions, in particular, imports could play an increasingly important role as a lower-cost alternative to domestic supplies.

China's impressive economic performance over the last 20 years, combined with sustained investment in coal production to meet demand growth, resulted in China overtaking the United States as the largest coal producer and consumer in the world in the late 1980s. This remains the case

25. This includes around 17 Mt of capacity already added since 2000 at the Dalrymple Bay coal terminal in Queensland.
26. Includes exports of around 15 Mt of coke oven coke.

today, despite a decline in production from a peak of 1,402 Mt in 1996 to 1,231 Mt in 2000. China has also become the second-largest exporter of coal after Australia.

The closure of the small town and village enterprise mines, as well as continued rationalisation of the coal industry into larger producers, have helped to boost the share of production of the largest state-owned coal enterprises, with the ten largest producers estimated to have a 22% share of total production in 2000, compared to around 16% in 1990.[27]

China's coal industry is benefiting from significant investment in rail and ports. A notable example is the dedicated double track line of around 600 km linking the Shendong coal field to the newly constructed Huanghua coal export harbour. This line is estimated to have a potential capacity of 100 Mt per annum, while the port facility has an initial annual capacity of 30 Mt.

The railways transported around 685 Mt of coal in 2000, or around two-thirds of China's reported production, over an average distance of 550 km. The coal export capacity of China's ports is estimated at around 257 Mt per annum, with throughput in 2000 of around 190 Mt.[28]

Mining and Port Investment

China will require cumulative investment in its coal mining and port infrastructure of $122.7 billion over the projection period.[29] This is equal to 34% of the world total and is the largest level of investment needed in a single country or region. Investment in coal mining will account for virtually all this investment, at around $120.6 billion. Even with growing imports, exports and internal seaborne coal trade, the investment need for coal-handling facilities at Chinese ports is only projected to be $2.1 billion over the projection period (Figure 6.19). China's mining investment needs will be driven by rapid growth in coal production, from 1,231 Mt in 2000 to 2,304 Mt in 2030. Metallurgical coal production will grow by 1.2% per annum and steam coal production by 2.2% per annum (from 1,107 Mt in 2000 to 2,127 Mt in 2030).[30]

Virtually all of the production growth is required to meet growing domestic demand. Exports fall as a share of production, from 5.7% in 2000 to 5.4% in 2030, although the absolute tonnage grows from 70 Mt in 2000 to 123 Mt in 2030. China's exports are concentrated in Asia, where it has a freight

27. RWE Rheinbraun (2002).
28. RWE Rheinbraun (2002), Ball, A, *et al.* (2003).
29. As discussed in Annex 2, these figures do not include investment in the existing rail network, only to reach the existing network. Unlike most countries, this investment is likely to be very large in China.
30. Chinese brown coal production is not reported separately from total thermal coal production.

Figure 6.19: **Cumulative Coal Investment in China**

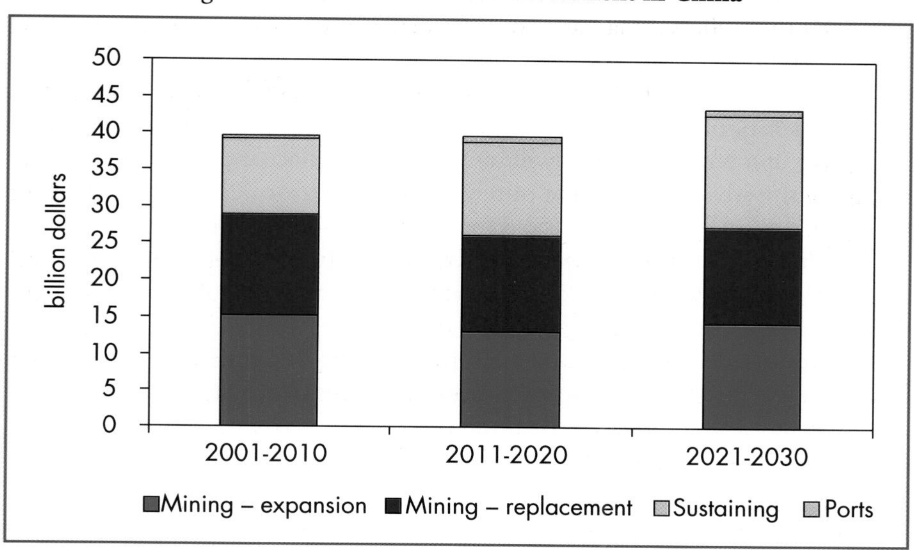

advantage. In 2030, 90% of its exports are expected to go to Japan, Korea and East Asia. Exports to Korea will grow steadily, accounting for 36% of China's exports in 2030, while exports to East Asia rise rapidly from 17% in 2000 to 28% in 2030. China's exports to Japan are projected to increase until around 2015, before declining slightly in line with Japan's reduced import needs.

Around 68%, or $82.4 billion, of the cumulative mining investment will be needed to add around 2,126 Mt of new capacity to meet production growth and replace depleted mines. This is equal to more than today's combined production from all of the OECD countries, plus Indonesia and India.

The closure of the small town and village enterprise mines, reports suggest that between 50,000 and 80,000 have been closed, is expected to continue and contributes to the high new capacity needs in China.[31] Expansion of these mines was encouraged by the government from 1983 to 1997 to accelerate rural development and solve some supply problems in the countryside. They were estimated to have reached a 45% share of production by 1996, and the decline in production after 1996 was mostly driven by government policy to close a large proportion of these small, inefficient and often unsafe mines.

The official reasons given for the new policy were a desire to improve the quality of production, eliminate unsafe and inefficient mining practices and improve reserve life. However, part of the reason for closing these small mines may be that the decline in consumption from 1996 meant that the smaller

31. Reuters (2002).

town and village mines, with lower costs, were gaining market share at the expense of key larger state mines.[32] The key state mines found themselves accumulating large stocks and in a deteriorating financial position.

The share of investment required just to replace capacity that closes over the *Outlook* period is around 33% ($39.7 billion) of total mining needs. Around 1,054 Mt of capacity will be needed to replace capacity that is either closed by the government or for purely economic reasons. New mine capacity to meet demand growth is expected to amount to 1,072 Mt between 2000 and 2030. The cumulative investment required for this expansion is around $42.6 billion or around 35% of Chinese mining investment requirements over the *Outlook* period.

Investment requirements are affected by the share of large state-owned enterprises compared to municipal and town mines in total production. The key state mines' share is projected to grow from around 26% in 2001 to around 40% in 2030. This is in line with China's ambitious Tenth Five-Year Plan for 2001 to 2005, which aims to concentrate around 35% of production in eight large coal corporations by 2005. The increase in the large state mining companies' share of total production will raise investment needs. Investment costs per tonne of capacity for large state mines are expected to be between 20% and 50% higher than those of smaller mines, because of large state mines being fully mechanised, adhering to social and safety regulations, etc.

Around 32%, or $38.2 billion, of the total cumulative mining investment is needed to maintain production and productivity (sustaining investment). This share is lower than in the export-focused, high-productivity regions, but it will increase over time as the industry becomes more mechanised, more advanced technologies are applied and higher safety standards are introduced.

Although exports are not a large proportion of China's total production, they account for a significant proportion of the international trade in coal. Although developments in China's coal exports are driven by hard-currency receipts, relatively minor changes in the domestic coal market balance of supply and demand could add or remove significant quantities of coal from export markets. This can have a noticeable impact on world trade patterns and the volatility of prices, especially since most of China's exports are destined for the Asian market. China's coal exports (including coke oven coke) increased from 37 Mt in 1995 to close to 100 Mt in 2001. Further growth is expected to be modest given the demands of the domestic market, with exports reaching 123 Mt in 2030.

32. These small mines have poorer safety standards and often do not contribute to worker welfare payments. They often also "free-ride" on the state-owned mines infrastructure development in a mining area, benefiting from road, rail, and power infrastructure to which they do not contribute financially.

China has significant internal trade in coal, much of which is carried by rail from the producing provinces in the north to ports and then shipped to the consuming provinces in the south. In 2000, this internal seaborne trade amounted to around 120 Mt.[33] Total port throughput (including exports) was around 190 Mt in 2000. The coal export capacity of China's ports in 2000 was around 220 Mt to 230 Mt.

Growth in the amount of coal both exported and shipped internally is expected to result in a need for an additional 69 Mt of export-facilities at ports, and an additional 61 Mt of receiving capacity to handle internal shipments and imports. This will require cumulative investment of around $2.1 billion over the *Outlook* period, or around 17% of the world total for ports.

Major reforms will need to be implemented in the coal industry if the investment projected here is to be achieved in a timely and efficient manner. Investments will be required to raise the mechanisation rate and productivity of mines in order to reduce production costs. Subsidies, social obligations and other market-distorting government interventions will need to be phased out. Ensuring that potential foreign investors are on a level playing field with domestic producers will encourage foreign investment and help China to bring its industry standards for safety, productivity and costs closer into line with international norms. An inability of the domestic industry to meet the investment requirements presented here could lead to increased foreign investment and/or substantially higher imports.

India

India is projected to remain the world's third-largest coal producer after China and the United States and will require cumulative investment of around $25 billion over the *Outlook* period. Production will increase from 329 Mt in 2000 to 652 Mt in 2030. India's primary energy demand for coal is projected to grow by 2.4% per annum from around 165 Mtoe (345 Mt) in 2000 to 341 Mtoe (696 Mt) in 2030. Imports grow from around 22 Mt to 46 Mt in 2030.

Coal Sector Profile

Coal provided around one-third of India's primary energy demand in 2000. However, the very high contribution of biomass obscures coal's vital role in the Indian economy. Coal provided 85% of all fuel consumed in the electricity generation sector in 2000.

India has around 84 billion tonnes (9% of the world total) of proven recoverable coal reserves. The coal is generally low in sulphur, but has a high

33. Ball, A, *et al.* (2003).

ash content and low calorific value. The high ash content raises costs to power generators, as it lowers boiler efficiency and increases ash disposal costs. Most of India's coal resources are to be found in the centre and east of the country, far from many consuming areas. As in China, this means large quantities of coal are transported over long distances by the rail network with some shipped by a combination of rail and sea, which significantly increases handling and transport costs.

The domestic coal industry provided around 95% of coal needs in 2000. Given the poor quality of Indian metallurgical coal, significant high-quality coking coal imports are required for direct use or for blending with local metallurgical coal to improve its performance. Steam coal imports have been growing in recent years and are generally competitive with local coal in the coastal states of Tamil Nadu, Kerala, Maharashtra and Gujarat, which are all some distance from local production areas of coal.

Indian coal is generally of low quality and is relatively expensive for consumers at a distance from the mines, as mine-mouth prices are high as a result of low productivity and transport is subject to high freight costs.[34] The generally poor ability of electricity generators to pay because of their own financial problems, and the price ceiling set by import prices, mean that the sales revenues received by coal companies often do not exceed costs.

Mining and Port Investment

India will need cumulative investment of $24.9 billion in its coal mining industry and ports over the period 2001-2030, or around 7% of the world total. Investment in coal mining, at around $24.1 billion, is by far the most important component. With growing imports and internal seaborne coal trade, investments of around $0.8 billion are required in coal loading and unloading facilities at ports (Figure 6.20).

In 2000, over 95% of hard coal production in India was produced by the government-owned companies, Coal India Ltd (CIL's) and Singareni Colleries Company Ltd (SCCL). Around 77% of the coal mined in India is from open-cast operations, and this percentage has been increasing steadily over the past 15 years.

Despite CIL's impressive growth in productivity in the 1990s, output at open-cast mines was only 1,675 tonnes per miner-year in 2000. This is significantly lower than the levels achieved by other major coal-producing countries, reflecting a shortage of modern equipment and overmanning.

34. Rail freight customers subsidise the passenger service, raising freight rates. As coal accounts for 42% (Tata Energy Research Institute, 2002) of freight transported (in tonne-km) on the rail network, this burden falls mainly on coal consumers. The low calorific value of coal transported also raises transport costs per unit of energy compared to imports.

Figure 6.20: **Cumulative Coal Investment in India**

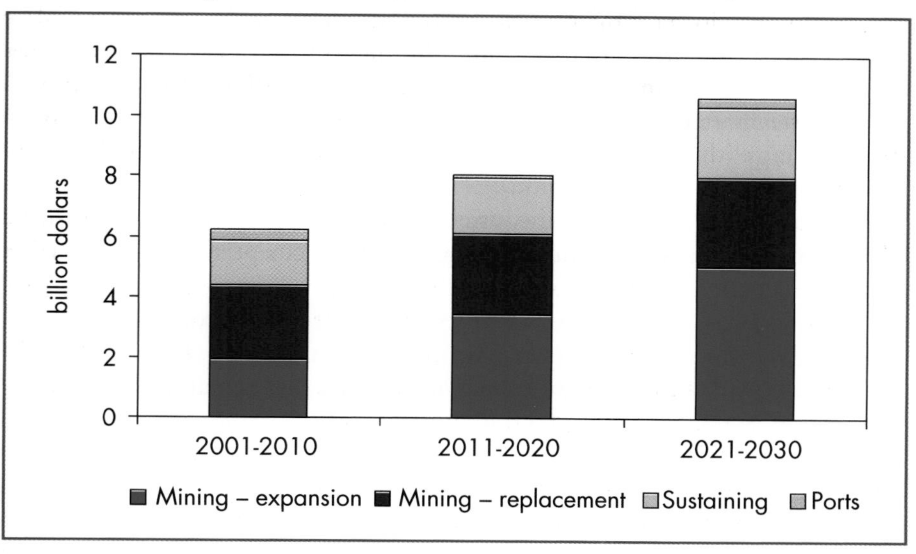

Productivity at underground mines is even worse, at just 200 tonnes per miner-year at CIL and 225 tonnes at SCCL. This low level of productivity is the result of low levels of mechanisation in underground mines in India. In 1999-2000 only around 40% of the underground production of hard coal came from mechanised mines.[35]

New investment needs to be focused on large-scale, efficiently designed mines, incorporating a greater degree of mechanisation in order to raise productivity and lower coal-production costs. If this is not achieved, CIL's financial position, and the sustainability of its operations, will not improve. Foreign direct investment could play a vital role in injecting the necessary capital and technological and operational expertise required for India's coal industry is to meet demand growth.

India's mining investment will allow production growth of 323 Mt, from 329 Mt in 2000 to 652 Mt in 2030. Metallurgical coal production has declined in recent years, but is expected to increase slightly from 2000 levels and stabilise at around 34 Mt. Steam coal production will grow strongly, at around 2.4% per annum, from 279 Mt in 2000 to 562 Mt in 2030. Lignite production will grow even more rapidly, at 3.0% per annum, but from the lower base of 23 Mt in 2000 to 56 Mt in 2030. All of the production growth is required to meet growing domestic demand.

35. IEA (2002b).

Of the total cumulative mining investment, around 77%, or $18.5 billion, is needed in order to add new capacity of 578 Mt to meet production growth and replace production capacity at depleted mines. This investment will need to include a greater proportion of investment in coal-washing machinery, in order to improve the competitiveness of Indian coal per unit of energy (the calorific value of Indian steam coal is projected to increase modestly over the *Outlook* period). This will also reduce the volume of coal transported by rail, because of the lower ash content of the coal transported. It will also benefit electricity generators by improving the performance of power stations.

Of the required new capacity of 578 Mt, around 256 Mt will be needed to replace the depleted capacity at existing mines that close over the projection period. This will require investment of $8.1 billion. The share of investment required to replace depleted mines is only slightly higher than in OECD countries, at around 34% of the total mining investment needs. The new productive capacity required to meet demand growth between 2000 and 2030 is around 323 Mt. The cumulative investment needed for this expansion is around $10.4 billion, or around 43% of total Indian mining investment requirements over the *Outlook* period.

Sustaining investment will amount to $5.6 billion, or around 23%, of total cumulative mining investment. This is relatively low compared to India's level of production and although this investment is crucial to maintaining productivity and keeping costs low, it is unlikely that the financial situation of the major coal mining companies will improve sufficiently to allow this investment to increase significantly.

Imports will more than double over the *Outlook* period, but still remain a relatively small percentage of total Indian coal demand. However, if investment in the Indian coal-mining industry were to be constrained and domestic coal prices were to rise, then import growth could be even higher than projected here.

Although not as large a share as in China, India's internal seaborne coal trade is still significant. It is carried by rail from the producing states in the north and east to ports and then shipped to the consuming states in the south. This internal seaborne trade, currently around 15 Mt, will grow modestly over the *Outlook* period owing to competition from imports. International coal imports and internal trade will together require an additional 55 Mt of coal-handling facilities. This will require investment of around $800 million over the period 2001-2030, or around 6% of the world total.

The ability of the Indian coal industry to finance the investment expected in this study is uncertain. At present, the problems in the power sector, the coal industry and the rail industry are interlinked. Much will therefore depend on the progress made in implementing reforms across the Indian economy. The financial viability of each of these three industries needs to be improved simultaneously if the constraints are to be overcome.

Power-sector reform will be particularly important as it is the principal consumer of Indian coal. Profitability needs to be improved in order to make the industry financially self-sustaining. This would not only increase coal demand, as more new power stations could be financed, but would also ensure coal producers are paid a fair price for the coal, improving their ability to invest. Reform of the coal industry would be complementary to this process. Similarly, the loss-making Indian railway system needs to be put on a commercial footing and cross-subsidisation phased out. This would lower freight costs and help remove logistical constraints.

Reform of the coal industry means freeing prices, allowing competition for access to coal reserves, reducing the constraints on foreign investment and removing bureaucratic hurdles and delays. The problems of restrictive employment policies and overmanning also need to be tackled if productivity is to rise, while rationalisation of loss-making mines is required. Without these reforms, the Indian coal industry will remain capital-constrained and will struggle to meet demand growth, as the government is unlikely to be able to meet any shortfall in funding requirements, given the constraints on its budget.

Africa

Cumulative coal investment of around $22 billion will be required in Africa over the *Outlook* period. Production, over 97% of which came from South Africa in 2000, is expected to increase from 230 Mt in 2000 to 415 Mt in 2030. Africa's primary energy demand for coal is projected to grow by 2.2% per annum, from around 91 Mtoe (167 Mt) in 2000 to 174 Mtoe (318 Mt) in 2030. Exports are expected to grow from around 71 Mt in 2000 to 110 Mt in 2030. Imports, most of which are intra-regional trade from South Africa to other African countries, grow from around 8 Mt to 13 Mt in 2030.

Coal Sector Profile

Coal met around one-fifth of Africa's primary energy needs in 2000 and provided around 48% of all the electricity generated in Africa (the percentage is higher in South Africa). A significant quantity of coal is also consumed by coal liquefaction plants in South Africa.

Africa has around 55 billion tonnes of proven recoverable coal reserves. This is around 6% of the world total. South Africa accounts for 50 billion tonnes of the total reserves and, similarly, dominates coal production, with 224 Mt of production in 2000 compared to the total for Africa of 230 Mt.

South Africa has 19 coalfields, with around 90% of production in 2000 coming from the Witbank, Highveld and Vereeniging-Sasolburg regions. The South African export coal industry is concentrated in the inland Witbank coalfield, with around 59 Mt of this coalfield's 108 Mt of production in 2000 carried by rail the 460 to 640 km to the export terminal at Richards Bay. The Highveld coalfield provides the bulk of coal used domestically, with around 68 Mt of the total production of 73 Mt used by local electricity generators, synfuels and metals plants.

South Africa's coal rail network comprises three main routes: from Witbank in the Transvaal to Richards Bay (580 km);[36] the Maputo rail line, which links the provinces of Gauteng, Mpumalanga and the Northern Province to Maputo in Mozambique (approximately 420 km); and the line from Mpumalanga to Durban. Freight rates rose significantly in rand terms during the 1990s in order to finance the large capital expenditures incurred by Spoornet, the railroad operator.[37]

South Africa's coal exports are predominantly shipped through the Richards Bay coal terminal (RBCT), which had a capacity of around 72 Mt per annum in 2000. Exports are also shipped from Durban (capacity of 2.5 Mt per annum) and Maputo in Mozambique (capacity of 5 Mt per annum). The RBCT is being expanded and 10 Mt per annum of capacity should be added by 2005.

Mining and Port Investment

It is projected that Africa will require cumulative investment in its coal mining and port infrastructure of $22.2 billion over the *Outlook* period, or around 6% of the world total. Investment in coal mining itself makes up $21.8 billion of this figure. With imports and exports projected to grow, investment of around $400 million in coal loading and unloading facilities at ports will be required (Figure 6.21).

Africa is projected to increase its coal production by 184 Mt, from 230 Mt in 2000 to 415 Mt in 2030. Almost all this increase in production will be of steam coal. Only 39 Mt of this increase will be for exports. South Africa is expected to account for around 179 Mt, or 97%, of this production increase.

South Africa's coal exports are expected to expand only moderately to 83 Mt in 2010, before climbing to around 103 Mt in 2020 and 110 Mt in 2030. Its exports to OECD Europe will increase in absolute terms, but decline as a share of Africa's exports from around 63% in 2000 to 51% in 2030.

36. A South African rand 201-million upgrade was approved in 2000, which will lift this line's capacity to 82 Mt per annum between 2006 and 2011.

37. For example the freight rate in rand per tonne for coal from Blackhill to Richards Bay increased by around 40% between 1994/95 and 2001/02.

Figure 6.21: **Cumulative Coal Investment in Africa**

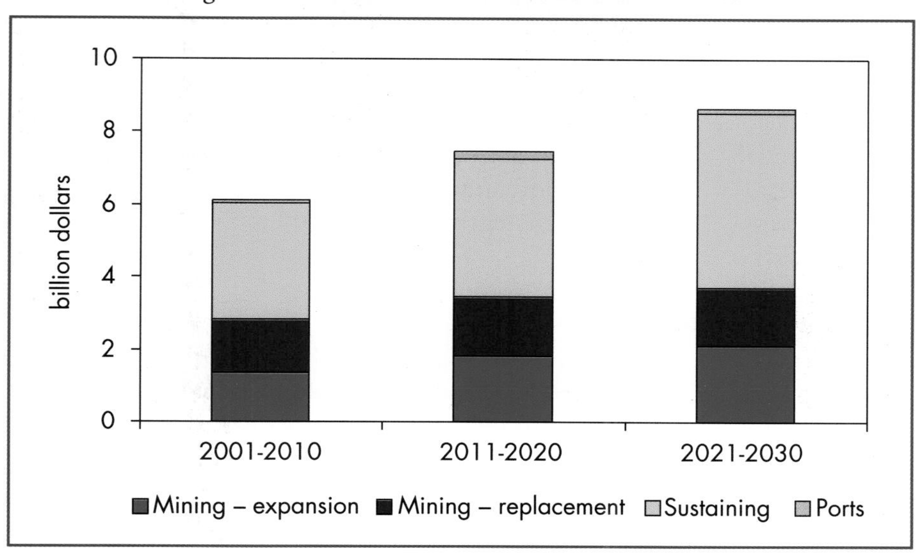

Exports to the Middle East will more than double, but from a low level, reaching 11% of exports by 2030.

Part of the reason for the modest growth in exports is that, although South Africa has abundant coal reserves, the depletion of many of the low-cost mines producing high-quality coal in the Witbank region will mean significant new investment is required in order to maintain and expand exports over the *Outlook* period. The development of new export reserves further from export ports will raise costs and limit export growth. Production for domestic consumption in electricity generation and industry (including coal liquefaction plants) is expected to grow significantly, but lower-quality reserves that are cheaper to mine should meet this requirement.

Of total cumulative mining investment, 46%, or $10.0 billion, is needed in order to add 366 Mt of new production capacity to replace capacity at depleted mines and meet demand growth. Of this capacity, 173 Mt will be required to replace production capacity from mines that will deplete their economic reserves over the *Outlook* period. The share of total mining investment required to replace depleted mines is 22% ($4.7 billion), or just less than half of the total needs for investment in new capacity. The new productive capacity to meet demand growth is around 193 Mt. The cumulative investment required will be $5.3 billion, or 24% of the total mining investment requirements over the *Outlook* period.

Around $11.8 billion, or 54% of the total cumulative mining investment, will be needed for sustaining capital investment in order to maintain production and productivity at mines. This level of investment will be an important component of South Africa's efforts to remain competitive in international markets.

Coal exports and imports will require an additional 40 Mt of coal-handling facilities. The 10 Mt per annum expansion of capacity at Richards Bay will probably be sufficient to meet export demand growth to around 2008-2009. The additional coal import and export-handling facilities (including at Richards Bay) will require investment of around $400 million, or around 3% of the world total.

Transition Economies

Cumulative coal investment of around $32 billion will be required in the transition economies over the *Outlook* period.[38] Production will increase from 528 Mt in 2000 to 645 Mt in 2030. Primary energy demand for coal in the transition economies is projected to grow by 0.7% per annum, from around 213 Mtoe (514 Mt) in 2000 to 260 Mtoe (628 Mt) in 2030. Production and primary energy demand for coal are both well down from their peaks prior to the collapse in demand in the transition economies that occurred in the 1990s. Exports will grow from around 70 Mt in 2000 to 94 Mt in 2030, while imports grow from around 51 Mt to 77 Mt.

Russia is the world's fifth-largest hard coal producer, with production in 2000 of around 242 Mt (46% of the total transition economies' figure). Russia's coal production is projected to grow to 290 Mt in 2030. This is well below the peak of 437 Mt achieved in 1988. Russia's primary energy demand for coal was around 111 Mtoe in 2000, or around 52% of the transition economies' total. Russia's exports are projected to grow from around 40 Mt in 2000 to 62 Mt in 2030.

Russia's modest growth in coal production, despite its large coal resources, is due to the large distances between Russia's major coal reserves and its population centres, industry and ports.

Coal Sector Profile

Coal met a little more than one-fifth of the transition economies' primary energy needs in 2000. This reflects both the current and historical relative price levels which favour oil and gas, particularly in Russia where coal was 6% more expensive than gas as late as 1999.[39]

38. The transition economies are separated into three regions in the World Energy Model: Russia, Annex B transition economies, and the other transition economies (see Annex 3 for regional definitions).
39. IEA (2002c).

and financial) likely to be expended on the continuing restructuring of the industry. This element is expected to be a higher share (48%) of Russia's investment requirements, with a higher priority being given to maintaining competitiveness in domestic and international markets.

The coal exports of the transition economies are projected to expand from around from 70 Mt in 2000 to 94 Mt in 2030. However, only around 43 Mt of these exports in 2030 will be to countries outside the transition economies. 51 Mt in 2030 will be exports from one transition economy to another.[42]

Russia will increase exports from around 40 Mt in 2000 to 62 Mt in 2030. OECD Europe will remain the largest market for Russian coal, with no significant change in Russia's pattern of trade expected over the *Outlook* period.

The prospects for exports could be better than projected here, as a number of new developments have been initiated in Russia, with the objective of using the latest "high-technology" engineering, in the most favourable geological conditions, in order to achieve highly profitable operations. The aim is to achieve high productivity (6,000 to 9,000 tonnes per miner-year, nearly twice the current maximum) and low production costs, estimated not to exceed $8 per tonne.

If exports were to be able to expand faster than projected here, significant problems with the transport and port infrastructure will need to be overcome. The Russian rail network is in a generally poor state of repair and there is no imminent prospect of improvement, given that subsidies fail to cover even the losses on current operations. This could become a significant constraint on the growth of exports and domestic consumption.

Russia's coal export capacity at ports handling vessels greater than 30,000 DWT is currently around 25 Mt per annum. There are options for export through the Baltic and Ukrainian ports, but these are limited. The growth in Russia's coal exports will require an additional 21 Mt of coal-handling facilities, requiring an investment of $300 million over the *Outlook* period, or around 2% of the world total. The Russian Ministries of Transport and Railways favour expansion at Vostochniy and Vanino on the Pacific and at Novorossiysk and Tuapse on the Black Sea, along with the Rosterminalugol project at Ust Luga on the Baltic. However, serious doubts exist about the plans for Vostochniy and Rosterminalugol.[43]

Even if the rail and port infrastructure constraints were to be adequately addressed, it is doubtful whether exports could grow much more than projected here in an already crowded and very competitive international market.

42. Only Russia exports coal to countries other than the transition economies.
43. McCloskey (2002b).

Latin America

Cumulative coal investment of around $10 billion will be required in Latin America over the *Outlook* period. Coal production in the region is expected to grow at 2.6% per annum, from almost 54 Mt in 2000 to 115 Mt in 2030. Primary energy demand for coal in Latin America is projected to grow at 2.3% per annum, from almost 23 Mtoe (33 Mt) in 2000 to around 44 Mtoe (70 Mt) in 2030. Exports will increase from 44 Mt in 2000 to more than 92 Mt in 2030.

Coal Sector Profile

Coal met 5% of primary energy demand in Latin America in 2000, of which 65% was used in Brazil. Latin America has proven recoverable coal reserves of 21.8 billion tonnes, of which 6.7 billion tonnes are in Colombia, 11.9 billion tonnes in Brazil and 0.5 billion tonnes in Venezuela.

Latin America exported 82% of its coal production in 2000. Coal production in Latin America in 2000 was headed by Colombia (71%), followed by Venezuela (15%) and Brazil (13%). Exports come mainly from Colombia (around 81%) and Venezuela (around 18%). The largest importers are Brazil (68%) and Chile (21%). The main destinations for Latin America's exports are the EU15 and North America, which together account for more than 90% of Latin America's export demand.

Colombia exported around 36 Mt in 2000, which accounted for around 93% of its production of mainly low-sulphur steam coal. The main producing areas are the Guajira peninsula (Cerrejón Norte) and the Cesar province. Cerrejón Norte is one of the world largest opencast mines and produced 19.4 Mt in 2001.

Mining and Port Investment

Latin America will need to invest around $9.8 billion in coal mining and port infrastructure over the projection period. Investments in Brazil will account for $600 million, or only 6% of the total. Production in Latin America will grow to 115 Mt by 2030, with exports accounting for around 92 Mt of the total. Country shares of production will remain constant, with Colombia still accounting for around three-quarters of the total.

Investments in coal mining will account for around 2.4% of the total world investment, or $8.6 billion. The new productive capacity that will need to be added is around 103 Mt, of which 61 Mt represents new capacity to meet demand growth and 42 Mt is needed to replace depleted mines. The corresponding investment needs will be $3.3 billion (39% of mining investment) for new capacity to meet demand growth and $2.3 billion (26% of mining investment) for new capacity to replace depleted capacity.

The remaining 35% of mining investment, or $3.0 billion, will be required for sustaining capital investment to maintain and increase the mine productivity.

The capital cost of new capacity additions rises over the projection period in order to reflect the increasing investment that will be required to improve Latin America's infrastructure if it is to support a doubling of exports. The average cost of new port capacity is also above the world average, reflecting the higher cost of developing predominantly greenfield port infrastructure in Latin America.

Around 13% of the region's total investment will be required for coal export and import-handling facilities at ports, corresponding to $1.2 billion (around 10% of the world total for coal ports). This relatively large investment is due to the more than 100% increase in exports and to the currently sparse existing infrastructure, in particular in Venezuela, which will require significant new investment if projections of export growth are to be met.

Indonesia

Cumulative coal investment of around $12 billion will be required in Indonesia over the *Outlook* period. Coal production will grow at an average annual rate of 4.0%, from 77 Mt in 2000 to 248 Mt in 2030. Indonesia's primary energy demand for coal will increase more than four-and-a-half times from 2000 to 2030, growing at an average annual rate of 5.2% over the next 30 years, from 14 Mtoe (22 Mt) to 63 Mtoe (102 Mt). Exports will grow more slowly than demand, at still healthy 3.3% per annum, from 55 Mt to 146 Mt in 2030.

Coal Sector Profile

Domestic coal demand is predominantly for power generation, with the state electricity utility, Perusahaan Listrik Negara (PLN), being Indonesia's largest coal consumer. Coal demand growth will be driven by demand from the power sector, because electricity demand grows rapidly at a time when coal is projected to increase its share of electricity generation from around 31% in 2000 to 54% in 2030.

Coal is produced by the state-owned company Tambang Batubara Bukit Asam (PTBA), 25 private companies (both foreign and domestic) that operate under the new Work Agreement on Coal Mining Enterprises (CCoW) and 7 Cooperative Units. The CCoW requires that the majority of coal companies operating in Indonesia should be owned by domestic investors. This has discouraged some foreign investors. However, when first introduced, the CCoW was seen as a model of how to attract investment.

The four biggest private companies are Adaro Indonesia, Kaltim Prima Coal, Arutmin and Kideco Jaya Agung. In 2000, these companies produced more than 44 million tonnes of coal (58% of total coal production), and more than 53 million tonnes in 2001.[44]

Indonesia has become a major coal exporter during the last 10 years, with exports increasing more than elevenfold, from just 5 million tonnes in 1990 to 55 million tonnes in 2000. Indonesian exports go mostly to East Asia, Japan, Korea and the EU15. Indonesia has 17 ports handling coal from vessels of more than 30,000 DWT, with coal-handling capacity of over 90 Mt per annum.[45]

Mining and Port Investment

Investment of around $11.6 billion will be needed in coal mining and ports over the *Outlook* period in Indonesia. Mining investment will need to be $9.8 billion, or an average of $330 million per year for the next three decades in order for coal production to expand from 77 Mt in 2000 to 248 Mt in 2030. Indonesia's share of world exports is projected to reach almost 14% by 2030. Of the total mining investment, 56% will be needed to meet demand growth, 24% to replace depleted mines and the remaining 20% for investment in sustaining production and productivity at mines (Figure 6.23).

The average investment need of $330 million per year compares to an average of around $200 million per year for the period 1997-2001.[46] The difference may not be as large as it seems, since the reported figure relates to only major mining companies and excludes a significant amount of mining equipment investment, as it is leased. Nevertheless, this level of investment remains a large challenge and will require significant foreign investment.

Although Indonesia's proven recoverable coal reserves, at 5.4 billion tonnes, are large, substantial investments will be needed in exploration, mine development and infrastructure if the production growth projected here is to be achieved. To reach the indicated levels of production, Indonesia will have to add 244 million tonnes of new capacity in the next three decades, of which only 30% will be needed to replace depleted mines.

Around 99% of Indonesian coal mines are opencast and their recent capital investment costs have been quite low ($22 per tonne on average). This is, in part, due to the fact that many recent developments have used leasing arrangements for the capital equipment, reducing the reported capital

44. Directorate of Mineral and Coal Enterprises (2002).
45. SSY Consultancy and Research Ltd (2002).
46. Ministry of Energy and Mineral Resources of Indonesia (2003).

Figure 6.23: **Cumulative Coal Investment in Indonesia**

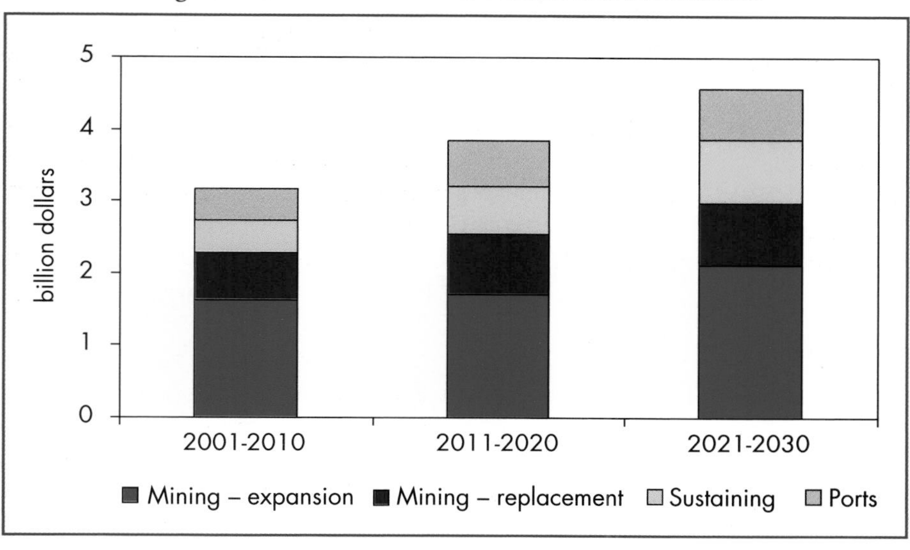

requirement. As a result, capital costs for new developments are assumed to grow over the *Outlook* period to reach the level of $35 per tonne by 2010. The depletion rate of existing mines is expected to be higher than in many other regions, reflecting the continuing trend for developments which are high-volume operations exploiting short-life deposits.

The big expansion in Indonesian exports will require investment of $1.8 billion in new coal-handling facilities at ports. Around one-third of this investment is needed to accommodate the growth in internal coal trade. Indonesian port capacity is expected to grow from around 104 Mt in 2001 to around 227 Mt in 2030, including facilities for internal trade.

East Asia

The investment in East Asia (excluding Indonesia) will amount to almost $8.3 billion in the next thirty years. One-third of this investment will be needed at ports in order to develop import-handling facilities to meet the growth in imports.

Coal production is expected to grow by less than 1% per annum over the *Outlook* period in East Asia. The Democratic People's Republic of Korea, Thailand and Vietnam will continue to account for almost all coal production.

The rapid growth in coal-fired electricity generation in this region and the relatively low percentage of this demand that can be met economically from domestic sources will result in rapid growth in coal import needs. East Asia

Figure 6.24: **Cumulative Coal Investment in East Asia**

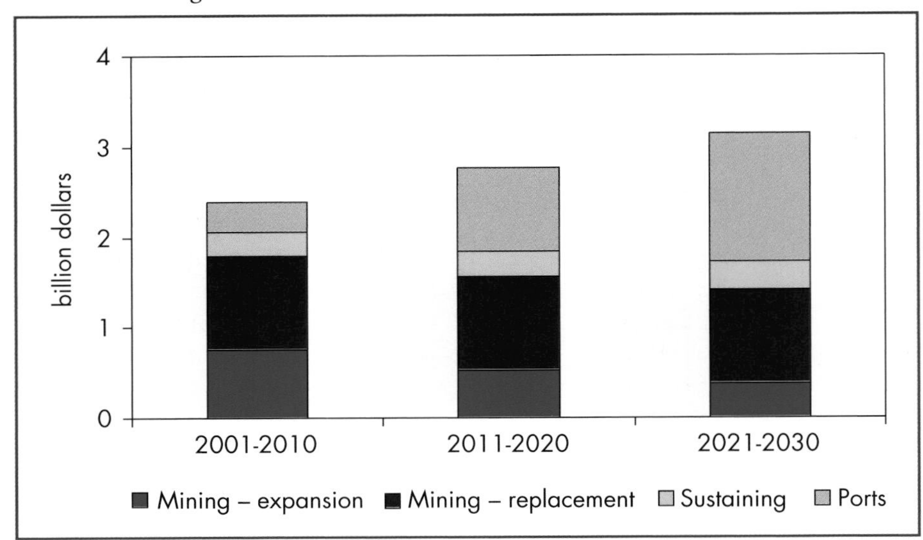

will become the largest coal-importing region in the world as imports increase from 60 Mt in 2000 to 251 Mt in 2030, accounting for almost 24% of world imports in 2030. To support this increase in coal imports, the region will need to add 184 Mt of coal import-handling facilities at ports, at a cost of $2.7 billion, over the *Outlook* period.

OECD Alternative Policy Scenario

As in previous chapters, the analysis up to this point has related to the world view contained in the *WEO-2002* Reference Scenario. However, that study also included, for the OECD countries, an Alternative Policy Scenario assessing the effect on OECD energy demand, supply and CO_2 emissions of more vigorous policies designed to address climate change and energy security.[47]

The *WEO-2002* Reference Scenario included all policies and measures that had been adopted by governments in mid-2002 in order to address energy security of supply and environmental issues. The OECD Alternative Policy Scenario analyses the impact of the policies and measures that OECD countries were considering in mid-2002, but had not adopted. Developments since then are of local significance, but leave the overall picture unchanged.

47 The OECD Alternative Policy Scenario covered EU15, United States and Canada, and Japan, Australia and New Zealand.

The policies in question include additional policies aimed at curbing CO_2 emissions, reducing local pollution and energy-import dependence. The basic assumptions about macroeconomic conditions and population are the same as in the Reference Scenario.[48] The OECD Alternative Policy Scenario does not take into account carbon sequestration technologies, which could alter the situation presented in this scenario by achieving CO_2 emissions cuts without reducing coal demand to the same extent.

Global coal investment is around $24.7 billion, or 6.2%, lower than in the Reference Scenario. Some $13.8 billion of this total occurs in the OECD regions examined. This results from a fall in coal demand in the OECD to 524 Mt below the Reference Scenario in 2030. A coal demand increase in these regions from 1,828 Mt in 2000 to 1,993 Mt in 2030 in the Reference Scenario is reversed to a decline to 1,480 Mt in 2030, a 26% decline below the Reference Scenario in 2030, or 19% below 2000 levels.

EU15 imports in 2030 are 75 Mt (35%) lower than in the Reference Scenario at only 141 Mt, while Japan's imports are 28 Mt (21%) lower, at only 105 Mt. World trade in coal is 108 Mt lower than the Reference Scenario in 2030, increasing by only 48% over the *Outlook* period, compared to 65% in the Reference Scenario.

For coal, the critical elements affecting demand are those policies and measures that reduce electricity demand, given coal's competitiveness for electricity generation after 2010, and policies that induce a switch away from fossil fuels in electricity generation. Policies that directly target coal demand in the industrial, residential and services sectors will have a much lower impact on demand, reflecting the relatively low coal consumption in these sectors in OECD countries.

Total electricity demand is reduced significantly below the Reference Scenario, with savings reaching 11%, or 107 Mtoe, in 2030. Around 41% of the savings are attributable to the residential sector and two-thirds to the residential and services sectors combined.

In the first decade, the most important reductions in fossil fuels are from natural gas, while the savings from coal accelerate after 2010. This is the result of savings in electricity demand and increased renewables growth, which displace new gas-fired generation to 2010 and mostly new coal-fired generation from then on. By 2030, the savings in coal exceed those for both oil and gas.

Global primary demand for coal is 524 Mt (7.5%) lower in 2030 than in the Reference Scenario, with around 96% of this reduction occurring in the electricity generation sector (Figure 6.25). The near-term impact on

48. See the *WEO-2002* for a complete description of the assumptions, scope and analytical framework of the OECD Alternative Policy Scenario.

Figure 6.25: **World Coal Production and Investment in the Reference and Alternative Policy Scenarios**

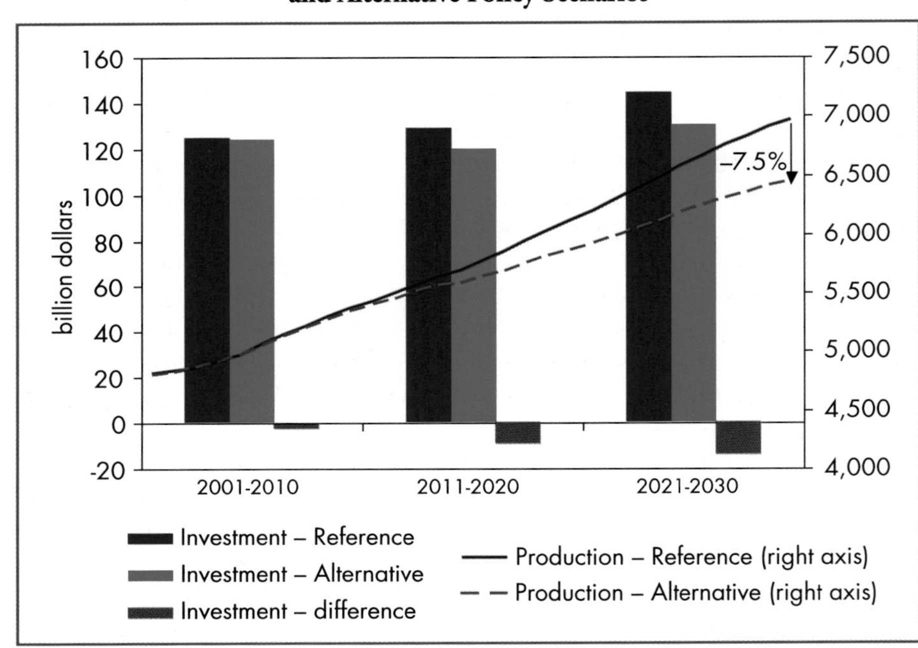

Note: Production is annual for the period 2000-2030, while investment is cumulative for each decade indicated.

production and investment is minimal: no premature scrapping of capital is assumed and the very long capital life of electricity generating units means that most of the reductions result from the abandonment of coal-fired plants that would have been built in the Reference Scenario but are not needed in the Alternative Policy Scenario.

Just over 60% of the reduction in demand of coal in 2030 comes from the United States and Canada. However, this has only a minor effect on their import requirements. Japanese demand is reduced by 21% below the Reference Scenario in 2030, to only 105 Mt. Japan's import requirements are, therefore, around 40 Mt per annum lower than 2000 levels in 2030.

The percentage declines in EU15 coal demand, production and imports are reasonably similar. Coal demand drops by 161 Mt below the Reference Scenario in 2030, to 272 Mt, or not much more than half of 2000 levels (Figure 6.26). Production in 2030 is around 90 Mt lower than the Reference Scenario, at 143 Mt, or less than half 2000 levels, as less brown coal is required for electricity generation. Imports are around 75 Mt (35%) lower than in the Reference Scenario in 2030, at just 141 Mt. Primary coal demand in the United States and Canada declines by around 303 Mt below the Reference

Scenario in 2030. Imports and exports also decline somewhat, while production is 317 Mt below the Reference Scenario in 2030. This means production in the United States and Canada is 73 Mt below 2000 levels in 2030.

The policies and measures considered in the OECD Alternative Policy Scenario would reduce cumulative global investment in coal by around $24.7 billion dollars, with $13.8 billion of this reduction occurring in the OECD regions examined, $6.1 billion in the rest of the world and a $4.8 billion reduction in shipping investment.

In addition to the direct impact of lower production levels in the OECD regions examined, investment by coal exporters around the world is affected by the reduction in global trade in coal of around 108 Mt below the Reference Scenario in 2030. This limits the increase in global coal trade between 2000 and 2030 to 307 Mt, representing a reduction in global trade of around 10% below the Reference Scenario in 2030.

Total cumulative investment in Japan, Australia and New Zealand is around $2.3 billion lower than in the Reference Scenario over the *Outlook* period, with around $2.0 billion of that attributable to the reduction in mining investment (Figure 6.26) and $0.3 billion less for export facilities at Australia's ports. Total cumulative investment in the United States and Canada is around $9.1 billion lower than in the Reference Scenario, with about $100 million less

Figure 6.26: **OECD Coal Production, Demand and Mining Investment**
(reduction below the Reference Scenario)

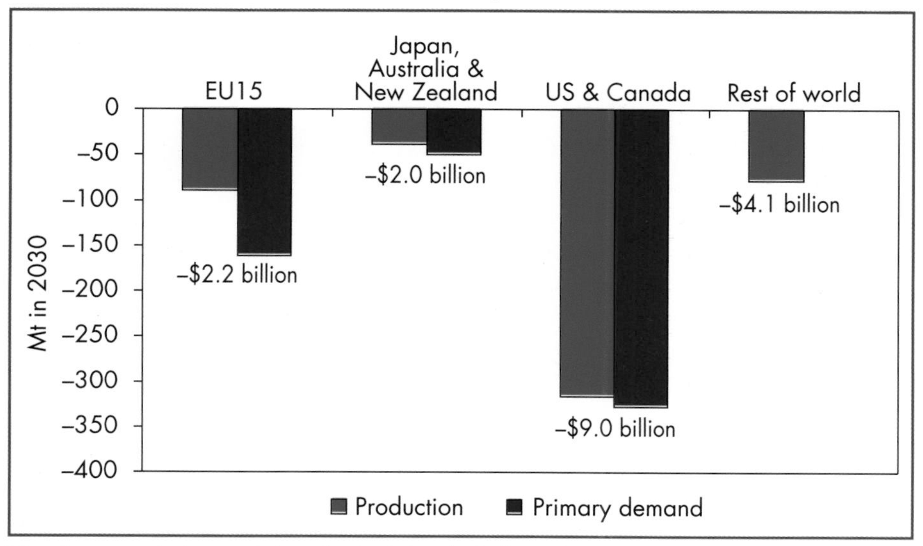

Note: Investment is the cumulative reduction below the Reference Scenario.

for ports. In the EU15 total cumulative investment is around $2.4 billion less than the Reference Scenario, with about $100 million less required in port development.

Although some of the reduction in import requirements in the OECD Alternative Policy Scenario of 108 Mt in 2030 is borne by OECD exporters, much of the reduction in exports to Japan and the EU15 affects other coal exporters. The biggest losers are Colombia and Venezuela, as most of their export growth into the EU15 does not eventuate, leaving their exports some 30% below Reference Scenario levels in 2030. Other notable exporters to suffer in the Atlantic or Pacific markets (or both) are China, South Africa, Indonesia and Russia.

The global reduction in mining capacity and port facilities below the Reference Scenario by 2030 means that global cumulative investment in mining and ports is around $19.8 billion lower in the OECD Alternative Policy Scenario. Around $2.5 billion of this saving comes from lower investment in export facilities at ports. In addition to the reductions in investment required in mining capacity and ports, around 14.8 Mt less of deadweight ship capacity is needed, given the reduction in world seaborne coal trade over the *Outlook* period. This results in investment in the bulk-dry fleet for coal being $4.8 billion less than in the Reference Scenario.

CHAPTER 7
ELECTRICITY

HIGHLIGHTS

- Global electricity-sector investment over the next three decades will amount to $10 trillion. This is 60% of total energy investment and nearly three times higher in real terms than investment in the electricity sector during the past thirty years. More than $5 trillion will go into transmission and distribution networks. Key factors that will determine investment in the power sector are competition and electricity-sector reform, environmental constraints and access to capital.
- OECD countries will require more that $4 trillion to expand and replace power production and delivery infrastructure. The new investment framework in liberalised electricity markets has created many new challenges and uncertainties. Concerns exist about the adequacy of investment as markets adapt to the new conditions, particularly with regard to electricity supply at times of peak load. The risks to investors for building peaking capacity are high, compared to baseload plant. Policy-makers need to address these concerns by providing a market framework that encourages adequate and timely investment. The value of security of supply needs to be recognised within the market framework.
- Investment in transmission networks requires particular attention. It has lagged behind investment in generation in some OECD countries, notably in the United States and some European countries. Liberalised electricity markets require increased levels of investment in transmission to accommodate greater volumes of electricity trade. Higher investments in transmission will also be required because of increased use of intermittent renewables. Although transmission and distribution remain largely regulated, the owner, operator and generator are increasingly distinct, making planning more complicated. Siting transmission lines and obtaining approval is also becoming increasingly difficult. These issues stress the need for policies that facilitate investment in networks.
- Environmental regulations, requiring power plants to reduce their emissions of pollutants such as sulphur dioxide and nitrogen oxides, are becoming tighter. Environmental legislation will increasingly address greenhouse gas emissions. Uncertainty about future legislation increases investors' risks. Environmental protection will increase investment requirements for both existing and new power plants.

- Developing countries together will require investment of the order of $5 trillion. Fuel costs will be of the same order of magnitude as investment in infrastructure, increasing the scale of the challenge. For most countries, investment needs to rise well above current levels to meet economic growth and social development goals, but there is no guarantee that the projected investment will be forthcoming. The uncertainty about whether developing countries will be able to mobilise this level of investment is significant, particularly for Africa and India. Overcoming these obstacles will require significant efforts to restructure and reform the electricity sector. A major challenge will be to make tariff structures more cost-reflective.
- More private sector involvement in developing countries will be required, but private investment has been declining since 1997. There are major uncertainties about when and to what extent private investment will rise again and where the new investors will come from. Renewed expansion of private-sector participation will take time and appropriate policies. This question represents one of the biggest uncertainties about future electricity-sector investment.
- The OECD Alternative Policy Scenario illustrates how government policies to address environmental concerns and to increase energy efficiency may affect investment over the next thirty years. With lower electricity demand and a more capital-intensive electricity mix, total power-sector investment in OECD markets in the Alternative Policy Scenario is about 20% lower than in the Reference Scenario. Investment in renewables in the Alternative Policy Scenario would amount to half the investment needed in total new capacity. Given the fact that other generating options are less expensive, investors in renewable energy projects will seek a guaranteed market for their electricity. To encourage renewables, governments will need to create a market framework that rewards those who invest in renewables.
- With present policies, about 1.4 billion people in developing countries will still have no access to electricity in 2030. The additional investment required to achieve 100% electrification is $665 billion, making the investment challenge even greater. This added investment would be needed mostly in the poorest regions of the world — sub-Saharan Africa and South Asia. If it can be mobilised — largely a matter of government priorities — a substantial contribution will be made to poverty alleviation.

The first section of this chapter summarises the results of the investment analysis, at each segment of the electricity-supply chain (generation, transmission and distribution). It looks at investment requirements in relation to GDP. It reviews trends in technology and costs. An analysis of the major challenges facing the sector in OECD and developing countries and the implications for policy follows. The second section of the chapter discusses investment trends and issues by major region. The third section examines the impact on investment of the additional environmental constraints within the Alternative Policy Scenario for OECD countries. It also analyses the investment requirements under a universal Electrification Scenario.

Global Investment Outlook

Over the next thirty years the world will need investment of $10 trillion in power-sector infrastructure.[1] This is nearly three times higher in real terms than investment in the sector during the past thirty years and 60% of total energy-sector investment. As demand for electricity increases, investment will gradually rise, from $2.6 trillion in the first decade, to $3.4 trillion during the decade 2011-2020, and $3.9 trillion in the last decade (Figure 7.1).

Figure 7.1: **World Electricity Sector Investment, 2001-2030**

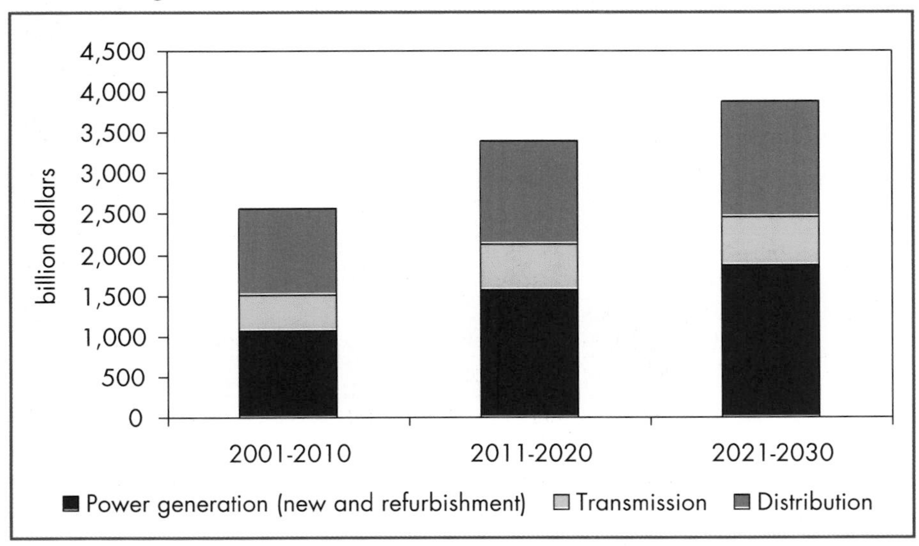

1. Details about the methodology on which the investment estimates are based are given in Annex 2.

Box 7.1: **The *WEO-2002* Reference Scenario**[2]

The projections of power-sector investment requirements presented in this study are based on the *World Energy Outlook 2002.* In the Reference Scenario, world electricity demand is projected to double between 2000 and 2030, growing at an annual rate of 2.4%. The main changes in the fuel mix are:

- Coal's share in total generation declines in the period from 2000 to 2020, but recovers slightly thereafter. Coal remains the largest source of electricity generation throughout the projection period.
- Oil's share in total generation, already small, will continue to decline.
- The share of natural gas will increase significantly, from 17% in 2000 to 31% in 2030. The rate of growth in power-sector demand for gas will slow as natural gas prices increase.
- Nuclear power production will increase slightly, but its share in total generation will be reduced by half because very few new plants will be built and many existing reactors will be retired.
- Hydroelectricity will increase by 60% over the projection period, but its share will fall.
- Generation from non-hydro renewable sources will increase almost six-fold over the period 2000 to 2030, providing 4.4% of the world's electricity in 2030.
- Fuel cells using hydrogen from reformed natural gas are expected to emerge as a new source of power generation after 2020. About 100 GW of fuel cells could be installed in OECD countries by 2030, 3% of total capacity.

Policies under consideration in OECD countries but not yet implemented were included in the Alternative Policy Scenario in *WEO-2002.* These policies could achieve a significant reduction in CO_2 emissions. They would also change investment requirements, which are given in the penultimate section of this chapter.

The countries of the OECD will need investment exceeding $4 trillion in power-sector infrastructure. This investment will take place in an increasingly competitive market environment. Electricity-sector investment now accounts for a small percentage of GDP in OECD countries, typically around 0.5%. The investment needed over the next thirty years will represent, on average, 0.3% of GDP in OECD countries.

2. IEA (2002a).

The power sector in developing countries will require more than half of the global investment, exceeding $5 trillion. Two-thirds of the total investment, some $3.5 trillion, must flow into developing Asia. China's investment needs will be the largest in the world, approaching $2 trillion (Figure 7.2). India will need investment close to $700 billion, while East Asia and Latin America each will need investment approaching $800 billion.

Figure 7.2: **Electricity Sector Investment by Region, 2001-2030**

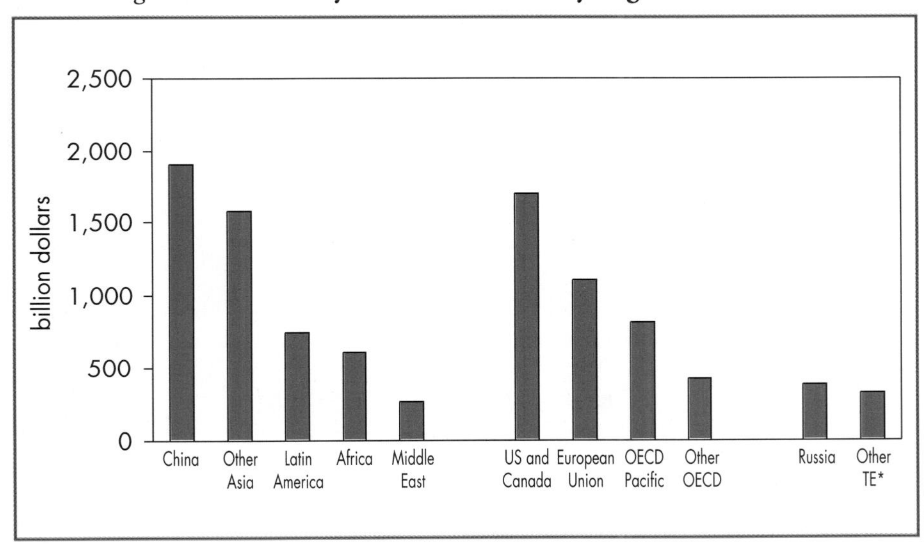

* Transition economies.
Note: Other Asia includes East Asia and South Asia.

Electricity-sector investment in developing countries generally accounts for a larger share of GDP than in OECD countries, often ranging between 1% and 3%. If the share is lower, it can indicate that existing levels of investment are insufficient. In many developing regions, this share is expected to rise, at least in the near term (Figure 7.3). In Indonesia, it will be lower in the decade 2001-2010, reflecting the current poor investment climate in the country.[3]

Investment in the transition economies will be $700 billion, with more than half of it going into the Russian power sector. These countries now have excess capacity, because electricity demand is still below the level reached before

3. See also World Bank (2003a).

the break-up of the former Soviet Union.[4] However, existing power plants and networks are old and poorly maintained and they need extensive refurbishment to be able to provide reliable supplies to national or export markets. Substantial new investment is needed to meet domestic demand only after 2010.

Figure 7.3: **Investment as a Proportion of GDP by Region**

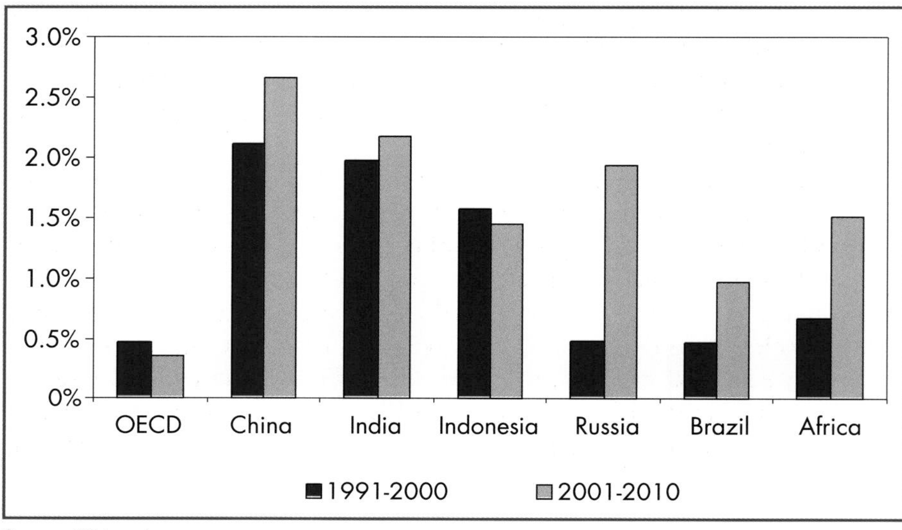

Source: IEA estimates.

The five largest countries in the world outside the OECD ("the big five") — China, Russia, India, Indonesia and Brazil — will need about a third of the global electricity investment (Table 7.1).

Table 7.1: **Investment in the Big Five Non-OECD Countries**

	GDP ($ billion) 2001	GDP ($ billion) 2030	Investment 2001-2030 ($ billion)	Ratio to GDP (%)
China	1,398	5,335	1,913	2.1
Russia	411	947	377	1.9
India	530	1,961	665	2.0
Indonesia	236	712	184	1.4
Brazil	872	2,025	332	0.8

Note: GDP at market exchange rates.

4. Aggregate electricity generation in the transition economies in 2000 was three-quarters of the 1990 level.

There are two key reasons that explain why the electricity sector will continue to need large investments:

- First, the electricity sector is very capital-intensive (Figure 7.4). All technology options available to generate electricity involve substantial investment in fixed assets. Nuclear, coal and renewable energy technologies are the most capital-intensive options. Gas turbines in combined or open cycle are at the low end of intensity. Capital intensity in the 1990s was lower compared to earlier decades. This can be explained to a large extent by the greater use of combined cycle gas turbine (CCGT) plants[5] and lower and often inadequate investment in networks.
- Second, the world will continue to shift from primary fuels to electricity; and demand for electricity increases as incomes increase. Over the next thirty years, global demand for electricity will double. The share of electricity in energy consumption will increase everywhere though the reasons for growth are different in developed and developing economies. In industrialised countries, electricity is the preferred energy source and a substitute for fossil fuels. In developing countries, increasing population, urbanisation, economic growth and rapid industrialisation constitute the major factors of growth.

Figure 7.4: **Capital Intensity by Industry**

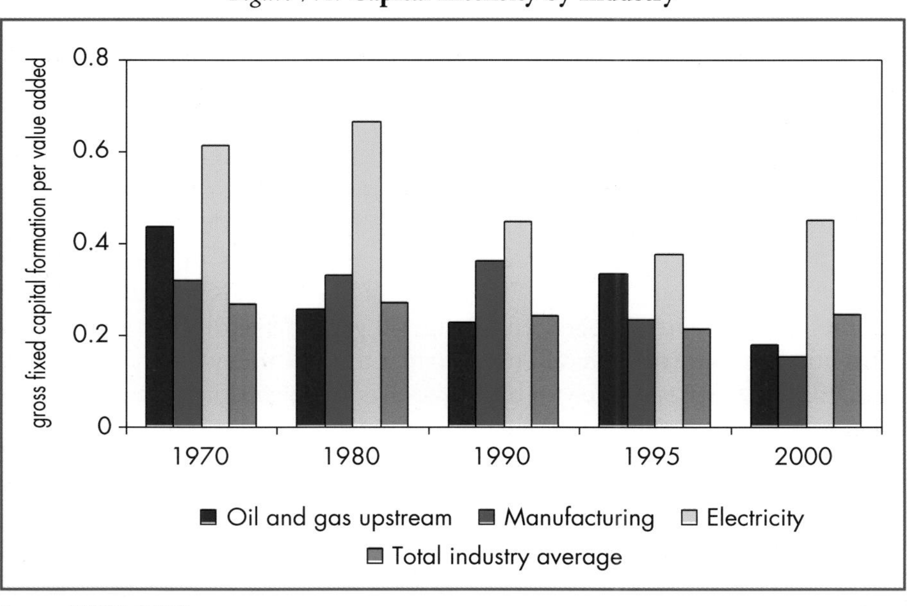

Source: OECD (2003).

Demand will be particularly strong in developing countries, where it is expected to increase by 4% per year on average. In OECD countries, demand is expected to progress at a much lower rate, 1.5% per year. However, a large part of the existing infrastructure in these countries will also need to be replaced and refurbished over the next three decades.

Generation is the largest single component of total electricity-infrastructure investment. Investment in new plant over the next thirty years will be more than $4 trillion, accounting for 41% of the total. Most of this investment will go into the development of gas- and coal-fired power plants. Refurbishment of existing power plants over the next thirty years will need investment of $439 billion.

Investment in transmission and distribution networks together will take 54% of the total. This amount includes investment in refurbishment and replacement of existing networks. Network extension, as a component of investment, is more important in developing countries, because of population growth and an increase in the rate of electrification. In OECD countries, where networks are more developed, most network investment will be needed for refurbishment and replacement of existing components, such as lines, substations and control centres. Distribution is the most important component of investment in networks. Investment in distribution networks over the period 2001-2030 will reach $3.8 trillion, while transmission networks will require investment of the order of $1.6 trillion.

Electricity Generation

New Plant

Over the period from 2001 to 2030, some 4,700 GW of generating capacity is expected to be built worldwide. About a third of the new capacity will be in developing Asia. OECD countries will require more than 2,000 GW.

Power plants in the OECD, as well as in many of the transition economies, are ageing (Figure 7.5). The *WEO-2002* Reference Scenario assumes a retirement age between 40 and 60 years depending on the type of plant and the country. Overall, more than a third of today's total capacity in the OECD countries is likely to be retired over the next thirty years. Most of the retired capacity will be coal-fired. About 40% of today's installed nuclear capacity in OECD countries will be retired, either because the plants will have

5. The capital cost of a CCGT plant is, on average, $500 per kW of installed capacity. The unit cost of a coal plant is twice as much, while that of a new nuclear plant is three to four times higher.

Figure 7.5: **Average Age of Power Plants in the OECD, 2003**

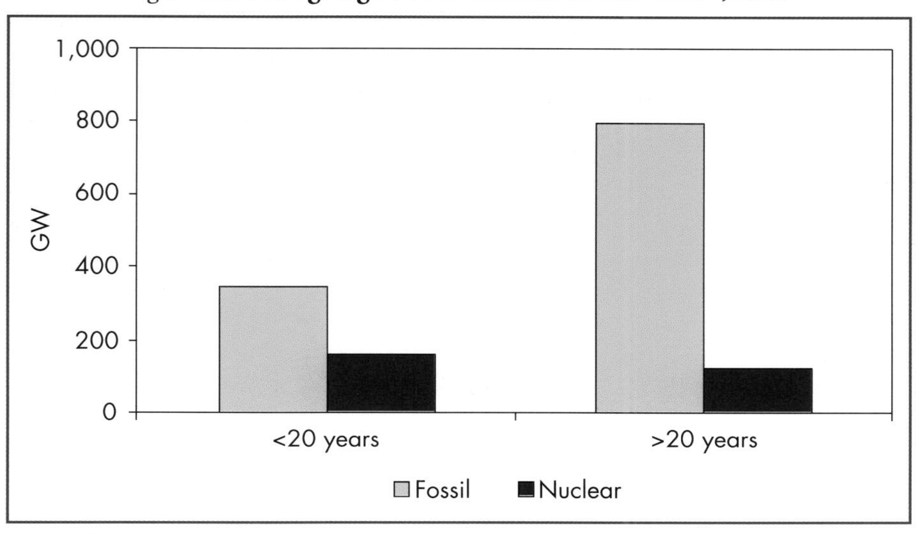

Sources: Platts (2001) and IEA analysis.

reached the end of their design lifetime or because of policies to phase out nuclear power. The retirement of fossil-fuel plants in the OECD will create opportunities to improve efficiency and to reduce CO_2 emissions. Older and inefficient coal plants are expected to be replaced, in most cases, either by gas-fired or by coal-fired plants. In either case, these new plants will be cleaner than the existing ones. The need to replace existing facilities will pose a particular investment challenge in the period 2015-2030.

Investment in the electricity sector has followed cyclical patterns, although the cycles were not as prominent or volatile as in some other industries (Figure 7.6). Orders reached their peak between the late 1960s and early 1970s, at about 150 GW a year, and then plummeted in the mid-1980s. They recently increased again, notably because of a substantial increase in the US market. It is possible that under liberalised energy markets, as electricity becomes more and more a commodity, the business cycle effect will be greater, resulting in fluctuating reserve margins, high electricity prices when margins are tight and, possibly, some threat to security of supply. However, evidence to support this theory is still limited. There was a major construction cycle in the United States between 1999 and 2002 and construction in the United Kingdom grew steadily throughout the 1990s, though it has now come to a halt. It is not yet clear how investment cycles will evolve in the future.

Figure 7.6: **World Annual Power Plant Orders, 1950-2002**

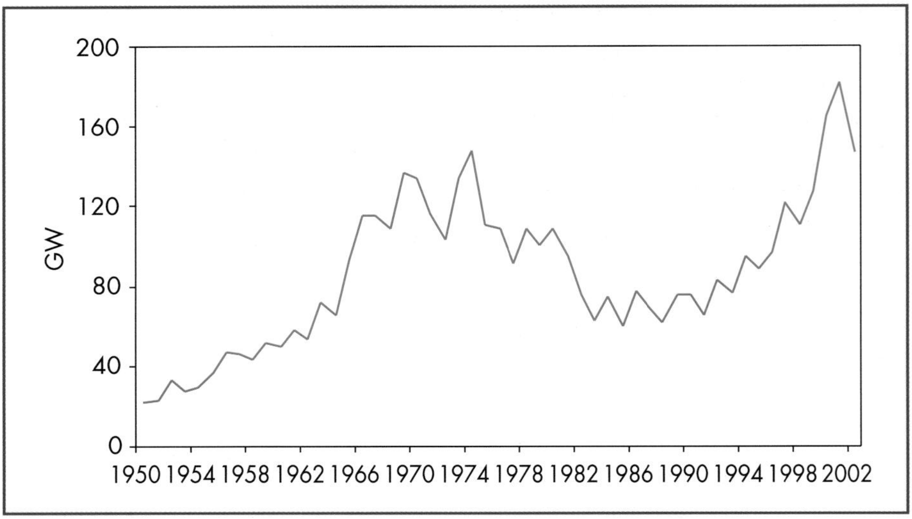

Source: Siemens.

The electricity market is organised around power companies whose size varies substantially. Table 7.2 shows the ten largest power companies in the world, ranked by their installed capacity. These companies account for about one-fifth of the world's installed capacity. In the past decade, many companies chose to invest in other countries in their region or overseas. Activity was particularly intense in Europe. A number of large power

Table 7.2: **The World's Ten Largest Electricity Companies**

Company	**Home-base**	**Capacity** (GW)
RAO-UES	Russia	156
EDF	France	121
TEPCO	Japan	59
E.ON	Germany	54
SUEZ	France	49
ENEL	Italy	45
RWE	Germany	43
AEP	United States	42
ESKOM	South Africa	42
ENDESA	Spain	40

Source: Company websites.

companies invested in power projects in developing countries. However, many of these companies are now withdrawing or selling their assets and interest in new projects in developing countries is very limited.

Investment in new generating capacity is expected to be of the order of $4 trillion in the period to 2030. The figure is based on a mix of technologies with different capital costs (Table 7.3). Investment, both in terms of capacity additions and money will rise in real terms over time (Figure 7.7).

Table 7.3: **Current Capital Cost Estimates**

Technology	**Capital Cost** ($ per kW)
Gas combined cycle	400-600
Conventional coal	800-1,300
Advanced coal	1,100-1,300
Coal gasification (IGCC)	1,300-1,600
Nuclear	1,700-2,150
Gas turbine — central	350-450
Gas turbine — distributed	700-800
Diesel engine — distributed	400-500
Fuel cell — distributed	3,000-4,000
Wind onshore	900-1,100
Wind offshore	1,500-1,600
Photovoltaic — distributed	6,000-7,000
Photovoltaic — central	4,000-5,000
Bioenergy	1,500-2,500
Geothermal	1,800-2,600
Hydro	1,900-2,600

Source: IEA analysis.

Nearly 2,000 GW of new plant will be gas-fired. Almost half of this increase occurs in the OECD, where gas is widely available and meeting the cost of environmental restrictions limits the use of coal. Big increases are expected in gas-fired power generation in the transition economies and developing countries too.

Coal-fired capacity additions over the period 2001-2030 will exceed 1,400 GW. Nearly half of these new plants will be developed in China and India. Nuclear plant construction will amount to 150 GW, concentrated in Asian countries. Hydropower will remain the most important renewable energy source, with 430 GW of new capacity over the next thirty years, while the generating capacity of other renewable energies will increase by nearly 400 GW.

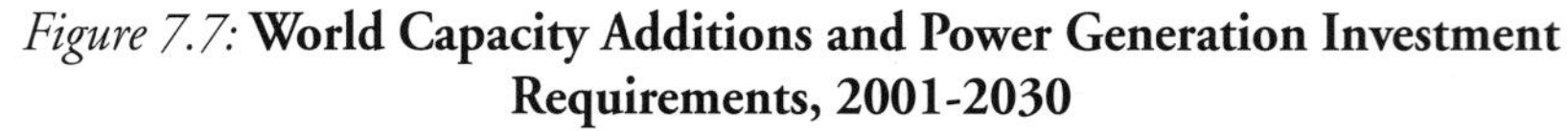
Figure 7.7: **World Capacity Additions and Power Generation Investment Requirements, 2001-2030**

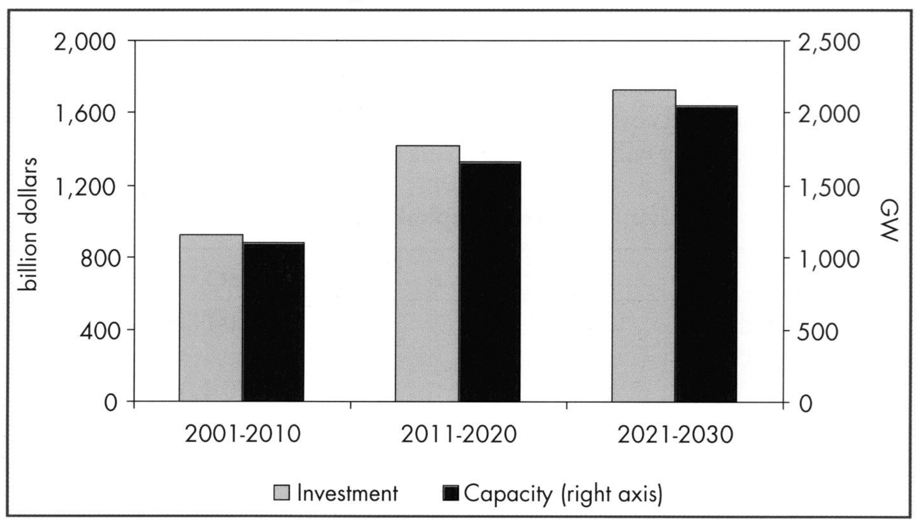

Beside the capital costs noted here, investment decisions take into account fuel costs and operation and maintenance costs. Figure 7.8 shows indicative electricity-generating costs for coal, gas and wind in current markets. CCGT plants now have the lowest generating costs. However, over the next thirty years, CCGT generating costs are expected to increase with higher natural gas prices.[6]

Figure 7.8: **Indicative Generating Cost Ranges, 2000**

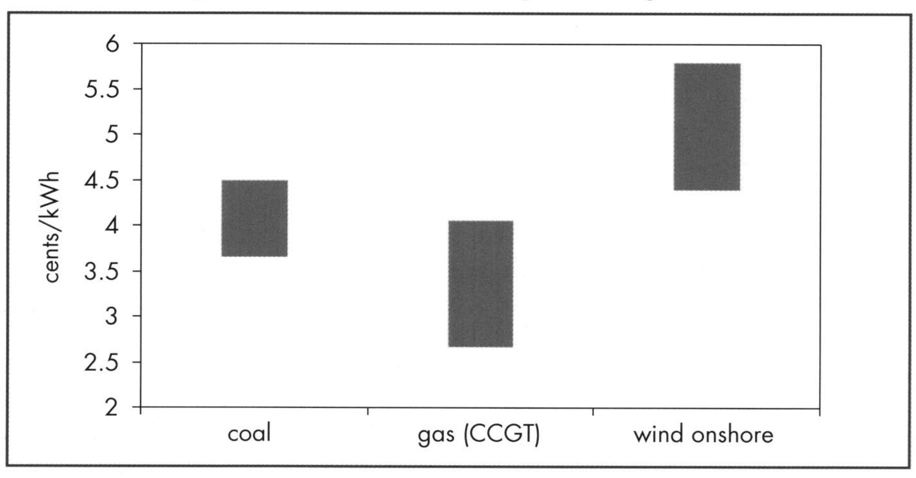

6. Although the expected increases in efficiency and reductions in capital costs of CCGTs will offset some of the impact of higher natural gas prices.

The generating cost of wind will decline with reductions in capital costs and improved performance of wind turbines.

Refurbishment of Existing Plants

Investment in power-plant refurbishment is expected to amount to $439 billion over the period 2001-2030. This estimate refers to major upgrades of existing power plants assumed to take place once in their lifetime. Most of this investment will be needed to refurbish fossil-fuel power plants in OECD countries. The transition economies will need about 10% of the global investment. Power infrastructure in these countries has been poorly maintained.

Liberalised electricity markets can bring about increased investment in cost-effective refurbishment of existing power-generation plants. By investing in plant refurbishment, power companies can improve plant output, reduce production costs and therefore become more competitive. There are, however, concerns that some companies may choose to cut costs by avoiding or deferring investment in refurbishment.

In many developing countries, insufficient investment in plant refurbishment — because of lack of resources — has contributed to the poor performance of power plants and the grid. Investment in refurbishment can increase electricity production in these countries and reduce somewhat the very large capacity additions needed to meet rising electricity demand.

Transmission and Distribution

Investment in transmission and distribution networks over the period 2001-2030 is expected to be of the order of $5 trillion — more than the total investment in power generation. About 30% of this amount will go into the development of transmission networks.

Liberalised electricity markets are likely to require increased levels of investment in transmission to accommodate greater volumes of electricity trade. Moreover, investment in efficient transmission and distribution reduces losses and can be an effective way to lower costs to consumers as well as to reduce emissions of power generation-related pollutants. At the same time, costs will increase if environmental issues related to power networks are addressed. Such issues include the need to reduce emissions of SF_6 (used in transformers and other equipment), which contribute to global warming, and the impact of electromagnetic fields around cables. Higher investment in transmission may also be required because of increased use of intermittent renewables. Government policies to promote renewables will have to address this issue.

Investment in transmission and distribution has lagged behind investment in generation in some OECD countries, notably in the United States and some European countries. There is a clear need across the OECD to develop electricity networks further and to reinforce cross-border links.[7] Investment in both transmission and distribution will have to rise in future if grids are to provide reliable power. Though formal inquiries have not yet concluded, grid failures in 2003, notably in North America, lend support to this analysis.

Although transmission and distribution remain largely regulated, the owner, operator and generator are increasingly no longer the same. While access to capital for investment in networks does not appear to be an issue in OECD countries, these changes have made transmission planning more complicated. Efficient investment in networks will be forthcoming only if there is appropriate network pricing that sends signals to the market to invest. But at the same time, siting transmission lines and obtaining approval is becoming increasingly difficult. These issues stress the need for policies that facilitate investment in networks.

Figure 7.9: **Savings in Transmission Investment from the Growth in Distributed Generation**

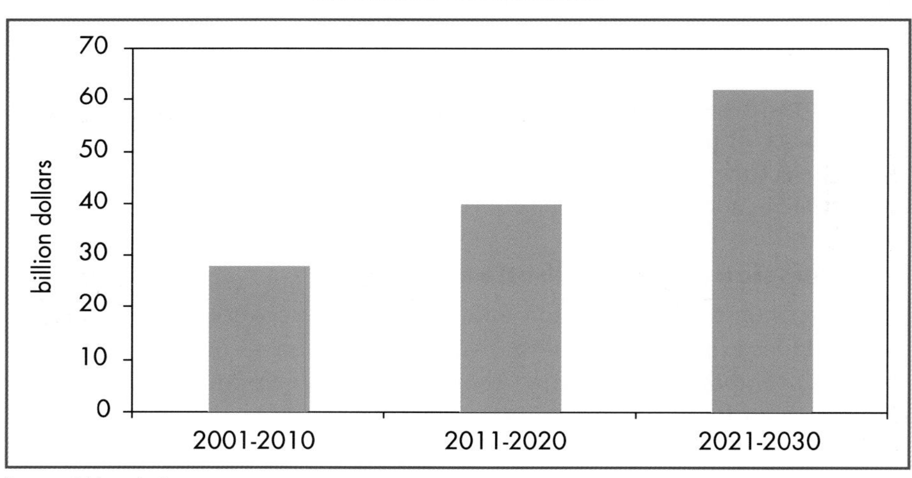

Source: IEA analysis.

Distributed generation, which is projected to increase its share over the next thirty years, can help reduce investment in transmission networks.[8] Most investment in distributed generation will be in OECD countries, supported by government policies. In developing countries, unreliable power supply will continue to be a key driver of distributed power. Over the period 2001-2030, increased use of distributed generation technologies in the Reference Scenario will avoid around $130 billion of investment in transmission networks (about 8% of the world total). Savings rise from $28 billion in the first decade to $62 billion in the last decade (Figure 7.9).

7. See also IEA (2002b).

8. Distributed generation includes photovoltaics. For definition, see Box 3.5 in *WEO-2002*, p.127.

Table 7.4: **Summary of Power Sector Investment Requirements, 2001-2030** ($ billion)

	Generation		Transmission	Distribution	Total
	New	Refurbishment			
OECD Europe	645	62	143	501	1,351
of which European Union	*525*	*52*	*120*	*413*	*1,110*
OECD North America	717	137	295	728	1,876
of which US and Canada	*654*	*130*	*261*	*649*	*1,694*
OECD Pacific	357	61	131	260	809
of which Japan, Australia, New Zealand	*274*	*48*	*98*	*185*	*606*
Total OECD	**1,719**	**260**	**569**	**1,488**	**4,036**
Transition economies	**297**	**41**	**82**	**280**	**700**
of which Russia	*157*	*21*	*45*	*154*	*377*
China	795	50	345	723	1,913
East Asia	344	22	133	301	799
of which Indonesia	*72*	*6*	*33*	*74*	*184*
South Asia	310	18	142	312	783
of which India	*268*	*15*	*119*	*262*	*665*
Latin America	317	19	128	281	744
of which Brazil	*149*	*7*	*54*	*122*	*332*
Middle East	92	15	47	103	258
Africa	206	13	123	266	609
Total developing countries	**2,064**	**138**	**918**	**1,987**	**5,106**
Total world	**4,080**	**439**	**1,568**	**3,755**	**9,841**

In developing countries, priority is often given to investment in generation. Investment in transmission and distribution must rise in the future. Among others, China and India have started to increase the amount of investment going into transmission and distribution.

Investment Uncertainties and Challenges

The major factors that will affect investment in the power sector over the next thirty years are competition, market reform, environmental constraints and access to capital. The issues are different between developed and developing countries.

Power-sector investment now accounts for less than 0.5% of GDP in most OECD countries (Figure 7.10). The share of investment has declined somewhat since the mid-1990s for a number of reasons, including high reserve margins in some countries, the lower capital costs of new power plants, low demand growth and uncertainty caused by environmental policies and market liberalisation. Competition between utilities has reduced profit margins, especially in markets with excess capacity and low demand growth (Figure 7.11).

Figure 7.10: **OECD Electricity Sector Investment Relative to GDP**

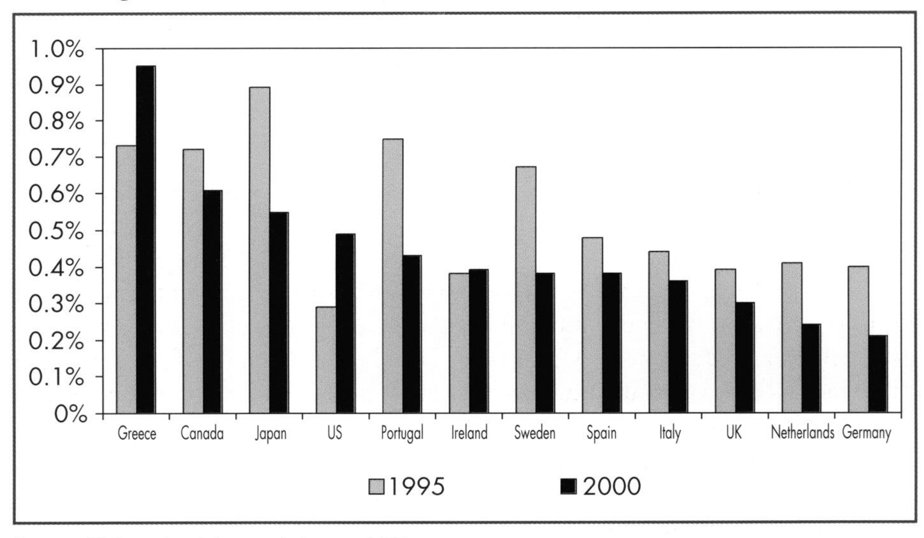

Sources: Various electricity associations and IEA.

Market liberalisation has created new challenges and uncertainties in OECD countries. There is new concern about the adequacy of investment as markets adapt to the new conditions. Policy-makers need to address this concern by providing a market framework that sends the right market signals to investors.

Investors in liberalised markets are more exposed to risk than they were in regulated markets and in different ways. Power companies will have to improve their risk management skills.

Figure 7.11: **Return on Investment of OECD Electricity Companies, 1993-2002**

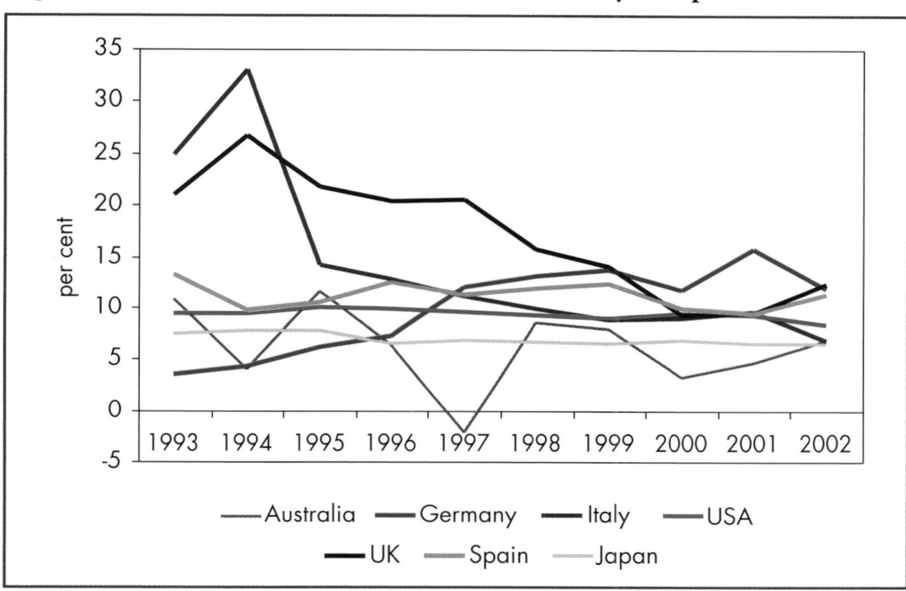

Source: Standard and Poor's (2003).

Developing and transition countries are, in many cases, also seeking to restructure their electricity industries by introducing new market structures to encourage competition. Many of their efforts have not brought about the expected results, and for many of these countries, attracting investment to meet rapidly growing demand and to improve the infrastructure for power production and delivery will be a major challenge. Some of these countries may want to delay the introduction of competition until their electricity sector is sufficiently mature and economically viable.

Environmental regulations, requiring power plants and other industrial facilities to reduce their emissions, are becoming tighter. Uncertainty about future environmental legislation increases investor risk.

Environmental protection will increase investment requirements for both existing and new power plants. Environmental costs may account for 10% to 40% of total plant costs in fossil-fuel plants and more in nuclear plants.[9]

9. IEA (1999).

Existing legislation is directed principally at emissions that have a local or regional impact, such as sulphur dioxide, nitrogen oxides and particulate matter. These emissions depend on the fuel mix used in power generation and tend to be higher in countries whose electricity generation is based heavily on coal. Emission standards for these pollutants are tight and are becoming tighter in many OECD countries. Developing countries will also be increasingly seeking to reduce these pollutants. This will increase further their already large needs for power sector investment.

The power generation sector accounts for 38% of total energy-related CO_2 emissions in the OECD countries and 40% worldwide. Environmental regulation may increasingly address carbon dioxide emissions in all countries; but in the medium term the impact will be greater in the countries that act to reduce their greenhouse gas emissions under the Kyoto Protocol. The impact on their investment paths is described in the section towards the end of the chapter that discusses investment in the OECD Alternative Policy Scenario.

Box 7.2: **Sources of Capital**

In OECD countries, electricity companies finance new projects by providing part of the project capital as equity (internally generated cash or equity issued as public shares), while the remainder is financed as debt – they may borrow money from the bank or issue bonds. Trends in the debt-equity structure of OECD power companies are shown in Figure 7.12. The chart shows that the power sector in Japan relies more on debt, while in the United States reliance on equity is larger. Companies with high-debt levels (Japan, France) have reduced their debt in preparing for competition. Countries with high levels of recent investment (United States, United Kingdom) have increased their debt, although the trend in the US now is to reduce debt. It is not clear how liberalised electricity markets will affect this structure in the future, *e.g.* whether the equity share will move towards the high levels typical of the oil market. Financing issues are discussed in more detail in Chapter 3.

In non-OECD countries, where utilities are often state-owned and revenue collection is insufficient, investment capital often comes from the government and in the form of loans from multilateral lending agencies (such as the World Bank or the Asian Development Bank). Over the past decade, many developing countries have attempted to attract private-sector (domestic or foreign) investment to meet part of their needs. Private participation declined after the Asian economic crisis. At the moment, there are many uncertainties about the scale of private-sector involvement in power projects in developing countries in the future.

In developing countries, achieving social equity will be a big challenge. Even if the huge electricity investment needs which arise in developing countries in the Reference Scenario are met in a timely fashion, there will still be 1.4 billion people without access to electricity in 2030. An Electrification Scenario has been developed to quantify the added investment required to achieve 100% access to electricity in the developing world by 2030.

Figure 7.12: **Debt-Equity Structure of OECD Power Companies**

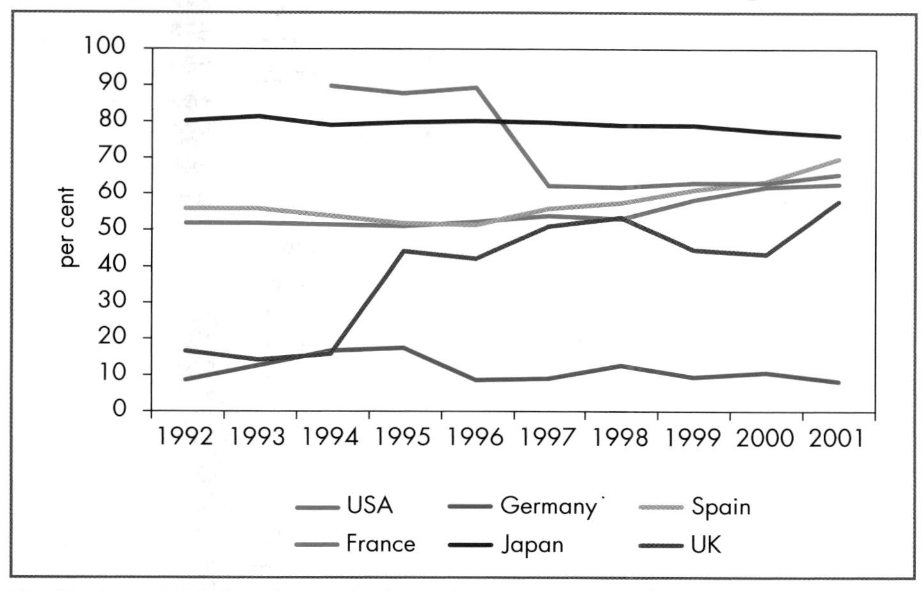

Note: For France, EDF's sharp change in its debt-equity structure in 1997 is due to the issuance of stocks.
Source: Standard and Poor's (2003).

OECD Countries[10]

Adequacy of Investment

OECD countries will require more that $4 trillion in power infrastructure investment over the next thirty years. Investment requirements will rise from $1.2 trillion in the period 2001-2010, to $1.4 trillion from 2011 to 2020 and will remain at this level during the third decade (Figure 7.13). Electricity markets are open to competition to varying degrees in nearly all OECD countries. This section discusses issues related to adequacy of investment in these liberalising electricity markets.

10. This section is based on IEA (2003a).

Figure 7.13: **OECD Power Sector Investment, 2001-2030**

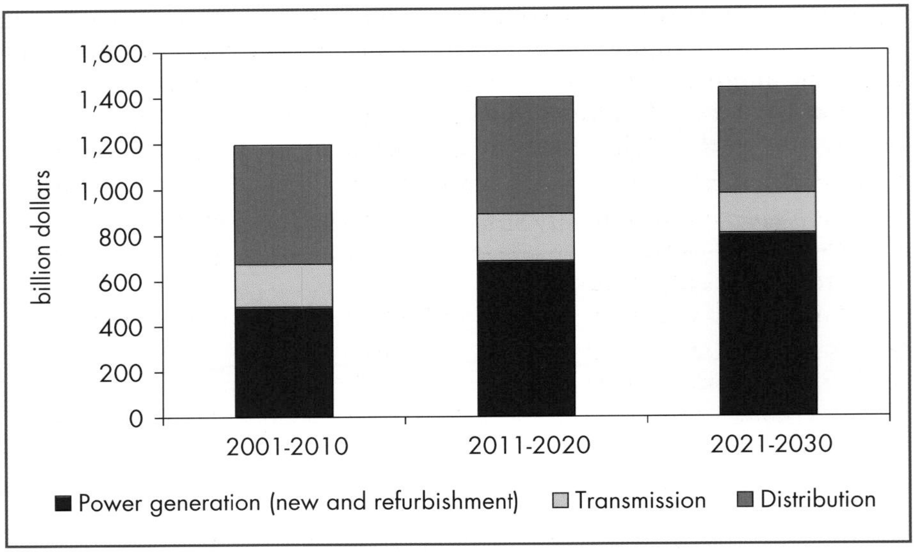

1,600
1,400
1,200
1,000
800
600
400
200
0
billion dollars
2001-2010
2011-2020
2021-2030
Power generation (new and refurbishment)
Transmission
Distribution

Source: Standard and Poor's (2003).

Figure 7.14: **Electricity Spot Prices in the Victoria Market**

250
200
150
100
50
0
weekly average price (AUD/MWh)
07-04-99
01-02-00
07-02-00
31-12-00
07-01-01
23-12-01
23-06-02
22-12-02
22-06-03
week

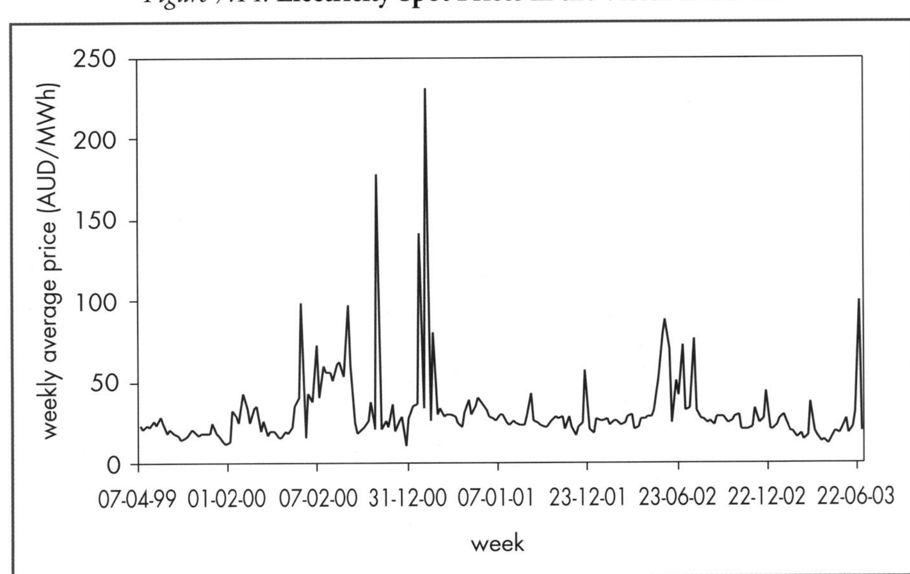

Source: National Electricity Market Management Company.

The main objective behind the past 15 years of electricity-market liberalisation has been to improve the economic efficiency of the electricity supply industry, improving productive efficiency by reducing operating costs and improving the efficiency of the allocation of capital by aligning prices with costs.[11] Electricity markets should stimulate the right level of investment at the right time. When surplus capacity exists, electricity markets can expect to see prices below long-run marginal costs (LRMC), not permitting producers of peaking capacity to recover any of their fixed costs. However, when capacity starts to become scarce during peak hours, prices will rise significantly (Figure 7.14).

Given the low elasticity of demand with price, cost-recovering prices for peak capacity may be needed over a relatively small number of hours.[12] Table 7.5 shows the average price needed over the most expensive 5% of hours (*i.e.* 438 hours) annually to recover the long-run marginal cost of a peaking plant. Under conditions likely to generate these prices, investors would make timely investments in new generating capacity.

Table 7.5: **Prices Needed to Recoup Peaking Plant LRMC in PJM* in Top 5% of Hours by Load**

LRMC ($ per MWh)	**Average price in top 5% of hours** ($ per MWh)
35	225
40	294
45	363
50	432
55	500

* Pennsylvania — New Jersey — Maryland.
Source: Hughes and Parece (2002).

A number of market and regulatory imperfections may lead to underinvestment in electricity markets.[13] Price signals may be distorted *e.g.* by government policies to protect small consumers. Lags between price signals and construction of new plant can cause boom-and-bust cycles in generating capacity. Very hydro-dependent power systems may have large system risks, *e.g.* with variations in rainfall that may not easily be accommodated. Box 7.3 describes the issues in the hydro-dependent Norwegian electricity market in the winter of 2002/2003.

11. Newbery (2000).
12. Hughes and Parece (2002).
13. For a discussion of capacity adequacy in US markets, see Joskow (2003).

Box 7.3: **Norway's Electricity Situation in Winter 2002/2003**

Norway deregulated its electricity market in 1991. It was later joined by Sweden, Finland and Denmark to open up a fully competitive Nordic market for trading electric power (Nord Pool). After the deregulation, Norwegian consumers generally enjoyed lower prices than before.

However, during the winter 2002/2003 prices increased dramatically; private consumers saw prices during the first quarter of 2003 go up by a factor of three compared to the same quarter one year earlier. The tight supply situation came after years of very low investment in new capacity and was triggered by an extraordinarily dry autumn in 2002. This left the 99% hydro-dominated Norwegian electricity system with very low reservoir levels at the beginning of the winter season, a winter that turned out to be colder than normal.

Electricity is the most common space heating choice in Norway and during the coldest days of the winter, peak load was very close to installed capacity. The low level of hydro reservoirs is a source not only of potential peak-load problems, but can also be a problem in the spring, before the melting of the mountain snows starts to fill up the reservoirs again. The high prices triggered increased imports — to a large extent from re-activation of coal-fired power plants in Denmark — and demand-side responses from both industry and private consumers, which together were enough to avoid both winter blackouts and spring rationing.

Although the electricity system was able to provide sufficient electricity services throughout the crisis, the situation has generated a debate in Norway about how reserve margins and investment in new capacity can be ensured in a deregulated market. The steep price hike did not make a significant impact on the four-year futures market of Nord Pool, indicating that the market regards the price increase as a temporary result of rather extreme weather conditions and not an indication that higher prices will be sustained. This means that prices are still too low to make investments in gas-fired power stations profitable. Further expansion of hydropower capacity is unlikely because of environmental constraints and agreeing on investment in new interconnections to other countries has proven difficult, because of unresolved issues about who should cover the costs and how grid operators should co-ordinate. In the short run, this leaves Norway with sharp price increases to consumers as the primary mechanism to avoid power shortages in dry and cold years. To reduce the system's vulnerability to rainfall variations in the longer run will require agreements on new interconnections, ways to get more peak capacity out of existing hydro plants and reduced dependence on electricity for space heating through energy efficiency and fuel switching.

Inconsistent investment performance across OECD countries has led a number to review the adequacy of their market arrangements for encouraging timely investment. Some countries have concluded that direct market intervention to stimulate investment in peaking capacity is unnecessary at this time. In others, measures are being taken to ensure peak capacity investment. Many markets are seeking ways to enhance the response of electricity demand to changes in price, as a means of decreasing the volatility of prices. Generally, policy-makers have concluded that current market designs do not guarantee an adequate level of security of supply.

Risk and Power Generation Investment

Prior to the liberalisation of electricity markets, electricity companies were usually operated as integrated monopolies, able to pass on their full costs to energy consumers. In such an environment, there was only limited risk in investment decisions. In today's market there are both concerns about wider commercial risks and new concerns about the environment and security of supply.

The level of future electricity prices in competitive electricity markets can be a major source of risk. Price volatility can greatly affect investors' revenues and profits. Uncertain electricity prices expose projects that have a long lead and construction time to additional risks. Economies of scale favour large power projects over small ones, as capital costs per kW for a given technology generally decrease with increasing scale. However, the combination of a long lead time for constructors, uncertain growth in demand for electricity and the cost of financing adds to the risks for these types of investments. Estimates of profitability for such projects rely principally on a long-term assessment, independent of the spot power market conditions. Very large projects that must effectively be built as a single large plant (*e.g.* a very large hydro dam) are more vulnerable to this type of risk than projects which can be developed as several smaller power plants, in response to market conditions.

There are a number of ways to manage electricity price risk, for example the use of long-term bilateral contracts, futures and forwards contracts, either through established or over-the-counter exchanges. The more liquid these markets become, the easier it will be to use these tools. Although fuel prices have always been uncertain, fuel price risks have been increased by the liberalisation of the natural gas market. Very long-term contracts are not generally available, with the exception of "take-or-pay" arrangements in LNG markets.

Regulatory uncertainty about future environmental legislation is another major source of risk. Existing coal plants are already subject to controls over emissions of three basic pollutants: sulphur dioxide, nitrogen oxides and particulates. However, today's investor faces a high risk of new constraints being imposed on emissions, particularly on carbon dioxide emissions.

Emissions trading is expected to be the most common mechanism employed to control emissions from larger power plants. Nuclear power plants may also be subject to additional safety regulations. Other regulatory risks include risks associated with gaining approval to construct a new power plant, changes in electricity market rules and market intervention. In general, a clear and stable regulatory framework is necessary to give investors confidence in markets.

Firms are adopting different strategies to manage the risks associated with investment in today's power-generation market, with varying success. One early trend was to develop so-called merchant plants that would rely on prices in spot electricity markets and not on long-term contracts. Recent experience with the merchant plant model in the US has driven investors away from this type of investment. Investment banks, concerned by their losses in the United States and in some European markets, are now paying greater attention to companies with stable revenue flows and customer bases. Consequently, it will be very difficult in the medium term to finance new power plant construction in the United States on a merchant plant basis.

Generating companies that have been able to retain contractually based customers are better able to withstand falls in wholesale power prices caused by excess capacity.

Companies with significant investments in natural gas power generation have been investing in companies with upstream natural gas assets in order to hedge the fuel cost risks associated with gas-fired power generation. However, many companies in the US, in an effort to improve their balance sheet, have now divested these upstream investments.

Mergers and acquisitions are one means to improve the prospects of stable cash flow as a source of finance for large capital-intensive investments in an environment where there is reduced access to debt capital. Indeed, the growth in mergers and acquisitions for this reason is not surprising. Mergers and acquisitions have led to the emergence in Europe of "Seven Brothers" (EDF, E.On, RWE, Vattenfall, Endesa, Electrabel, ENEL) — very large electricity firms which are expected to finance a significant portion of new investment from internal resources. At the same time, this consolidation has raised concerns about undue concentration.[14]

Customers may also become more involved in power-generation investment. Falling capital costs for small power plants (particularly reciprocating engines or small turbines) combined with various incentives to encourage power generation on a smaller scale, are creating opportunities for economic combined heat and power generation (CHP) and contributing to higher system reliability. While successful in high electricity price jurisdictions, such as Japan, distributed generation technologies have made only limited progress elsewhere in the OECD

14. Thomas (2002).

as a result of recent increases in natural gas prices and lower electricity prices, as well as institutional barriers.[15]

Box 7.4: Peak Capacity in Liberalised Electricity Markets

Since electricity markets opened to competition, reserve margins have been declining in most OECD countries. While reducing excessive reserve margins is in line with the objectives of market liberalisation, this situation has provoked a debate about the appropriate level of reserve margin to ensure that electricity demand will be met during peak demand periods.

Supplying electricity at times of peak demand requires adequate total generating capacity or purchases from another market with a different peak, along with adequate transmission capacity. Peak demand is most economically met with power plants of low capital cost, since fixed expenses can be recovered only over relatively short annual periods of operation. The risks to investors building this type of peaking capacity may be high, especially when compared to baseload plant. Such risks include:

- *Market risk:* Peak demand is greatly influenced by weather conditions. Unusual weather patterns such as very warm winters or cool summers could result in zero annual revenues to certain peaking plants.
- *Fuel-supply risk:* In systems where the demand for natural gas for space heating and for peak electricity generation coincide, gas supply for space heating will generally be given a higher priority. Thus, there is a risk that fuel supply to gas-fired peaking plants could be interrupted or curtailed during cold periods.
- *Regulatory risk:* Because peaking plants are called into service when prices are highest, they are disproportionately exposed to the risk of government-imposed electricity price caps.

Investors can mitigate risks through appropriate hedging mechanisms, such as power purchase agreements. Governments can also reduce investment risks by ensuring that the value of security of supply is adequately recognised within the market framework. But until experience proves otherwise, concerns about meeting peak demand are likely to persist.

Demand response measures are one way to reduce the need for peaking capacity and moderate price volatility. Such measures include campaigns to increase consumer awareness of the threat of supply disruption when demand peaks, demand-side bidding to induce industrial customers to reduce their load at peak times, or the use of advanced technology, such as advanced meters, to reduce or reschedule peak load.

15. IEA (2002c).

Developing Countries

Today, developing countries account for a little over a quarter of global electricity production. By 2030, this share is expected to rise to 44% and these countries would be producing as much electricity as the OECD. To provide for this rapid increase, they will need to invest over $5 trillion in electricity infrastructure (Figure 7.15). For most countries, this means that investment should rise well above current levels. The uncertainty about whether developing countries will be able to mobilise this level of investment is significant, particularly for Africa and India. Poorly developed domestic financial markets are a major constraint in developing countries.[16] The exchange rate risk is an important factor limiting their access to international financial markets.[17] Investment in power-sector infrastructure in developing countries has traditionally been the responsibility of governments, though the 1990s saw an increasing number of countries turning to the private sector for part of the investment needed to finance the electricity sector.

Figure 7.15: **Power Sector Investment in Developing Countries, 2001-2030**

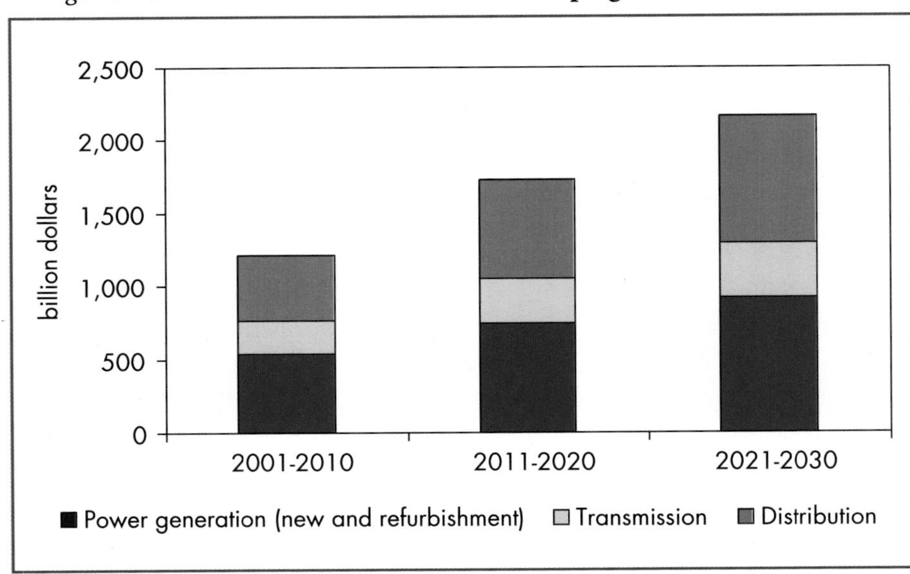

Source: OECD (2003).

16. Bacon *et al.* (2001).
17. See also the discussion in Chapter 3.

Figure 7.16 shows how additions in generating capacity have evolved over the thirty-year period 1971-2000. China's increase in capacity during the 1990s stands out and can be attributed to the reforms initiated in the 1980s. Between 1991 and 2000, China increased installed capacity by as much as all other developing Asian countries taken together. Nonetheless, Indonesia and other Asian countries saw continuous expansion throughout the thirty-year period, despite the set-back in the late 1990s attributable to the Asian economic crisis.

Figure 7.16: **Capacity Additions in Developing Regions, 1971-2000**

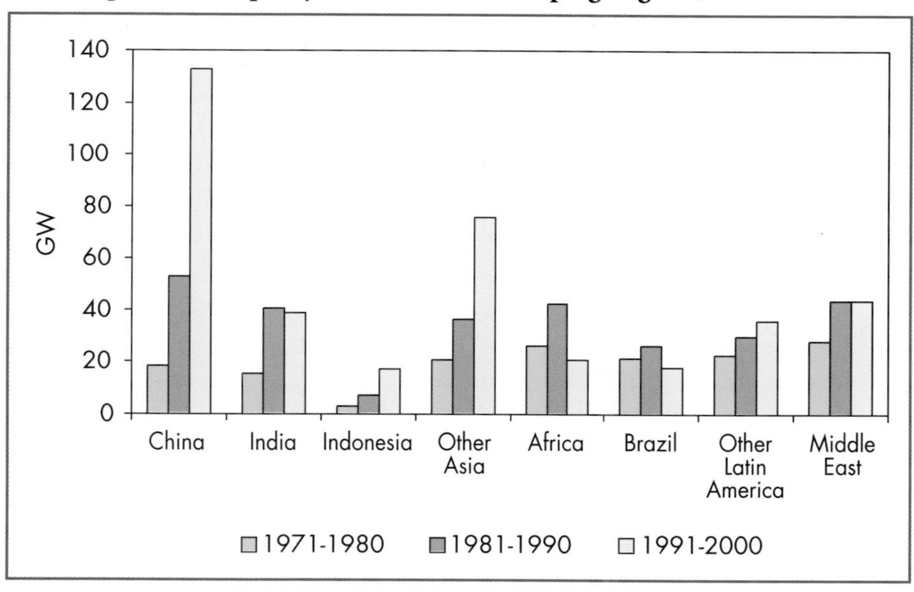

Source: Platts (2001).

The rate of capacity expansion in India, the Middle East and Latin America in the decade 1981-1990 did not continue in the 1990s. In the Middle East, this can be explained, to some extent, by the high levels of per capita electricity generation achieved in some countries in the region. In India and Latin America, particularly in Brazil, market reforms aiming at encouraging private investment did not bring the anticipated results.

In Africa, the rate of investment in power infrastructure declined in the 1990s. Economic growth was modest and per capita income remained flat throughout the 1990s. Spending on infrastructure relative to GDP appears to have been particularly low, less than 1%. This compares with 2% to 3% of GDP spent on electricity in Asian countries. The decline was particularly pronounced in sub-Saharan Africa, where only 5 GW of new capacity were added in the 1990s.

Capacity additions are assumed to accelerate in the future (Figure 7.17). Total capacity additions during the period 2001-2030 will need to be three times higher than in the past thirty years. Mobilising the capital needed to build those plants and to add sufficient transmission and distribution capacity may prove an insurmountable challenge for some developing economies.

Figure 7.17: **Capacity Additions in Developing Countries, 1971-2030**

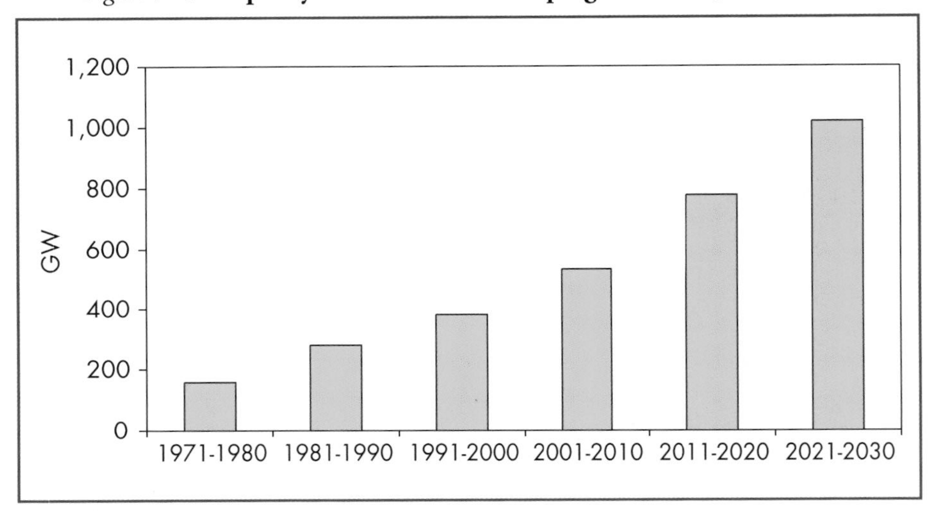

Sources: Platts (2001) and IEA (2002a).

Public utilities in several developing countries are not profitable and are not able to finance new projects themselves. Moreover, investing in new plant is only part of the challenge. Utilities must also purchase fuel to run their power plants. Expenditure on fuel in power stations in developing countries over the next thirty years is expected to be of the same order of magnitude as the investment in infrastructure (Figure 7.18).

The poor financial health of public utilities in these countries results from a series of factors:

- Low revenues because of low electricity tariffs (Figure 7.19). On average, electricity tariffs in developing countries are not high enough for the public utility to be profitable and they may not even cover the utility's short-run marginal costs. Revenue collection can also be inadequate because of non-payment or theft.
- High production cost, which increases the challenge of raising tariffs to cover costs. The cost of producing electricity in many developing

Figure 7.18: **Power Sector Fuel Expenditure in Developing Countries, 2001-2030**

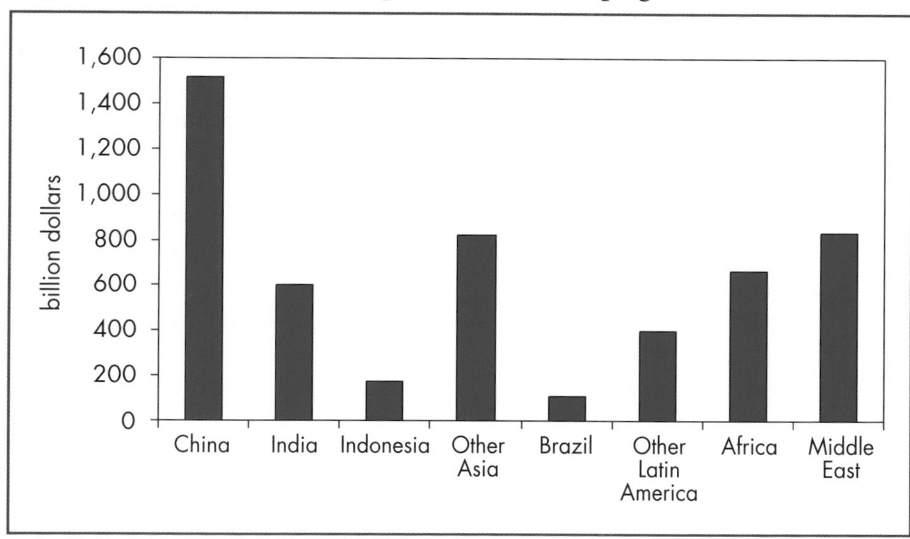

Source: IEA.

Figure 7.19: **Current Electricity Prices, Selected Countries**

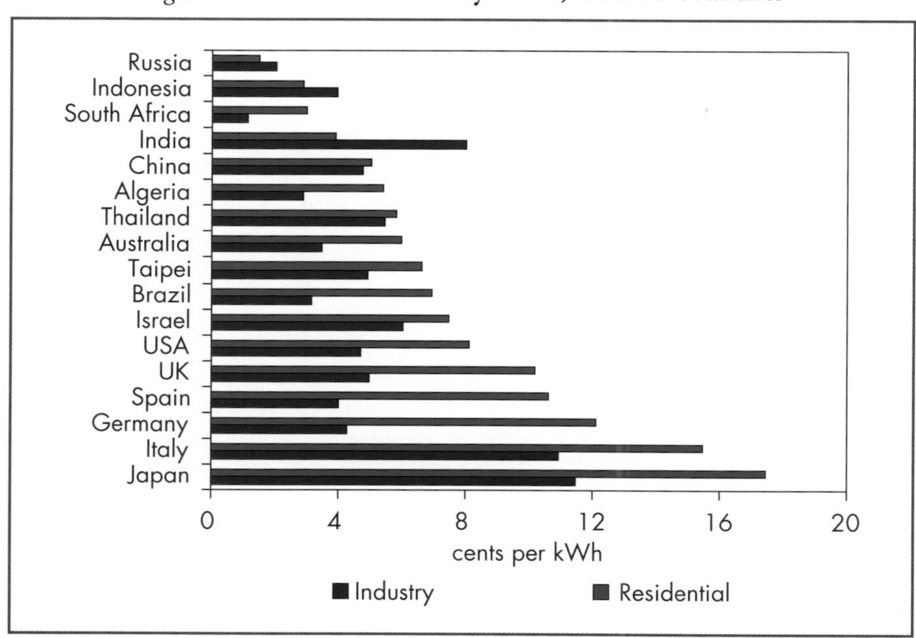

Source: IEA data survey.

countries can be high compared to the OECD for reasons that include low plant efficiency (because of poorly maintained equipment), poor fuel quality, high grid losses (because of poor grid performance or theft), high capital costs (because of non-competitive and non-transparent purchases of equipment), high transmission and distribution costs (because of low consumption density) and high operating costs (because of poor management and low productivity). Exchange rates also adversely affect a utility's costs when loan servicing and purchases of fuel and equipment have to be made in a foreign currency. Utilities have accumulated debt and interest charges which increase their costs and are not usually passed on to consumers.

While much of the funding for investment in developing countries in the past has come from government budgets, this may not continue in the future for a number of reasons:

- Government revenues in many developing countries are low and volatile. They account, on average, for between 15% and 17% of GDP in low- and middle-income countries, while this share is more than 25% in OECD countries, where the electricity sector is mostly privately-owned.[18] Slower economic growth may weaken a government's ability to invest, even though demand for electricity may still be rising.
- Governments may feel obliged to free up resources to increase spending in other sectors (for example in health and social security).
- Governments may wish to increase the economic efficiency of the sector (often along with other sectors of the economy) by encouraging competition and opening up the sector to private investors.

Attracting private investment can be challenging. The private sector, while in principle welcoming business opportunities in rapidly growing developing economies, will respond only if it perceives a sufficiently stable and adequate legal framework and can expect returns high enough to compensate for the risks.[19]

Many countries initiated reforms in the 1990s, aimed at attracting private investment. The initial response was encouraging, but private investment declined rapidly after 1997 (Figure 7.20).[20] Total private-sector investment in electricity between 1990 and 2002 in developing countries amounted to $193 billion. Brazil and other Latin American countries attracted half of it. However, much of it was spent on existing assets that were privatised rather than on new projects.

18. World Bank (2003b).
19. See also, World Bank (2003c).
20. World Bank (2003d).

Reasons for the decline in private investment include badly designed economic reforms, economic crisis or bad business judgements. Many private companies are now selling their assets in developing countries. The reasons for this are diverse and include poor returns on investment, loss of position in their home markets (notably in the case of US investors) and mergers and take-overs under corporate retrenchment policies (in the case of European investors). The result is a drastic reduction in the number of active international investors in developing countries. There are great uncertainties about when and to what extent private investment will revive and where the new investors might come from. One possible answer is local conglomerates, especially in Asia. But development of this source will take time and appropriate policies. These uncertainties create large doubts in attempting to estimate future private investment.[21]

Another handicap for developing countries is growing constraints on their ability to borrow money in international markets. Traditionally, part of the power-sector funding has come from international lending institutions and export credit organisations. Funds from these sources are becoming less and less available. Competition between countries for global investment funds is likely to get fiercer in the future, underlining the need for developing countries to create attractive investment climates.

Figure 7.20: **Power Sector Private Investment in Developing Countries, 1990-2002**

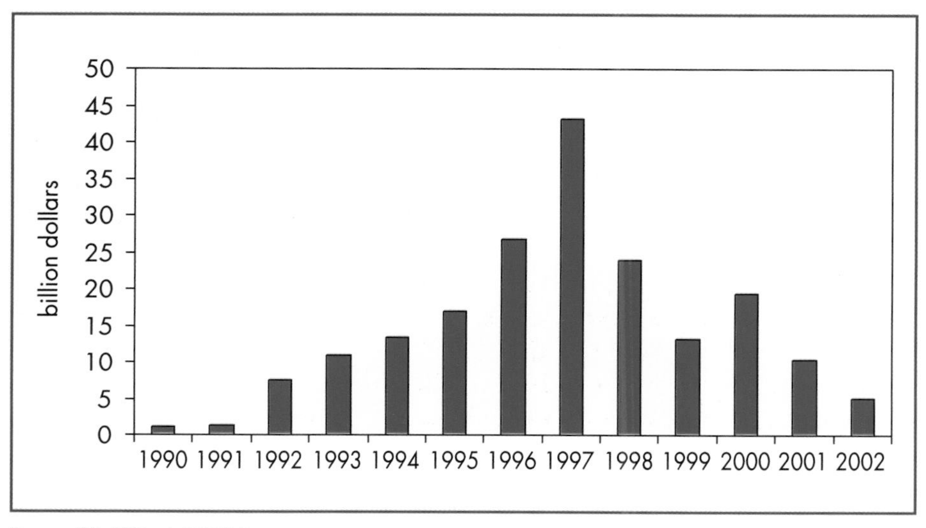

Source: World Bank (2003e).

21. See ADB (2000) for a discussion of best practices in promoting private investment.

Box 7.5: **Potential Impact of CDM on Investment in Developing Countries**

The Clean Development Mechanism (CDM) is one of the three "flexible mechanisms" defined under the Kyoto Protocol (the other two are Joint Implementation and Emissions Trading). It allows countries with binding greenhouse gas emission reduction targets (Annex-I Parties) to implement projects that reduce emissions in non-Annex I Parties.

Although the Kyoto Protocol has not yet come into force, it allows for projects undertaken since 2000 to generate Certified Emission Reductions (CERs). A CDM Executive Board was established at the end of 2001 to supervise such projects. Energy efficiency and renewable energy projects are most likely to form the bulk of projects related to power generation.

The market value of CERs will determine the development of CDM projects. Such projects will only be undertaken if they yield CERs at a cost lower than the market value and below the cost of domestic emission reductions. If the Kyoto Protocol comes into force, the development of future emission reduction commitments beyond the first commitment period (2008-2012) will affect CER prices and therefore CDM projects. Whether countries that have not ratified the Kyoto Protocol decide to join a broader emissions-trading regime will also have an important impact on the CER market. This is particularly true for the United States, the world's single largest emitter of GHG emissions. The fungibility of CERs with other emission reduction accounting units will also affect the market price of CERs and the subsequent development of CDM projects.

Defining emissions baselines for CDM projects is an inherently difficult process. The time and costs involved in the approval of CDM projects could limit the pace of development of these projects. The development of a clear, efficient and affordable approval process will increase the appeal of CDM projects to Annex-I Parties and companies.

CDM projects could lead to significant investment and technology transfers from developed to less developed countries. Given the substantial investment in power infrastructure needed in the developing world over the next thirty years, CDM projects could be one important component in mobilising the necessary investment.

While most investment in the developing world goes towards expanding public utilities, another source has been direct investment by private electricity consumers in their own electricity generating capacity, either as back-up to the public supply or as a replacement for the public supply. This response to underinvestment in public supply is most notable in those countries where the quality of electricity supplied by public utilities is poor and deteriorating, such as India, Nigeria and Indonesia. In Indonesia, for example, autonomous electricity producers own 15 GW out of the country's 40 GW of total installed capacity. This trend could become more significant in the future, if shortfalls in utility investment persist.

Overcoming these obstacles will not be easy. It will require significant improvements in governance and continued restructuring and reform in the electricity and associated sectors, which will test the institutional capacity of developing countries. A threshold challenge will be reform in tariff structures to make prices cost-reflective and to improve revenue collection.[22] It is a major challenge to do this in a way that does not unduly hurt low-income consumers who are not able to afford even basic electric services.

Reforms will be necessary in all segments of the power sector. Given the significant level of investment needed in distribution networks (about $2 trillion, or 40% of total power-sector investment), reform of electricity distribution will be critical. In many developing countries, the priority is to reduce non-technical losses from theft of electricity and from non-payment of bills. Such reforms are difficult and take considerable time, at least five years and more likely ten or so. The gap between investment needs and actual investment is likely to continue for some time in the worst-affected countries.

Regional Analysis

United States and Canada

Overview

The United States and Canada will need to invest $1.7 trillion in the electricity sector over the next thirty years, of which $900 billion will have to go towards the transmission and distribution system, particularly in the United States. The power blackout across the north-east of the region in August 2003, has directed renewed attention to the importance of maintaining electricity reliability. Adequate and timely investment in infrastructure is a key element of a reliable electricity system.

The North American electricity market is the largest in the world, accounting for about 30% of the world's electricity production. Electricity sales in North America comprise almost 3% of GDP, which is more important than the contribution of telecommunications, airlines or the natural gas industries.

22. See also Jamasb (2002).

Figure 7.21: **Electricity Sector Investment in the United States and Canada, 1992-2001**

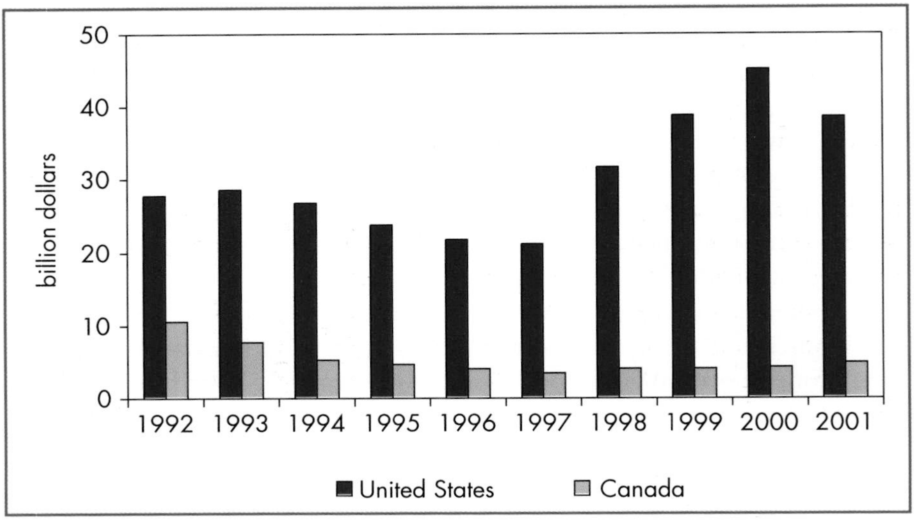

50
40
30
20
10
0
billion dollars
1992
1993
1994
1995
1996
1997
1998
1999
2000
2001
United States
Canada

Sources: Edison Electric Institute (2002) and Canadian Electricity Association (2000).

Investment in the sector has increased dramatically over the past few years and reached $43 billion in 2000/2001. Most of the increase in investment reflects increased power plant construction in the United States (Figures 7.21 and 7.22). Investment in transmission and distribution did not keep pace with investment in generation.

Figure 7.22: **Generating Capacity Additions in the United States, 1995-2002**

60
50
40
30
20
10
0
GW
1995
1996
1997
1998
1999
2000
2001
2002

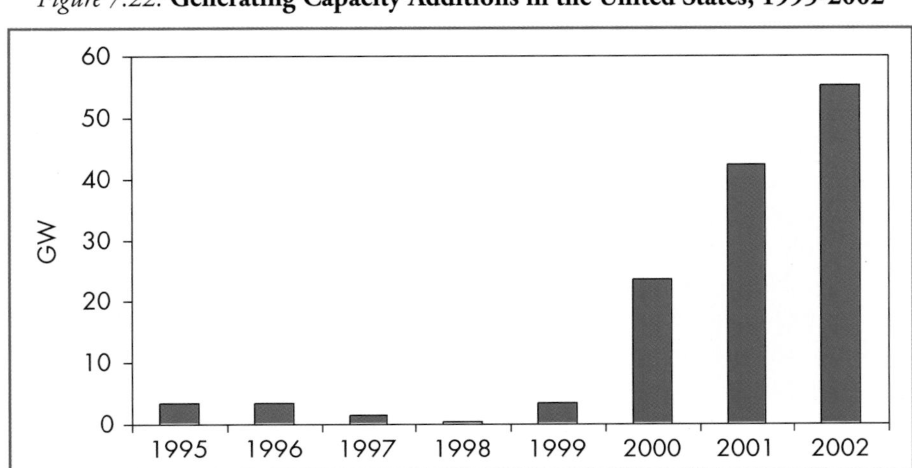

Sources: IEA and US DOE Energy Information Administration (2003).

A competitive environment is being developed in the region's wholesale and retail markets. In the United States, about three-quarters of generating capacity was still operating in a regulated environment in 2000, while some 70% of utilities were privately-owned. Since the mid-1990s, successive regulatory actions have encouraged or required the development of competitive electricity supply. Nearly all the power plants constructed recently in the United States have been unregulated plants.

Deregulation has also prompted an increase in mergers and acquisitions, resulting in substantial industry consolidation. By the end of 2000, the ten largest companies held more than half of the country's privately-owned capacity, compared to 36% in 1992.

Confidence in the new competitive market model is at a low ebb. The California energy crisis caused the pace of deregulation to slow in many states and increased regulatory uncertainty. The trading scandal and ensuing financial collapse of Enron has prompted close financial scrutiny of the industry. These events, combined with a slowing economy (and resulting lower demand growth for electricity) and volatile fuel and wholesale electricity prices, have resulted in poor financial performance, high debt and downgrades by credit-rating agencies, most significantly for merchant companies. While regulated utilities can expect to recover their costs, they too have seen their profit margins shrink (Figure 7.23).

Figure 7.23: **US Privately Owned Electricity Companies' Profit Margin**

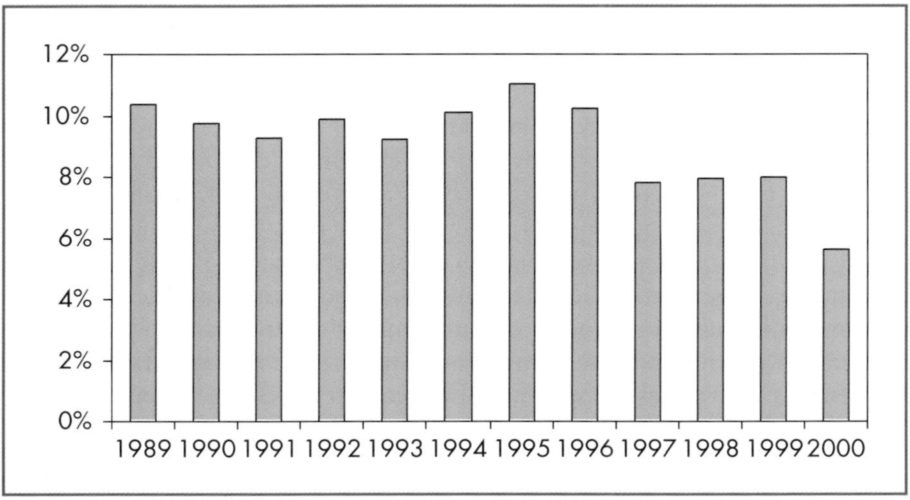

Source: US DOE Energy Information Administration (2002).

In Canada, most generating capacity is publicly-owned. Two provinces — Alberta and Ontario, accounting for half of the country's electricity consumption — have introduced wholesale and retail competition. The events in the United States have shaken confidence in Canadian electricity market reform.

Investment Outlook

Following the substantial capacity increases in the 1980s, few power plants were built in the region in the early and mid-1990s. However, starting in 1999 large amounts of new capacity came on line in the United States. About 144 GW of new capacity, mostly merchant plants spurred by market liberalisation, were constructed in the period 1999-2002, resulting in capacity overbuild. After 2002, given the uncertainties about economic growth and the current poor financial situation of the electricity sector, many additional projects that had been announced have been delayed or cancelled. Investment in new plant is expected to be relatively slow in the near term. The power plants that were constructed recently, if operated in baseload, could meet rising electricity demand for about a decade. Some regions, however, will need earlier investment in power generation. Several gas-fired power plants are under construction in Canada. Six closed nuclear reactors in Ontario, four at Pickering A and two at Bruce A, are expected to reopen soon.

Over the period 2001-2030, the region will need about 830 GW of new power plant. More than half of the new capacity will be gas-fired, continuing a recent surge of gas-fired power plants. In the longer term, as gas prices rise, more coal-fired plants could be constructed. About 15% of the capacity increase could come from non-hydro renewable energy sources. The total cost of these new plants will be $654 billion and an additional $130 billion will be needed to refurbish ageing units. North America has some of the oldest power plants in the world. A significant portion of this amount will go towards extending the lifetime of existing nuclear generating facilities.

An important issue currently under discussion in the United States is the low level of investment in the country's transmission grid.[23] Until the late 1980s, transmission capacity was somewhat higher than generating capacity.[24] In the 1990s, however, investment in transmission did not keep pace with investment in generation for a number of reasons that include siting and approval difficulties, a focus on unregulated activities and regulatory uncertainty arising from the transition to competitive markets. Transmission bottlenecks emerged as the substantial increase in electricity demand over the

23. US DOE (2002a) and US DOE (2002b).
24. See also Hirst and Kirby (2001).

past decade, combined with new generation in wholesale markets, increased electricity flows.

Investment in transmission systems has increased somewhat over the past three years. Private investment in transmission was $3.7 billion in 2001[25], compared to $2.6 billion a decade earlier, and is likely to stay at this level at least until 2004.[26] It will need to more than double to provide adequate transmission capacity in the long run.

North American investment in transmission in the period 2001-2030 is projected to reach $260 billion, amounting to $73 billion in the decade 2001-2010 and $99 billion in the second decade and then falling to $89 billion in the last decade. Along with transmission, investment in distribution should also rise above present levels. The investment needed in distribution networks in the period 2001-2030 is of the order of $650 billion.

Investment Issues and Implications

The North American power sector is expected to become increasingly reliant on natural gas. The share of natural gas in electricity generation could rise from about 15% now to 30% by 2030. Increased interdependence between natural gas and electricity will require investment in gas-supply infrastructure to move closely in step with investment in gas-fired power plants.

Natural gas-fired power plants continue to be constructed close to the fuel supply, rather than close to demand centres, with very little investment in transmission lines.[27] This issue can be addressed through transmission pricing that accurately reflects the costs of transmission and allows the market to determine whether it is more appropriate to expand transmission capacity or to install new generating capacity closer to the load.

The North American electricity sector faces increasingly stringent environmental regulation, both on the generation and on the transmission sides. Stricter emission limits for sulphur dioxide, nitrogen oxides, particulate matter and new limits on mercury and possibly CO_2 will increase investment requirements in both existing and new facilities. For example, in the United States, less than 30% of coal-fired capacity was fitted with FGD in 2000. Additional investment in scrubbers may be required to meet the Clear Skies Initiative. The US Environmental Protection Agency (EPA) estimates that some 130 GW of power plant will need to be fitted with scrubbers by 2020. The average cost of installing such equipment is of the order of $200 per kW.

25. Edison Electric Institute (2002).
26. By comparison, Japanese investment in transmission lines was between $4 billion and $7 billion per annum in the late 1990s. Japan's electricity generation is a quarter that of the United States.
27. NERC (2002).

Canada has ratified the Kyoto Protocol. The implementation of policies to achieve its target could alter the country's investment outlook by encouraging low or zero carbon technologies.

European Union

Overview

With demand for electricity rising by nearly 50% over the next thirty years, an ageing infrastructure of which a large part will have to be replaced, and converging electricity markets that will be more and more interdependent through connected electricity networks, the countries of the European Union will need to invest over a trillion dollars in power infrastructure, divided equally between generation and networks.

The electricity sector in the European Union has an annual turnover of about $150 billion and contributes about 1.5% to EU GDP. Investment in the sector is now about $30 billion per year. Investment in electricity in the 1990s was low compared to the 1980s because the amount of capacity built was lower and because investment was directed towards low capital cost plant, notably CCGTs. Figure 7.24 shows investment in European Union countries over the five-year period 1993-1997.

Figure 7.24: **Electricity Sector Investment in EU Countries, 1993-1997**

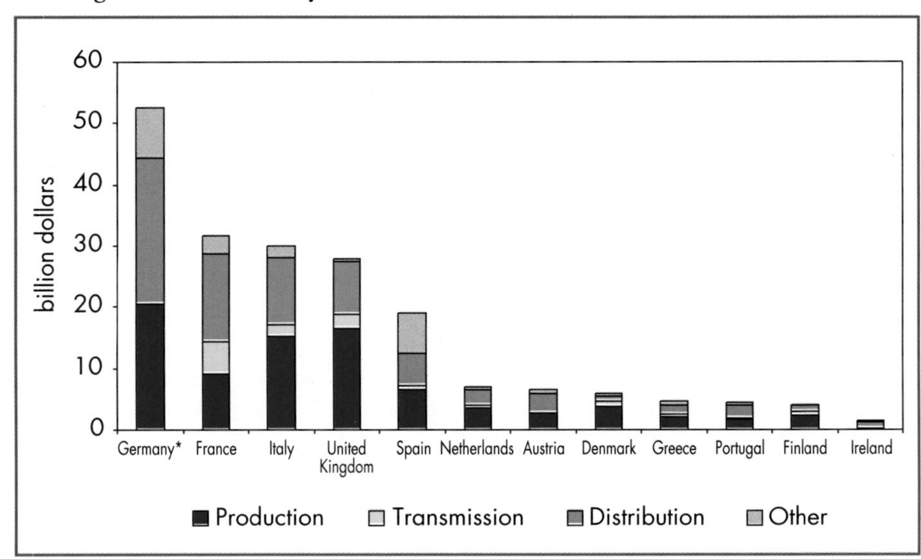

* Transmission in Germany is included in distribution.
Source: Eurelectric (1999).

The European electricity sector is undergoing profound restructuring, from state- owned utilities to private generators in competitive markets. The process of market liberalisation began in the United Kingdom and Sweden in the early 1990s. EU-wide liberalisation commenced with the adoption of an electricity directive in 1996. Most future power-sector investment will be private and the private sector will determine the levels of capacity and the fuel mix. This creates an obligation on policy-makers to monitor carefully how much and when new capacity and network infrastructure is brought on line, especially in the longer term, when demand for new plant is expected to be substantially higher than now.

The opening of electricity markets to competition has reduced the size of national companies, but at the same time it has triggered a series of mergers and acquisitions among electricity companies. This trend is likely to continue in the near term, with companies trying to strengthen their position in the changing market environment. Mergers and acquisitions may, in certain cases, improve economic efficiency through economies of scale. Importantly, they create large companies with the deep pockets needed to finance investment. However, large-scale mergers could impede competition. At the beginning of 2003, seven companies controlled more than half of the European Union's capacity, although there are more than 5,000 producers of electricity in the region, including autonomous producers of electricity (Figure 7.25).

Figure 7.25: **Capacity Ownership by Major European Power Companies, 2002**

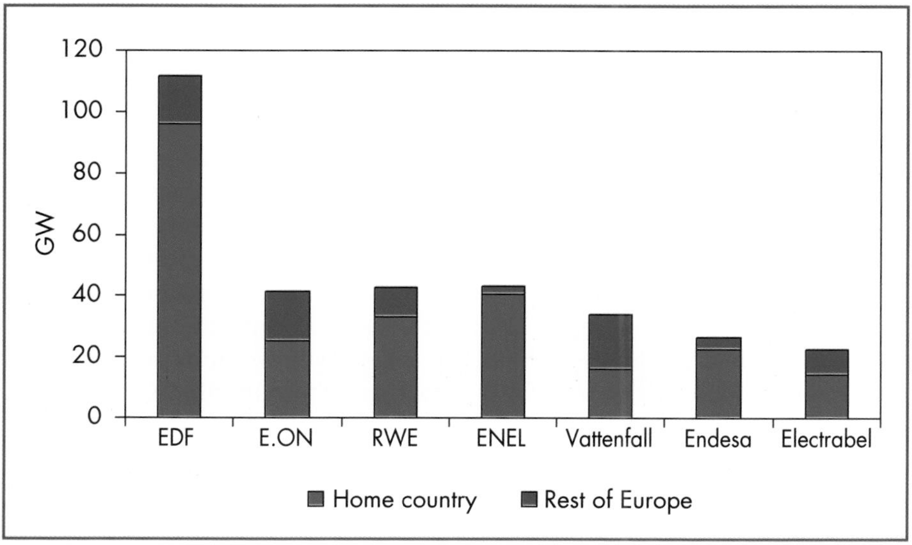

Source: IEA and company websites.

European power plants had an installed capacity of 584 GW at the end of 2000 and produced 2,572 TWh of electricity. Substantial increases in generating capacity took place in the 1980s, partly because of high electricity demand expectations. About 30% of today's installed capacity was built in that one decade. This resulted in high reserve margins in most countries throughout the 1990s. The supply situation appears to be tightening in many countries, although some still enjoy fairly high reserve margins. New capacity is being continually added almost everywhere, but in many cases the capacity additions have not kept up with the increase in the load.

Investment Outlook

Over the next thirty years, the countries of the European Union will need nearly 650 GW of new capacity to meet rising electricity demand and to replace about 330 GW of existing power stations. These capacity additions are greater than the current total installed capacity. The total finance needed for these new generating projects is in the order of $525 billion.

Investment in power generation is likely to rise over the long term for two reasons: first, the capacity requirements will increase as the rate of retirement of existing power plants increases; second, the average cost of adding a new power plant is expected to increase over time. The 1990s saw a decrease in average cost per kW of new plants, because of the introduction of low capital cost CCGT plants. Favoured by market liberalisation and low gas prices, this trend is likely to continue in the medium term. However, in the long term, the projected increase in coal and renewables use for electricity generation will increase the average plant cost, making the electricity sector more capital-intensive.

More than half of the new capacity is expected to be gas-fired and more than 20% of new plant will be based on non-hydro renewable energy sources, particularly wind and biomass. A significant amount of new capacity, some 116 GW, is expected to be coal-fired, using advanced technology. Construction of nuclear plants is expected to be limited to France and Finland, making investment in nuclear capacity a small part of the total. Whether more new nuclear plant will be built in Europe is uncertain and depends on a number of factors, including the CO_2 emissions cap-and-trade EU scheme.

The existing transmission network will need to be extended and refurbished to accommodate increased trade and new patterns of energy flows brought about by competition. Consequently, some $120 billion will be invested in transmission over the next three decades. The transmission networks in EU countries are relatively mature and most investment is expected to go towards upgrading and replacement of existing lines.[28]

Much effort is being directed towards making more efficient use of the existing infrastructure.[29] This involves the extensive use of existing interconnections between EU countries, as well as the expansion of interconnection capacity. Import capacity is used intensively in some countries, already creating congestion. There were some 30 GW of excess generating capacity in early 2003 in continental Europe, but current interconnection capacity does not allow all this capacity to be reflected in imports.

The European Union attaches particular importance to the development of transmission networks. Key areas of transmission network investment include the connection of isolated electricity networks, the development of interconnections between the European Union member States, the development of interconnections within countries and the development of interconnections with countries outside the EU.[30]

Investment will be necessary in distribution networks for network reinforcement, asset replacement and new connections. Investment in distribution has been a substantial part of electricity-sector investments and it is likely to remain high in the future. It will amount to $413 billion over the period 2001-2030.

Investment Issues and Implications

The outlook described above shows that future power-sector investment needs are expected to be much greater. Current trends in investment differ between EU countries, but there is evidence that capacity margins are getting tighter in many of them. Power plant retirements, expected to increase over time, will add to the need for new plants. The development of interconnections between countries will help improve competition and security of supply but these interconnections should not be seen as a means of avoiding additional investment in generating plants. Governments, while providing a stable regulatory framework that allows investors to be rewarded, will need to monitor developments in investment to ensure that adequate infrastructure is built.

28. For example, the National Grid Company, which operates the transmission system in England and Wales, spends some 10% of its annual investment on new lines, while more than 80% of the investment is used to replace and refurbish existing lines.
29. Commission of the European Communities (2001).
30. *Official Journal of the European Union* (2003).

The current pattern of investment points to a substantial increase in gas consumption. Gas will increasingly be imported from countries outside Europe. Adequate and timely investment in gas production, transport and storage facilities will therefore be of crucial importance.

The next thirty years should see substantial investment in renewable energy sources. This growth will be encouraged by the EU Renewables Directive, which calls for a substantial increase in the share of renewables by 2010. A significant amount of new capacity will come from wind farms. As an intermittent source of energy, wind is not always available to produce electricity. Integrating wind into the network requires investment beyond that necessary for conventional generation technologies, involving greater voltage regulation power and network reinforcement.

With most future investment directed towards natural gas, renewables and cleaner coal, CO_2 emissions per unit of electricity produced will decline. However, annual power-sector emissions will rise by a third by 2030 and the power-sector's share in total energy-related CO_2 emissions will increase. Governments will need to intervene to encourage greater investment in technologies that emit less CO_2. The EU-wide cap-and-trade scheme, which is expected to be put in place in 2005, should influence power-sector investment in this direction over the longer term.

The projected retirement of a substantial number of fossil-fuel plants will create great opportunities for replacement by cleaner technologies. At the same time, the increased investment requirements might give rise to pressure to defer decommissioning decisions. If fossil-fuel plants remain in operation longer than assumed here, CO_2 emissions will be much higher.

Japan, Australia and New Zealand

Overview

These three countries in the OECD Pacific region will need investment approaching $600 billion over the next three decades. Japan will account for the largest part of this investment, because of the size of its market — it accounts for 80% of the electricity produced in the region — and because of the higher capital intensity of its power sector, a reflection of its nuclear programme.

In Japan, generating capacity amounted to 262 GW in 2001. Most electricity is produced and supplied by ten private-sector electricity companies, listed in Table 7.6. Electricity-sector investment by the major companies in 2001 was nearly $22 billion, equivalent to 0.5% of the country's GDP. Revenues from electricity sales contributed 2.7% to GDP.

Investment in Japan has been declining since the mid-1990s. The level of investment in 2001 was about half the 1994 level (Figure 7.26). This decline reflects, to some extent, a pause in electricity demand growth: demand has been almost constant over the past five years.

Table 7.6: **Japanese Power Companies**

Power company	**Total assets** ($ million)	**Capacity** (MW)	**Revenues** ($ million)
Tokyo	116,665	60,375	42,219
Kansai	57,971	35,585	20,723
Chubu	50,446	32,231	17,680
Kyushu	32,796	19,336	11,395
Tohoku	32,681	16,076	12,814
Chugoku	22,298	12,179	8,004
Hokuriku	12,534	6,759	3,971
Shikoku	11,750	6,877	4,550
Hokkaido	11,396	5,904	4,279
Okinawa	3,284	1,676	1,140
Total	**351,821**	**196,998**	**126,775**

Note: Data refer to March 2002.
Source: Federation of Electric Power Companies of Japan (2002).

Figure 7.26: **Electricity Sector Investment in Japan, 1988-2001**

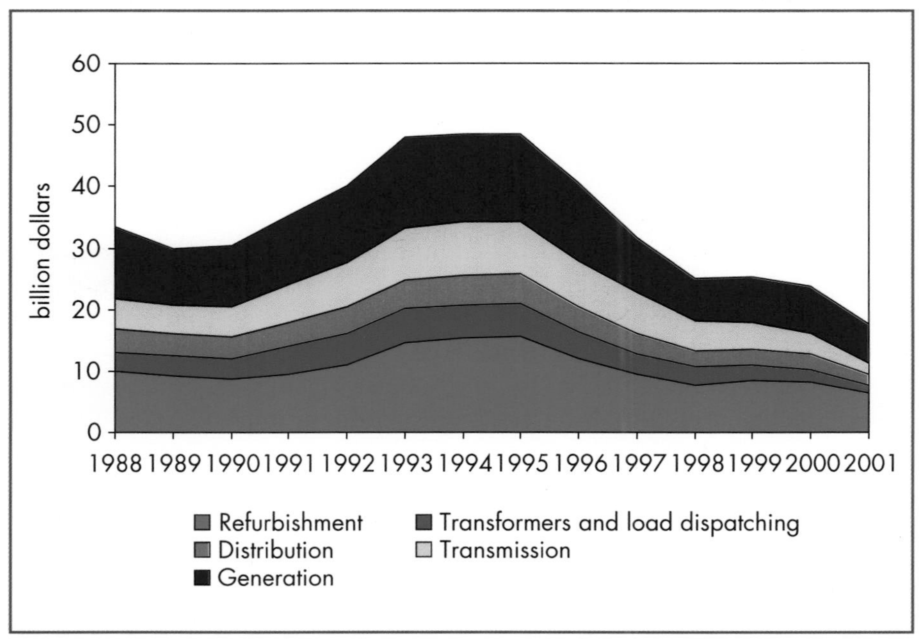

Source: Federation of Electric Power Companies of Japan (2002).

Australia had 47 GW of installed capacity in 2001, producing 217 TWh of electricity. Market reforms began in the early 1990s, but it was not until 1998 that a national wholesale market was established. Box 7.6 describes how investment has evolved in Australia's liberalised electricity market.

Box 7.6: **Investment Performance in the Australian National Electricity Market**

The National Electricity Market (NEM) commenced operation in December 1998. It includes the states of New South Wales (which incorporates the Australian Capital Territory), Victoria, Queensland and South Australia. The NEM uses a regional pricing model to approximate full nodal pricing.

Demand has been growing at around 3.7% per year compound across the NEM since the start of the market, with the highest growth rates recorded in Queensland (12.3%) and South Australia (9.4%).

Pre-existing surplus capacity has largely been absorbed and over 3,300 MW of new dispatchable generation has been commissioned since the start of the market, representing a 9.6% increase in capacity over the period. New generation investment has typically been well timed and has emerged in those regions with the tightest supply-demand balances and highest average spot prices. Nearly 80% of this new generation is located in the Queensland or South Australian region. Over 8,000 MW of new generation projects have been proposed for the next few years, but it is unlikely that all will proceed in this time frame.

Since the market began, there have also been some significant additions to the interconnected network. Overall, around 1,800 MW of interconnection capacity has been added, representing an increase in NEM interconnection capacity of over one-third. In addition, a further two interconnection projects, with up to 850 MW of new capacity, have recently been approved.

New generating investment, combined with new interconnection investment, has helped substantially reduce regional spot prices in South Australia and Queensland. Average annual spot prices in South Australia have nearly halved, falling from AUD 61/MWh in 1999/2000 to AUD 32/MWh in 2001/2002, while in Queensland, average annual spot prices fell by over 20% during the same period, from AUD 45/MWh to AUD 35/MWh.

The National Electricity Code Administrator noted in the Triennial Review of the NEM (June 2002) that average spot prices were beginning to converge across the NEM at or near the threshold for new entry, with some seasonal variations reflecting weather patterns. The Triennial Review of the NEM also indicated that new investment has been sufficient to meet reliability requirements.

New Zealand's installed capacity was 8.6 GW in 2001, most of which is hydro. There has been hardly any increase in the country's capacity over the past few years, although electricity demand has been rising. Concerns over future gas supplies have delayed decisions on new projects. There are serious concerns about the adequacy of reserve margins to meet demand in a very dry year, when output from hydro plants can be very low.

Investment Outlook

Japan, Australia and New Zealand will need to invest over $600 billion in generation, transmission and distribution. New capacity in the region is expected to amount to 236 GW, requiring investment of $274 billion, while the refurbishment of existing plants will require about $48 billion.

The *WEO-2002* projections assume that nuclear plant construction in Japan will proceed more or less as planned. However, the recent turmoil in Japan's nuclear operations has undermined public confidence in nuclear power and may hinder future development. If fewer nuclear plants are built, the industry is likely to build coal- and gas-fired plants. The investment requirements in this case will be lower than the estimate given here.

Investment in transmission is expected to be around $100 billion, while distribution will take $185 billion. In Japan, both the share and the absolute level of investment in electricity transmission and distribution have declined substantially since 1998. They will have to rise again in the future to meet supply requirements. In Australia, increased investment will be needed in the country's interconnection capacity in order to facilitate trade.

Investment Issues and Implications

In Japan, investment in gas-fired plants will increase the country's dependence on imported sources of energy, particularly if fewer nuclear plants are built than assumed here. Japan has ratified the Kyoto Protocol, which calls for a 6% reduction of the country's GHG emissions relative to the 1990 level. If coal and gas plants are built instead of nuclear, it will be much harder for Japan to meet this target. Regarding other emissions, Japanese power plants have low SO_2 and NO_x emissions. These are higher in Australia and any effort to reduce these emissions will require increased investment in pollution control or may shift investment towards cleaner sources of electricity. In the near term, New Zealand will need urgent investment in new capacity. Concerns about supply shortages may encourage distributed generation.

China

Overview

China's investment requirements in electricity infrastructure over the next three decades will amount to nearly $2 trillion, of which more than $1 trillion will be for transmission and distribution. This means that 2.1% of the country's GDP should be invested every year in power infrastructure. Although, the country's vigorous economy should attract substantial private investment, whether such capital will come depends on the framework created by the new electricity authorities established earlier this year.

The Chinese electricity system is now the world's second-largest after the United States, with 322 GW installed capacity in 2000. Most of this capacity is coal-fired. China has been very successful in developing its power infrastructure. This may be attributed to two main factors: structural reforms opening up new capital sources and high domestic savings. Over the past decade, the country added 13 GW of new capacity and 30 thousand kilometres of transmission every year. Investment in 2000 amounted to $25 billion.

China initiated its power-sector reforms in the mid-1980s. Since then, and through various steps in the reform process, China has successfully attracted increasing sums of new capital to supplement government funding in power generation. As new investors came on board, the central government's share of financing declined from 100% to 45% in the second half of the 1990s. However, private investment, both domestic and foreign, still amounts to less than a fifth of the total (Figure 7.27).

Figure 7.27: **China's Capital Construction Financing, 1996-2000 (9th Five-year Plan)**

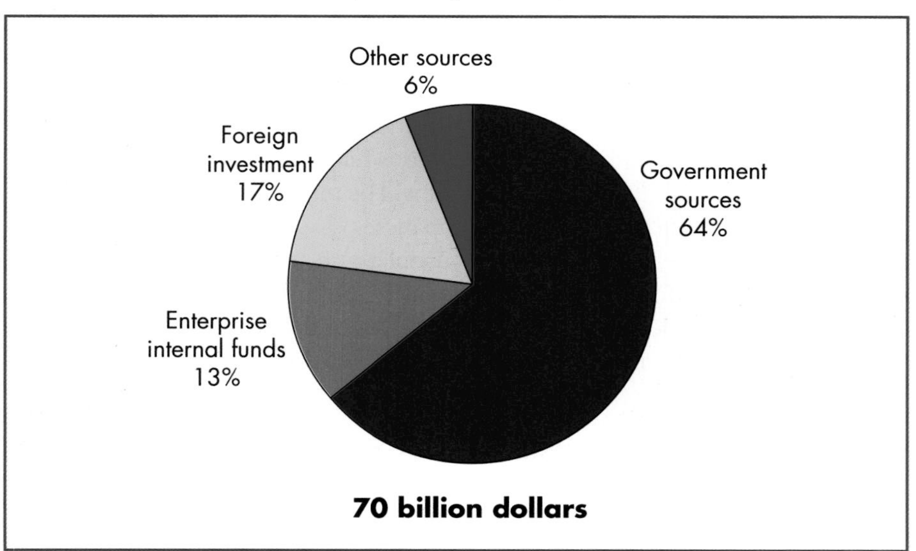

Note: Construction expenditure does not include refurbishment and distribution.

The reforms might not have been so successful without abundant domestic savings. China has one of the highest domestic saving rates in the world, at around 40% of GDP. Domestic savings financed $89 billion out of $101 billion invested in the electricity industry over the period 1996-2000, through bank loans, government bonds and power company equities, with foreign investment accounting for 12.2% (and 17.4% of investments dedicated to capital construction).[31]

In China, transmission and distribution remain the responsibility of the central government. In order to bring electricity to over 98% of its population, China has built an extensive distribution network over the past 15 years. China is now putting more emphasis on the development of transmission networks and interconnections. Lack of adequate transmission prevents low-cost generation in one province or region from reaching a neighbouring area.

Investment Outlook

China will need to add about 800 GW of new capacity over the next thirty years. Most of this new capacity will be based on coal and hydropower but the adverse impact of coal on the environment is pushing China to increase the use of natural gas for electricity generation. Over 100 GW of natural gas-fired capacity is expected to be added by 2030.

To meet the country's rapidly growing electricity demand, a total of nearly $2 trillion will be needed for electricity generation, transmission and distribution over the projection period. The investment requirements for new power plants will amount to $795 billion over 30 years. Refurbishment of existing power plants will add $50 billion.

China's corresponding transmission network extension will require investment of $345 billion, 18% of the overall investment. The investment in distribution will be double that in transmission. Investment in transmission and distribution, as a share of total power-sector investment, will be higher in the future. Investment in transmission has not matched investment in generation in the past and will need to rise substantially in the future. Investment in distribution will be driven by the residential and services sectors, where electricity demand is projected to rise by 6% per year.

Power-sector investment in China accounted for 1.3% of GDP in the mid-1980s and has been constantly rising since then. It now stands at around 2.5% of GDP. This share will rise slightly to 2.7% in the decade 2001-2010, and then, as the sector reaches maturity, fall to 2.3% and 1.7% in 2011-2020 and 2021-2030, respectively (Figure 7.29).

31. China State Planning Commission (2001).

Figure 7.28: **Average Annual Investment Requirements in China**

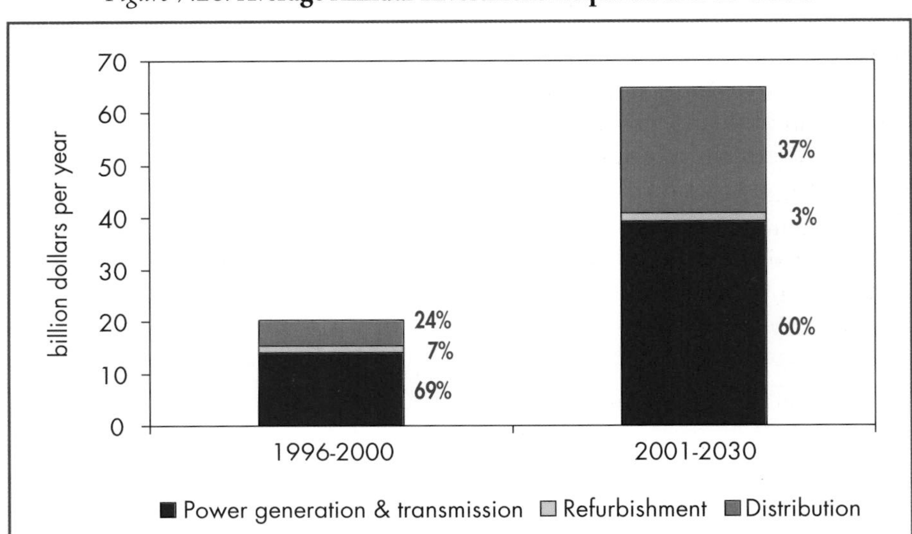

Figure 7.29: **China Power Sector Investment as a Proportion of GDP**

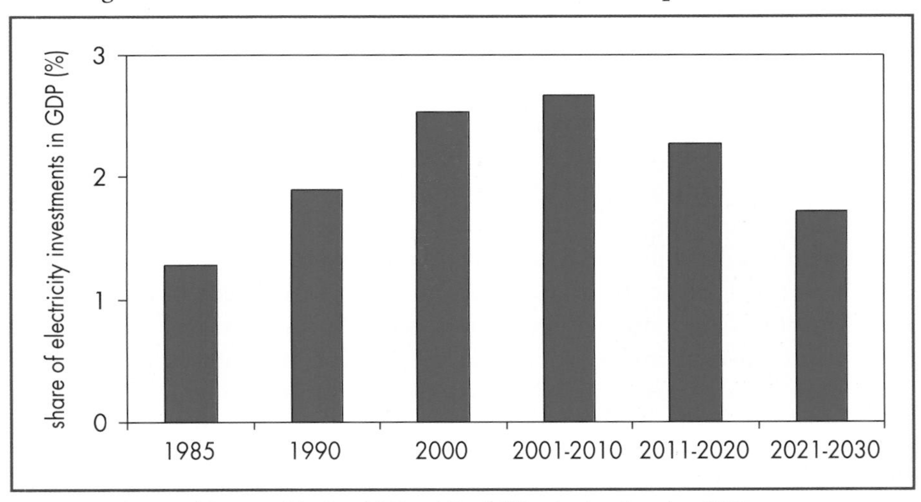

Source: State Statistical Bureau of the People's Republic of China (various issues) and IEA.

Investment Issues and Implications

As China's economy expands to meet the growing needs of its population and industry, the country will continue to face great investment challenges. China is currently undertaking a third set of reforms aimed at achieving a market economy. While the Chinese government's effort to raise capital in the past has been successful, future investment funding may prove more difficult.

Although the economy is growing fast and domestic savings are high, relying on financial markets for investments in a competitive generation market may prove to be difficult in China, where the private sector is weak and financial markets are still underdeveloped. For the moment, risks in power project financing are often low owing to government underpinning and the monopolistic nature of the industry. As the government withdraws from business and competition is introduced, private-sector investors will take due account of the increased risk.

Much will depend on the reform of fuel and electricity prices. Until now, investors in the power sector have benefited from sales contracts based on a cost-plus pricing regime. Though there are sound arguments for change, new pricing arrangements and the lack of firm power purchase agreements could deter investment.

The price reform policy seeks to let the wholesale market determine the tariffs on the generation side, while government will regulate transmission and distribution prices as well as relative prices to end-users. China has tested competitive power pricing in Shanghai and five other provinces, but that pilot programme covered less than 10% of the electricity generated in those areas. The government has promised freer power pricing, but has pursued the issue slowly and power executives say they do not expect major progress before 2005 because of the political sensitivity of the issue and concerns about social stability. Whether China will be able to attract the necessary investment depends very much on the nature of the future electricity pricing regime.

Loans from multilateral agencies are not expected to play a big role in financing the Chinese power sector. International financial loans to the sector have never exceeded $1 billion a year over the past ten years, while multilateral aid has been declining worldwide. China has been the World Bank's largest borrower of investment financing since 1992, but most of this lending has gone towards environmental protection, including some power projects with the requisite environmental standards. China's energy sector has been the principal recipient of loan support from the Asian Development Bank.

Along with sizeable financial constraints, there are also major threats to environmental sustainability. Producing 13.5% of global CO_2 emissions, China is the world's second-largest emitter, after the United States. This share will rise to 18% by 2030. The power sector accounts for 44% of the total now and this will grow to 55% in 2030. Many of China's environmental problems stem from the increasing and inefficient use of coal, much of it in power plants.

These plants emit substantial amounts of SO_2, accounting for about 44% of the country's total SO_2 emissions. NO_x and dust emissions from coal-fired power plants are also key pollutants.

In order to tackle these environmental issues, China is putting more emphasis on fuel diversification and on clean energy development. However, these new policies were designed in an era of surplus. The recent electricity shortages could shift the focus back to capacity expansion and to coal (although more stringent emission standards apply to new coal plants). Given that China is expected to rely on domestic coal for the major part of its power supply, the clean and more efficient use of coal in the power sector will be critically important for both the local and global environment.

India

Overview

India's electricity demand will increase more than threefold over the next thirty years. To meet this increase, the country will need to invest $665 billion in the power sector (Figure 7.30). This will require investment equivalent to 2% of Indian GDP every year in power projects. Given the extremely poor financial situation of the Indian power sector, the availability of the necessary finance remains very uncertain.

Figure 7.30: **India Power Sector Investment, 2001-2030**

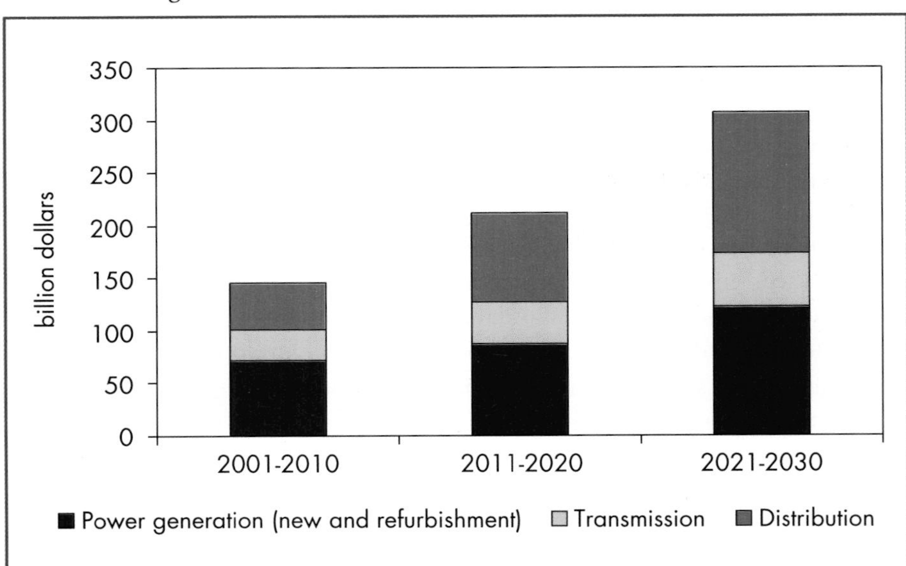

India's installed capacity was 111 GW at the end of 2000. Most of India's power plants are coal-fired, producing more than three-quarters of the country's electricity. Hydropower is the second-largest source, with 23 GW of installed capacity and a 14% share in electricity generation.

Most power plant capacity is publicly-owned. The central government, through public companies, owns one-third of capacity. Most of the remaining capacity is owned by the State Electricity Boards (SEBs). The high-voltage transmission grid is operated by a central transmission utility, Powergrid, while the rest of the grid is under the responsibility of the states.

Public-sector power development is the joint responsibility of the central government and the states. The power sector is funded mainly through budgetary support and external borrowings. Until the early 1990s, it received between 15% and 20% of the total budget. This share has declined since economic reforms were introduced in 1991, in the expectation that part of the required investment would come from the private sector. Despite government desire to attract private and foreign investment in new independent power producers, most of the projects proposed have not proceeded, because of an inadequate legal and commercial framework and delays in obtaining regulatory approvals. The dispute over the Dabhol power plant, the largest single foreign investment in India, has drawn attention to the inadequacy of the current framework.

The electricity industry faces enormous challenges in providing reliable service and meeting rising demand. Inadequate reserve margins and the poor performance of the transmission and distribution systems cause frequent, widespread blackouts and brownouts. Plant capacity factors are often low, owing to the age of generating units, poor quality coal, defective equipment and insufficient maintenance. The lack of inter-regional grid connections accentuates local power shortages. Power theft, the non-billing of customers and non-payment of bills are common.

The most problematic area in the electricity sector has been the operational and financial performance of the SEBs. The financial health of many of India's state electricity boards has been deteriorating because of high operating costs and pricing policies that keep tariffs to most customers well below the cost of supply. Insufficient revenues have been driving up debt and discouraging investment. The following points illustrate the critical situation of the SEBs:[32]

- Revenues from electricity sales are 70% of costs. This share has fallen more than 10 percentage points since the early 1990s (Figure 7.31).
- About 40% of revenues come from subsidies.

32. Data are for 2000, except for revenue arrears which refer to 1999.

- The electricity boards' rate of return on capital is -44%, almost four times worse than in 1992.
- Revenue arrears as a share of total revenue have been increasing, reaching 40%.

Because of the poor reliability of India's power sector and because supply does not match demand, many customers, particularly industrial facilities, have invested in small generators for on-site production of electricity. About 10% of India's electricity generation now comes from such installations.

The government has taken a number of steps in recent years to restructure the electricity industry, to reform pricing and to introduce more market-based mechanisms. In April 2003, the Electricity Bill 2001 was approved by the lower house of the Parliament. Once approved by the upper house, the new law will replace the 1910 Indian Electricity Act, the 1948 Electricity Supply Act and the 1998 Electricity Regulatory Commission Act. Key measures of the new bill include fewer licensing restrictions for fossil-power projects, open access in transmission, an obligation on states to establish regulatory commissions which would set retail tariffs on the basis of full costs and promote competition, and a requirement that any subsidies on electricity retail sales be paid out of state budgets rather than through cross-subsidisation.

Figure 7.31: **Electricity Revenue and Return on Capital of the State Electricity Boards in India**

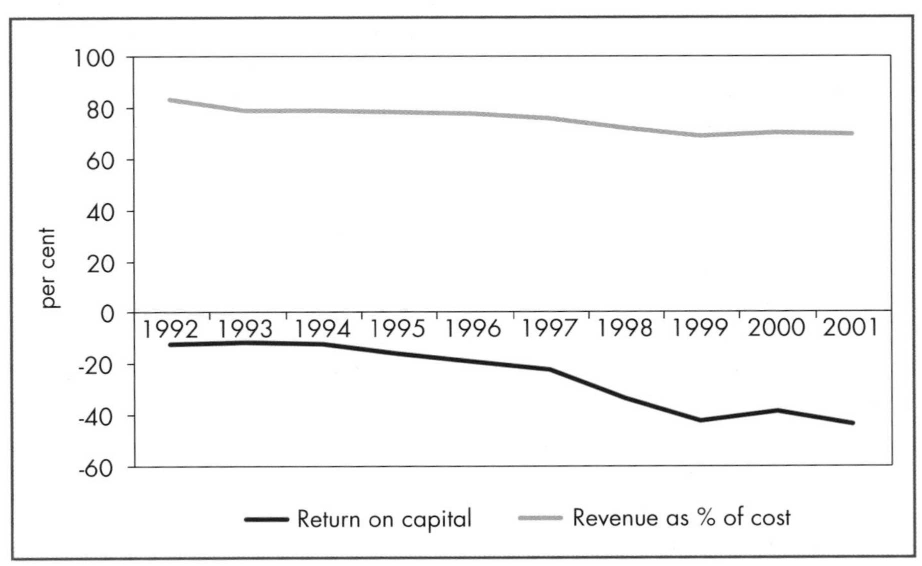

Note: Return on capital is defined as the ratio of commercial profits/losses to net fixed assets, expressed as a percentage.

Source: Government of India (2002).

Investment Outlook

Over the next thirty years, electricity production in India needs to rise at an average rate of 4% a year. Although this rate is low by historical standards, it remains one of the highest in the world. The Indian power sector will need $665 billion in electricity generation, transmission and distribution over the period 2001-2030.

Power plant additions should amount to 272 GW. The total cost of these new plants will be in the order of $268 billion. The refurbishment of existing plants will require about $15 billion. Most new capacity is expected to be coal-fired, while natural gas could become the second-largest source of electricity generation, surpassing hydro. India has set ambitious targets to increase its hydropower capacity but, given the current difficulties the sector is facing and the high initial costs, it is unlikely that the target will be met.

Gas-fired generation is also faced with uncertainty, particularly regarding the availability of infrastructure to supply natural gas to power stations. It could, however, be favoured by private developers as market reforms proceed, encouraging more private participation in the power sector. Recent increases in domestic gas reserves could further boost investment in gas-fired projects.

Development of India's transmission and distribution systems will require investment of the order of $380 billion, 30% for transmission. Investment in transmission and distribution facilities has lagged behind investment in generation, and this explains to a degree the high level of power losses. Network investment now stands at about 35% of total investment, having increased from about 28% in the early 1990s. Over the next thirty years, network investment will need to exceed investment in generation by about 40%.

Investment Issues and Implications

India's security of electricity supply over the next thirty years will largely depend on the country's ability to mobilise the funds to build new power plants and to expand the transmission and distribution networks. While reforms in the sector commenced in 1991, the financial situation of the electricity sector has deteriorated and private-sector participation has been limited. The gap between supply and demand has declined, but there is no guarantee at the moment that this trend will continue, unless effective reforms are implemented.

India hopes to attract more private capital to finance new projects. Although the government has encouraged private and foreign investment in new independent power producers, most of the projects proposed have encountered financing problems and delays in obtaining regulatory approvals. Uncertainties about the profitability of projects, along with political problems, bureaucracy and corruption, have turned many potential investors away. Substantial efforts will be necessary to improve the confidence of potential developers and banks.

Price reforms will help improve the financial health of the electricity sector, but reducing the cost of electricity supplied is also important. The cost of every kWh of electricity sold by the SEBs in 2000 was equal to the average retail electricity price in the United States (Figure 7.32). This high cost can be attributed to high transmission and distribution losses, theft, the high cost of coal,[33] the low efficiency of power plants and high administrative costs.

Figure 7.32: **Electricity Cost and Price Comparison for India**

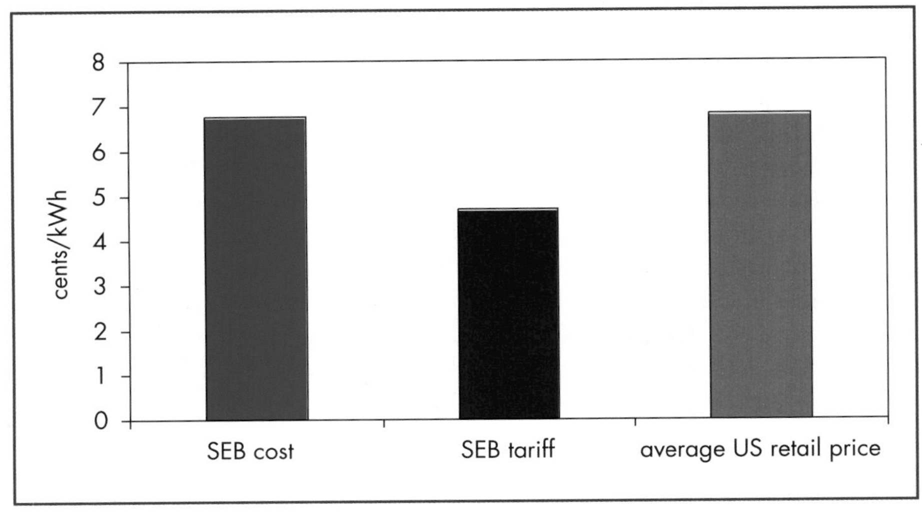

Source: TERI (2002) and IEA.

The power sector, which will continue to depend on coal, will remain a major emitter of CO_2. It now accounts for 54% of total energy-related CO_2 emissions and is likely to remain at this level in 2030. India's power plants emit substantial quantities of SO_2, NO_x and particulate matter, since few plants are equipped with pollution-control equipment.[34] Investment in more efficient plant and adequate maintenance of generation, transmission and distribution systems can prevent these problems from getting worse.

33. Coal in India is mined far from the power plants and the cost of moving it to the generation point is high.
34. Indian coal has low sulphur content but low generation efficiency drives emissions up. Indian thermal plants emit 7.4 g SO_2 per kWh. This is higher than in most OECD countries: for example, it is 5.3 g/kWh in the United States.

Brazil

Overview

Brazil's electricity demand will increase by two-and-a-half times from 2000 to 2030, growing at an average annual rate of 3.2%. To meet this big increase, the country will need to invest more than $330 billion in the power sector, more than half in transmission and distribution networks (Figure 7.33).

Figure 7.33: **Electricity Sector Investment in Brazil, 2001-2030**

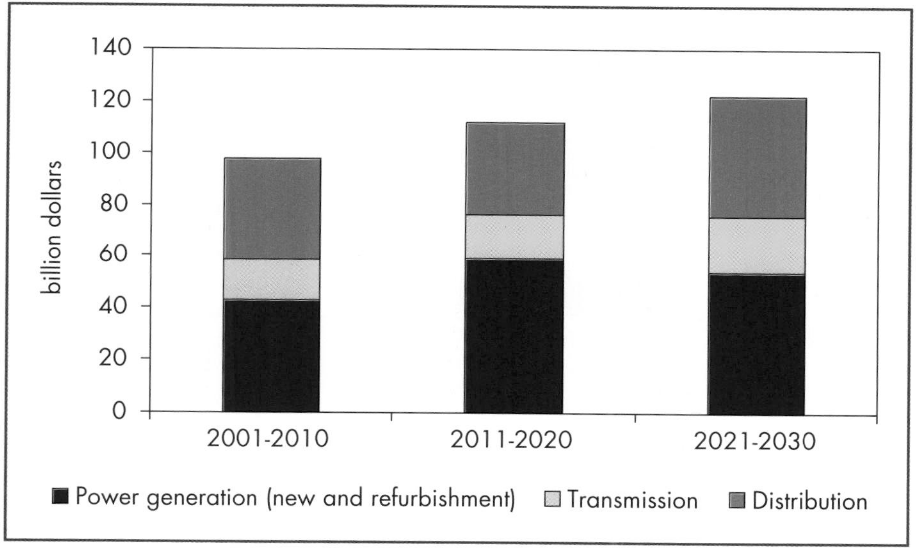

Source: Eletrobras.

Installed capacity was 71 GW at the end of 2000, with 85% of it hydropower. Brazilian power plants produced 349 TWh of electricity in 2000, while imports amounted to 44 TWh.

Transmission and distribution losses are among the highest in the world — around 15-16% of total domestic supply — because of the long distances that characterise Brazil's power networks (with hydro resources located far from demand centres), old and poorly maintained systems with high losses and power theft.

While the distribution sector is 80% privatised, generation is still mainly publicly-owned. Eletrobras alone controls over 45% of the total Brazilian capacity through its subsidiaries. The previous Brazilian government planned to reduce the influence of Eletrobras by privatising its subsidiaries. The new Labour Party government gave up the plans of privatising the state-owned generators and intends to increase the role of Eletrobras as the major promoter of new large investments. The revenues of Eletrobras were over $6 billion in 2002 (about the same as in 2001) but net income fell dramatically (Figure 7.34) mainly because of the devaluation of Brazilian currency, which increased the weight of debts in dollars.

Figure 7.34: **Eletrobras Operating Revenue and Net Income, 2001-2002**

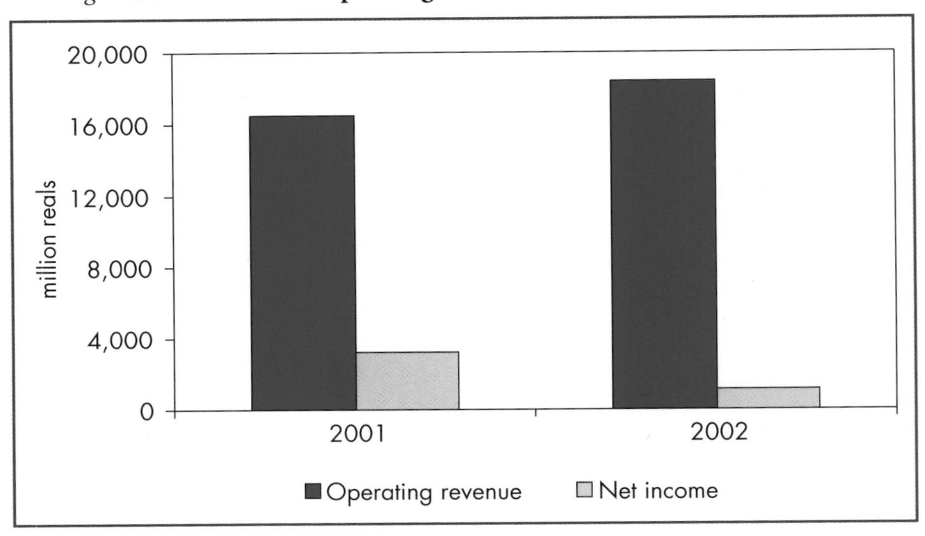

Source: Eletrobras.

Brazil started deregulating its electricity market in 1995. A regulatory agency (ANEEL) was established in 1996 and a national transmission system operator was created in 1998. A wholesale electricity market was created in September 2000 and was put under ANEEL's authority in 2002. ANEEL has had a primary role in promoting the construction of new transmission lines, having awarded contracts for a total of more than 6,700 miles of lines since September 2000.

Tractebel Energia (part of Suez/Tractebel) became the biggest privately-owned company in Brazil, with an installed capacity of more than 5 GW, following the acquisition of a former federal generator that was privatised (Electrosul). Several other foreign utilities (mainly US and European) are also present in the Brazilian electricity market.

In 2001, Brazil had a severe electricity crisis, caused by low rainfall for hydro generation and the lack of investment in generation and transmission capacity.[35]

35. See also World Bank (2002a).

Between 1990 and 2000, electricity production increased at an average annual rate of 4.6%, while installed capacity grew only by 3.1% per annum. As a consequence, reserve margins were low and the whole system became too dependent upon the annual rains. The crisis was solved through a rationing programme during a period of 10 months, which had a profound effect on the electricity sector and the Brazilian economy in general. The crisis highlighted the need for Brazil to diversify the fuel mix to reduce dependence on hydro.

Investment Outlook

Investment in power generation over the next thirty years should reach $156 billion, most of which will go into the construction of 120 GW of new plants. Development of Brazil's transmission and distribution systems will require investment in the order of $175 billion. Insufficient investment in the transmission and distribution network was one of the causes of the electricity crisis in 2001 and this will be one of the major challenges during the next 30 years.

Investment Issues and Implications

New capacity will be almost equally divided between capital-intensive hydro and low capital cost gas-fired plant. Construction of hydro plants will gradually slow down in order to reduce reliance on hydropower and because the remaining hydro resources are located far from the most populated areas, requiring huge investments in transmission lines. Environmental considerations may also have an impact on hydro expansion, since much of the remaining potential is in the Amazon.

Private investors are unlikely to construct new hydropower plants because of their high initial cost and long construction time. New hydro development is therefore likely to remain the government's responsibility. In the long term, investment in hydro will probably be more focused on the upgrading of existing plants, construction of medium-size plants or reactivation of small hydropower plants.

In February 2000, the Brazilian government launched the Thermoelectric Priority Program, which consists of a series of measures to increase and stimulate investment in thermal power plants — mainly CCGT plants — based on preferential fuel prices and financing terms. Following the 2001 electricity crisis, the deadline for the programme has been extended to December 2004. Several of these projects are co-financed by the state-owned company, Petrobras, and foreign investors. The national development bank, BNDES, provides finance on favourable terms.

Whether enough gas-fired power plants will be built is very uncertain and will depend on the cost of natural gas, the development of the gas infrastructure system and the tariffs and contracts for the supply of natural gas. Natural gas investors seek long-term contracts to protect their investments. But in an electricity market dominated by hydropower, electricity prices will be highly dependent on the rainfall levels. The economic attractiveness of gas-fired power plants for foreign investors will depend critically on the type of contracts established.[36]

Brazil was the world's largest recipient of private investment in electricity in the period 1990-2001. Private investment in new projects grew constantly in the late 1990s, following the privatisation programme, but declined in absolute terms in 2001 (Figure 7.35).

The future growth of private investment in Brazil will depend on the capacity of its electricity and gas market to provide stable and reliable conditions. Much of investors' concern has to do with uncertainty about the import price of gas, mainly because of the exchange rate and the possible devaluation of the national currency.

Figure 7.35: **Private Investment in Brazil Greenfield Projects**

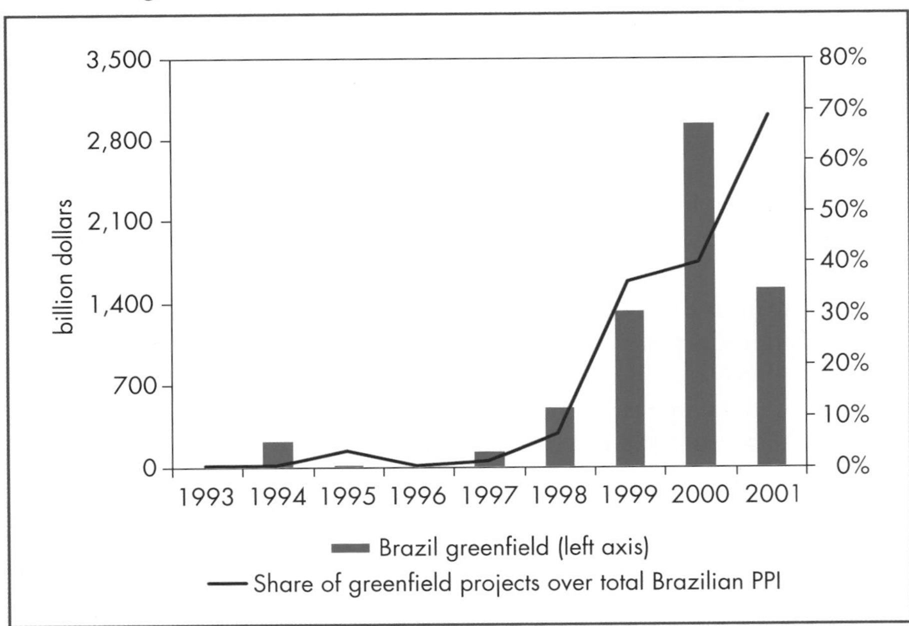

PPI: private participation in investment.
Source: World Bank (2003e).

Private investment in the distribution sector will be highly dependent on the regulatory framework. Difficulties in evaluating the risks linked to the

36. See IEA (2003b) for a discussion of natural gas markets in South America.

progress of reforms and the lack of clear rules have been a cause of uncertainty and could adversely influence investments in the future.

Indonesia

Overview

Electricity demand in Indonesia is expected to rise fivefold over the next thirty years requiring investment just over $180 billion in electricity infrastructure. This corresponds to 1.4% of the cumulative GDP over the same period. The country urgently needs new capacity and has again taken up reform of the electricity sector in order to improve its performance and to attract private investment.

Indonesia's installed capacity was close to 40 GW in 2000. PLN, the state-owned electricity company, owns and operates some 21 GW of the total capacity, as well as the transmission and distribution grid. Private power producers owned 1.6 GW in five power plants at the beginning of 2000. An interesting feature of the Indonesian electricity market is the high share of autonomous power producers, who owned 15 GW of the country's total capacity in 2000. Growth in this sector was driven by high electricity tariffs and unreliable power supplies, which encouraged industry to install its own generation plants.

Indonesia's electricity demand was one of the fastest growing in the world until the Asian economic crisis of 1997. Demand doubled between 1990 and 1997, growing at an average rate of 11% per year. The impact of the crisis on electricity demand was short-term, as demand began to rise again within two years. However, there were serious implications for investment in new capacity. As a result of the crisis, PLN's financial performance deteriorated, projects were cancelled and private investors abandoned their plans. Shortly after the beginning of the crisis, PLN cancelled several projects, with a combined capacity of about 15 GW. Moreover, some of the foreign investors involved in independent power producer (IPP) projects abandoned partially constructed plants. Between 1997 and 2001, PLN's installed capacity increased modestly from 19 GW to 21 GW. Reserve margins were high before the crisis and this helped meet the additional post-crisis demand. But there have been electricity shortages recently in some areas and this situation could worsen, indicating the urgent need for investment.

PLN's losses amounted to $1.9 billion in the period 1998-2001. Its debt reached $5.6 billion in 2001, of which more that 40% was short-term debt. PLN's expenses increased substantially over this period, to a large extent because of the sharp fall in the rupiah exchange rate, involving 300% devaluation between 1997 and 1998. Although tariffs in rupiah have increased since then, the increase has not been high enough to compensate for the change in the exchange rate (Figure 7.36). The case of PLN is a clear example of foreign exchange risk for both the domestic power sector and foreign investors in developing countries. Foreign currencies, especially the US dollar, are necessary for a large part of PLN's expenditure covering investment in new plants (with imported equipment, payment for imported spare parts and maintenance, and repayment of loans), fuel purchases and electricity purchases from IPPs.[37] Because of the economic crisis, PLN was not able to pay for all of the power for which it had signed contracts with IPPs. PLN's financial performance started to improve in 2001, because electricity tariffs and revenues went up. PLN has begun to increase electricity prices on a quarterly basis. The aim is to bring prices to about 7 US cents per kWh by 2005.

Figure 7.36: **Indonesia Average Tariff, Rupias versus US Cents**

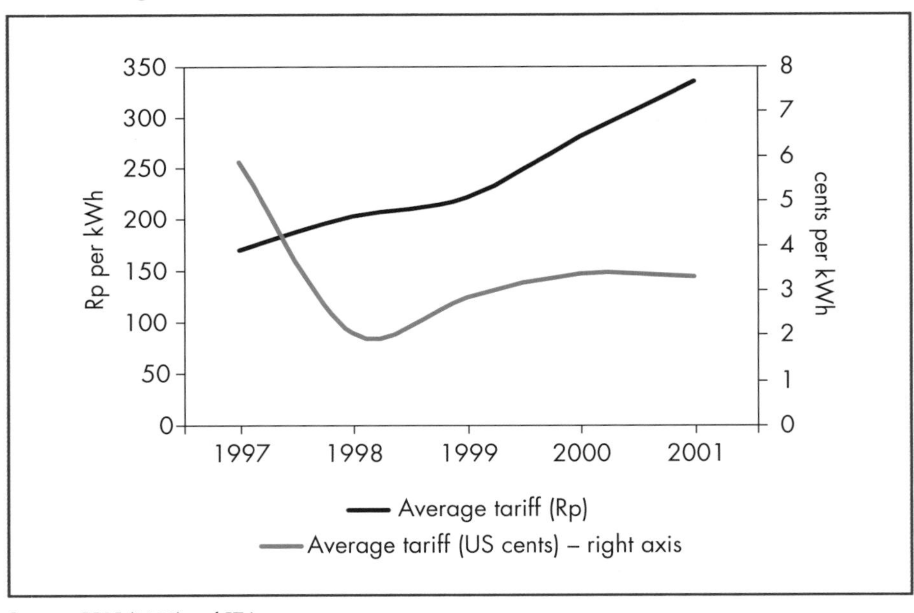

Sources: PLN (2002) and IEA.

Reforms in the Indonesian electricity sector were introduced as early as 1985, but the first private project was approved in 1990. This was the Paiton I coal-fired power plant, built under a build-own-operate scheme.[38]

37. PLN (2002).
38. APEC (1995).

The Paiton I project, a 1,230 MW coal-fired plant, reportedly cost between $2 billion and $2.6 billion to build. This corresponds to a unit cost of between $1,600 and $2,000/kW, which is at least 60% higher than the capital cost of a coal-fired power station in OECD countries. PLN agreed to pay a high price for the electricity it purchased from the Paiton plant. The electricity tariff was set at 8.6 cents per kilowatt-hour for the first six years of the thirty-year power purchase agreement. This is about three times the average wholesale price of electricity and twice the generating cost of a new coal-fired power station in OECD countries. Although this is an extremely high price, it reflects, to some extent, the higher risk of investing in a power project in a developing country and the investor's consequent requirements for high rates of return. Following PLN's inability to honour its contract with the power plant owners (because of its financial situation), an agreement was reached in 2002 to reduce the tariff.

In September 2002, the Indonesian Parliament passed a new law to allow for deeper reforms in the electricity sector. Under the current system, only PLN is authorised to sell power to the public. Private companies are allowed to produce power, but they are only able to use the power themselves or sell it to PLN. The new electricity law will introduce competition and will gradually reduce PLN's monopoly. Under the new law, the liberalisation of Indonesia's electricity sector will commence in 2007. Competition is expected to be introduced first in areas where prices are higher.

Investment Outlook

Indonesia's electricity demand is expected to rise at an average annual rate of 5.4% in the period to 2030. The country will need 90 GW of new capacity over this period. Coal and gas will account for most of the incremental capacity. Total investment in generation, transmission and distribution is estimated to be in the order of $180 billion. Investment in generation is expected to amount to about $77 billion, of which $6 billion in refurbishment. To meet the fivefold increase in electricity demand over the next three decades, over a hundred billion dollars must be invested in transmission and distribution networks.

Investment Issues and Implications

Indonesia's major challenge over the next three decades will be successfully to reform its power sector so as to attract the investment needed to meet the country's growing electricity needs. The severe financial problems that the Indonesian power sector faces have resulted in the cancellation of many plans.

There are concerns as to whether there will be adequate, timely investment to meet the country's rising electricity demand, particularly in the near term. Given the long construction times of power plants (four to five years for a coal plant, two to three years for a CCGT plant), it appears to be impossible to raise the country's capacity sufficiently over the next few years to avoid power shortages.

In the longer term, the challenges are equally daunting. Private investors had shown strong interest in Indonesia's power sector before the crisis. At present, Indonesia's high-risk rating and poor investment climate make international investors cautious. Substantial efforts are needed to rebuild investor confidence. This will require a stable and transparent regulatory structure, improved corporate governance and a growing sense of national political harmony.

At the same time, it will be necessary to improve the financial health of the public sector. This will require improvements in production costs, as well as electricity prices that are cost-reflective. Moreover, investment in generating plant must be accompanied by the appropriate investment in transmission and distribution networks to avoid bottlenecks and to improve power quality.

Russia

Overview

The Russian electricity sector will require investment in the order of $380 billion over the next thirty years (Figure 7.37). This corresponds to 1.9% of the country's cumulative GDP during the period.

Figure 7.37: **Russian Electricity Sector Investment, 2001-2030**

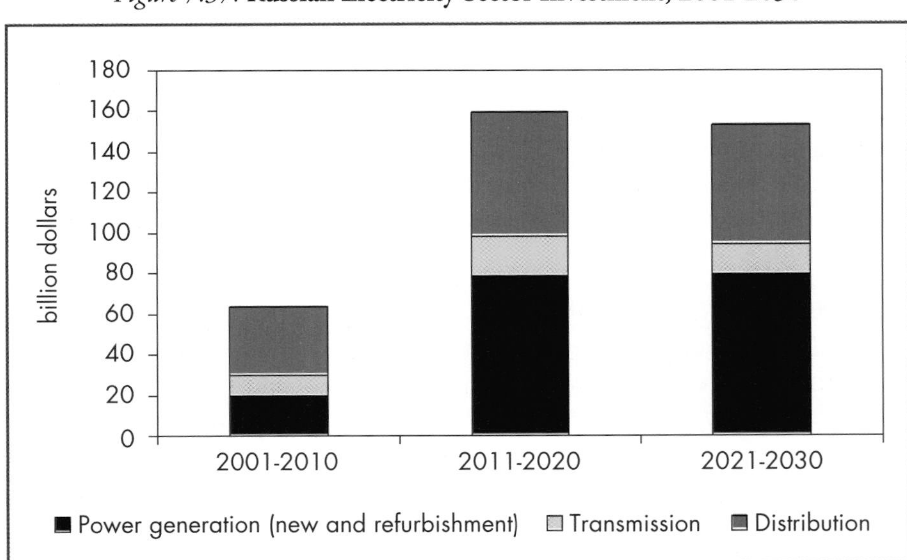

Russia's installed capacity was 215 GW in 2001, the fourth-largest in the world. Electricity production, which began to rise again in 1999 after eight consecutive years of decline, reached 889 TWh in 2001. This is still 18% less than the 1990 level. The decline can be almost exclusively attributed to much lower demand by industry, reflecting the economic decline after the break-up of the Soviet Union.

Russia has about 23 GW of nuclear capacity. Since 1992, all nuclear power stations have operated under the state company, RosEnergoAtom (with the exception of the Leningrad NPP, which is independent), which is controlled by the Ministry of Atomic Energy.

Russia's electricity sector is dominated by RAO UES, which was created in 1992 as a joint-stock corporation, with the federal government being the company's major shareholder. RAO UES owns nearly three-quarters of Russia's generating capacity and is the largest power company in the world. There are only two other energy companies, in which UES has no shareholdings: Irkutsenergo (13 GW) and Tatenergo (7 GW).

Russian electricity prices are low compared to OECD countries.[39] Electricity tariffs are set by the government. Those for residential consumers and for government organisations continue to be subsidised, with average tariffs of less than 49 kop/kWh (1.6 US cents/kWh). Industry tariffs are significantly higher, averaging 65 kop/kWh (2.2 cents/kWh) in 2002. However, the level of cross-subsidisation is decreasing, as residential tariffs continue to increase faster than industrial ones.

Investment Outlook

Over the next thirty years, electricity demand in Russia will increase by 2.3% per year on average. The capacity additions to match this demand will be 204 GW. Nearly half of that will be needed in the last decade of the projection period, as existing capacity can meet most of the additional demand to 2010, provided that the necessary upgrading is carried out.

Total investment in the Russian electricity sector over the next thirty years will amount to $377 billion. Near-term investment needs are relatively low, amounting to less than $6.5 billion per year to 2010. They will be about $16 billion per year in the longer term. Higher prices to consumers and the enforcement of payment will be critical in attracting investment. Significant progress has been made in recent years in improving payment.

The investment needed in new generating plant is estimated to be $157 billion. About $21 billion will be needed to refurbish existing power plants. Over 80% of new generating capacity in the next thirty years will be gas-fired. As Russia's electricity sector becomes more market-oriented, investors

39. Although, in terms of purchasing power parity, they are high for Russian consumers.

will be looking for the most economic way to generate electricity. Natural gas-fired CCGT plants are expected to be the lowest-cost option, so long as natural gas prices are low. Investment in nuclear plant will account for nearly a quarter of total investment in new plant. There are significant uncertainties about whether enough funds will be available to finance these new plants.

Investment in the transmission and distribution networks will be higher than investment in generation, approaching $200 billion. Much of this investment will go towards improving the existing infrastructure.

Investment Issues and Implications

In terms of the adequacy of capacity, Russia is in a fairly comfortable position to meet rising electricity demand for the next decade, although there is a need for investment to refurbish power plants, transmission and distribution networks. The need to add new capacity will be higher in the longer term. Much of the investment needed could come from the private sector, but its extent and timeliness will largely depend on the success of current market reform efforts.

The Ministry of Economic Development and Trade (MEDT), which is leading the electricity reform drive (Electricity Reform Laws were passed in April 2003), is keen to create well-functioning markets in electricity and to provide an economically sensible governance and regulatory structure for the operation of the grid and for dispatch, which are to remain state-controlled in view of their natural monopoly character. However, there are substantial challenges ahead — as in any country liberalising its electricity market. Issues critical to effective implementation of reforms are:

- Creating as competitive a market structure as possible by ensuring that transmission between regions is not limited.
- Cost-reflective tariffs, to ensure an efficient market and the attractiveness of the sector to investors.
- Strong and credible legislative and regulatory frameworks, well received by market players, to underpin electricity markets and inspire confidence.
- Credible, transparent, predictable and effective administration of market rules and regulations by effective regulatory bodies.

Given the important share of natural gas as an input to electricity generation, it will be critical to ensure access to gas supply at a "fair and reasonable" price. This is essential for the development of an efficient and competitive wholesale electricity market. Distorted input fuel pricing, especially during the period in which electricity reform are implemented, could cause distorted investment decisions, which would continue to be a problem over the life of the generation asset.[40]

40. IEA (2002d).

Nearly half of Russia's CO_2 emissions come from the power sector. CO_2 emissions will continue to rise in the future and, in 2030, they are likely to be 27% above the 2000 level. CO_2 emissions will rise at a much lower pace than electricity generation, because new plants will substantially improve power generation efficiency. Repowering of existing gas-fired facilities — not provided for in the investment requirements — could be another way to save gas and reduce emissions.

OECD Alternative Policy Scenario

Investment decisions in liberalised markets are no longer taken only by governments. However, government policies still strongly affect these decisions. The Reference Scenario of the *WEO-2002* included only policies that were in place before mid-2002. An Alternative Policy Scenario was also developed to analyse the impact on energy markets, fuel consumption and energy-related CO_2 emissions of the policies and measures that OECD countries had under consideration but had not adopted or implemented. These included more aggressive policies principally aimed at curbing CO_2 emissions as well as reducing energy-import dependence. The OECD Alternative Policy Scenario applies to the European Union, the United States and Canada, and Japan, Australia and New Zealand.

The Alternative Policy Scenario focuses on policies with the potential to have a major impact on energy use. It makes detailed assumptions on the impact on each major consuming sector (residential, services, industry, transport, power generation). Indicative stronger policies in the power sector include promoting the use of low-carbon or no-carbon fuels and increasing efficiency (Table 7.7). The power generation sector is affected in two ways: first, electricity demand is lower because of demand-side policies to improve efficiency (such as efficiency standards for appliances); and second, the electricity fuel mix is different.

The level of nuclear electricity generation in the Alternative Policy Scenario remains the same as in the Reference Scenario. At present, there are no new policies supporting nuclear power in the majority of OECD countries. However, a debate exists in some of them about the future role of nuclear power in combating climate change and enhancing security of supply. Future policies could affect nuclear power generation in two ways: First, through support for existing nuclear plants by not phasing them out, notably, in Europe. And second, through the construction of new plant in Europe and possibly in North America.

The Alternative Policy Scenario illustrates that if existing policies were strengthened and new policies adopted to curb emissions and reduce electricity

Table 7.7: **Policies Considered in the Power Generation Sector under the Alternative Policy Scenario**

Policy type	Programme/measure	Impacts on power-generation sector
Increased renewables	Renewable Energy Directive (EU) Renewable Portfolio Standard (United States and Canada)	Increased share of renewables
	Renewable energy targets (Japan, Australia and New Zealand)	
Increased CHP	Policies to promote CHP in end-use sectors	Increased share of electricity generation from CHP plants
Improved efficiency	Various policies and R&D to accelerate the penetration of even higher-efficiency coal and gas plants and new technologies such as fuel cells	Higher efficiency for new gas, coal and fuel cells plants

consumption, the reduction in CO_2 emissions would be considerable. Under the Alternative Policy Scenario, emissions continue to increase, initially at a lower rate, but by 2030 they fall back to their 2000 level.

Total power-sector investment in the Alternative Policy Scenario is $2.7 trillion, about 20% less than the investment in the Reference Scenario (Figure 7.38). The difference between the two scenarios becomes more pronounced in the long term. In the period 2021-2030, investment in the Alternative Policy Scenario is nearly a third less than in the Reference Scenario. However, the reliance on more expensive generating options in the Alternative Policy Scenario is likely to result in higher electricity prices.

Investment in electricity generation in the Alternative Policy Scenario is about the same as in the Reference Scenario, while there are major savings in transmission (40% less) and distribution (36% less). The reduction in generation investment is small because the technologies in the Alternative Policy Scenario are more capital-intensive, particularly renewables and distributed generation, which have a much higher share of total generation in the Alternative Policy Scenario.

For the private sector to invest in more expensive generation, there must be confidence that electricity prices will be high enough to ensure an adequate return on investment. Given the fact that other generating options

Figure 7.38: **OECD Investment in the Reference and Alternative Policy Scenarios, 2001-2030**

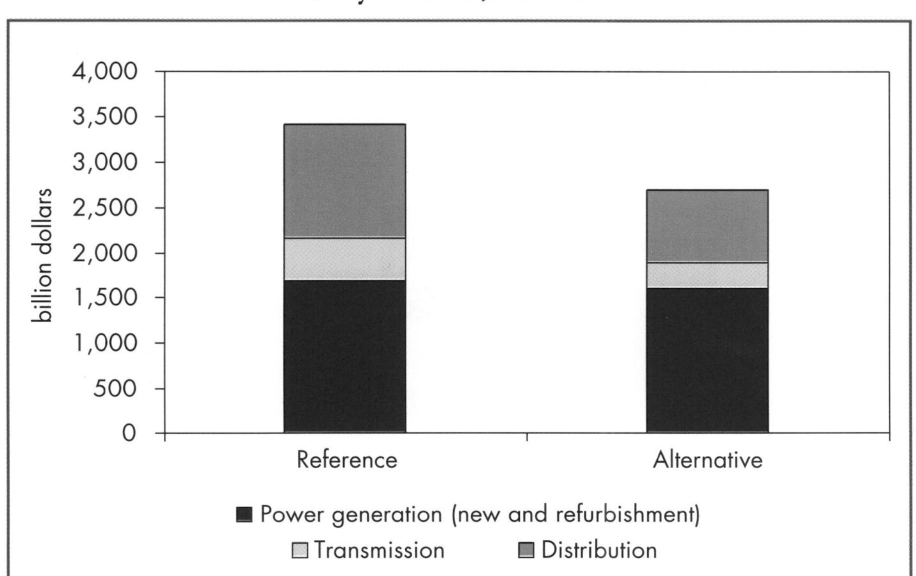

are less expensive, investors in renewable energy projects frequently seek a guaranteed market for their electricity. To encourage renewables, governments may need to help create a market framework that rewards those who invest in renewables. A price for emitting carbon will raise the price of fossil-based electricity and thereby create a more favourable environment for renewables.

In the Alternative Policy Scenario, investment in transmission and distribution is lower than in the Reference Scenario because the amount of electricity demanded by final consumers is lower. There are additional savings in transmission investment because of greater use of distributed generation. Investment in energy efficiency to achieve reductions on the demand side will be greater in the Alternative Policy Scenario.[41]

Implications for Investment in Renewables

The implementation of new policies to promote renewables will have considerable implications on investment in this source of electricity. Policies under consideration are projected to achieve a 25% share of

41. Demand-side investments are not included in these projections.

Figure 7.39: **OECD Share of Renewables in Electricity Generation in the Reference and Alternative Policy Scenarios**

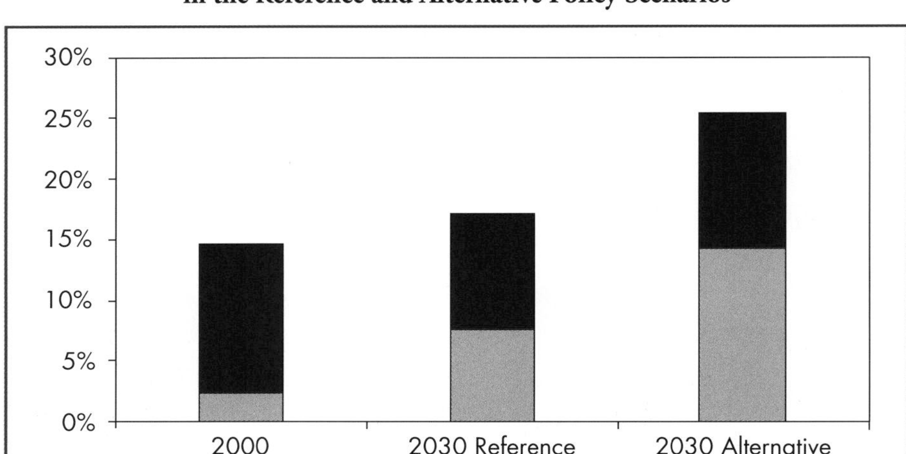

renewables by 2030 across the OECD, compared to 17% in the Reference Scenario (Figure 7.39).[42]

In the Reference Scenario, investment in OECD renewables electricity plants amounts to $477 billion. This is nearly a third of investment in new power generation. In the Alternative Policy Scenario, because of the higher share of renewables, this total reaches $724 billion and corresponds to half the investment needed in new plants (Table 7.8).

Table 7.8: **OECD Shares of Renewables in Generation, Capacity Additions and Investment**

	Reference Scenario (%)	**Alternative Policy Scenario** (%)
Electricity generation		
2000	15	15
2030	17	25
Capacity additions, 2001-2030	19	32
Investment in new power plant, 2001-2030	33	52

42. Renewable energy sources, including hydropower, accounted for 15% of OECD electricity generation in 2000.

Figure 7.40: **Changes in Capital Costs of Renewables, 2001-2030**

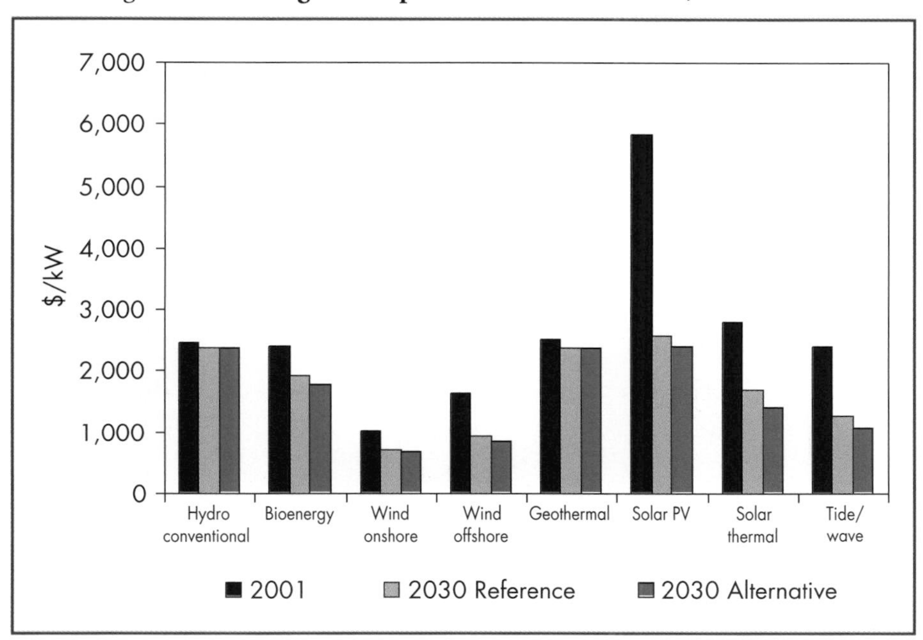

Source: IEA.

The capital costs of renewables are assumed to continue to decline in the future. The ratio of decline will depend on the rate at which they are deployed and on the maturity of the technology (Figure 7.40).[43] The highest rate of decline will be in the capital cost of photovoltaics, which drops by 63% between 2001 and 2030 in the Alternative Policy Scenario. This is the most capital-intensive of the renewable energy technologies considered here. Substantial decreases are also expected in the capital cost of offshore wind, solar thermal and tidal and wave technologies.

To achieve investment in renewables at the level expected in the Alternative Policy Scenario, OECD governments will have to develop vigorous incentive strategies. A number of countries have achieved substantial increases in recent years by using feed-in tariff mechanisms. Another approach is to impose portfolio quotas, with or without accompanying tradable certificates. This strategy can increase investment in renewables in a market-oriented way. A third approach is through tax incentives, such as the US production tax credit. An additional instrument to foster the demand for renewables and attract investment is green pricing, although this voluntary type of measure has not proven to have a significant impact.

43. The capital cost reductions in the two scenarios have been estimated using learning curves. Learning relates cost reductions to cumulative production. The learning rate is time- and technology-dependent.

Universal Electricity Access

In the Reference Scenario, electrification in developing countries reaches 78% of the population in 2030, leaving 1.4 billion people without access to electricity. The Electrification Scenario projects an added investment of $665 billion in order to reach 100% access by 2030. This would increase global electricity supply by 3% (7% in developing countries).

Box 7.7: **Methodology for Developing the Electrification Scenario**[44]

The methodology followed to develop the 100% Electrification Scenario has three main determinants:

- The urban/rural breakdown by region under the Reference Scenario and the distribution of the extra connections under the Electrification Scenario.
- Basic consumption needs and the way consumption might evolve over the period during which access is extended to the balance of the population.
- The associated costs for generation, transmission and distribution.

The model underlying the Electrification Scenario projects the number of people connected each year, the related supply requirements and the corresponding investments. These projections are broken down between urban and rural areas and by regions.

Supply Need and Options

Under the Electrification Scenario, supply will need to be increased by nearly 1,000 TWh, 6% of the world total, to satisfy the added demand.[45] In the Reference Scenario, the average consumption per capita of those with electricity in 2030 in developing countries is 2,136 kWh, while for the 1.4 billion people gaining access, the average consumption would be 526 kWh/capita.

The Electrification Scenario assumes that the electrification process will bring electricity to the poorest and the more remote progressively over time. Each person gaining access is at first going to use electricity only as a substitute for the traditional fuels (candles, LPG, kerosene) used to cover basic needs. Basic electricity consumption in rural areas is estimated to be 50 kWh per person per year. The minimum urban consumption is set at 100 kWh per person per year. This higher consumption in urban areas reflects specific urban consumption patterns. This basic consumption is a starting point.

44. The methodology is described in more detail in Annex 2.
45. This generation figure takes into account transmission and distribution losses.

Consumption changes over time, reflecting the income-generating effects of electricity on the populations concerned. The exponential process means that by 2015, only 12% of the additional 1.4 billion will have access to electricity. Nearly 90% of the investment will accordingly be required in the second half of the projection period. If more people were connected earlier, consumption and investment needs would be higher.

Although the overall increase needed in supply is relatively low, it hides great regional disparities. Most of the added supply will be required in just two regions, Africa (437 TWh) and South Asia (377 TWh). For these two regions, the increase in supply compared to the Reference Scenario is substantial — over 25% in Africa and 18% in South Asia. India alone would need 15% more electricity to supply all its people by 2030.

The picture is also quite different as between rural and urban areas. Only sub-Saharan Africa and South Asia will need to add supply in urban areas, as they are the only two regions, under the Reference Scenario, which do not reach 100% electrification in urban areas by 2030. The additional urban electricity supply amounts to 320 TWh, of which two-thirds will be needed in sub-Saharan Africa (Figure 7.41). Urban supply will still represent more than a third of the total.

Figure 7.41: **Additional Electricity Generation in the Electrification Scenario Compared to the Reference Scenario, 2030**

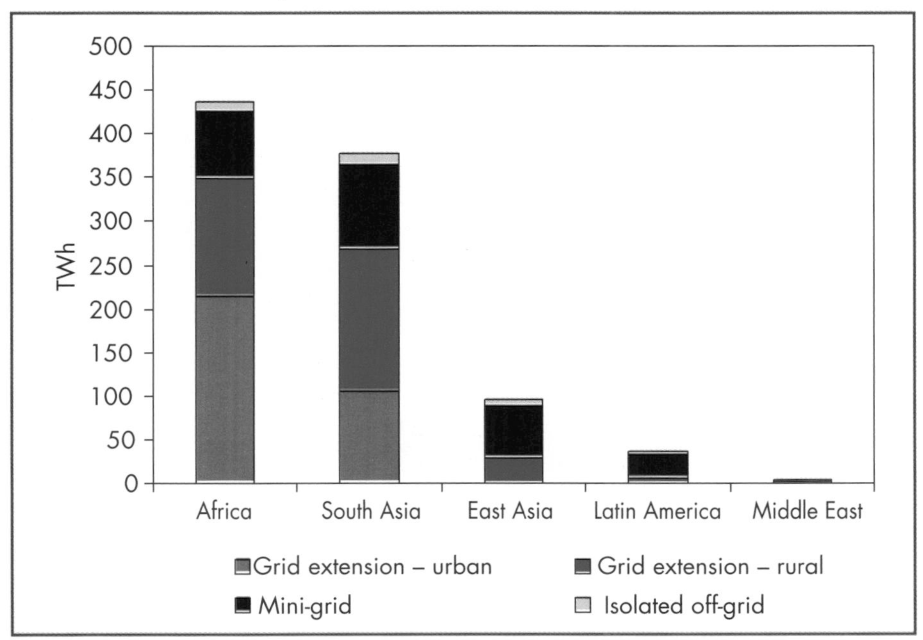

Options to increase electrification include extension of existing grids, creation of mini-grids and isolated off-grid generation. Two-thirds of the additional supply would be based on grid extension, as this is the most economic way to increase electrification. Mini-grid generation would make up 27%, leaving isolated off-grid systems to account for 4%. South Asia, with its lower urbanisation trends, has a greater share of rural and thus off-grid supply. Regions with already high rates of rural electrification, such as the Middle East and Latin America, have a bigger share of mini-grids and isolated off-grid generation as they are reaching the most far-flung populations. Sub-Saharan Africa still needs to bring electricity to its cities as well as to its rural population.

Assuming no change in the fuel mix, the impact of bringing electricity to 100% of the world population would increase world oil consumption by only 0.2% in 2030. This could be higher if the 31% of supply achieved through decentralised generation assumed in the Electrification Scenario were higher, or if the share of mini-grids fuelled by oil were higher.

Investment Issues and Implications

The investment needed by 2030 to supply electricity to 1.4 billion people is estimated to be around $665 billion, adding 6.7% to the global investment required over the projection period in the Reference Scenario. Nearly 80% of this investment will be needed in Africa and South Asia (Figure 7.42). Their investment needs are of similar magnitude. Although

Figure 7.42: **Additional Cumulative Investment Requirements in the Electrification Scenario Compared to the Reference Scenario, 2001-2030**

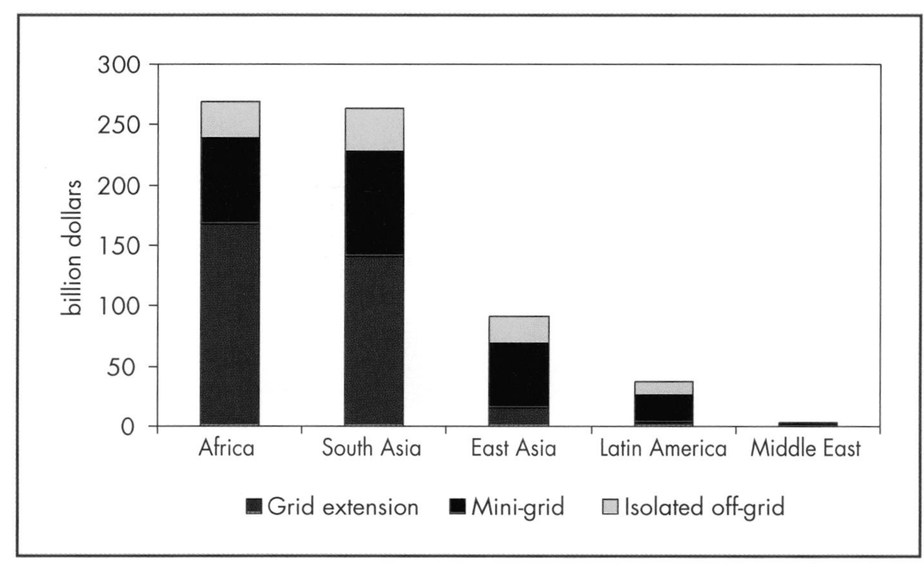

Africa will require a greater volume of additional supply, South Asia's high rural population increases the cost there per added kWh.

In Africa, electrification will be achieved mostly through grid extension, while in East Asia most of the additional supply will be provided through autonomous off-grid systems. Decentralised solutions in rural areas are more capital-intensive per GW installed.

The prospects of financing this additional investment in the Electrification Scenario need to be seen in the wider context of the challenge which exists even to achieve the results of the Reference Scenario. In 2000, over 77% of the sub-Saharan population and nearly 60% of South Asians had no access to electricity. In the poorest regions of the world, the unelectrified population is growing every year and is not expected to start to decline for a long time (the sub-Saharan unelectrified population as a whole is projected to start declining in the mid-2020s according to the Reference Scenario). About $5 trillion will be required in developing countries as a whole to meet the increase in electricity demand projected in the Reference Scenario. The $665 billion needed under the Electrification Scenario is additional to this. The task is, clearly, formidable. But increasing world electricity investment by the relatively modest figure of 6.7% could contribute substantially to the alleviation of poverty by ending electricity deprivation.

What priority the world community should give to this challenge is beyond the scope of this analysis. Access to electricity *per se* will not alleviate poverty, but

Figure 7.43: **Additional Power Generation and Investment in the Electrification Scenario Compared to the Reference Scenario**

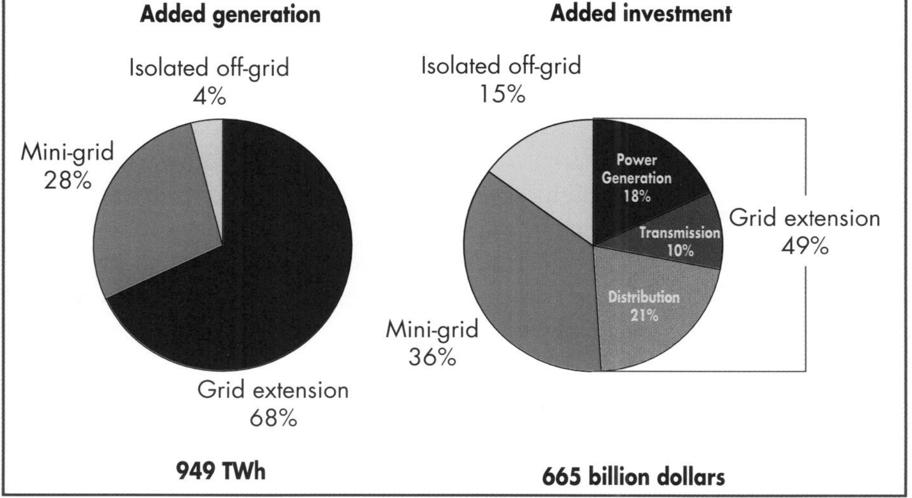

Note: Added generation is in 2030; added investment is cumulative over the period 2001-2030.

as part of a bigger scheme (*e.g.* the Millennium Development Goals) it can make a substantial contribution. Electricity helps alleviate poverty directly and by contributing to income-generating activities. The least developed countries suffer from a general lack of infrastructure, and electricity access would need to be complemented by investment in roads, telecommunications, rail and other basic infrastructure. The investment requirement for electricity must accordingly be seen in this wider context.

The required investment is most unlikely to be taken up by the private sector. In sub-Saharan Africa, for example, where 77% of the population still needs access to electricity at an estimated cost of $270 billion, there is no market and there are no guarantees. Population density is low and distances immense. Urban circumstances are more favourable to prospective private finance, but there are still formidable obstacles.

Local communities can practice self-help.[46] Although Africa and South Asia do not have the savings rates of China, there are ways of garnering local investments. Micro-credit lending, first initiated in the 1970s by the Grameen Bank and through the African tontines, has spread and has been quite successful in reaching the poorest communities. At the end of 2001, there were 3,000 micro-financing institutions in the world, which have granted more than 55 million credits; 27 million of these went to populations living on less than $2 per day. The micro-credit world summit held in 2002 projected that the number of loans will reach 100 million by 2005.

The general international perspective has not been favourable to developing countries in recent years. Since 1997, private investment has plunged and international aid for infrastructure has diminished substantially. In the 1990s, international organisations, such as the World Bank, reduced the proportion of lending devoted to infrastructure and imposed conditions linking aid with reforms favouring the private sector. There is now evidence (in India, for example) that the intended effect was not achieved. There was a very negative effect on investment, especially in the poorest countries. The international community has recently again decided to put infrastructure back to centre stage.[47] In this light, the prospects for electricity investment in developing countries may not be as gloomy as past trends suggest.

But there is a need for realism. Universal electricity access means providing electricity to those who are so poor that paying would be out of the question. For these people, the only solution is for the service to be provided

46. African energy networks and partnerships across countries, such as the African Energy Commission (AFREC), may also help link more adequately resources with demand throughout the continent and thus help provide a better framework for local and foreign investors.
47. In July 2003, the World Bank announced that it planned to pay increased attention to infrastructure.

by governments or the international community as an investment in future social and income benefits. Subsidising the basic needs of 1.4 billion people, assuming a price of 7 cents per kWh (higher than the average in the countries concerned) would require expenditure of $1.1 billion per year ($600 million in sub-Saharan Africa and $500 million in South Asia)[48].

Environmental Implications

Bringing electricity to the remaining 1.4 billion people, assuming no change in the fuel mix, would increase CO_2 emissions by 1.4% in 2030. Even if all added generation were fossil fuel-fired, the increase of CO_2 emissions would be only 1.6% worldwide (corresponding to the total emissions of the United Kingdom in 2000). The data on emissions stemming from biomass use in developing countries is not readily quantifiable. If unsustainable biomass emissions were to be taken into account, the switch to electricity may change the level of emissions given here.

For South Asia and Africa, where the added generation is greatest, CO_2 emissions would increase by 9% and 13% respectively in 2030. Per capita CO_2 emissions in Africa, at 1.4 Mt, would still be nearly ten times lower than the average in the OECD in 2000.

48. In most countries, the poorest spend many times more on candles and kerosene than they would for the kilowatt needed for a light bulb. The challenge is the initial hook-up and ensuring that the kWh is paid for.

CHAPTER 8: ADVANCED TECHNOLOGIES

HIGHLIGHTS

- Advanced energy technologies could change the long-term energy investment outlook. Of the currently known areas of technology development, those which appear most likely to modify the investment picture painted in the previous chapters are carbon sequestration, hydrogen and fuel cells, advanced nuclear and advanced electricity transmission and distribution technology.
- Carbon sequestration technologies are relatively mature but face unresolved environmental, safety, legal and public acceptance issues. If these issues are resolved, very large reductions of CO_2 emissions could be achieved through carbon sequestration in the electric power sector. For example, a reduction of 3 gigatonnes of CO_2 emissions in 2030 in the OECD countries would result from equipping 250 GW of coal-fired plants and 500 GW of gas-fired plants with carbon sequestration by that date.
- Carbon sequestration increases power plant investment costs by between 30% and 120%. In the above illustrative example, investment costs are estimated to increase by between $350 billion and $440 billion. This would increase OECD investment in power generation by 20% to 25%. Commensurate incentives would be required.
- Hydrogen can be produced from fossil fuels, but production without CO_2 emissions depends on successful application of renewable energies, carbon sequestration technologies or nuclear-generated electricity.
- In the electricity sector, the fuel cells that are expected to achieve commercial viability first will involve the reforming of natural gas. In the Reference Scenario, about 100 GW of fuel cells are expected to be constructed in OECD countries by 2030 for distributed power generation. Production from other sources will require larger cost reductions.
- Because of high distribution costs in an emerging market, the fuel costs of travel in hydrogen fuel cell vehicles would, initially, be substantially higher than those of the alternative conventional fuels, but they could ultimately become similar to those of gasoline vehicles. Fuel cell vehicles face significant technology development hurdles, however, and

large cost reductions would be needed in vehicle costs before fuel cell vehicles become attractive to the general public. Fuel storage is also a problem. The widespread use of hydrogen is likely to require strong government intervention in order to create the conditions necessary for the co-ordination of the availability of fuel supply, refuelling facilities and vehicles.

- Advances in nuclear technology could lead to designs with lower capital costs, shorter construction time and flexible operation, so improving the economics of nuclear electricity. There is widespread interest worldwide in a fourth generation of nuclear power reactors, though public opposition can be expected to persist.
- There is substantial scope for improvement in electricity transmission and distribution losses, which account for 6.5% of electricity generation in OECD countries and are significantly higher in developing countries. An improvement of one-third of a percentage point in the efficiency of distribution transformers would save more than 200 TWh in 2030. Technologies such as gas-insulated transmission lines have higher capacities, lower losses and lower electro-magnetic fields, but also entail higher investment costs.

Overview

New technologies can significantly alter world fuel markets and, with them, the requirements for energy investments. One significant driver of technology is the threat of global climate change and the resulting need to reduce energy-sector greenhouse gas emissions. Concerns over energy supply security, particularly dependence on imported fuels, may also accelerate the development of new technological solutions. Governments and industry are pursuing programmes to develop cheaper renewable energy technologies, more efficient fuel combustion technologies, coal-gasification technologies, carbon capture and storage, hydrogen and fuel cell technologies, advanced nuclear technologies and advanced transmission and distribution technologies.

Beyond the allowances already made in the Reference Scenario, these technologies are not competitive on a full cost basis but continuous research could overcome the current technological barriers and bring costs down. These technologies could be encouraged through incentives or other measures. Nonetheless, they are not expected to play an important role in energy supply before 2020. In general, these technologies are more capital-intensive than those expected to be used in the Reference Scenario and therefore would require increased levels of investment.

CO_2 Capture, Transport and Storage

CO_2 capture and storage involves the *separation* of CO_2 produced during fossil fuel use, its *transport* and its *storage* in the earth or the ocean. All three elements have been implemented on a commercial scale in certain applications. For example, CO_2 capture is widely used in the chemical industry. Likewise, pipeline transport of CO_2 is an established technology, and about 44 Mt of CO_2 per year is already injected for enhanced oil recovery. Part of this CO_2 is stored underground permanently. However, for capture to become widespread in the power generation sector, these technologies need to be even further developed and demonstrated.

Separation: This technology is used today in the production of hydrogen in chemical production processes and refineries, in which CO_2 is a by-product. For capture technology to be suitable for power generation, technological improvements to reduce efficiency losses will be necessary. Losses can be mitigated further if highly efficient power generation technology is used. Taken together, carbon separation technologies are relatively mature and, given sufficient incentive to develop them further for use in electric power or hydrogen production, they could be available in a relatively short period of time compared to other technologies that can provide deep cuts in CO_2 emissions.

Transport: After capture, CO_2 must be transported by high-pressure pipelines or tankers to land-based or offshore geological sites or the deep sea. Relevant internationally recognised standards of CO_2 transportation exist; problematic issues include commercialisation of CO_2 shipping technologies and obtaining public acceptance of increased CO_2 transport. No major technological obstacles exist to the development of CO_2 transport, but substantial capital investment would be required.

Storage and Utilisation: CO_2 disposal options include depleted oil and gas fields, deep saline aquifers, or the ocean, mineralisation and use of CO_2 for enhanced oil, gas or coal-bed methane recovery. An example of the first storage option is a planned US project which will produce electricity and hydrogen using coal-gasification technologies, capture carbon dioxide and store it in a depleted oil reservoir. CO_2 utilisation in deep coal seams with enhanced coal-bed methane recovery is a tested technology. CO_2 utilisation in oil reservoirs with enhanced oil recovery is also a demonstrated technology. The Canadian Weyburn project, for example, imports CO_2 from a US synthetic gas plant and uses it for enhanced oil recovery. Underground storage of CO_2 in deep saline aquifers has been demonstrated in one commercial scale project, the Norwegian Sleipner project, where CO_2 associated with the produced natural gas is injected into a saline aquifer below the gas field to avoid CO_2 taxes. Two more projects using deep saline aquifers are planned in the Barents Sea (2006) and in Natuna in Indonesia (2010). While further effort is needed to demonstrate safety and to improve understanding of geological formations, deep saline aquifers represent a potentially huge and widely dispersed medium for CO_2 storage.

Some CO_2 may potentially leak into the atmosphere. Determining the potential for such leakage from reservoirs will depend on careful analysis of underground geological structures, cap rock integrity and well-capping methods. Small leaks over a period of tens or hundreds of years would reduce the effectiveness of CO_2 storage as an emission mitigation strategy. Natural CO_2, oil and gas reservoirs have contained these compounds for millions of years, but this does not provide conclusive evidence that underground storage is permanent. Monitoring of leakage would be required and would add to cost. A greater understanding of specific reservoirs is required before reliable assessments can be made.

Sequestration in the deep ocean poses much greater environmental uncertainties, particularly with regard to changes in pH and the effects on marine life, as well as issues related to the legality of storing CO_2 in the ocean.

More efforts will be needed to reduce current uncertainties if carbon sequestration is to be accepted by the scientific and environmental community and the general public. For CO_2 capture and storage to be used to combat climate change in the foreseeable future, governments would need to actively and in a diverse manner promote its development and to facilitate industry participation. Financial and regulatory incentives will be required, within a sufficiently stable long-term policy and legal framework, to win confident industry participation.

Given the early stage of development, there are significant uncertainties surrounding the investment cost for capture and storage projects. Expected improvements of cost and efficiency for future CO_2 capture depend on the development of power generation technologies, like integrated gasification combined cycle (IGCC), oxyfuel combustion chemical looping and solid oxide fuel cell, as well as the development of associated CO_2 capture options. The additional investment costs arise from:

- Additional power plant costs (including CO_2 pressurisation for storage).
- Additional investment in the fuel supply chain (because of higher fuel consumption stemming from lower efficiency in the generating plant).
- The need for CO_2 transportation pipelines.
- Drilling of injection wells and the provision of auxiliary injection installations.

Our cost estimates (Table 8.1) suggest that, compared to power plants without CO_2 capture, investment costs increase by 28% to 78% for coal plants and 75% to 113% for gas plants (though the total capital costs for a gas plant remain substantially below those for a coal plant).

A new IGCC coal plant with CO_2 capture would have lower investment costs (total plant cost) than a steam-cycle coal plant similarly equipped, though an IGCC plant with CO_2 capture would be 57% to 62% more costly than a steam-cycle plant without capture. Considering also that fuel efficiency loss in an IGCC coal plant with CO_2 capture (15% to 21% increase in fuel needs) is lower than the loss of efficiency in steam-cycle coal plants with capture (22% to 39% increase in fuel needs), use of CO_2 capture will encourage the use of IGCC technology in coal plants.

Significant efficiency losses also occur in gas-fired plants with CO_2 capture, resulting in an increased fuel need of 16% to 19%. Because fuel cost is a more significant element of total generating cost in a gas plant than a coal plant, CO_2 capture will increase the cost-competitiveness of coal generation relative to gas. Nonetheless, because of its capital cost advantage, gas may well remain the most competitive option overall, depending on local gas prices. The prospect of improving the market for coal is a major incentive to develop carbon sequestration and capture technology for countries with important coal reserves, such as the United States.

In the Reference Scenario, over the next thirty years, the OECD countries will need about 2,000 GW of new capacity. As an illustrative example, if 250 GW of coal-fired plants and 500 GW of gas-fired combined cycle gas turbine (CCGT) plants were equipped with CO_2 capture by 2030, CO_2 capture would amount to 3 gigatonnes.[1] Because of the additional fuel requirements, global coal supply would need to increase by 1%, and gas supply by 2%. The additional investment for CO_2 capture in coal plants would amount to $188 billion for steam-cycle power plants or $153 billion for IGCC plants. For gas-fired power plants, the additional investment would amount to $200-$250 billion (Figure 8.1). Assuming the use of IGCC coal-fired plants, the total incremental investment in power plants would be $350-$440 billion. This would increase investment in power generation in OECD countries by 20% to 25%.

The increased use of coal and gas that would arise from carbon capture and sequestration would require additional investment in gas and coal production capacity and infrastructure of the order of $35 billion, or about 5% of the total incremental investments needed for carbon capture.

CO_2 transportation costs depend on the distance and on the energy costs for pressurisation. The costs for pipeline transportation can range from $1 to

1. Global CO_2 emissions in the Reference Scenario in 2030 reach 38 gigatonnes.

Table 8.1: **Characteristics of Power Plants without and with CO_2 Capture**

Plant type	Year available	Reference cost ($/kW)	Additional investment ($/kW)	Efficiency without capture (%)	Efficiency loss with capture (percentage points)	Additional fuel requirements (%)
Coal steam cycle	2010	1,075	750-825	43	-12	39
Coal steam cycle – advanced	2020	1,025	700-800	44	-8	22
Coal IGCC	2010	1,455	650-750	46	-8	21
Coal IGCC – advanced	2020	1,260	350-400	46	-6	15
Gas CCGT	2010	400	350-450	56	-9	19
Gas CCGT – advanced	2020	400	300-450	59	-8	16

Note: The IGCC data for 2010 refer to a European highly integrated plant based on a Shell gasifier, while the 2020 data refer to a less integrated US design based on an E-gas gasifier. The efficiency remains at the same level because new gas turbines will become available in the 2010-2020 period (the so-called "H-class"), that result in an increase of the efficiency. The different gasifier type reduces capture efficiency losses and reduces investment cost penalties.

$3 per tonne of CO_2 per 100 km. The costs for injection into storage wells are small in comparison to the capture and transportation costs ($1 to $2 per tonne of CO_2), although the longer-term costs for containment, monitoring and verification have yet to be determined.

Figure 8.1: **OECD CO_2 Capture: Illustrative Capacity and Cumulative Investment Requirements through 2030**

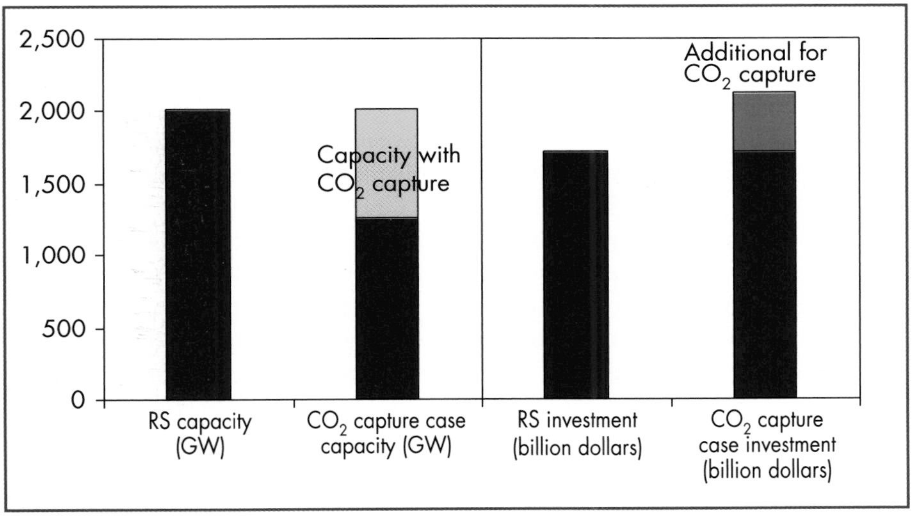

2,500
2,000
1,500
1,000
500
0
Capacity with CO_2 capture
Additional for CO_2 capture
RS capacity (GW)
CO_2 capture case capacity (GW)
RS investment (billion dollars)
CO_2 capture case investment (billion dollars)

RS: Reference Scenario.

Overall, the use of CO_2 capture and storage in electricity generation would increase electricity costs by about 2 to 3 cents per kWh, with some optimistic estimates as low as 1 cent per kWh. Long-term policy targets and government incentives would be required in order to induce power generators in liberalised markets to undertake investment in this technology.

Hydrogen and Fuel Cells

Power Generation

Fuel cells convert oxygen and hydrogen into electricity. Hydrogen can be extracted from hydrocarbon fuels using a process known as reforming and from water by electrolysis. While the use of fossil fuels releases CO_2 emissions into the atmosphere, expected improvements in the efficiency of fuel cells will result in much lower emissions compared to conventional coal or gas plants. Moreover, the CO_2 released is in concentrated form, which makes its capture and sequestration much easier. A major advantage of fuel cells is their flexibility. They come in different sizes, from a few watts for specific applications to many megawatts, suitable for larger-scale electricity generation. Factors that limit their use now are their high capital cost compared to conventional alternatives, their relatively unproven status and limited commercialisation, and the fuel choice for hydrogen production and its cost.

There are many fuel cell technologies suitable for power generation, but the most prominent are:

- *Phosphoric Acid Fuel Cells:* These were the first fuel cells to be commercialised, with more than 200 units in operation worldwide.[2] Phosphoric acid fuel cells use liquid phosphoric acid as the electrolyte and operate at temperatures between 150° C and 200° C. Their electricity generation efficiency is relatively low, around 40% or less. If used in combined heat and power (CHP) mode, the efficiency can rise to 80%. Hydrogen comes from an external source, typically natural gas. These fuel cells now cost around $4,000 per kilowatt. Because of their low efficiency, these systems are likely to be replaced in the future by more advanced technologies, offering much higher efficiencies.
- *Molten Carbonate Fuel Cells:* These fuel cells use lithium-potassium carbonate salts, which are heated to around 650° C to conduct the ions to the electrodes. Because of this higher operating temperature, molten carbonate fuel cells can achieve much higher electricity-generating efficiencies, approaching 60%, and 85% if they produce heat along with electricity. The reform process takes place inside the cell, which eliminates the need for an external reformer and therefore reduces costs. Another advantage is that the electrodes can be made of nickel, which is cheaper than the platinum used in phosphoric acid systems. The main disadvantages are related to the durability of the stack, which is the electricity production unit of the fuel cell. Commercially available molten carbonate fuel cells are expected to have a stack lifetime of five years with 25 years for the balance of plant. Some experts estimate that they could become commercially available as early as 2004.[3]
- *Solid Oxide Fuel Cells:* These fuel cells use ceramic materials, which can achieve very high operating temperatures, reaching 1,000° C. The electricity-generating efficiency of these fuel cells can reach 50% and, combined with a gas turbine, efficiencies can reach 60% to 70%. The conversion of fuel to hydrogen takes place inside the cell. The use of solid

2. US Department of Energy, Office of Fossil Energy (www.fe.doe.gov).
3. See Rocky Mountain Institute at www.rmi.org.

materials is advantageous because it avoids electrolyte leakage and offers greater stability. The high operating temperature requires costly ceramic materials. Research is continuing to produce materials that would reduce costs. There are several projects at the demonstration stage.

The major challenge facing fuel cells is their high initial cost. As noted earlier, the cost of a fuel cell today is in the order of $4,000 per kW or more. A diesel generator or gas turbine would cost three to ten times less. The development of less costly materials will help reduce costs. The conversion of fuel to hydrogen inside the cell will also lower costs. Moreover, higher operating temperatures allow for the exhaust heat to be used for space heating, water heating or additional power production.

In the *World Energy Outlook 2002*, fuel cells emerge as a new source for electricity generation around 2020. The fuel cells that are expected to achieve commercial viability first will involve the reforming of natural gas. Almost all the fuel cells in use for electricity generation by 2030 will be for distributed power generation. Fuel cells are expected to become competitive in distributed generation when capital costs fall below $1,000 per kW, just over a quarter of current costs, and their efficiency approaches 60%. In the Alternative Policy Scenario, fuel cells start increasing their market share around 2015 and achieve a higher share by 2030 relative to the Reference Scenario. More substantial market penetration can be achieved with additional R&D efforts, additional incentives or more stringent environmental policies than envisaged in either of the two scenarios.

Hydrogen Fuel Cell Vehicles

Fuel cell vehicles incorporate a number of technologies not used in conventional vehicles. Several of these technologies are common to electric and hybrid vehicles, including electric drives, electronic controls, higher-voltage direct current electrical circuits, regenerative braking and others. The two most notable new technologies unique to the hydrogen fuel cell vehicles are the fuel cell system and on-board hydrogen storage.

Proton exchange membrane (PEM) fuel cells are widely considered the technology of choice for passenger cars. Such fuel cells currently have a life span of about 50,000 km, meaning that the fuel cell would have to be changed several times during the life span of a car, which adds to the cost compared to an internal combustion engine (ICE) vehicle. Producers are aiming for longer life fuel cells, with the objective of avoiding a fuel cell change altogether during the vehicle life. The efficiency of PEM fuel cells compared to ICEs is not yet clear. Some studies suggest that their energy efficiency will be two to three times higher than that of current ICEs. These differences can be attributed to the assumptions regarding the reference car type, acceleration capability, drive cycles and baseline ICE efficiency trends. The efficiency gains offset, to some extent, the additional fuel costs per unit of energy compared to ICEs using gasoline or diesel.

A number of hydrogen on-board storage systems have been proposed, including:

- Liquid H_2.
- Gaseous H_2 at up to 800 bars.
- Binary metal hydrides.
- Carbon nanotubes.

Liquid H_2 on-board storage is likely to be more expensive than gaseous H_2 and suffers from significant energy losses. Gaseous H_2 has lower energy losses, but may result in lower vehicle range. While liquid and gaseous storage are proven technologies, binary metal hydrides and carbon nanotubes are still at a laboratory stage.

Hydrogen Production and Distribution

Hydrogen is an energy carrier that can be produced from a range of sources. Current hydrogen production technologies include steam reforming of natural gas (used in more than 90% of total hydrogen production), partial oxidation (gasification) of heavy oil products and coal, and electrolysis of water. Steam reforming and oxidation of fossil fuels involve significant CO_2 emissions. Therefore, the focus has been on producing hydrogen using emerging technologies that do not emit CO_2. These include:

- Coal gasification with CO_2 capture and storage.
- Natural gas reforming with CO_2 capture and storage.
- Electrolysis of water with carbon-free sources of electricity.
- Cogeneration in a high temperature gas-cooled nuclear reactor.
- Biomass gasification.

Breakthrough technologies, such as photoelectrochemical water splitting and algal systems for water production, are speculative and unlikely to be practical before 2050.

Cost Estimates for Hydrogen Production and Distribution

Unit supply cost estimates for nine sources of hydrogen are provided in Table 8.2. These estimates include production and distribution to the retail customer. They reflect costs for a system with full economies of scale and cost reductions achieved through progressive improvements in commercial scale production (technology learning). The technology learning effects depend on there being sufficient installed capacity. If installation of each type of capacity were to be limited, these cost reductions could not be counted upon. Natural gas or coal with CO_2 capture and storage is the least costly source ($12-$18/GJ). The next least costly group of technologies consists of

Table 8.2: **Hydrogen Supply Costs**

	Future fuel/ electricity resource price	Fuel cost ($/GJ)	Other production cost ($/GJ)	Transport cost ($/GJ)	Refuelling ($/GJ)	Future supply cost ($/GJ LHV)
Gasoline/diesel	Crude oil $25-$29/bbl	4-5	2	<1-1	2	8-10
Natural gas	Import price $3-4/GJ	3-4	n.a.	<1-1	4	7-9
H_2 from natural gas – CO_2	$3-5 /GJ	3.8-6.3	1.2-2.7	2	5-7	12-18
H_2 from coal – CO_2	$1-2 /GJ	1.3-2.7	4.7-6.3	2	5-7	13-18
H_2 from biomass (gasification)	$2-5 /GJ	2.9-7.1	5-6	2-5	5-7	14-25
H_2 from onshore wind	3-4 cents/kWh	9.8-13.1	5	2-5	5-7	22-30
H_2 from offshore wind	4-5.5 cents/kWh	13.1-18.0	5	2-5	5-7	27-37
H_2 from solar thermal electricity	6-8 cents/kWh	19.6-26.1	5	2-5	5-7	32-42
H_2 from solar PV	12-20 cents/kWh	39.2-65.4	5	2-5	5-7	52-82
H_2 from nuclear	2.5-3.5 cents/kWh	8.2-11.4	5	2	5-7	20-27
H_2 from HTGR cogeneration	n.a.	n.a.	8-23	2	5-7	15-32

Note: Electrolysis efficiency is assumed to be 85%. Costs include technology learning effects that are dependent on installed capacity. It is also assumed that the hydrogen transportation pipelines serve as storage, eliminating the need for additional storage at the production site or the refuelling station. In a transition period, this assumption would not be valid, because transportation pipelines would either not exist or be of insufficient capacity. Hydrogen storage systems may add $5-10 per GJ.
LHV: Lower heating value.
HTGR: High temperature gas-cooled reactor.

biomass, onshore wind and nuclear, all within a range of $14-$30/GJ. The most costly technologies ($27-$82/GJ) are hydrogen production from electrolysis using offshore wind, solar thermal, and solar PV.

Hydrogen is more expensive than conventional fuels partly because of relatively high distribution and retail costs. These costs depend on the configuration of the hydrogen supply system and on the scale of hydrogen demand. If there is low demand (such as in a transition period), decentralised production and/or delivery of hydrogen by trucks to refuelling stations may be the best option: the cost during the transition period to a full-scale hydrogen economy is one of the key problems. Over the longer term, a hydrogen pipeline distribution system would be a less costly solution; and it would also be necessary for CO_2 capture. The energy efficiency of central production would also be significantly higher and the investment costs lower (Figure 8.2). Since a comparison between petroleum-based fuels, which benefit from an established distribution system, and small-scale hydrogen supply would result in an unduly negative assessment of the future potential of hydrogen, the estimates in Figure 8.2 cover only centralised production and distribution of hydrogen.

Pipeline transport of hydrogen involves relatively high pressures (between 10 and 100 bars) and may require the use of special materials. Pressure at the fuel station needs to be increased to 600-800 bars for gaseous on-board storage. Such pressurisation is energy-intensive, requiring between 0.05 and 0.1 GJ electricity per GJ hydrogen (and up to 0.15 GJ if the

Figure 8.2: **Long-term Investment Costs for Alternative Hydrogen Production and Supply Systems**

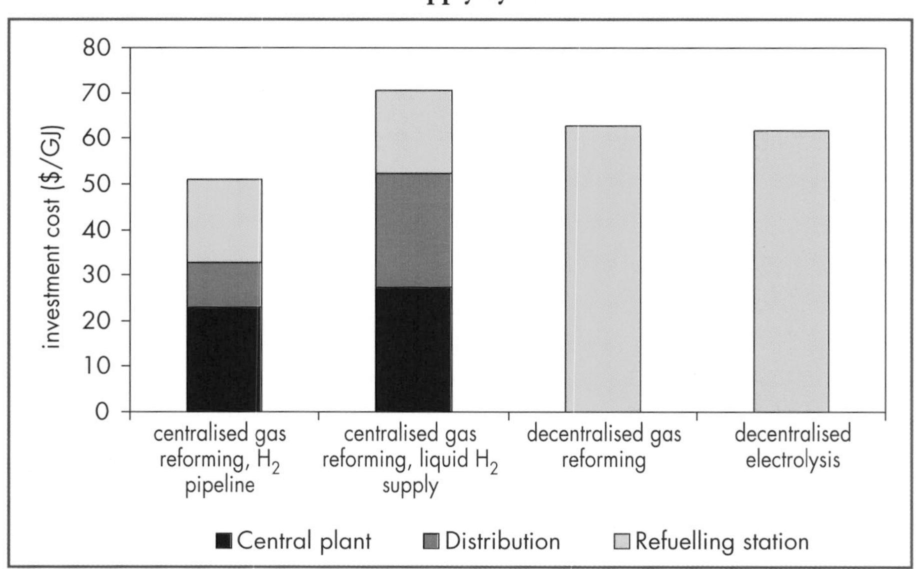

hydrogen were delivered at atmospheric pressure). At an electricity price of 7 cents per kWh, this equals 1 to 2 $/GJ. Moreover, there are capital costs and operating costs for the refuelling station. The total retail cost depends on the fuel station throughput. For a station that serves 300 cars per day, the retail element of the total cost would amount to 5-7 $/GJ.

Distribution and retail sales are an important planning problem in a transition to a hydrogen energy system. A difficult transition period will exist where there are either too few vehicles to justify widespread refuelling, or strong consumer resistance to purchase hydrogen vehicles because of insufficient fuel availability. Without government intervention, it is unlikely that this dilemma would be overcome. Even then, substantial investment will be needed to produce the large volumes of hydrogen needed for an expanding fleet of fuel cell vehicles. Thus, such a system will require huge investments in three major components: fuel cell vehicles, hydrogen production facilities, and a system for transporting hydrogen between production facilities and vehicles. Investors in each component may be reluctant to commit themselves if it is uncertain whether the other components will be developed in time to achieve profits.

Hydrogen fuel cell vehicles are more efficient than conventional vehicles using gasoline or diesel. The fuel cost per kilometre of travel would be lower than a current ICE vehicle but higher than an advanced ICE vehicle (Figure 8.3). But, as discussed below, the transition costs to achieve a system with full economies of scale may be quite large and these are not reflected in the long-term consumer costs presented in Table 8.2 and Figure 8.3.

Figure 8.3: **Fuel Cost per 1,000 km of Travel**

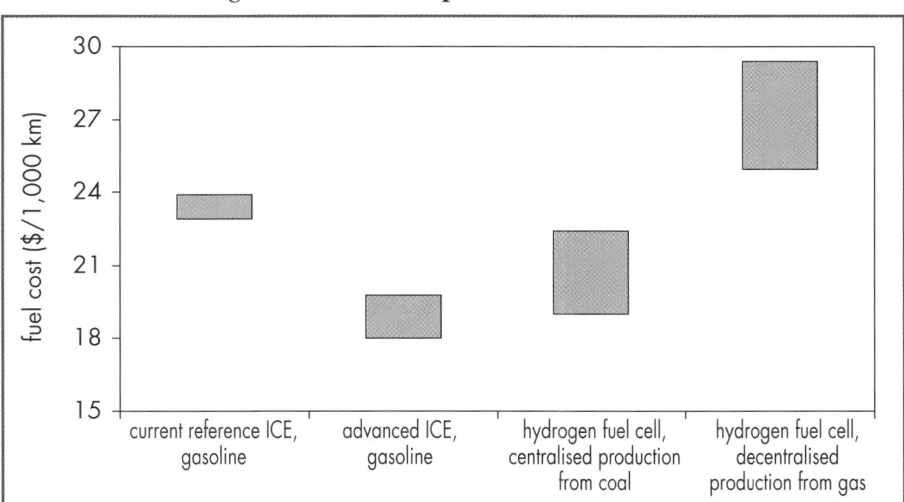

Note: The efficiency of the reference gasoline ICE is 2.6 MJ/km by 2020. The fuel consumption of the advanced gasoline ICE and the fuel cell vehicle are 81% and 50% of this reference.

Cost of Hydrogen Fuel Cell Vehicles

The costs of fuel cell systems and vehicles constitute a major uncertainty. Recently, Toyota began offering hydrogen fuel cell vehicles at $10,000 per month on a 30-month lease. If this fully covers the vehicle production costs, it would imply fuel cell system costs of around $6,000/kW. Most studies suggest that fuel cell cost will decline substantially with mass production and learning. The required fuel cell output in kW terms is still unclear, but may be far less than current cars and light trucks (for example, if they become lighter and more efficient). As mentioned above, the life span of the fuel cell system currently is shorter than a normal vehicle life span. Although its durability is expected to improve, the fuel cell may require replacement twice or three times during the vehicle life. These various uncertainties imply a wide range of cost estimates. At the low end, a fuel cell vehicle might eventually be no more expensive than a comparable ICE vehicle, except for the hydrogen fuel tank (expected to cost between $500 and $1,500 per vehicle.) At the high end, the incremental cost of fuel cell cars could remain greater than $10,000, probably too expensive to achieve commercial success. A recent IEA study estimated a long-run cost of $6,000 to $6,500 in excess of that of an ICE vehicle, even with fuel cell retail costs of $100 per kW (including on-board hydrogen storage) — more than fifty times lower than current costs.[4] There are large uncertainties associated with any estimate in which future technological progress could be so large.

The total investment required in order to achieve a substantially hydrogen-powered transport system will depend both on the rate at which costs are reduced and the timing of the development of this system. Fast transition would be more expensive, particularly if it occurred so fast as to render current investments obsolete before they were fully amortised. Another key factor is the rate of learning (cost reduction) achieved though increased production. Different "learning rates" could result in widely different future costs. With optimal learning and cost reduction, a mature fuel cell vehicle market could be reached at an incremental cost of several hundred billion dollars, but if cost reductions are slow, the incremental costs of achieving a mature market could be around $5 trillion, which would clearly reduce the likelihood of widespread hydrogen use in transport.

Even under a favourable cost scenario, significant government co-ordination and intervention will be needed in the transition period in which there will be too few vehicles to justify the commercial erection of a widespread refuelling network and strong consumer resistance to hydrogen vehicle purchase while fuel is insufficiently available. The potential benefits of widespread hydrogen use have spurred several governments to begin to develop strategies to overcome these transition barriers.

4. IEA (2001).

Advanced Nuclear Reactors

The scenarios developed in the *WEO-2002* show a limited role for nuclear power over the next thirty years, as a result of unfavourable economics and government policies which constrain use in response to public opposition.

Advances in nuclear technology could improve the competitive position of nuclear power against fossil fuels, enhance safety and solve the waste disposal problem. In terms of economics, some argue that the emphasis should be on small (100-600 MW) and modular (100-300 MW) designs with lower capital costs, a shorter construction time and flexible operation.[5] However, very large units, like those used today, will still be more suitable for large and dense grids.

Even now, proven nuclear technologies could be used indirectly to provide hydrogen, by providing energy for electrolysis. This process is more costly than natural gas steam reforming (now the most economic way to produce hydrogen) but it does not produce GHG emissions. Carbon dioxide emission constraints will make this option more competitive, although carbon sequestration offers competing potential for fossil fuels.

Many argue that entirely new nuclear reactor designs are needed if there is to be a major nuclear expansion. There is widespread interest in a fourth generation of nuclear power reactors. Ten countries have pooled their efforts to develop candidate systems.[6] Public opposition, however, can be expected to persist. Table 8.3 lists the six systems included in this programme and their potential best deployment date.

Table 8.3: **Generation IV Systems and Best Deployment Date**

System	Best deployment date
Sodium-Cooled Fast Reactor (SFR)	2015
Very-High-Temperature Reactor (VHTR)	2020
Gas-Cooled Fast Reactor (GFR)	2025
Molten Salt Reactor (MSR)	2025
Supercritical-Water-Cooled Reactor (SCWR)	2025
Lead-Cooled Fast Reactor (LFR)	2025

Source: US Department of Energy (2002).

5. Duffey *et al.* (2001).
6. Argentina, Brazil, Canada, France, Japan, Korea, South Africa, Switzerland, the United Kingdom and the United States have formed the Generation IV International Forum (GIF).

Advanced Electricity Transmission and Distribution Technologies

Global electricity transmission and distribution losses amounted to 1,342 TWh in 2000. Network-related losses account for 6.5% of OECD electricity generation and are higher in developing and transition countries, because of less efficient and poorly maintained equipment and, in a number of countries, theft (Figure 8.4).

Figure 8.4: **Transmission and Distribution Losses as Percentage of Total Electricity Production, 2000**

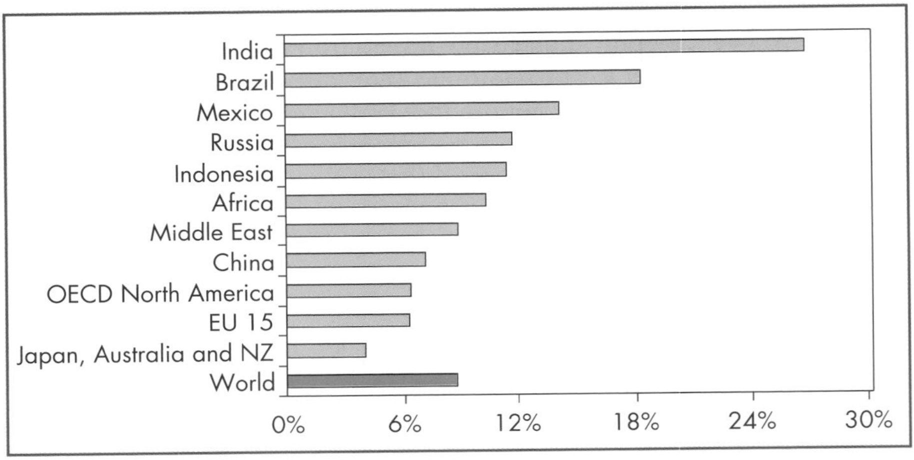

Source: IEA.

World electricity production will increase at an average growth rate of 2.4% over the next three decades. Transmission and distribution losses, in the Reference Scenario, increase at about the same rate and could amount to about 2,700 TWh in 2030.[7]

Losses in transmission and distribution systems arise at every step of the way, *e.g.* conductors and cables, transformers and network protection equipment.

In OECD countries, transformers have high efficiencies, 99.75% on average for transmission transformers and around 99% for distribution transformers;[8] but there is room for further improvement. Efficiency varies according to a number of factors, including the transformer material, the load,

7. At a regional level, losses increase at a lower rate than electricity demand. However, because losses are a higher percentage of electricity production in developing countries and electricity demand there grows faster than in OECD countries, losses at world level increase at the same rate as electricity production.
8. European Copper Institute (1999).

and the age of the transformer. The shares of the different components of transmission and distribution losses in the European electricity networks are shown in Figure 8.5. The importance of the contribution of transformer losses — in particular on the distribution side — is clear.

Important savings could accordingly be achieved through investment in more efficient distribution transformers. Since distribution transformers account for about a quarter of total losses, an improvement of 0.33 percentage points in their efficiency (*i.e.* from 99% to 99.33%, reducing related losses by a third) would result in an overall reduction of around 8%. This corresponds to more than 100 TWh in 2000 and possibly more than 200 TWh in 2030, which is equivalent to the current electricity generation of Australia.

There are other factors driving change, apart from economic efficiency. While the technology for overhead lines in the transmission and distribution networks is relatively mature, there are many environmental and health concerns related to them, especially to high-voltage lines. Concerns range from a possible link between health and exposure to magnetic fields, deforestation in some regions (for example in the Amazon in Brazil), or merely the high visual impact of overhead lines.

Figure 8.5: **Transmission and Distribution Losses in European Electricity Networks**

Conductors and cables 15%
LV conductors and cables 25%
System transformers 10%
Meters, unbilled consumption, theft, etc. 5%
HV transformers 10%
HV conductors and cables 10%
Distribution transformers 25%

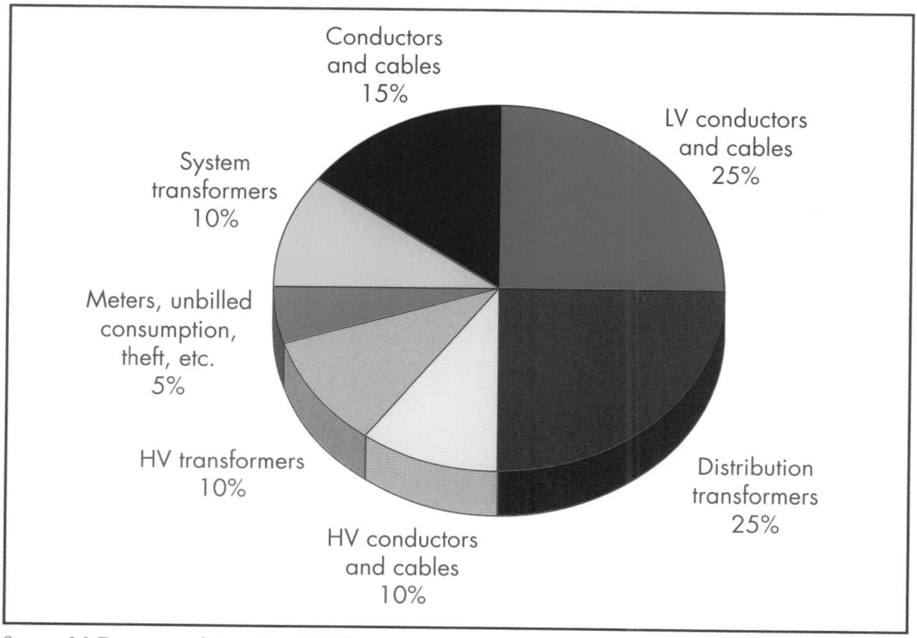

Source: McDermott and Associates (2000).

Improvements are taking place for cables and newer technologies such as gas-insulated transmission lines (GILs). These systems have higher capacities, lower losses and lower electro-magnetic fields, but investment costs are higher. When making a comparison, the value of the land dedicated to transmission and distribution systems should be considered.[9] Building restrictions because of electro-magnetic fields can extend over 200-300 metres for an overhead line of 400 kV, while the area affected is around 15 metres for gas-insulated transmission lines.

For gas-insulated transmission lines, the main environmental concern relates to the use of sulphur hexafluoride (SF_6), which is a very strong greenhouse gas (global warming potential is 23,900 times higher than CO_2). However, GILs of the second generation use mixtures of N_2/SF_6, with a 10-20% content of SF_6, and have much reduced losses due to leakages and consequent overall environmental impact.[10]

9. Benato *et al.* (2001).
10. Koch (2003).

ANNEX 1:
TABLES OF INVESTMENT, SUPPLY AND INFRASTRUCTURE PROJECTIONS

General Note to the Tables

The tables show projections of energy investment, supply and infrastructure for the following regions:

- World
- OECD
- United States and Canada
- EU15
- OECD Pacific
- Transition economies
- Russia
- Developing countries
- China
- India
- Latin America
- Middle East
- Africa

The definitions of regions and fuels are given in Annex 3.

Investment numbers are cumulative for the period indicated. All monetary values are in real billion US dollars using year 2000 prices and market exchange rates.

Non-conventional oil investment numbers include gas-to-liquids (GTL) plants, while gas production includes the gas for GTL.

Supply and infrastructure numbers are expressed in levels for the years given. The column 2000 for the capacities of oil tankers, LNG liquefaction, LNG regasification, and LNG ships shows data for 2001. Oil production includes crude oil, NGL and non-conventional oil. Other renewables capacities include bioenergy, wind, geothermal, solar, tide and wave power. For trade, negative values are exports and positive values imports. Unlike oil and gas, coal trade includes all cross-border coal trade between nations. Energy per capita only takes into account commercial energy.

Both in the text of this book and in the tables, rounding may cause some differences between the total and the sum of the individual components.

Reference Scenario: **World**

	Investment (billion dollars)	**2001-2010**	**2011-2020**	**2021-2030**	**2001-2030**
	Total Investment	**4,551**	**5,610**	**6,320**	**16,481**
Oil	**Total**	**916**	**1,045**	**1,136**	**3,096**
	Exploration and development	689	740	793	2,222
	Non-conventional oil	49	60	96	205
	of which GTL	*6*	*14*	*20*	*40*
	Refining	122	143	147	412
	Tankers	37	79	76	192
	Pipelines	20	23	23	65
Gas	**Total**	**948**	**1,041**	**1,157**	**3,145**
	Exploration and development	478	575	678	1,731
	LNG liquefaction	46	32	38	116
	LNG regasification	21	21	25	67
	LNG ships	30	16	22	69
	Transmission	201	196	182	579
	Distribution	135	160	194	489
	Underground storage	36	41	17	94
Coal	**Total**	**125**	**129**	**144**	**398**
	Total mining	113	113	125	351
	new mining capacity	*71*	*66*	*71*	*208*
	sustaining mining capacity	*41*	*47*	*54*	*143*
	Ports	2.5	4.5	6.0	12.9
	Shipping	10	11	13	34
Electricity	**Total**	**2,562**	**3,396**	**3,883**	**9,841**
	Generating capacity	926	1,422	1,731	4,080
	of which renewables	*401*	*497*	*496*	*1,394*
	Refurbishment	145	152	142	439
	Transmission	439	548	581	1,568
	Distribution	1,052	1,274	1,429	3,755

Reference Scenario: **World**

	Supply and Infrastructure	2000	2010	2020	2030
Oil	Oil production* (mb/d)	75	89	104	120
	Net trade (mb/d)	–	–	–	–
	Refinery capacity (mb/d)	82	92	105	121
	Tanker capacity (million DWT)	271	365	425	522
Gas	Gas production (bcm)	2,513	3,407	4,362	5,280
	Net trade (bcm)	–	–	–	–
	LNG liquefaction capacity (bcm)	163	441	678	997
	LNG regasification capacity (bcm)	353	607	882	1,252
	LNG shipping capacity (bcm)	141	365	566	838
	Transmission pipelines (thousand km)	1,139	1,453	1,769	2,058
	Distribution pipelines (thousand km)	5,007	6,214	7,377	8,523
	Underground storage working volume (bcm)	328	471	621	685
Coal	Coal production (Mt)	4,595	5,354	6,099	6,954
	Exports (Mt)	637	785	905	1,051
	Imports (Mt)	637	785	905	1,051
	Port capacity** (Mt)	2,212	2,316	2,545	2,879
	for exports	*1,072*	*1,138*	*1,265*	*1,445*
	for imports	*1,140*	*1,178*	*1,280*	*1,434*
	Shipping capacity (million DWT)	67	82	99	116
Electricity	Generating capacity (GW)	3,498	4,408	5,683	7,157
	Coal	*1,078*	*1,277*	*1,599*	*2,090*
	Oil	*501*	*547*	*540*	*507*
	Gas	*729*	*1,162*	*1,865*	*2,501*
	Hydrogen fuel cell	*0*	*0*	*4*	*100*
	Nuclear	*354*	*379*	*362*	*356*
	Hydro	*776*	*911*	*1,080*	*1,205*
	Other renewables	*61*	*133*	*233*	*399*
	Urban population without electricity (million)	250	291	321	316
	Rural population without electricity (million)	1,395	1,275	1,212	1,110
Indicators	GDP (billion dollars)	37,087	47,970	60,664	74,605
	Population (million)	6,035	6,778	7,518	8,196
	Population density (persons/km^2)	45	51	56	61
	Electrification rate (%)	73	77	80	83
	Energy/capita (toe/capita)	2	2	2	2

* World oil production includes processing gains.

** The port capacity figures in this table do not include the capacity required for internal coal trade in China, India and Indonesia (cumulative additions for internal trade are 192 Mt of capacity).

Reference Scenario: **OECD**

	Investment (billion dollars)	2001-2010	2011-2020	2021-2030	2001-2030
	Total Regional Investment	**2,092**	**2,228**	**2,231**	**6,552**
Oil	**Total**	**353**	**285**	**251**	**888**
	Exploration and development	287	224	173	684
	Non-conventional oil	29	33	54	115
	Refining	37	28	24	89
Gas	**Total**	**501**	**499**	**496**	**1,496**
	Exploration and development	269	275	282	826
	LNG liquefaction	5	4	4	13
	LNG regasification	19	19	23	61
	Transmission	95	80	63	238
	Distribution	98	103	117	318
	Underground storage	15	17	7	39
Coal	**Total**	**41**	**44**	**46**	**131**
	Total mining	41	43	44	128
	new mining capacity	*20*	*20*	*19*	*58*
	sustaining mining capacity	*21*	*23*	*25*	*70*
	Ports	0.2	1.4	1.8	3.4
Electricity	**Total**	**1,197**	**1,400**	**1,438**	**4,036**
	Generating capacity	390	595	734	1,719
	of which renewables	*173*	*167*	*221*	*561*
	Refurbishment	98	90	71	260
	Transmission	188	209	172	569
	Distribution	520	507	461	1,488

Reference Scenario: **OECD**

	Supply and Infrastructure	2000	2010	2020	2030
Oil	Oil production (mb/d)	22	21	19	18
	Net trade (mb/d)	24	29	35	41
	Refinery capacity (mb/d)	44	46	49	52
Gas	Gas production (bcm)	1,099	1,251	1,340	1,391
	Net trade (bcm)	274	552	833	1,091
	LNG liquefaction capacity (bcm)	10	42	75	106
	LNG regasification capacity (bcm)	347	578	840	1,182
	Transmission pipelines (thousand km)	791	948	1,080	1,185
	Distribution pipelines (thousand km)	3,626	4,302	4,807	5,245
	Underground storage working volume (bcm)	192	266	335	368
Coal	Coal production (Mt)	2,019	2,084	2,258	2,408
	Exports (Mt)	321	342	396	483
	Imports (Mt)	453	516	548	550
	Port capacity (Mt)	1,503	1,510	1,601	1,731
	for exports	*658*	*658*	*716*	*826*
	for imports	*845*	*853*	*885*	*905*
Electricity	Generating capacity (GW)	2,063	2,430	2,847	3,294
	Coal	*611*	*608*	*659*	*723*
	Oil	*274*	*282*	*237*	*173*
	Gas	*418*	*674*	*1,006*	*1,238*
	Hydrogen fuel cell	*0*	*0*	*4*	*100*
	Nuclear	*298*	*313*	*284*	*269*
	Hydro	*413*	*445*	*468*	*485*
	Other renewables	*49*	*109*	*189*	*306*
	Urban population without electricity (million)	0	0	0	0
	Rural population without electricity (million)	8	1	0	0
Indicators	GDP (billion dollars)	30,142	37,742	45,839	53,965
	Population (million)	1,117	1,173	1,217	1,248
	Population density (persons/km^2)	31	33	34	35
	Electrification rate (%)	99	100	100	100
	Energy/capita (toe/capita)	5	5	5	6

Reference Scenario: **United States and Canada**

	Investment (billion dollars)	**2001-2010**	**2011-2020**	**2021-2030**	**2001-2030**
	Total Regional Investment	**979**	**1,071**	**1,113**	**3,164**
Oil	**Total**	**208**	**172**	**165**	**545**
	Exploration and development	164	130	102	397
	Non-conventional oil	28	32	53	114
	Refining	15	9	10	35
Gas	**Total**	**288**	**286**	**282**	**855**
	Exploration and development	166	172	171	509
	LNG liquefaction	–	–	–	–
	LNG regasification	10	11	11	32
	Transmission	49	39	30	119
	Distribution	58	59	65	182
	Underground storage	5	5	3	13
Coal	**Total**	**21**	**24**	**25**	**70**
	Total mining	21	24	25	70
	new mining capacity	*7*	*10*	*9*	*26*
	sustaining mining capacity	*13*	*14*	*16*	*44*
	Ports	0.1	0.1	0.1	0.2
Electricity	**Total**	**463**	**589**	**642**	**1,694**
	Generating capacity	111	234	309	654
	of which renewables	*41*	*52*	*104*	*197*
	Refurbishment	46	44	39	130
	Transmission	73	99	89	261
	Distribution	232	212	204	649

Reference Scenario: **United States and Canada**

	Supply and Infrastructure	**2000**	**2010**	**2020**	**2030**
Oil	Oil production (mb/d)	11	12	11	12
	Net trade (mb/d)	10	11	14	16
	Refinery capacity (mb/d)	19	20	21	23
Gas	Gas production (bcm)	718	824	856	842
	Net trade (bcm)	24	109	228	371
	LNG liquefaction capacity (bcm)	–	–	–	–
	LNG regasification capacity (bcm)	17	136	284	449
	Transmission pipelines (thousand km)	539	629	700	755
	Distribution pipelines (thousand km)	2,023	2,405	2,660	2,850
	Underground storage working volume (bcm)	129	165	190	209
Coal	Coal production (Mt)	1,045	1,077	1,214	1,289
	Exports (Mt)	86	54	65	72
	Imports (Mt)	38	41	44	48
	Port capacity (Mt)	328	330	333	336
	for exports	*309*	*309*	*309*	*309*
	for imports	*19*	*22*	*25*	*28*
Electricity	Generating capacity (GW)	942	1,086	1,265	1,473
	Coal	*332*	*328*	*379*	*422*
	Oil	*80*	*79*	*58*	*46*
	Gas	*234*	*362*	*488*	*577*
	Hydrogen fuel cell	*0*	*0*	*2*	*35*
	Nuclear	*107*	*109*	*95*	*84*
	Hydro	*166*	*170*	*175*	*177*
	Other renewables	*22*	*39*	*68*	*133*
	Urban population without electricity (million)	0	0	0	0
	Rural population without electricity (million)	0	0	0	0
Indicators	GDP (billion dollars)	10,560	13,469	16,188	19,061
	Population (million)	306	334	361	387
	Population density (persons/km^2)	16	17	18	20
	Electrification rate (%)	100	100	100	100
	Energy/capita (toe/capita)	8	9	9	9

Reference Scenario: **EU15**

	Investment (billion dollars)	**2001-2010**	**2011-2020**	**2021-2030**	**2001-2030**
	Total Regional Investment	**496**	**564**	**542**	**1,603**
Oil	**Total**	**53**	**37**	**27**	**117**
	Exploration and development	47	31	22	99
	Non-conventional oil	0.0	0.1	0.1	0.2
	Refining	7	7	4	18
Gas	**Total**	**137**	**123**	**105**	**365**
	Exploration and development	63	54	44	161
	LNG liquefaction	–	–	–	–
	LNG regasification	7	4	7	19
	Transmission	32	27	16	74
	Distribution	28	31	36	95
	Underground storage	7	7	1	15
Coal	**Total**	**5**	**3**	**3**	**10**
	Total mining	5	3	3	10
	new mining capacity	*3*	*2*	*2*	*6*
	sustaining mining capacity	*2*	*1*	*1*	*4*
	Ports	0.0	0.0	0.1	0.1
Electricity	**Total**	**302**	**401**	**408**	**1,110**
	Generating capacity	108	190	227	525
	of which renewables	*57*	*58*	*65*	*181*
	Refurbishment	20	19	13	52
	Transmission	42	44	34	120
	Distribution	131	149	133	413

Reference Scenario: **EU15**

	Supply and Infrastructure	2000	2010	2020	2030
Oil	Oil production (mb/d)	3	2	2	1
	Net trade (mb/d)	9	11	12	13
	Refinery capacity (mb/d)	13	13	14	15
Gas	Gas production (bcm)	241	221	191	150
	Net trade (bcm)	187	350	510	632
	LNG liquefaction capacity (bcm)	–	–	–	–
	LNG regasification capacity (bcm)	43	130	188	300
	Transmission pipelines (thousand km)	182	227	266	288
	Distribution pipelines (thousand km)	1,189	1,370	1,509	1,633
	Underground storage working volume (bcm)	54	80	106	111
Coal	Coal production (Mt)	340	276	234	233
	Exports (Mt)	16	16	15	17
	Imports (Mt)	183	209	223	216
	Port capacity (Mt)	452	452	452	460
	for exports	*30*	*30*	*30*	*30*
	for imports	*422*	*422*	*422*	*430*
Electricity	Generating capacity (GW)	584	679	792	901
	Coal	*146*	*134*	*122*	*136*
	Oil	*78*	*77*	*55*	*33*
	Gas	*98*	*176*	*310*	*372*
	Hydrogen fuel cell	*0*	*0*	*1*	*30*
	Nuclear	*124*	*118*	*88*	*76*
	Hydro	*118*	*124*	*129*	*134*
	Other renewables	*19*	*50*	*87*	*120*
	Urban population without electricity (million)	0	0	0	0
	Rural population without electricity (million)	0	0	0	0
Indicators	GDP (billion dollars)	10,663	13,360	16,284	19,005
	Population (million)	378	378	374	367
	Population density (persons/km^2)	117	117	115	113
	Electrification rate (%)	100	100	100	100
	Energy/capita (toe/capita)	4	4	5	5

Reference Scenario: **OECD Pacific**

	Investment (billion dollars)	**2001-2010**	**2011-2020**	**2021-2030**	**2001-2030**
	Total Regional Investment	**381**	**333**	**287**	**1,000**
Oil	**Total**	**20**	**14**	**10**	**44**
	Exploration and development	8	6	6	19
	Non-conventional oil	0.6	0.3	0.3	1.2
	Refining	12	8	4	24
Gas	**Total**	**33**	**36**	**42**	**111**
	Exploration and development	12	15	20	46
	LNG liquefaction	4	4	4	12
	LNG regasification	1	4	4	9
	Transmission	7	5	7	19
	Distribution	7	7	7	21
	Underground storage	1	1	1	3
Coal	**Total**	**10**	**12**	**14**	**36**
	Total mining	10	11	13	33
	new mining capacity	*5*	*5*	*6*	*16*
	sustaining mining capacity	*5*	*5*	*7*	*16*
	Ports	0.0	1.3	1.6	3.0
Electricity	**Total**	**319**	**270**	**220**	**809**
	Generating capacity	119	114	124	357
	of which renewables	*43*	*35*	*31*	*110*
	Refurbishment	27	21	13	61
	Transmission	58	46	28	131
	Distribution	114	90	56	260

Reference Scenario: **OECD Pacific**

	Supply and Infrastructure	**2000**	**2010**	**2020**	**2030**
Oil	Oil production (mb/d)	0.9	0.6	0.5	0.5
	Net trade (mb/d)	8	10	10	10
	Refinery capacity (mb/d)	8	9	10	10
Gas	Gas production (bcm)	41	65	94	125
	Net trade (bcm)	81	104	109	121
	LNG liquefaction capacity (bcm)	10	37	69	100
	LNG regasification capacity (bcm)	283	301	354	414
	Transmission pipelines (thousand km)	28	37	43	52
	Distribution pipelines (thousand km)	133	161	177	194
	Underground storage working volume (bcm)	2	6	11	14
Coal	Coal production (Mt)	318	386	460	553
	Exports (Mt)	181	241	301	389
	Imports (Mt)	209	241	263	258
	Port capacity (Mt)	652	654	741	861
	for exports	*289*	*289*	*347*	*457*
	for imports	*364*	*365*	*395*	*404*
Electricity	Generating capacity (GW)	367	453	523	591
	Coal	*83*	*97*	*107*	*103*
	Oil	*91*	*97*	*92*	*62*
	Gas	*67*	*97*	*130*	*169*
	Hydrogen fuel cell	*0*	*0*	*2*	*35*
	Nuclear	*57*	*77*	*94*	*105*
	Hydro	*63*	*70*	*76*	*81*
	Other renewables	*6*	*15*	*23*	*36*
	Urban population without electricity (million)	0	0	0	0
	Rural population without electricity (million)	0	0	0	0
Indicators	GDP (billion dollars)	7,429	8,930	10,772	12,718
	Population (million)	197	203	205	203
	Population density (persons/km^2)	22	23	23	23
	Electrification rate (%)	100	100	100	100
	Energy/capita (toe/capita)	4	5	6	6

Reference Scenario: **Transition Economies**

	Investment (billion dollars)	2001-2010	2011-2020	2021-2030	2001-2030
	Total Regional Investment	**438**	**612**	**622**	**1,672**
Oil	**Total**	**124**	**154**	**170**	**448**
	Exploration and development	114	144	163	422
	Non-conventional oil	–	–	–	–
	Refining	10	10	6	26
Gas	**Total**	**154**	**174**	**164**	**492**
	Exploration and development	75	93	103	272
	LNG liquefaction	2	1	0	4
	LNG regasification	–	–	–	–
	Transmission	47	45	33	125
	Distribution	12	17	22	51
	Underground storage	17	17	6	40
Coal	**Total**	**13**	**9**	**10**	**32**
	Total mining	13	9	10	32
	new mining capacity	*10*	*5*	*6*	*21*
	sustaining mining capacity	*4*	*4*	*4*	*11*
	Ports	0.2	0.0	0.1	0.3
Electricity	**Total**	**147**	**276**	**277**	**700**
	Generating capacity	35	123	139	297
	of which renewables	*22*	*31*	*27*	*80*
	Refurbishment	11	16	14	41
	Transmission	21	33	28	82
	Distribution	80	103	97	280

Reference Scenario: **Transition Economies**

	Supply and Infrastructure	**2000**	**2010**	**2020**	**2030**
Oil	Oil production (mb/d)	8	13	14	15
	Net trade (mb/d)	–4	–7	–8	–8
	Refinery capacity (mb/d)	11	11	11	12
Gas	Gas production (bcm)	722	914	1,143	1,222
	Net trade (bcm)	–112	–167	–267	–277
	LNG liquefaction capacity (bcm)	0	13	19	22
	LNG regasification capacity (bcm)	–	–	–	–
	Transmission pipelines (thousand km)	241	311	379	420
	Distribution pipelines (thousand km)	877	1,059	1,248	1,425
	Underground storage working volume (bcm)	132	189	247	266
Coal	Coal production (Mt)	528	638	621	645
	Exports (Mt)	70	92	90	94
	Imports (Mt)	51	67	66	77
	Port capacity (Mt)	107	120	121	129
	for exports	*62*	*75*	*76*	*79*
	for imports	*45*	*45*	*45*	*50*
Electricity	Generating capacity (GW)	406	421	526	624
	Coal	*112*	*110*	*110*	*142*
	Oil	*38*	*39*	*33*	*19*
	Gas	*126*	*134*	*232*	*300*
	Nuclear	*41*	*38*	*35*	*31*
	Hydro	*88*	*97*	*110*	*119*
	Other renewables	*1*	*3*	*6*	*13*
	Urban population without electricity (million)	0	0	0	0
	Rural population without electricity (million)	2	0	0	0
Indicators	GDP (billion dollars)	726	987	1,378	1,814
	Population (million)	353	344	338	327
	Population density (persons/km^2)	15	15	15	14
	Electrification rate (%)	99	100	100	100
	Energy/capita (toe/capita)	3	4	4	5

Reference Scenario: **Russia**

	Investment (billion dollars)	**2001-2010**	**2011-2020**	**2021-2030**	**2001-2030**
	Total Country Investment	**269**	**391**	**389**	**1,050**
Oil	**Total**	**97**	**111**	**120**	**328**
	Exploration and development	90	104	114	308
	Non-conventional oil	–	–	–	–
	Refining	7	7	6	20
Gas	**Total**	**103**	**117**	**111**	**332**
	Exploration and development	52	65	70	187
	LNG liquefaction	2	1	0	4
	LNG regasification	–	–	–	–
	Transmission	33	34	24	92
	Distribution	7	11	14	32
	Underground storage	8	7	2	17
Coal	**Total**	**6**	**4**	**4**	**13**
	Total mining	5	4	4	13
	new mining capacity	*3*	*1*	*2*	*7*
	sustaining mining capacity	*2*	*2*	*2*	*6*
	Ports	0.2	0.0	0.0	0.3
Electricity	**Total**	**64**	**159**	**153**	**377**
	Generating capacity	15	69	72	157
	of which renewables	*7*	*15*	*8*	*30*
	Refurbishment	5	9	7	21
	Transmission	10	20	15	45
	Distribution	34	61	59	154

Reference Scenario: **Russia**

	Supply and Infrastructure	2000	2010	2020	2030
Oil	Oil production (mb/d)	7	9	9	10
	Net trade (mb/d)	–4	–5	–5	–5
	Refinery capacity (mb/d)	7	7	7	7
Gas	Gas production (bcm)	583	709	872	914
	Net trade (bcm)	–188	–223	–288	–280
	LNG liquefaction capacity (bcm)	0	13	19	22
	LNG regasification capacity (bcm)	–	–	–	–
	Transmission pipelines (thousand km)	150	196	245	275
	Distribution pipelines (thousand km)	540	633	764	888
	Underground storage working volume (bcm)	73	101	124	130
Coal	Coal production (Mt)	242	296	285	290
	Exports (Mt)	40	56	57	62
	Imports (Mt)	26	32	29	29
	Port capacity (Mt)	45	57	59	61
	for exports	*31*	*44*	*45*	*48*
	for imports	*14*	*14*	*14*	*14*
Electricity	Generating capacity (GW)	217	225	298	360
	Coal	*53*	*52*	*50*	*64*
	Oil	*12*	*12*	*11*	*9*
	Gas	*88*	*90*	*159*	*205*
	Nuclear	*20*	*23*	*22*	*21*
	Hydro	*44*	*47*	*53*	*54*
	Other renewables	*1*	*2*	*3*	*7*
	Urban population without electricity (million)	0	0	0	0
	Rural population without electricity (million)	0	0	0	0
Indicators	GDP (billion dollars)	389	519	732	947
	Population (million)	145	137	130	121
	Population density (persons/km^2)	9	8	8	7
	Electrification rate (%)	100	100	100	100
	Energy/capita (toe/capita)	4	5	6	8

Reference Scenario: **Developing Countries**

	Investment (billion dollars)	**2001-2010**	**2011-2020**	**2021-2030**	**2001-2030**
	Total Regional Investment	**1,923**	**2,641**	**3,332**	**7,897**
Oil	**Total**	**382**	**505**	**616**	**1,502**
	Exploration and development	288	372	457	1,116
	Non-conventional oil	20	28	42	89
	Refining	75	105	117	297
Gas	**Total**	**263**	**352**	**474**	**1,089**
	Exploration and development	134	207	292	633
	LNG liquefaction	38	27	34	99
	LNG regasification	2	1	3	6
	Transmission	59	70	86	216
	Distribution	25	40	56	120
	Underground storage	4	7	4	15
Coal	**Total**	**61**	**65**	**74**	**200**
	Total mining	59	62	70	191
	new mining capacity	*42*	*42*	*46*	*129*
	sustaining mining capacity	*16*	*20*	*25*	*62*
	Ports	2.1	3.0	4.0	9.2
Electricity	**Total**	**1,218**	**1,720**	**2,168**	**5,106**
	Generating capacity	501	704	859	2,064
	of which renewables	*206*	*299*	*247*	*752*
	Refurbishment	35	46	57	138
	Transmission	230	307	382	918
	Distribution	452	664	871	1,987

Reference Scenario: **Developing Countries**

	Supply and Infrastructure	2000	2010	2020	2030
Oil	Oil production (mb/d)	43	53	68	84
	Net trade (mb/d)	–21	–22	–28	–33
	Refinery capacity (mb/d)	26	35	45	57
Gas	Gas production (bcm)	692	1,241	1,879	2,667
	Net trade (bcm)	–162	–385	–566	–814
	LNG liquefaction capacity (bcm)	153	386	584	869
	LNG regasification capacity (bcm)	6	28	42	70
	Transmission pipelines (thousand km)	107	195	311	452
	Distribution pipelines (thousand km)	504	852	1,323	1,853
	Underground storage working volume (bcm)	4	17	39	51
Coal	Coal production (Mt)	2,047	2,632	3,220	3,901
	Exports (Mt)	246	351	420	474
	Imports (Mt)	133	202	292	425
	Port capacity* (Mt)	602	686	823	1,019
	for exports	*352*	*405*	*473*	*540*
	for imports	*250*	*281*	*350*	*479*
Electricity	Generating capacity (GW)	1,029	1,558	2,311	3,238
	Coal	*354*	*559*	*830*	*1,224*
	Oil	*190*	*226*	*271*	*315*
	Gas	*185*	*354*	*627*	*963*
	Nuclear	*15*	*29*	*42*	*56*
	Hydro	*274*	*368*	*502*	*601*
	Other renewables	*11*	*22*	*39*	*79*
	Urban population without electricity (million)	250	291	321	316
	Rural population without electricity (million)	1,385	1,275	1,212	1,110
Indicators	GDP (billion dollars)	6,219	9,240	13,447	18,826
	Population (million)	4,565	5,261	5,964	6,621
	Population density (persons/km²)	61	70	79	88
	Electrification rate (%)	64	70	74	78
	Energy/capita (toe/capita)	1	1	1	1

* The port capacity figures in this table do not include the capacity required for internal coal trade in China, India and Indonesia (cumulative additions for internal trade are 192 Mt of capacity).

Reference Scenario: **China**

	Investment (billion dollars)	**2001-2010**	**2011-2020**	**2021-2030**	**2001-2030**
	Total Country Investment	**578**	**787**	**888**	**2,253**
Oil	**Total**	**39**	**41**	**39**	**119**
	Exploration and development	27	23	20	69
	Non-conventional oil	0	0	0	0
	Refining	12	19	19	50
Gas	**Total**	**22**	**31**	**45**	**98**
	Exploration and development	7	10	14	31
	LNG liquefaction	–	–	–	–
	LNG regasification	1	1	2	4
	Transmission	8	8	11	26
	Distribution	5	11	18	35
	Underground storage	0	1	1	2
Coal	**Total**	**40**	**40**	**43**	**123**
	Total mining	39	39	43	121
	new mining capacity	*29*	*26*	*27*	*82*
	sustaining mining capacity	*10*	*13*	*15*	*38*
	Ports	0.5	0.8	0.8	2.1
Electricity	**Total**	**478**	**675**	**761**	**1,913**
	Generating capacity	199	285	311	795
	of which renewables	*68*	*122*	*79*	*270*
	Refurbishment	13	17	20	50
	Transmission	90	119	136	345
	Distribution	175	254	294	723

Reference Scenario: **China**

	Supply and Infrastructure	2000	2010	2020	2030
Oil	Oil production (mb/d)	3	3	2	2
	Net trade (mb/d)	2	4	7	10
	Refinery capacity (mb/d)	4	6	8	10
Gas	Gas production (bcm)	30	55	90	115
	Net trade (bcm)	2	6	20	47
	LNG liquefaction capacity (bcm)	–	–	–	–
	LNG regasification capacity (bcm)	0	8	16	31
	Transmission pipelines (thousand km)	11	18	32	48
	Distribution pipelines (thousand km)	89	150	269	434
	Underground storage working volume (bcm)	0	1	5	8
Coal	Coal production (Mt)	1,231	1,606	1,941	2,304
	Exports (Mt)	70	100	112	123
	Imports (Mt)	8	18	27	40
	Port capacity* (Mt)	162	162	168	181
	for exports	*119*	*119*	*124*	*137*
	for imports	*43*	*43*	*43*	*44*
Electricity	Generating capacity (GW)	322	517	787	1,087
	Coal	*212*	*342*	*499*	*696*
	Oil	*21*	*22*	*22*	*23*
	Gas	*7*	*27*	*69*	*113*
	Nuclear	*2*	*11*	*21*	*31*
	Hydro	*79*	*112*	*171*	*209*
	Other renewables	*1*	*3*	*5*	*15*
	Urban population without electricity (million)	0	0	0	0
	Rural population without electricity (million)	18	0	0	0
Indicators	GDP (billion dollars)	1,313	2,291	3,627	5,335
	Population (million)	1,263	1,363	1,442	1,481
	Population density (persons/km^2)	132	142	150	154
	Electrification rate (%)	99	100	100	100
	Energy/capita (toe/capita)	1	1	1	1

* The port capacity figures in this table do not include the capacity required for internal coal trade (cumulative additions for internal trade are 111 Mt of capacity).

Reference Scenario: **India**

	Investment (billion dollars)	**2001-2010**	**2011-2020**	**2021-2030**	**2001-2030**
	Total Country Investment	**172**	**247**	**347**	**766**
Oil	**Total**	**9**	**11**	**12**	**32**
	Exploration and development	5	4	3	12
	Non-conventional oil	–	–	–	–
	Refining	4	7	9	20
Gas	**Total**	**11**	**16**	**17**	**44**
	Exploration and development	5	8	9	22
	LNG liquefaction	–	–	–	–
	LNG regasification	1	0	1	2
	Transmission	3	4	3	11
	Distribution	1	3	4	8
	Underground storage	0	1	0	2
Coal	**Total**	**6**	**8**	**11**	**25**
	Total mining	6	8	10	24
	new mining capacity	*4*	*6*	*8*	*18*
	sustaining mining capacity	*1*	*2*	*2*	*6*
	Ports	0.3	0.1	0.3	0.8
Electricity	**Total**	**145**	**212**	**307**	**665**
	Generating capacity	69	83	116	268
	of which renewables	*30*	*28*	*27*	*85*
	Refurbishment	4	5	6	15
	Transmission	29	39	51	119
	Distribution	44	85	134	262

Reference Scenario: **India**

	Supply and Infrastructure	2000	2010	2020	2030
Oil	Oil production (mb/d)	0.7	0.5	0.4	0.3
	Net trade (mb/d)	1	2	4	5
	Refinery capacity (mb/d)	2	2	3	4
Gas	Gas production (bcm)	22	38	57	58
	Net trade (bcm)	0	9	19	38
	LNG liquefaction capacity (bcm)	–	–	–	–
	LNG regasification capacity (bcm)	0	10	12	23
	Transmission pipelines (thousand km)	4	10	17	22
	Distribution pipelines (thousand km)	36	62	111	167
	Underground storage working volume (bcm)	0	1	4	5
Coal	Coal production (Mt)	329	395	503	652
	Exports (Mt)	1	1	2	2
	Imports (Mt)	22	26	35	46
	Port capacity* (Mt)	35	42	42	54
	for exports	*2*	*2*	*2*	*2*
	for imports	*33*	*40*	*40*	*52*
Electricity	Generating capacity (GW)	111	174	259	366
	Coal	*65*	*93*	*132*	*197*
	Oil	*5*	*7*	*7*	*7*
	Gas	*13*	*30*	*59*	*83*
	Nuclear	*3*	*4*	*6*	*8*
	Hydro	*23*	*37*	*49*	*59*
	Other renewables	*1*	*3*	*6*	*11*
	Urban population without electricity (million)	77	79	64	46
	Rural population without electricity (million)	503	484	438	382
Indicators	GDP (billion dollars)	508	831	1,310	1,961
	Population (million)	1,016	1,164	1,291	1,409
	Population density (persons/km^2)	309	354	393	429
	Electrification rate (%)	43	52	61	70
	Energy/capita (toe/capita)	0.3	0.4	0.4	0.5

* The port capacity figures in this table do not include the capacity required for internal coal trade (cumulative additions for internal trade are 35 Mt of capacity).

Reference Scenario: **Latin America**

	Investment (billion dollars)	**2001-2010**	**2011-2020**	**2021-2030**	**2001-2030**
	Total Regional Investment	**339**	**440**	**558**	**1,337**
Oil	**Total**	**91**	**112**	**133**	**336**
	Exploration and development	70	81	90	241
	Non-conventional oil	15	17	27	59
	Refining	6	14	17	37
Gas	**Total**	**54**	**78**	**115**	**247**
	Exploration and development	28	45	68	141
	LNG liquefaction	7	3	4	15
	LNG regasification	–	–	–	–
	Transmission	10	16	23	49
	Distribution	9	12	19	39
	Underground storage	0	1	1	2
Coal	**Total**	**3**	**3**	**4**	**10**
	Total mining	3	3	3	9
	new mining capacity	*2*	*2*	*2*	*6*
	sustaining mining capacity	*1*	*1*	*1*	*3*
	Ports	0.4	0.3	0.5	1.2
Electricity	**Total**	**191**	**247**	**306**	**744**
	Generating capacity	86	111	120	317
	of which renewables	*63*	*78*	*69*	*211*
	Refurbishment	5	6	8	19
	Transmission	32	41	55	128
	Distribution	69	89	124	281

Reference Scenario: **Latin America**

	Supply and Infrastructure	2000	2010	2020	2030
Oil	Oil production (mb/d)	7	9	10	12
	Net trade (mb/d)	–2	–3	–3	–3
	Refinery capacity (mb/d)	6	6	8	9
Gas	Gas production (bcm)	116	217	339	516
	Net trade (bcm)	–11	–45	–68	–103
	LNG liquefaction capacity (bcm)	4	48	74	111
	LNG regasification capacity (bcm)	–	–	–	–
	Transmission pipelines (thousand km)	30	47	74	114
	Distribution pipelines (thousand km)	121	217	320	472
	Underground storage working volume (bcm)	0	1	4	6
Coal	Coal production (Mt)	54	77	94	115
	Exports (Mt)	44	63	77	92
	Imports (Mt)	22	29	36	48
	Port capacity (Mt)	94	113	133	162
	for exports	*55*	*74*	*90*	*109*
	for imports	*39*	*39*	*42*	*53*
Electricity	Generating capacity (GW)	180	256	357	492
	Coal	*6*	*9*	*12*	*19*
	Oil	*30*	*32*	*32*	*26*
	Gas	*28*	*68*	*124*	*219*
	Nuclear	*3*	*3*	*4*	*4*
	Hydro	*109*	*139*	*175*	*205*
	Other renewables	*4*	*6*	*11*	*19*
	Urban population without electricity (million)	6	1	0	0
	Rural population without electricity (million)	50	35	28	23
Indicators	GDP (billion dollars)	1,748	2,339	3,171	4,219
	Population (million)	416	477	534	584
	Population density (persons/km^2)	22	26	29	32
	Electrification rate (%)	87	93	95	96
	Energy/capita (toe/capita)	1	1	1	2

Reference Scenario: **Middle East**

	Investment (billion dollars)	**2001-2010**	**2011-2020**	**2021-2030**	**2001-2030**
	Total Regional Investment	**268**	**332**	**444**	**1,044**
Oil	**Total**	**123**	**166**	**234**	**523**
	Exploration and development	87	129	193	408
	Non-conventional oil	3	7	7	16
	Refining	34	30	34	99
Gas	**Total**	**73**	**83**	**106**	**263**
	Exploration and development	33	46	60	140
	LNG liquefaction	19	12	16	46
	LNG regasification	–	–	–	–
	Transmission	17	19	25	61
	Distribution	3	5	4	12
	Underground storage	1	2	1	4
Coal	**Total**	**0.0**	**0.1**	**0.1**	**0.2**
	Total mining	0	0	0	0
	new mining capacity	*0*	*0*	*0*	*0*
	sustaining mining capacity	*0*	*0*	*0*	*0*
	Ports	0.0	0.0	0.1	0.1
Electricity	**Total**	**71**	**83**	**103**	**258**
	Generating capacity	24	29	40	92
	of which renewables	*9*	*5*	*4*	*18*
	Refurbishment	5	5	6	15
	Transmission	14	15	18	47
	Distribution	29	34	40	103

Reference Scenario: **Middle East**

	Supply and Infrastructure	2000	2010	2020	2030
Oil	Oil production (mb/d)	23	28	40	53
	Net trade (mb/d)	–19	–23	–33	–46
	Refinery capacity (mb/d)	6	10	13	16
Gas	Gas production (bcm)	223	421	619	861
	Net trade (bcm)	–21	–139	–229	–365
	LNG liquefaction capacity (bcm)	33	147	232	368
	LNG regasification capacity (bcm)	–	–	–	–
	Transmission pipelines (thousand km)	23	47	74	110
	Distribution pipelines (thousand km)	82	145	217	243
	Underground storage working volume (bcm)	0	5	12	15
Coal	Coal production (Mt)	2	2	2	2
	Exports (Mt)	0	0	0	0
	Imports (Mt)	11	13	17	21
	Port capacity (Mt)	17	17	19	23
	for exports	*0*	*0*	*0*	*0*
	for imports	*17*	*17*	*19*	*23*
Electricity	Generating capacity (GW)	120	160	214	277
	Coal	*4*	*6*	*8*	*11*
	Oil	*53*	*65*	*79*	*96*
	Gas	*56*	*77*	*113*	*154*
	Nuclear	*0*	*1*	*1*	*1*
	Hydro	*6*	*10*	*12*	*14*
	Other renewables	*0*	*1*	*1*	*3*
	Urban population without electricity (million)	2	0	0	0
	Rural population without electricity (million)	13	8	6	3
Indicators	GDP (billion dollars)	633	915	1,317	1,839
	Population (million)	165	218	272	327
	Population density (persons/km²)	33	44	55	66
	Electrification rate (%)	91	96	98	99
	Energy/capita (toe/capita)	2	2	2	2

Reference Scenario: **Africa**

	Investment (billion dollars)	**2001-2010**	**2011-2020**	**2021-2030**	**2001-2030**
	Total Regional Investment	**248**	**393**	**567**	**1,208**
Oil	**Total**	**82**	**130**	**149**	**360**
	Exploration and development	75	112	125	311
	Non-conventional oil	1	3	4	7
	Refining	7	15	20	42
Gas	**Total**	**48**	**69**	**99**	**216**
	Exploration and development	28	50	75	153
	LNG liquefaction	10	8	9	27
	LNG regasification	–	–	–	–
	Transmission	9	10	14	33
	Distribution	1	1	1	3
	Underground storage	0	0	0	1
Coal	**Total**	**6**	**7**	**9**	**22**
	Total mining	6	7	9	22
	new mining capacity	*3*	*3*	*4*	*10*
	sustaining mining capacity	*3*	*4*	*5*	*12*
	Ports	0.1	0.2	0.1	0.4
Electricity	**Total**	**112**	**187**	**310**	**609**
	Generating capacity	40	62	104	206
	of which renewables	*12*	*17*	*28*	*57*
	Refurbishment	3	4	7	13
	Transmission	24	39	61	123
	Distribution	46	82	138	266

Reference Scenario: **Africa**

	Supply and Infrastructure	**2000**	**2010**	**2020**	**2030**
Oil	Oil production (mb/d)	6	9	12	13
	Net trade (mb/d)	–4	–6	–8	–8
	Refinery capacity (mb/d)	3	4	5	7
Gas	Gas production (bcm)	130	246	389	589
	Net trade (bcm)	–74	–145	–212	–299
	LNG liquefaction capacity (bcm)	43	103	160	235
	LNG regasification capacity (bcm)	–	–	–	–
	Transmission pipelines (thousand km)	20	38	58	86
	Distribution pipelines (thousand km)	40	64	85	97
	Underground storage working volume (bcm)	0	0	1	2
Coal	Coal production (Mt)	230	272	338	415
	Exports (Mt)	71	83	103	110
	Imports (Mt)	8	9	10	13
	Port capacity (Mt)	99	109	127	140
	for exports	*87*	*96*	*115*	*125*
	for imports	*12*	*12*	*12*	*14*
Electricity	Generating capacity (GW)	103	160	254	400
	Coal	*36*	*46*	*64*	*99*
	Oil	*24*	*36*	*54*	*81*
	Gas	*19*	*49*	*96*	*165*
	Nuclear	*2*	*2*	*2*	*2*
	Hydro	*21*	*27*	*32*	*39*
	Other renewables	*1*	*2*	*5*	*15*
	Urban population without electricity (million)	111	150	188	206
	Rural population without electricity (million)	412	449	465	445
Indicators	GDP (billion dollars)	646	833	1,092	1,406
	Population (million)	795	997	1,231	1,489
	Population density (persons/km^2)	26	33	40	49
	Electrification rate (%)	34	40	47	56
	Energy/capita (toe/capita)	0.3	0.3	0.4	0.5

ANNEX 2: METHODOLOGY

Overview

This annex explains the methodology used to estimate the investment requirements presented in this report. The time frame of the analysis is to 2030. The study estimates the supply-side investments in the oil, gas, coal and electricity sectors. Investment is defined as capital expenditure only; it includes expenditure necessary to sustain production levels, but does not include spending that is usually classified as operational and maintenance. Investments in equipment and infrastructure involving the use of final energy are not included.

Except where alternative scenarios are being explicitly addressed, the calculation of the investment requirements is based on the supply and demand projections presented in the Reference Scenario of the *World Energy Outlook 2002*. The methodology adopted for calculating the investment required in each supply chain element involved, for each fuel and region,[1] the following steps:

- New-build capacity needs for production, transportation and (where appropriate) transformation were calculated on the basis of projected supply trends, estimated rates of retirement of the existing supply infrastructure and decline rates for oil and gas production.
- Unit cost estimates were compiled for each component in the supply chain. These costs were then adjusted for each year of the projection period using projected rates of change based on a detailed analysis of the potential for technology-driven cost reductions and on country-specific factors.

Incremental capacity needs were multiplied by unit costs to yield the amount of investment needed. The methodology used for each fuel and each step of the supply chain is explained below.

Oil Sector Investment

The investment requirements in Chapter 4 cover capital cost associated with the following sectors of the industry:

- Conventional oil exploration and development.
- Non-conventional oil and gas-to-liquids.
- Oil tankers and pipelines.
- Oil refining.

1. See Annex 3 for the definition of WEO regions.

Investment associated with retail and storage activities has not been included.

A detailed description of the methodology utilised for each component of the supply chain is provided below.

Exploration and Development

Global oil and gas exploration and development investment has been calculated as the sum of upstream spending in each of the 22 world regions identified in the Oil and Gas Supply Module of the *WEO-2002*.[2] For each of these regions, the costs of exploration and development activities have been assessed separately and on the basis of whether they are occurring onshore or offshore to reflect the different unit costs incurred. Exploration costs include all investment that is needed before a discovery is confirmed, including geophysical and geological analysis, and drilling of exploration wells. Development costs cover all spending after a discovery is confirmed, such as drilling of production wells and purchase and installation of surface equipment.

Exploration Investment

For each region, investment in exploration activities in onshore and offshore acreages is calculated as the product of the estimated unit cost for exploration and the volume of new discoveries made. Unit costs for exploration are calculated as a function of unit development costs, and take account of exploration success rates.

Discoveries are estimated as a logistic function over time, to reflect that in new areas discoveries increase rapidly but then slow down as the area matures. The maximum values for discoveries are constrained by the estimates for the regional level of ultimate recoverable resources.

$$CD_i = URR_i / [1 + \exp(a_i * t + b_i)]$$

Where:
CD_i = cumulative discoveries in each region
URR_i = ultimate recoverable resources
a_i, b_i = parameters

The parameters of the discoveries function are based on the regional historical evolution of discoveries, from 1962 to 2002. The parameters take into account the level of exploration activity and regional differences in production policy.

Development Investment

For each region, investment in development activities in onshore and offshore fields is calculated as the product of the estimated unit cost for development and the volume of new capacity additions.

2. See IEA (2002).

Capacity additions are calculated as the sum of capacity added to meet production growth and capacity added to replace decline in existing capacity:

new capacity = production (t) – production (t-1)
+ capacity (t-1) * (1-decline rate/100)

Additional capacity to meet production growth is calculated as the difference between new production and previous existing capacity and draws on the regional oil and gas production projections of the *WEO-2002*. Capacity needed to replace decline is calculated as a function of existing production capacity and decline rate. Assumed decline rates range from 5% per year to 11% per year. These vary regionally and over time, according to the maturity of the region.[3]

To estimate unit development costs in the base year, this study drew on data from a wide range of sources, including commercial databases, private and national oil and gas companies, international organisations — including OPEC — and literature surveys. The resulting database includes costs for different types of onshore and offshore locations and different field sizes on a regional and, in some cases, country-by-country basis. The assumed evolution of unit costs over time depends on two counterbalancing factors:

- The evolution of average field sizes over time: the smaller the fields, the higher the unit costs. Projections of average field size in any given region are based on the historical evolution of field size and the geology.
- Technology improvement, which tends to lower unit costs over time.

Non-conventional Oil and Gas-to-Liquids

The investment assessment for the non-conventional oil (NCO) and gas-to-liquids (GTL) sectors includes spending for new developments and for sustaining production at existing and new projects. The *WEO-2002* projections for incremental supply of NCO and GTL form the basis of the investment calculations in this sector. Unit costs for NCO and GTL projects, detailed in Chapter 4, have been derived from a review of historical investment spending and discussion with relevant industry experts and government officials. These costs vary by region and by project type and decline over time (at an assumed learning rate) reflecting the potential for further technological progress in these sectors.

Oil Tankers and Pipelines

The assessment of the investment required in inter-regional oil transportation is underpinned by a matrix of trade projections, through to

3. See Chapters 4 and 5 for decline rates assumptions.

2030, prepared from the regional oil and refined products demand and supply projections contained in the *WEO-2002* as well as other IEA databases. A matrix of the share of trade, for oil tankers and pipelines through to 2030 was also prepared, based on the likely evolution of the different markets. Volumes transported by pipelines and tankers were then calculated as the product of these two matrices. Investment associated with domestic and intra-regional oil transportation and port capacity expansion is not included.

Investment required for oil tankers comprises spending on vessels greater than 25,000 DWT for crude oil and refined products. This includes spending to increase capacity to accommodate growth in trade and to replace tankers phased out of service because of environmental regulations or age constraints.

Required oil-tanker capacity was derived as a function of inter-regional trade projections and other variables related to tanker transportation, such as voyage distances, cruising speeds and time spent in port and on routine maintenance. The trade projections utilised in this calculation reflect the changing trade patterns expected through to 2030 and the subsequent impact of this on the structure of the oil-tanker fleet.

A schedule for removing tankers from service to meet environmental regulations and age constraints was formulated. This schedule progressively phases single-hull tankers out of service by 2015 and applies a maximum lifespan of 25 years for all tankers.

The amount of oil-tanker capacity required was calculated as the sum of capacity needed to meet new demand and to replace tankers phased out of service. The associated investment was derived from this figure by applying unit costs for each new tanker.

As for oil tankers, the investment calculated for oil pipelines comprises inter-regional trade flows. Investment for domestic and inta-regional trade has not been considered. Short-term pipeline-investment requirements have been determined through a review of expected spending on currently gazetted projects. Long-term pipeline investment has been calculated as a function of the projected growth in trade by pipeline and average unit pipeline costs.

Oil Refining

Investment projections for the refining sector comprise spending on increasing refining capacity, on improving conversion and product quality treatment capability.

Investment for increasing refining capacity has been determined as a function of the *WEO-2002* refining capacity projections and unit refinery costs. The *WEO-2002* refining capacity projections are based on demand for refined products, past trends in refinery construction, currently announced plans for additional capacity and existing surplus capacity. The assumed unit refinery costs vary regionally reflecting differences in refinery types, construction costs and differing reliance on expansion through capacity creep as opposed to construction of completely new refineries.

Investment for refinery conversion capacity has been derived from the *WEO-2002* projections for the evolution of the refinery product slate and unit costs for conversion processes. These unit costs vary on the basis of the severity of the conversion required – they are highest in regions with high existing conversion capability as deeper conversion processes are necessary.

Investment requirements for product quality improvements have been derived from published cost estimates for meeting, worldwide, the fuel quality standards currently mandated (but not necessarily yet introduced) in various major markets including Europe, North America and Asia-Pacific. This assessment assumes that through the projection period, other world regions will adopt similar standards and that similar unit costs will be involved. The unit costs applied to different regions have been adjusted to reflect the relative quality of current refinery output. Investment associated with reducing environmental emissions from refineries and further fuel quality improvements beyond those already gazetted in major OECD markets was not considered.

Natural Gas Sector Investment

The natural gas-supply investment requirements in this report include the following gas supply chain elements:

- Exploration and development.
- New gas transmission pipelines.
- New LNG liquefaction plants, ships and regasification terminals.
- New gas distribution pipelines.
- New underground storage facilities.

They do not include investment in downstream refurbishment, because in many cases these investments are classified as operating expenditures not as capital. Moreover, region-specific unit costs and regional figures for capital assets depreciation proved in most cases unavailable or unreliable.

The methodology adopted for calculating the investment required in each gas supply chain element is provided below. The methodology for exploration and development investment is the same as that used for oil (see above).

Trade Flows

The calculation of investment needs in the liquefied natural gas (LNG) chain and in gas transmission pipelines required the estimation of future evolution of gas trade patterns and volumes. Gas trade flows among the WEO regions are based on the *WEO-2002* gas supply and demand projections. The breakdown of the trade flows into LNG and pipeline is done according to these rules:

- Accordance with existing long-term contracts, planned or under construction LNG and pipeline projects.
- Determination of the least costly option between LNG and pipeline transportation in relation to distance and terrain.
- Minimisation of transportation distances.

The additional capacity required for LNG and transmission pipelines over the projection period is estimated as a function of the increase in volumes, utilisation rates and installed capacity in the base year.

Gas Transmission Pipelines

Regional investment requirements in gas transmission pipelines are calculated by multiplying additional pipeline kilometres by a unit cost.

The lengths of additional transmission networks in each region are estimated yearly, with a dynamic model, as a function of the following variables: incremental gas demand, incremental transit flows and exports, installed capacity in the base year and utilisation rates.

Gas transmission pipeline capacities and lengths in place in the base year were compiled by gathering country data from companies, governments and literature sources. For regional aggregates, a database on the past evolution of gas transmission capacities and lengths was also compiled.

Investments in recent and planned gas pipelines were gathered, together with diameters, capacities and lengths. Country average unit costs were then estimated.

Table A2.1 shows some cost assumptions used in this report. Unit costs in the United States and Korea, where safety standards are high and labour is expensive, are higher than in developing countries. Costs in Ukraine and Russia are lower because of lower-quality materials, engineering standards and labour costs. Unit onshore pipeline costs are assumed to remain constant over the projection period. Unit offshore pipeline costs are expected to continue to decline, but at a lower rate than in the past.

Table A2.1: **Average Unit Cost of Gas Transmission Pipelines in Selected Countries**

	Weighted average* **diameter** (inches)	**Total length** (km)	**Unit cost** (M$/inch/km)
United States	24	463,000	26
Korea	28	2,066	21
Russia	36	150,000	14
Ukraine	32	36,700	16
Brazil	23	7,700	20
Argentina	30	12,800	21

* Diameters are weighted by the lengths.

LNG

LNG investments are calculated separately for liquefaction, ships and regasification. Investments in each element of the chain are calculated by multiplying the additional capacity by a unit cost.

Projections of LNG liquefaction capacities from 2001 to 2030 are a function of regional LNG incremental gas export volumes, utilisation rates and base year installed capacity. Projections of LNG regasification capacities are similarly a function of regional LNG incremental gas imports, utilisation rates and base year installed capacity. The number of additional LNG ships is estimated as a function of incremental LNG trade volumes, average ship capacity and cruising speeds. Average ship capacity is assumed to rise steadily over time. Cruising speed for a given trajectory is assumed to remain constant at its base year value.

The methodology used to estimate the evolution of unit costs is the experience curve method.[4]

$$UC(t) = UC(t_0) * CC^{(-a)}$$

Where:
$UC(t)$ = unit cost at time t
$UC(t_0)$ = unit cost in the base year
CC = cumulative capacity
a = experience parameter

Experience parameters were estimated for each element of the LNG chain. The unit cost in the base year for each link of the chain was estimated from literature surveys and companies' information. Four technology learning scenarios were developed to bracket the effects of technology learning on the investment estimates. The most probable value was chosen for the LNG investment presented in this report.

Gas Distribution

The methodology used for distribution is similar to that used for transmission. Regional projections of additional distribution pipelines are derived from a dynamic model based on residential gas demand projections and network densities.

4. Learning (experience) curve is the analytic tool which describes how unit cost declines with cumulative capacity. The learning rate is expressed as a percentage, namely, the percentage cost reduction for each doubling of capacity. For a detailed discussion, see IEA (2000).

Gas distribution pipeline capacities and lengths in place in the base year were compiled by gathering country data from companies, governments and literature sources. For regional aggregates, a database on the past evolution of gas transmission capacities and lengths was also compiled. Table A2.2 shows some cost assumptions used in this report. Distribution unit costs are assumed to increase slightly over time, as a result of stricter environmental regulation, increasing urbanisation and tougher safety standards.

Table A2.2: **Average Unit Cost of Gas Distribution Pipelines in Selected WEO Regions**

	Average cost (M$/km)	Total length (km)
EU15	0.07	1,189,327
United States and Canada	0.08	2,022,596
Russia	0.03	540,000
Indonesia	0.07	26,814
India	0.04	36,126
Brazil	0.14	11,767

Underground Gas Storage (UGS)

The quantification of investment requirements in UGS involved the projection of future regional UGS working volume requirements. These projections are based on the ratios of storage capacity to gas demand and to the development of the gas market. If the residential-commercial sector share rises, for instance, this implies a greater seasonal demand, which creates demand for new storage facilities. The strategic position of some countries in terms of transit or exports, where gas storage is also used to optimise long-distance gas line management, requires an additional ratio to take into account these incremental storage needs.

In countries where no existing UGS facilities are currently installed, UGS capacity projections are based on planned and potential projects, expert judgements and country-specific factors.

Investment projections are estimated by multiplying the additional working volume needed by unit capital costs. Unit costs were compiled by region. They are a function of the average size of storage facility and the storage types (see Chapter 5).

Coal Sector Investment

The investment requirements for the coal sector presented in this report include:

- Investment in new production capacity, comprising two components:
  – new capacity to meet demand growth; and
  – new capacity to replace depleted production capacity.
- Investment required to sustain production capacity at new and existing mines.
- Investment in coal import- and export-handling facilities at ports.
- Investment in the dry-bulk cargo fleet for coal trade.

They do not include investment in the existing transport networks of rail, road and waterways as insufficient data were available to permit the calculation of reliable investment figures.

Coal Demand and Trade

The investments required in the coal industry over the next thirty years depend on the level of production and the pattern of coal trade. These two factors determine the level of investment needed in new capacity and the infrastructure requirements of moving the coal to its end-users.

The *WEO-2002* Reference Scenario included detailed projections of the primary energy demand for coal. This report has expanded on these projections of demand and includes detailed projections of production, imports and exports by major producing and consuming country (not just by region). For the more important producers, production is also broken down by coal type.

The trade patterns for the next thirty years therefore contain all coal trade between countries and not simply inter-regional trade flows. This allows a more detailed assessment of the location of production to be joined with country-specific investment cost assumptions in order to arrive at a more accurate estimate of investment needs.

World seaborne coal trade was derived from projections of total world coal trade. In addition, separate projections of the internal seaborne trade in coal in China, India and Indonesia were also made. These are not included in the total world trade figures.

Mining Investment

The projections of production by country/region are input into a model framework that determines the new mining capacity required and calculates the level of sustaining production investment required.

Figure A2.1: **Investment Model for Coal**

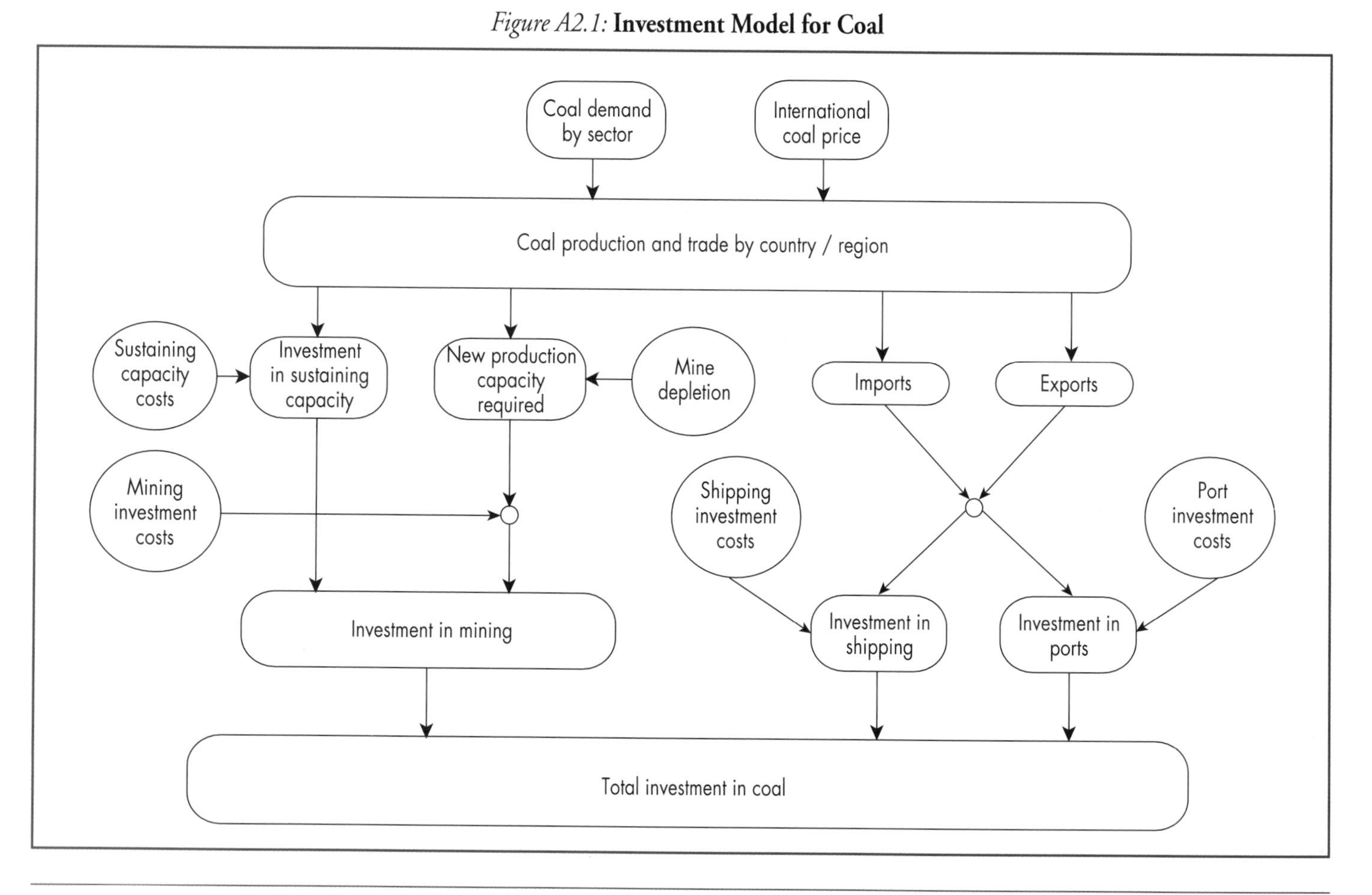

New Production Capacity

The need for new production capacity is split into two components:

- Replacement capacity investment – this is investment required in order to replace mine capacity that exhausts its economic reserves in a given year.
- Expansion capacity investment – this is the investment needed in new capacity in order to meet demand growth from 2000 levels.

In both cases the model does not distinguish from which type of mine the new capacity will come (greenfield, brownfield/expansion or productivity investment). It simply identifies the level of capacity that needs to be built. It is modelled assuming a one-off cost in dollars per tonne of capacity.

The one-off cost in dollars per tonne of new production capacity is an average development cost for recent new greenfield and brownfield/expansion developments reported for a country/region.

The calculation of replacement capacity requires assumptions on the rate at which existing capacity is closed. This is an important component of overall investment and is modelled as a percentage of the year 2000 capacity retired in each year.[5] For modelling purposes, mine depletion rates are defined as the amount of productive capacity retired for economic reasons in each year which requires investment in new replacement capacity.

This does not necessarily mean that coal reserves are exhausted in a given mine area, or that a new mine in a geographically separate location is needed. It simply reflects circumstances where, in order to continue production, a level of investment equivalent to a new capacity investment is required.

The rate of closure of 2000 capacity is determined by recent and near-term mine closure information, adjusted where necessary if the rate of closure is expected to be higher or lower in the future.

The total of new capacity needed to meet demand growth and to replace capacity that is closed is then multiplied by the specific investment cost for new capacity per tonne to arrive at the mining investment required for new capacity.[6]

In general, the investment costs for new capacity are assumed to be constant over the *Outlook* period. Exceptions are made where recent experience is not likely to be repeated over the *Outlook* period. In some countries, where data are available and there are significant differences in new mine development costs,

5. This is different from the decline rates used in the gas and oil chapters. The output of coal mines generally does not decline in a systematic way from when peak output is reached. The economics of coal mining means that mines tend to operate at design capacity throughout most of their anticipated life and close when production potential starts to decline.

6. The exception is for the EU15, where investment is calculated by using an average total mining investment figure per tonne of production, rather than per tonne of new capacity. This different approach is used in order to allow for the difficulty of modelling retirement rates relative to new capacity needs when production is declining.

separate assumptions are used for different coal types and/or regions. A summary of the specific assumptions for coal can be found in Table 6.4.

The new capacity cost assumptions, in general, include the investment required to connect the new development to the existing transport infrastructure. This might include access roads, conveyor systems, a branch rail line, a river port facility, etc.

Sustaining Investment

The second component of mining investment, sustaining investment, is that required to maintain productive capacity at all mines, including those built over the *Outlook* period. It can also in some cases result in improvements in productivity. It is investment in accessing new reserves within a mine, replacing worn capital equipment, extending infrastructure, etc. This investment is closely linked to annual production, rather than capacity. It is therefore modelled using an average dollar per tonne of production in each year.

This investment reflects the fact that mining equipment has a much shorter life than the mine itself and will need to be replaced over the life of the mine. It also includes the often significant additional development expenditure required over the life of the mine to access additional reserves. This investment does not include depreciation of mining equipment or expenditure of an operational nature.

The level of sustaining investment required per tonne of production varies by region depending on a number of factors, including mechanisation levels, geological conditions, predominant mine type, financial health of the industry and the intensity of competition faced domestically and internationally.

Ports

Investments in ports are calculated by taking the required additional capacity of import- or export-handling facilities and multiplying it by a specific investment cost per tonne of annual capacity.

Regional coal import and export capacities at ports are established from the IEA's Coal Information and other sources. These are then compared to export or import levels to determine if new capacity is required. The model maintains a capacity "operating margin" of around 10% above imports and exports, although this varies by region.

In addition, the port requirements for the internal shipping of coal in China, India and Indonesia are also allowed for. The quality of data available is, however, not as high as for the capacity of ports for international trade. There is therefore a higher uncertainty surrounding these estimates of investment, than for ports for international trade.

The cost of port handling facilities is taken from publicly available information on port developments. It is an average of the cost of both expansions of existing port facilities and of new port developments. The costs lie within a reasonably narrow range for the major importing and exporting regions of around $13 to $18 per tonne of annual capacity.

Shipping

International seaborne trade in coal is calculated by removing from total cross-border trade coal that is projected to be railed or shipped by inland waterways to neighbouring countries.

An estimate of the total dry-bulk cargo fleet required in order to service this trade in coal was calculated for the year 2000. This was done by calculating the capacity in tonnes of coal per year that one dead weight tonne of ship capacity could transport between different regions (see below). This required making assumptions regarding average distance between each import and export region, the useable capacity of a ship's tonnage, steaming speeds, number of days in service per annum, as well as the days required for loading and unloading.

$$SI = \Sigma\,(ST_c \times SP_c) \times DWT$$

$$DWT = [(\Sigma\,EX_i \div CDWT_i)_{2030} - (\Sigma\,EX_i \div CDWT_i)_{2000} \times CDF]$$

$$CDWT_i = (365 - L) \div (D_i \div MT + 2 \times UT)$$

Where:

SI	shipping investment	[$ million, 2000 values]
ST_c	share of ship	[Capesize, Panamax, Hanymax, Handysize]
SP_c	ship price	[Capesize, Panamax, Hanymax, Handysize]
DWT	additional DWT	[million tonnes]

And:

EX_i	exports	[by trade flow]
$CDWT_i$	annual capacity of one dead weight tonne	[by trade flow]
CDF	cumulative losses and disposal factor	[fraction]

And:

L	annual service/down time	[days]
D_i	distance	[nautical miles by trade flow]
MT	distance travelled in one day	[nautical miles]
UT	loading/unloading time	[days]

When combined with the level of trade between regions, these factors allow the total dead weight tonnage of capacity required to service trade for that year to be calculated. When summed over all exports, this yielded an estimate of the dry-bulk cargo fleet required for the coal trade in each year of the *Outlook* period. With an assumption regarding the rate of losses and disposals of this fleet, a total additional fleet capacity figure in dead weight tonnes was arrived at for the *Outlook* period.

The share of this fleet between different classes of ship was assumed to remain constant over time. Assumptions regarding the cost of each ship type were then used to calculate the total investment required over the *Outlook* period. Table A2.3 details the assumptions used.

Table A2.3: **Assumptions for Shipping Investment for Coal** ($ million)

	Capesize (170,000 DWT)	**Panamax** (75,000 DWT)	**Hanymax** (51,000 DWT)	**Handysize** (30,000 DWT)
Cost	38.9	22.7	20.6	15.0

Electricity Sector Investment

The electricity sector investment requirements in this report include:

- Investment in new plant (including CHP plant).
- Investment in refurbishment of existing plant.
- Investment in transmission networks (including replacement and refurbishment).
- Investment in distribution networks (including replacement and refurbishment).

They do not include:

- Investment in nuclear fuel processing, waste and plant decommissioning.
- Investment in CHP distribution networks.

These areas of investment have not been included in the projections presented in this report because of lack of data allowing for reliable investment estimates. However, they are only a small component of electricity-sector investment and therefore their exclusion does not have any significant impact on total electricity-sector investment.

Investment in New Power Plant

These investments are largely based on the *WEO-2002* and are derived from the WEM. The Power Generation Module of WEM takes into account capital costs, operating and maintenance costs, fuel costs and efficiency to calculate new plant costs. These costs are used to calculate the future mix of technologies and the corresponding new capacity for the following types of plant:

- Coal, oil and gas steam boilers
- Combined cycle gas turbine
- Open cycle gas turbine
- Integrated gasification combined cycle
- Oil and gas internal combustion
- Fuel cell
- Nuclear
- Pumped storage hydro
- Conventional hydro
- Bioenergy
- Geothermal
- Wind onshore
- Wind offshore
- Solar photovoltaic
- Solar thermal
- Tide and wave

The investment requirements are derived by multiplying the capacity requirements (by technology) by the corresponding capital cost. Capital costs for a given technology are region- and time-dependent. They decline over time, depending on the technology, its maturity and its market penetration. Current capital cost estimates are given in Table 7.3.

Renewables

The calculation of investment needs for renewables in electricity generation combines the *WEO-2002* capacity additions with specific investment costs by technology.[7] The technologies considered are bioenergy, hydro, wind onshore, wind offshore, geothermal, photovoltaic (PV), solar thermal, and tide/wave. The capacity additions include replacement capacity for those installations that reach the end of their lifetime some time during the forecast period.

The investment costs per unit of installed capacity take into account learning effects by incorporating learning curves into the methodology.[8]

The learning rates used in the investment calculations are shown in Table A2.4. These figures are based on a comprehensive literature survey. Different rates have been assumed for each decade of the 30-year projection period. In

7. The methodology was developed by the Technical University of Vienna for the IEA.
8. See footnote 4 in this annex.

general, learning rates are assumed to decline over time. Experience has shown that the rate of technological learning is often closely linked to the development stage of a certain technology. At an early stage of development, learning rates are high. In later stages of development, when the technology matures, learning rates slow down. No learning has been assumed for hydro plants, since the technology is already mature.

Table A2.4: **Learning Rate Assumptions** (%)

	2001-2010	2011-2020	2021-2030
Hydro	0	0	0
Wind onshore	10	10	5
Wind offshore	10	10	5
Geothermal	2	2	2
Bioenergy	15	10	10
PV	15	10	10
Solar thermal	15	10	10
Tide/wave	20	15	15

In addition to learning effects, regional differentiation of investment costs has also been taken into account. Assumed project investment costs include a fraction of country-specific labour and other region-specific cost components. This fraction is assumed to be 20% for hydro and 15% for all other technologies. These local costs are adjusted using the per capita GDP of the specific region.

Power Plant Refurbishment

Investment in refurbishment of existing plants is based on the assumption that major refurbishment will take place once during the lifetime of a plant during the next thirty years. Only plants that will not be retired during this period are considered. The cost assumptions used in this calculation are shown in Table A2.5.

Table A2.5: **Refurbishment Cost Assumptions** ($ per kW)

	Boiler	Turbine	Nuclear	Hydro
OECD	200-300	100	500	100
Transition economies	150-300	50-100	300	50-100
Developing countries	200	100	500	100

Electricity Transmission and Distribution[9]

The methodology applied in this study is based on historical investments in the transmission and distribution grid. These investments include replacement and refurbishment. Using these historical investments, a set of models was developed to calculate specific investment costs in the transmission and distribution grid and to project future investment needs. The specific transmission and distribution costs are calculated over a period of time using investment during this period and the corresponding increase in electricity generation during this period.

Long time series of investments in the transmission and distribution grid are available for the following countries:

- Austria
- Germany
- Japan
- Norway
- Switzerland
- United States

Historical information for investment in transmission and distribution exists for a number of other countries (mostly OECD countries and a few developing countries), but only for recent years.

Transmission

Total transmission investment costs (TTMIC) are a function of:

- Specific transmission investment costs
- Electricity demand increase
- Share of central generation
- Transmission grid usage
- Share of labour
- Labour cost factor
- Topology factor
- Supply security factor of transmission

The specific investment costs are corrected by the labour cost factor, topology factor, supply security factor transmission and the change in the usage of the grid. For each factor, except the supply security factor, transmission in Europe is the baseline. The labour cost factor and topology factor are assumed to be one for Europe and the supply security factor for transmission is assumed to be one for North America. The specific transmission costs are corrected using the following equations:

9. The models for estimating investment in transmission and distribution were developed by the Technical University of Vienna for the IEA.

$STMCC =$

$$STMIC \times [(1 - SL/100) + LCF \times SL/100] \times TF \times SSFT \times (1 - \Delta TMU/100)$$

$\Delta\mathrm{TMU} = \mathrm{TMU}_{2030} - \mathrm{TMU}_{2000}$

Where:

STMIC	specific transmission investment costs	[Mill. €/GWh, 2000 values]
STMCC	specific transmission costs corrected	[Mill. €/GWh]
SL	share of labour	[%]
LCF	labour cost factor	[/]
TF	topology factor	[/]
SSFT	supply security factor transmission	[/]
TMU	transmission grid usage	[%]

Total transmission investment costs are calculated for the period 2001-2030 using the following equations:

$CGF = 1 + \Delta SCG/100$

$\Delta SCG = SCG_{2030} - SCG_{2000}$

$TTMIC = STMCC \times \Delta D \times 1000 \times CGF$

Where:

ΔD	electricity demand increase	[TWh]
ΔSCG	change of share of central generation	[%]
CGF	central generation factor	[/]

The central generation factor (CGF) reflects the change of costs with respect to the share of central generation over a given period of time.

Distribution

The methodology to calculate investment in distribution is similar to that used for transmission. Total distribution investment costs (TDIC) are a function of:

- Specific distribution investment costs
- Electricity demand increase
- Share of distributed generation
- Distribution grid usage
- Share of labour
- Labour cost factor
- Topology factor
- Supply security factor distribution

Specific distribution costs are corrected using the following equations:

$$SDCC = SDIC \times [(1 - SL/100) + LCF^{*}SL/100] \times TF \times SSFD \times (1 - \Delta DU/100)$$
$$\Delta DU = DU_{2030} - DU_{2000}$$

Where:

SDIC	specific distribution investment costs	[Mill. $/GWh]
SDCC	specific distribution costs corrected	[Mill. $/GWh]
SL	share of labour	[%]
LCF	labour cost factor	[/]
TF	topology factor	[/]
SSFD	supply security factor distribution	[/]
DU	distribution grid usage	[%]

Total investment in distribution is based on the following equations:

$$DGIF = 1 - \Delta DGI/100)$$
$$\Delta DGI = DGI_{2030} - DGI_{2000}$$
$$TDIC = SDCC \times \Delta D \times 1000 \times DGIF$$

Where:

ΔD	electricity demand increase	[TWh]
ΔDGI	change in distributed generation	[%]
DGIF	isolated distributed generation factor	[/]
TDIC	total distribution investment costs	[Mill. $]

The distributed generation factor (DGIF) reflects the change of costs regarding the change of share of distributed generation from 2000 to 2030. An increase of distributed generation in 2030 compared to 2000 leads to a $\Delta DGI > 0\%$.

Universal Electricity Access

Model Description

The Electrification Scenario investment requirements calculations are based on three main determinants:

- Number of people gaining access to electricity over time.
- Consumption patterns over time.
- Costs.

The number of people gaining access over time under the Reference Scenario was estimated in the *WEO-2002*. In order to reach 100% electrification by 2030, the Electrification Scenario projects an exponential addition of people gaining access over time. This exponential distribution was chosen to be consistent with the system for building electricity network extensions.

Basic Consumption

The Electrification Scenario is demand-based and assumes a basic consumption for each person gaining access, as well as its evolution over time. The basic electricity consumption in rural areas corresponds to the minimum vital consumption, estimated to be 50 kWh per person per year. The minimum urban electricity consumption is set at 100 kWh per person per year. This higher consumption in urban areas reflects specific urban consumption patterns.

This basic consumption is a starting point which changes over time, reflecting the overall income-generating effects of electricity on the populations considered. People gaining access to electricity will start by consuming electricity for their basic needs (lighting and communication) and will then increase their consumption to meet new needs (cooling, water pumps, mills, etc.). Of course, this is an average, as the effects of electricity are not immediate and will not affect all the newly electrified population – electricity alone does not guarantee an income-generating process. The Electrification Scenario assumes that the consumption per capita of an individual catches up the average consumption of his/her region over time. Therefore, although the starting consumption is the same across regions, there is a regional difference in the "catching-up" process, depending on the average consumption of each region under the Reference Scenario and the rural/urban disparity. The model assumes that generation in urban areas is made through grid options. The breakdown for rural generation is given in the table below.

Table A2.6: **Rural Additional Generation Breakdown** (%)

	Grid extension	Mini-grid	Isolated off-grid	Total
Africa	60	35	5	100
South Asia	60	35	5	100
East Asia	30	61	9	100
Latin America	15	74	11	100
Middle East	5	83	12	100
Developing countries	**53**	**41**	**6**	**100**

Associated Investment Costs

Once the new level of demand is determined, and therefore the generation needs are estimated, the Electrification Scenario assumes a breakdown by power generation options. The generation options have been chosen according to the 100% electrification target in 2030. The Electrification Scenario, as a global plan, sets the most cost-efficient and likeliest picture for 2030 and defines generation addition patterns accordingly.

For urban-area electricity demand, the less costly choice is electricity grid extension. That part of the rural area – around 50% of total rural demand – closest to urban areas and/or likely to become more densely populated by 2030 is also projected to be supplied through the grid, as this will be the most economic option.

To evaluate the cost associated with grid generation, the Electrification Scenario feeds the added required generation to 2030 into the power generation module of the WEM and the transmission and distribution model of the Reference Scenario.

The remaining rural generation is off-grid, divided between mini-grids, which will constitute the bulk of off-grid generation, and isolated off-grid generation for the remotest populations. The average base costs per kW installed are given in Table A2.7.

Table A2.7: **Average Costs per kW Installed**

	Average cost ($/kW)
Mini-grid	4,000
Isolated off-grid	8,000

ANNEX 3: DEFINITIONS, ABBREVIATIONS AND ACRONYMS

This annex provides definitions of the energy, economic and financial terms and the regional groupings used throughout the study.

FUEL AND PROCESS TERMS

Readers interested in obtaining more detailed information should consult the annual IEA publications *Energy Balances of OECD Countries*, *Energy Balances of Non-OECD Countries*, *Coal Information*, *Oil Information*, *Gas Information* and *Electricity Information*.

API Gravity

Specific gravity measured in degrees on the American Petroleum Institute scale.

Associated Gas

Natural gas found in a crude oil reservoir, either separate from or in solution with the oil.

Biomass

Biomass includes solid biomass and animal products, gas and liquids derived from biomass, industrial waste and municipal waste.

Coal

Coal includes all coal: both coal primary products (including hard coal and lignite) and derived fuels (including patent fuel, coke-oven coke, gas coke, coke-oven gas and blast-furnace gas). Peat is also included in this category.

Electricity Generation

Electricity generation shows the total amount of electricity generated by power plants. It includes own-use and transmission and distribution losses.

Gas

Gas includes natural gas (both associated and non-associated with petroleum deposits, but excluding natural gas liquids) and gas works gas.

Heat

Heat is heat produced for sale. The large majority of the heat included in this category comes from the combustion of fuels, although some small amounts are produced from electrically-powered heat pumps and boilers.

Heavy Petroleum Products

Heavy petroleum products include heavy fuel oil and bitumen.

Hydro

Hydro refers to the energy content of the electricity produced in hydropower plants, assuming 100% efficiency.

Hydrogen Fuel Cell

A hydrogen fuel cell is a high-efficiency electrochemical energy conversion device that generates electricity and produces heat, with the help of catalysts.

International Marine Bunkers

International marine bunkers cover those quantities delivered to sea-going ships of all flags, including warships. Consumption by ships plying in inland and coastal waters is not included.

Light Petroleum Products

Light petroleum products include liquefied petroleum gas, naphtha and gasoline.

Liquefied Natural Gas (LNG)

Natural gas which has been liquefied by reducing its temperature to minus 258 degrees Fahrenheit (minus 162 degrees Celsius) at atmospheric pressure. As 625 cubic feet of natural gas can be contained in one cubic foot of space when liquefied, the space requirements for storage and transport are significantly reduced.

Middle Distillates

Middle distillates include jet fuel, diesel and heating oil.

Non-conventional Oil

Non-conventional oil includes oil shale, oil sands-based extra-heavy oil and bitumen and derivatives such as synthetic crude products, and liquids derived from natural gas (GTL).

Nuclear

Nuclear refers to the primary heat equivalent of the electricity produced by a nuclear plant with an average thermal efficiency of 33%.

Oil

Oil includes crude oil, natural gas liquids, refinery feedstocks and additives, other hydrocarbons and petroleum products (refinery gas, ethane, liquefied petroleum gas, aviation gasoline, motor gasoline, jet fuel, kerosene, gas/diesel oil, heavy fuel oil, naphtha, white spirit, lubricants, paraffin waxes, petroleum coke and other petroleum products).

Other Petroleum Products

Other petroleum products include refinery gas, ethane, lubricants, bitumen, petroleum coke and waxes.

Other Renewables

Other renewables include geothermal, solar, wind, tide, and wave energy for electricity generation. Direct use of geothermal and solar heat is also included in this category. For OECD countries, other renewables include biomass. Biomass is indicated separately for non-OECD regions, except for electricity output, which includes biomass for all regions.

Other Transformation, Own Use and Losses

Other transformation, own use and losses covers the use of energy by transformation industries and the energy losses in converting primary energy into a form that can be used in the final consuming sectors. It includes energy use and loss by gas works, petroleum refineries, coal and gas transformation and liquefaction. It also includes energy used in coal mines, in oil and gas extraction and in electricity and heat production. Transfers and statistical differences are also included in this category.

Power Generation

Power generation refers to fuel use in electricity plants, heat plants and combined heat and power (CHP) plants. Both public plants and small plants that produce fuel for their own use (autoproducers) are included.

Renewables

Renewables refer to energy resources, where energy is derived from natural processes that are replenished constantly. They include geothermal, solar, wind, tide, wave, hydropower, biomass, and biofuels and hydrogen derived from renewable resources.

Total Final Consumption

Total final consumption (TFC) is the sum of consumption by the different end-use sectors. TFC is broken down into energy demand in the following sectors: industry, transport, other (includes agriculture, residential, commercial and public services) and non-energy use. Industry includes manufacturing, construction and mining industries. In final consumption, petrochemical feedstocks appear under industry use. Other non-energy uses are shown under non-energy use.

Total Primary Energy Supply

Total primary energy supply (TPES) is equivalent to primary energy demand. This represents inland demand only and, except for world energy demand, excludes international marine bunkers.

Underground Storage Working Volume

The amount of gas in a storage facility above the amount needed to maintain a constant reservoir pressure (the latter is known as cushion gas).

ECONOMIC AND FINANCIAL TERMS

Corporate Finance

Companies use their overall creditworthiness, backed by their equity and assets, in order to raise money in the financial market to finance their operations and investment. This might be in the form of debt or equity. Funds for the repayment of debt for a specific project are expected to derive primarily from the revenues generated by the project concerned, but may be derived also from cash flows generated by other assets, especially when the project revenues prove to be insufficient to cover the interest and principal payments on the loan. Corporate finance is the traditional method of financing investment.

Debt Equity Ratio (Leverage)

This is the ratio of total debt (the sum of short-term and long-term debts) to the sum of shareholders' equity (common and preferred shareholders' interest in the company, plus any reserves reported as equity) and total debt. The ratio, often called leverage, is a key measure of a company's capital structure. A company with a high ratio is perceived to be riskier and its ability to raise new capital could be constrained. However, there is no definite level beyond which the ratio should not go.

Debt Maturity

Debt maturity is measured by dividing short-term debt, including long-term debt due in one year (debt in current liabilities), by total debt. The ratio measures the maturity structure of debt: the lower the ratio, the more a company is dependent on long-term debt.

Domestic Credit by Banking Sector

This refers to all credit extended by the banking sector to all other sectors. The calculation is done on a gross basis (*i.e.* disregarding repayments) except in relation to credit extended to central government, which is net. The banking sector includes monetary authorities, depository banks and other banking institutions, such as savings and mortgage loan institutions and loan associates. Domestic credit as a percentage of GDP indicates the size of the banking sector's activity in channelling savings to investors in the economy.

Earnings before Interest, Taxes, Depreciation and Amortisation (EBITDA)

This is a company's income before deduction of interest expenses, taxes, and non-cash charges (depreciation and amortisation). It is an approximate measure of a company's operating cash flow and is sometimes used to evaluate its capacity to raise loans or to derive its value in merger and acquisition deals.

Equity

See Debt Equity Ratio (Leverage).

Foreign Direct Investment (FDI)

FDI is international investment made by a resident entity in one economy in another economy with the objective of acquiring a lasting management interest in an enterprise. A lasting management interest implies the existence of a long-term relationship between the investor and the enterprise and a significant degree of influence by the investor on the management of the direct investment enterprise. It usually involves investment in at least 10% of the voting stock of the enterprise or project. FDI includes not only equity capital but also reinvested earnings and the borrowing or lending of funds between direct investors and the enterprise.

Futures Contract

A supply contract between a buyer and a seller, traded in an Exchange, whereby the buyer is obligated to take delivery and the seller is obligated to provide delivery of a fixed amount of a commodity at a predetermined price at a specified location. Futures contracts are traded exclusively on regulated exchanges and are settled daily, based on their current value in the marketplace.

Henry Hub

The delivery point for the largest New York Mercantile Exchange (NYMEX) natural gas contract by volume.

Leverage

See Debt Equity Ratio.

Levelised Cost

The present value of a cost, including capital, financing and operating costs, expressed as a stream of equal annual payments.

Official Aid

Flows which meet the conditions of eligibility for inclusion in Official Development Assistance (see below), but whose recipients are on Part II of the OECD's Development Assistance Committee (DAC) List of Aid Recipients. See http://www.oecd.org/dac/htm/glossary for details.

Official Development Assistance (ODA)

Grants or loans to countries and territories listed in Part I of DAC List of Aid Recipients, which are undertaken by the official sector with the promotion of economic development and welfare as the main objective, on concessional financial terms (to qualify, loans must have a grant element of at least 25%).

Official Capital Flows

Official capital flows to developing countries represent the total disbursements by the official sector of the creditor country to the recipient country. The term includes official development assistance (ODA), official aid (OA) and other official flows (OOF).

Portfolio Investment

Portfolio investment involves the purchase (or sale – divestment) of new and existing debt and equity. This term includes a variety of instruments which are traded or tradable in organised and other financial markets: bonds, equities and money market instruments. It includes derivatives or secondary instruments, such as options, but excludes FDI (see above). Portfolio investment is more passive in nature than FDI because the portfolio investor has little ability to influence the managerial decision-making of the enterprise in which investment takes place or the return on that investment.

Private Capital Flows

Private capital flows consist of flows financed out of private-sector resources (*i.e.* changes in holdings of long-term assets held by residents of the reporting county, such as bank lending, the purchase of bonds, shares and real estate) and private grants (*e.g.* grants by non-governmental organisations).

Project Finance

For the repayment of their loans, lenders of project finance have recourse principally to the revenues expected to be generated by the project and to the assets of the project which are provided as the collateral, not to the general assets of the project sponsors. Project finance is commonly used as a financing method in infrastructure-related investment, including energy projects. It may allow less creditworthy sponsors to obtain more favourable terms than those available to them under traditional corporate financing by allocating risks among a number of parties. By keeping project costs "off balance sheet", companies involved in the project retain greater flexibility to raise finance for other purposes on the strength of their balance sheets.

Purchasing Power Parity (PPP)

The rate of currency conversion that equalises the purchasing power of different currencies, *i.e.* makes allowance for the differences in price levels between different countries.

Return on Investment (ROI)

ROI, one of the main indicators of profitability, is defined as operating income in a fiscal year divided by invested capital (all types of long-term finance for the company, the majority of which is shareholders' equity and long-term debt). This definition of ROI provides a measure of how effectively the company is using the financial resources invested in its operations, in terms of the return generated.

Sovereign Risk

This refers to the risk that a government will default on its financial obligations, such as loans, or fail to honour other business commitments, owing to political circumstances or economic events. The risk is reflected in the cost of capital to the government, as a risk premium, when it raises debt in the international financial market. This cost of capital to the government often serves as a benchmark for the cost of international capital to the non-government sector in that country.

Standard and Poor's Global 1200 Index

This global equity index covers the performance of stocks in seven major markets in the world: the United States, Canada, Japan, Europe, Australia, Asia and Latin America. The base value of the index, 100, was established at 31 December 1997.

Take-or-Pay (ToP)

In a buyer's contract, take-or-pay is the obligation to pay for a specified amount of gas, whether this amount is taken or not. Depending on the contract terms, it may be possible to take divergences from the specified guarantees as make-up or carry forward quantities in the next contract period. When such divergences are credited into another contract period, they are called make-up gas.

Total External Debt

This is debt owed by an economy to non-residents repayable in foreign currency, goods, or services. It is the sum of public, publicly guaranteed, and private non-guaranteed long-term debt, use of International Monetary Fund (IMF) credit, and short-term debt.

Total Shares Traded

Total shares traded refer to the total value of shares traded on a stock market during a given period. The ratio of total shares traded to GDP measures the overall activity or liquidity of the stock market in the economy.

Value of Listed Shares (Stock Market Capitalisation)

The market value of all listed shares, also known as the stock market capitalisation, is the value reached by multiplying the share prices by the number of shares outstanding. This value, as a percentage of GDP, gives the overall size of a stock market in the economy.

Volatility (measured by standard deviation)

Volatility is a measurement of change in price over a given period. Markets are never free from volatility. The standard deviation measures how widely actual values are dispersed from the average. The larger the difference between the actual value and the average value, the higher the standard deviation will be and the higher the volatility.

REGIONAL GROUPINGS

OECD Europe

OECD Europe consists of Austria, Belgium, the Czech Republic, Denmark, Finland, France, Germany, Greece, Hungary, Iceland, Ireland, Italy, Luxembourg, the Netherlands, Norway, Poland, Portugal, Spain, Sweden, Switzerland, Turkey and the United Kingdom.

OECD North America

OECD North America consists of the United States of America, Canada and Mexico.

OECD Pacific

OECD Pacific consists of Japan, Korea, Australia and New Zealand.

Transition Economies

The transition economies include: Albania, Armenia, Azerbaijan, Belarus, Bosnia-Herzegovina, Bulgaria, Croatia, Estonia, the Federal Republic of Yugoslavia, the former Yugoslav Republic of Macedonia, Georgia, Kazakhstan, Kyrgyzstan, Latvia, Lithuania, Moldova, Romania, Russia, the Slovak Republic, Slovenia, Tajikistan, Turkmenistan, Ukraine and Uzbekistan. For statistical reasons, this region also includes Cyprus, Gibraltar and Malta.

Developing Countries

Developing countries include: China and countries in East Asia, South Asia, Latin America, Africa and the Middle East (see below for countries included in each regional grouping).

China

China refers to the People's Republic of China.

East Asia

East Asia includes: Afghanistan, Bhutan, Brunei, Chinese Taipei, Fiji, French Polynesia, Indonesia, Kiribati, Democratic People's Republic of Korea, Malaysia, Maldives, Myanmar, New Caledonia, Papua New Guinea, the Philippines, Samoa, Singapore, Solomon Islands, Thailand, Vietnam and Vanuatu.

South Asia

South Asia consists of Bangladesh, India, Nepal, Pakistan and Sri Lanka.

Latin America

Latin America includes: Antigua and Barbuda, Argentina, Bahamas, Barbados, Belize, Bermuda, Bolivia, Brazil, Chile, Colombia, Costa Rica, Cuba, Dominica, the Dominican Republic, Ecuador, El Salvador, French Guiana, Grenada, Guadeloupe, Guatemala, Guyana, Haiti, Honduras, Jamaica, Martinique, Netherlands Antilles, Nicaragua, Panama, Paraguay, Peru, St. Kitts-Nevis-Anguilla, Saint Lucia, St. Vincent-Grenadines and Suriname, Trinidad and Tobago, Uruguay, and Venezuela.

Africa

Africa comprises Algeria, Angola, Benin, Botswana, Burkina Faso, Burundi, Cameroon, Cape Verde, the Central African Republic, Chad, Congo, the Democratic Republic of Congo, Côte d'Ivoire, Djibouti, Egypt, Equatorial Guinea, Eritrea, Ethiopia, Gabon, Gambia, Ghana, Guinea, Guinea-Bissau, Kenya, Lesotho, Liberia, Libya, Madagascar, Malawi, Mali, Mauritania, Mauritius, Morocco, Mozambique, Niger, Nigeria, Rwanda, Sao Tome and Principe, Senegal, Seychelles, Sierra Leone, Somalia, South Africa, Sudan, Swaziland, the United Republic of Tanzania, Togo, Tunisia, Uganda, Zambia and Zimbabwe.

Middle East

The Middle East is defined as Bahrain, Iran, Iraq, Israel, Jordan, Kuwait, Lebanon, Oman, Qatar, Saudi Arabia, Syria, the United Arab Emirates and Yemen. It includes the neutral zone between Saudi Arabia and Iraq.

In addition to the WEO regions, the following groupings are also referred to in the text.

European Union (EU15)

Austria, Belgium, Denmark, Finland, France, Germany, Greece, Ireland, Italy, Luxembourg, the Netherlands, Portugal, Spain, Sweden and the United Kingdom.

Northwest Europe Continental Shelf

Denmark, the Netherlands, Norway and the United Kingdom, and comprises the North Sea, the Norwegian Sea and the Norwegian sector of the Barents Sea.

Annex B Countries

Australia, Austria, Belgium, Bulgaria, Canada, Croatia, the Czech Republic, Denmark, Estonia, Finland, France, Germany, Greece, Hungary, Iceland, Ireland, Italy, Japan, Latvia, Lithuania, Luxembourg, the Netherlands, New Zealand, Norway, Poland, Portugal, Romania, Russia, Slovakia, Slovenia, Spain, Sweden, Switzerland, Ukraine, the United Kingdom and the United States of America.

Asia

China, East Asia and South Asia.

Asia-Pacific

East Asia, South Asia and OECD Pacific.

Other Asia

East Asia and South Asia excluding India, unless other grouping is specified in the main text.

Other Latin America

Latin America excluding Brazil, unless other grouping is specified in the main text.

Other Transition Economies

Transition economies other than Russia, unless other grouping is specified in the main text.

Caspian Region

Azerbaijan, Kazakhstan, Turkmenistan and Uzbekistan.

Sub-Saharan Africa

Includes all African countries except North Africa (Algeria, Egypt, Libya, Morocco and Tunisia).

Organization of Petroleum Exporting Countries (OPEC)

Algeria, Indonesia, Iran, Iraq, Kuwait, Libya, Nigeria, Qatar, Saudi Arabia, the United Arab Emirates and Venezuela.

ABBREVIATIONS AND ACRONYMS

In this book, acronyms are frequently used. This glossary provides a quick and central reference for the abbreviations used.

ADB	Asian Development Bank
API	American Petroleum Institute's specific gravity measure
bcm	billion cubic metres
b/d	barrels per day
boe	barrels of oil equivalent
CCGT	combined cycle gas turbine
CCT	clean coal technology
CDM	Clean Development Mechanism
cm	cubic metre
CO_2	carbon dioxide
DOE	Department of Energy
DWT	dead weight tonne
E&D	exploration and development
EIB	European Investment Bank
EU	European Union
FDI	Foreign Direct Investment
GDP	gross domestic product
GHG	greenhouse gas
GTL	gas-to-liquids
GW	gigawatt (1 watt $\times 10^9$)
IEA	International Energy Agency
IFC	International Finance Corporation
IGCC	integrated gasification combined cycle
IMF	International Monetary Fund
kb/d	thousand barrels per day
kcm	thousand cubic metres
kW	kilowatt (1 watt $\times$ 1,000)
kWh	kilowatt-hour
LNG	liquefied natural gas
LPG	liquefied petroleum gas
mb/d	million barrels per day
MBtu	million British thermal units
mcf/d	million cubic feet per day

mcm/d	million cubic metres per day
Mt	million tonnes
Mtoe	million tonnes of oil equivalent
MW	megawatt (1 watt × 10^6)
MWh	megawatt-hour
n.a.	not applicable
NGL	natural gas liquid
NO_x	nitrogen oxides
NSW	New South Wales
OECD	Organisation for Economic Co-operation and Development
OPEC	Organization of Petroleum Exporting Countries
PV	photovoltaic
SO_2	sulphur dioxide
SF_6	sulphur hexafluoride
tce	tonne of coal equivalent
tcm	trillion cubic metres
toe	tonne of oil equivalent
tonne	metric ton
TPES	total primary energy supply
TW	terawatt (1 watt × 10^{12})
TWh	terawatt-hour
WEC	World Energy Council
WEM	World Energy Model
WEO	World Energy Outlook

REFERENCES AND DATA SOURCES

Chapter 1

International Energy Agency (2002), *World Energy Outlook 2002*, Paris: OECD.

Chapter 2

References

International Energy Agency (2001), *Oil Price Volatility: Trends and Consequences*, Economic Analysis Division Working Paper (not published).

International Energy Agency (2002), *World Energy Outlook 2002*, Paris: OECD.

Data sources

World Bank, *Global Development Indicators*, various issues, Washington D.C.: World Bank.

Chapter 3

References

Asian Development Bank (1999), Challenges and Opportunities in Energy, in *Proceedings from Workshop on Economic Cooperation in Central Asia: Challenges and Opportunities in Energy*, Manila: Asian Development Bank.

Boone, Jeff (2001), Empirical Evidence for the Superiority of Non-US Oil and Gas Investments, in *Energy Economics 23*, London: Elsevier Science.

Bosworth, B and Collins, S (2000), *From Boom to Crisis and Back Again: What Have We Learned?*, ADB Institute Working Paper 7, Tokyo: Asian Development Bank Institute.

Claessens, S, Djankov, S and Nenova, T (2000), *Corporate Risk Around the World*, World Bank Country Economics Department Papers, Washington D.C.: World Bank.

Caprio, G and Demirguc-Kunt, A (1997), *The Role of Long Term Finance: Theory and Evidence*, World Bank Policy Research Paper, Washington D.C.: World Bank.

Dailami, M and Leipziger, D (1997), *Infrastructure Project Finance and Capital Flows: A New Perspective*, Presentation at the Conference on Financial Flows and World Development, the University of Birmingham.

Demirguc-Kunt, A and Maksimovic, V (1996), *Institutions, Financial Markets, and Firms' Choice of Debt Maturity*, the World Bank Policy Research Working Paper 1686, Washington D.C.: World Bank.

Dunkerley, Joy (1995), Financing the Energy Sector in Developing Countries, in *Energy Policy*, Vol.23, No.11, London: Elsevier Science Ltd.

Energy Intelligence (2003), *Petroleum Intelligence Weekly*, September 1, 2003, New York: Energy Intelligence.

Harrison, Ann, Love, Inessa and MacMillan, Margaret (2002), *Global Capital Flows and Financing Constraints*, NBER Working Papers, Cambridge: National Bureau of Economic Research.

Humphries, Michael (1995), The Competitive Environment for Oil and Gas Financing, in *Energy Policy*, Vol.23, No.11, London: Elsevier Science Ltd.

International Energy Agency (2001), *World Energy Outlook 2001 Insights: Assessing Today's Supplies to Fuel Tomorrow's Growth*, Paris: OECD.

International Energy Agency (2002), *World Energy Outlook 2002*, Paris: OECD.

International Monetary Fund (2002), *World Economic Outlook September 2002 – Trade and Finance*, Washington D.C.: IMF.

Izaguirre, Ada Karina (2000), Private Participation in Energy, in *Public Policy for the Private Sector*, Washington D.C.: World Bank.

Izaguirre, Ada Karina (2002), Private Infrastructure, in *Public Policy for the Private Sector*, Washington D.C.: World Bank.

Jechoutek, Karl and Lamech, Ranjit (1995), New Directions in Electric Power Financing, in *Energy Policy*, Vol.23, No. 11, London: Elsevier Science Ltd.

Lamech, Ranjit and Saeed, Kazim (2003), *What International Investors Look for When Investing in Developing Countries*, World Bank Energy and Mining Sector Board Discussion Paper No. 6, Washington D.C.: World Bank.

Merrill Lynch (2003), *Global Oils: Capital Investment/F&D Study*, London: Merrill Lynch.

OECD (2002), *Foreign Direct Investment for Development: Maximizing Benefits, Minimizing Costs*, Paris: OECD.

Razavi, H (1998), *Investment and Finance in the Energy Sectors of Developing Countries*, Abu Dhabi: The Emirates Centre for Strategic Studies and Research.

Schmukler, Sergio and Vesperoni, Esteban (2001), *Globalization and Firms' Financing Choices: Evidence from Emerging Economies*, William Davidson Institute Working Paper Series, Ann Arbor: William Davidson Institute.

Société Générale (2003), *Current Financing Issues: the Banker's Perspective*, Presentation to the World Energy Outlook team at IEA.

United Nations Conference on Trade and Development (1999), *World Investment Report*, Geneva: United Nations.
United Nations Conference on Trade and Development (2003), *World Investment Report*, Geneva: United Nations.
World Bank (2003b), *Global Development Finance*, Washington D.C.: World Bank.

Data sources

Moody's Investors Service (2003), *Rating List: Government Bonds & Country Ceilings*, www.moodys.com.
Standard & Poor's (2003a), *Compustat Global Database*, London: MacGraw-Hill Companies.
Standard & Poor's (2003b), *S&P 1200 Indices*, www.spglobal.com
United Nations Conference on Trade and Development (2003), *Foreign Direct Investment*, Data Extract Service, Geneva: United Nations.
World Bank (2003a), *Global Development Indicators*, Washington D.C.: World Bank.
World Bank (2003b), *Global Development Finance*, Washington D.C.: World Bank.
World Bank (2003c), *The Private Participation in Infrastructure Project Database*, Washington D.C.: World Bank.

Chapter 4

References

Alberta Energy and Utilities Board (AEUB) (2002), *Alberta's Reserves 2002*, Alberta: AEUB.
Arab Petroleum Research Centre (APRC) (2002), *Arab Oil and Gas Directory 2002*, Paris: APRC.
Beck, R (2002), *Worldwide Petroleum Industry Outlook 19th Edition*, Oklahoma: PennWell Corporation.
Canadian National Energy Board (NEB) (2000), *Canada's Oil Sands: A Supply and Market Outlook to 2015*, Alberta: NEB.
Concawe (2000), *Impact of a 10 ppm Sulphur Specification for Transport Fuels on the EU Refining Industry*, Brussels: Concawe.
Energy Information Administration (EIA) (2003), *Annual Energy Outlook*, Washington D.C.: US DOE.
IEA (2001), *World Energy Outlook Insights 2001: Assessing Today's Supplies to Meet Tomorrow's Growth*, Paris: OECD.
IEA (2002a), *World Energy Outlook 2002*, Paris: OECD.

IEA (2002b), *Russia Energy Survey 2002*, Paris: OECD.

IEA (2002c), *Oil Information*, Paris: OECD.

Institut Français du Pétrole (IFP) (2002), *Recherche et Production du Pétrole et du Gaz*, Paris: IFP.

Institut Français du Pétrole (IFP) (2003), *Panorama 2003*, Paris: IFP.

Ismail, I (1995), Raising Oil Output in Major Producing Regions: the Financial Implication, in *OPEC Bulletin*, November, Oxford: Blackwell Publishers.

Government of the Russian Federation (2003), *Energy Strategy of the Russian Federation for the Period up to 2020*, Moscow: Ministry of Energy.

Kellas, G and Castellani, M (2002), Recognising and Mitigating E&P Fiscal Risk, in *17th World Petroleum Congress Proceedings*, Rio: World Petroleum Congress.

Kemp, A and Kasim S (2002), *An Analysis of Production Decline Rates in the UK Continental Shelf*, Aberdeen: University of Aberdeen.

Norwegian Petroleum Directorate (2003), *The Petroleum Resources on the Norwegian Continental Shelf*, Oslo: Norwegian Petroleum Directorate.

Natural Resources Canada (2003), Canada's Oil Sands, in *Proceedings of SMI Conference on Non Conventional Oil and Gas*, 4-5 June, London: SMI.

Organization of the Petroleum Exporting Countries (OPEC) (2003), *Proceedings of the Joint OPEC-IEA Workshop on Oil Sector Investment*, 25 June, Vienna: OPEC.

Simmons, M (2002), The World's Giant Oil Fields, in *Hubert Centre Newsletter*, January, Colorado: Colorado School of Mines.

Shihab-Eldin, A, Hamel, M and Brennand, G (2003), Oil Outlook to 2020, in *OPEC Review*, Vol. XXVII, No. 2, June, Oxford: Blackwell Publishers.

Wene, C (2003), *Oil Upstream Investments and Technology Learning* (not published).

Wood Mackenzie Consultants (2002), *Global Oil and Gas Risks and Returns*, Edinburgh: Wood Mackenzie Consultants.

Chapter 5

References

Cedigaz (2003), *Major Trends in the Gas Industry in 2002*, Rueil Malmaison: Institut Français du Pétrole.

CERA (2003), *LNG Shipping: Boom or Bust?*, Decision Brief April 2003.

Commission of the European Communities (CEC) (2002), *Energy dialogue with Russia - progress since the October 2001 EU-Russia Summit*, Commission Staff Working Paper, SEC(2002)333.

Commission of the European Communities (CEC) (2003), *Long-term Gas Supply Security in an Enlarged Europe*, Brussels: CEC (forthcoming).

Energy Information Administration (EIA) (1999), *Corporate Realignments and Investments in the Interstate Natural Gas Transmission System*, Washington D.C.: US DOE.

Energy Information Administration (EIA) (2001), *The Majors' Shift to Natural Gas,* Washington D.C.: US DOE.

Energy Information Administration (EIA) (2003a), *Monthly Gas Report* (various issues), Washington D.C.: US DOE.

Energy Information Administration (EIA) (2003b), *US LNG Markets and Uses*, Washington D.C.: US DOE.

Energy Information Administration (EIA) (2003c), *Annual Energy Outlook*, Washington D.C.: US DOE.

Energy Information Administration (EIA) (2003d), *Expansion and Change on the US Natural Gas Pipeline Network*, Washington D.C.: US DOE.

Energy Information Administration (EIA) (2003e), *International Energy Outlook*, Washington D.C.: US DOE.

ERI/SDPC, Energy Research Institute of the State Development Planning Commission (2001), *A Prospect of Gas Supply and Demand in China*, presentation by Liu Xiaoli to IEA Gas Experts Mission, Beijing, 13 November.

FACTS (2003a), *Iran Oil and Gas Industry: Outlook for 2003 and Beyond*, Confidential Memo, February, Honolulu: Fesharaki Associates Commercial and Technical Services, Inc.

FACTS (2003b), *Iran LNG: Recent Developments*, May, Honolulu: Fesharaki Associates Commercial and Technical Services, Inc.

Gazprom (2003), *Preliminary Offering Circular*, 14 February.

IEA (2002a), *World Energy Outlook 2002*, Paris: OECD.

IEA (2002b), *Russia Energy Survey 2002*, Paris: OECD.

IEA (2002c), *Developing China's Natural Gas Market: the Energy Policy Challenges*, Paris: OECD.

IEA (2003), *South American Gas*, Paris: OECD.

IHS Energy/Petroconsultants (2003), *Gulf of Mexico Deep Shelf Production Performance Study*, Colorado: IHS Energy.

Jensen, J (2003), The LNG Revolution, in *Energy Journal*, Volume 24 (2).

National Energy Board (2002), *Short-term Natural Gas Deliverability from the Western Canada Sedimentary Basin 2002-2004*, Calgary: NEB.

National Petroleum Council (2003), *Balancing Natural Gas Policy – Fueling the Demands of a Growing Economy*, Volume 1, Washington D.C.: NPC.

Observatoire Méditerranéen de l'Energie (2003), *Financing Energy Projects in the Southern and Eastern Mediterranean Countries*, Sophia Antipolis: OME.

Renaissance Capital (2002), *Gazprom: The Dawning of a New Valuations Era*, Moscow, April.

Royal Institute of International Affairs (2001), *China Natural Gas Report*, London.

Simmons, Matthew R (2002), *Unlocking the Natural Gas Riddle*, Houston: Simmons & Company International.

UNDP/World Bank ESMAP (2003), *Cross-Border Oil and Gas Pipelines: Problems and Prospects*. Washington D.C.: UNDP.

UNECE (2000), *Study on Underground Gas Storage in Europe and Central Asia*, Geneva: UNECE.

Chapters 4 and 5

Data sources

Asia Pacific Energy Research Centre (2000a), *Natural Gas Pipeline Development in Southeast Asia*, Tokyo: Institute of Energy Economics.

Asia Pacific Energy Research Centre (2000b), *Natural Gas Pipeline Development in Northeast Asia*, Tokyo: Institute of Energy Economics.

BP (2003), *BP Statistical Review of World Energy 2003*, London: BP.

Canadian Association of Petroleum Producers (CAPP), *Annual Oil Sands Investment and Production 1958-2001*, Calgary: CAPP.

Cedigaz (2000), *Natural Gas in the World*, Rueil Malmaison: Institut Français du Pétrole.

Clarkson Research Studies (2003), *World Shipyard Monitor*, Vol. 10, No. 6, June, London: Clarkson Research Studies.

Deutsche Bank AG (2003), *Oil and Gas Abacus*, 30 May, London: Deutsche Bank AG.

Eurogas (2001), *Annual Report 2001*, Brussels: Eurogas.

FACTS-EWCI (2003), *Gas Databook: Asia-Pacific Natural Gas and LNG*, Honolulu: Fesharaki Associates Commercial and Technical Services, Inc.

Gas Technology Institute (2001), *LNG Source Book*, Des Plaines, IL: GTI.

Gas Transmission Europe (2003), *The European Gas Network*, Brussels: GTE.

IHS Energy (2003), *PEPS Database - Exploration and Production Costs Module*, London: IHS Energy.

International Group of LNG Importers (2001), *The LNG industry*, Clichy: GIIGNL.

Korea Maritime Institute (KMI), *World Shipping Statistics 2002*, Seoul: KMI.

Lehman Brothers (2002), *The Original E&P Spending Survey*, New York: Lehman Brothers.

Merrill Lynch (2003), *Global Oils: Capital Investment/F&D Study*, 29 May, New York: Merrill Lynch.

Natural Resources Canada (2001), *Natural Gas Transportation and Distribution*, Ottawa: StatCan.

Norwegian Ministry of Petroleum and Energy (2003), *Facts 2003 - The Norwegian Petroleum Sector*, Oslo: Norwegian Ministry of Petroleum and Energy.

OLADE (2001), *Study for Natural Gas Market Integration in South America*. Quito: OLADE.

Organization of the Petroleum Exporting Countries (OPEC) (2002), *Annual Statistical Bulletin 2002*, Oxford: Blackwell Publishers.

Organization of Arab Petroleum Exporting Countries (2002), *Annual Statistical Report 2002*, Safat: OAPEC.

PennWell Corporation (1991-2003), *Oil and Gas Journal* (various issues), Oklahoma: PennWell Corporation.

UK Offshore Operators Association (UKOOA) and Department of Trade and Industry (DTI) (2002), *Activity Survey 2002*, London: UKOOA.

United States Department of Energy/Energy Information Agency (DOE/EIA) (2001a), *Performance Profiles of Major Energy Producers*, Washington: DOE/EIA.

United States Department of Energy/Energy Information Agency (DOE/EIA) (2001b), *International Energy Annual 2001* (World Crude Oil Refining Capacity), Washington: DOE/EIA.

United States Department of Energy/Energy Information Agency (DOE/EIA) (2003), *Financial Reporting System (Financial and Operating Data Series, 1977-2001)*, Washington: DOE/EIA.

United States Geological Survey (USGS) (2000), *World Petroleum Assessment 2000*, Washington: USGS.

Chapter 6

References

Ball, A, Hansard, A, Curtotti, R and Schneider, K (2003), *China's Changing Coal Industry – Implications and Outlook*, Canberra: ABARE.

Energy Information Administration (EIA) (2003), *Annual Energy Outlook 2003*, Washington: D.C.: US DOE.

Energy Information Administration (EIA) (2003a), *Annual Coal Report 2001*, Washington: D.C.: US DOE.

International Energy Agency (2001), *Coal Information 2001*, Paris: OECD.

International Energy Agency (2002a), *World Energy Outlook 2002*, Paris: OECD.
International Energy Agency (2002b), *Coal in the Energy Supply of India*, Paris: OECD.
International Energy Agency (2002c), *Russia Energy Survey 2002*, Paris: OECD.
International Energy Agency (2003), *Coal Information 2003*, Paris: OECD
International Energy Agency Clean Coal Centre (2002), *Coal upgrading to reduce CO_2 emissions*, London: International Energy Agency Clean Coal Centre.
International Energy Agency Coal Research (2001), *Comparative Environmental Standards – Deep Mine and Opencast*, London: International Energy Agency Clean Coal Centre.
McCloskey (2002a), *Australia's Coal Exports: Prospects to 2015*, Hampshire: The McCloskey Group Ltd.
McCloskey (2002b), *Russia's Coal Exports: producers, specs and ports*, Hampshire: The McCloskey Group Ltd.
McCloskey (2003), *Coal Report 65*, Hampshire: The McCloskey Group Ltd.
Reuters (2002), China's thermal and coking coal exports to rise to 95 Mt, in *Coal Week International*, 24 June: Reuters.
RWE Rheinbraun (2002), *World Market for Hard Coal*, Köln: RWE Rheinbraun Aktiengesellschaft.
Tata Energy Research Institute (2002), *Teri Energy Data Directory & Yearbook 2001/02*, New Delhi: Tata Energy Research Institute.
UK Department of Trade and Industry (2003), *Coal Investment Aid*, London: UK DTI.
UNECE (2001), *Restructuring of the Coal Industry in Economies in Transition*, ENERGY/GE.1/2001/4 and Add.1-8, September 2001.
US Census Bureau (2001), *1997 Economic Census: Mining*, Washington: D.C.: Department of Commerce.

Data sources

Australian Bureau of Agricultural and Resource Economics (ABARE) (1999), *Australian Commodities*, Vol. 6, No. 4, Canberra: ABARE.
Australian Bureau of Agricultural and Resource Economics (2000), *Australian Commodities*, Vol. 7, No. 2, Canberra: ABARE.
Australian Bureau of Agricultural and Resource Economics (2001), *Australian Commodities*, Vol. 8, No. 4, Canberra: ABARE.
Clarkson Research Studies (2002), *Dry Bulk Trade Outlook*, London: Clarkson Research Studies.
Clarkson Research Studies (2003), *World Shipyard Monitor*, Vol. 10, No. 6, June, London: Clarkson Research Studies.

Connell Wagner Ltd (2001), *India Coal Port Infrastructure Study*, Canberra: Dept. of Industry, Science and Resources.

Department of Natural Resources and Mines (2001), *Queensland Coal Industry Review 2000-2001*, Brisbane: Queensland Government.

Directorate of Mineral and Coal Enterprises, *Indonesia Mineral and Coal Mining: Company Profile 2002*, Jakarta: Ministry of Energy and Mineral Resources.

Energy Information Administration (EIA) (1999), *The US Coal Industry in the 1990's: Low Prices and Record Production*, Washington: D.C.: US DOE.

Energy Information Administration (EIA) (2003), *US Coal Supply and Demand: 2002 Review*, Washington: D.C.: US DOE.

Energy Information Administration (EIA), *Annual Coal Report (various editions)*, Washington: D.C.: US DOE.

European Coal and Steel Community (2001), *Investment in the Community coal and steel industries*, Luxembourg: European Commission.

Fearnleys (1999), *World Bulk Trades 1998*, Oslo: Fearnresearch.

Gruss, H (2002), *New coal mining and coal ports investment projects* (not published).

IMC Group Consulting Ltd (2003), *A Review of the Remaining Reserves at Deep Mines*, London: UK DTI.

Lloyd's Register of Shipping (2001), *World Fleet Statistics 2001*, London: Lloyd's Register.

Simpson Spence and Young Ltd. Consultancy and Research (2002), *Coal Port Survey 2002*, London: SSY Consultancy and Research.

Statistics Canada (2003), *Private and Public Investment in Canada*, Ottawa: Statistics Canada.

Statistics South Africa (2001), *Mining, Financial Statistics*, Pretoria: Statistics South Africa.

World Energy Council (2001), *Survey of Energy Resources*, London: WEC.

Chapter 7

References

APEC (1995), *Regional Cooperation for Power Infrastructure*, Australia: APEC.

Asian Development Bank (2000), *Developing Best Practices for Promoting Private Sector Investment in Infrastructure – Power*, Philippines: ADB.

Bacon, R W, Besant-Jones, J (2001), Global Electric Power Reform, Privatization and Liberalization of the Electric Power Industry in Developing Countries, *Energy and the Environment, Annual Reviews*, 26:331-359.

China State Planning Commission (2001), *Main Report: Summary of the Electricity Industry Performance during the 9th Five-year Plan*, Beijing: China Electric Power Press.

Commission of the European Communities (2001), *Communication from the Commission to the European Parliament and the Council - Infrastructure*, Brussels: CEC.

Dube, I (2002), *Energy Services for the Urban Poor*, Nairobi: African Energy Policy Research Network (AFREPREN).

Edison Electric Institute (2002), *Financial Review 2001*, Washington, D.C.: EEI.

EDF (2002), *Electricity for All*, Paris: EDF.

Hirst, E and Kirby, B (2001), *Transmission Planning for a Restructuring US Electricity Industry*, Washington, D.C.: Edison Electric Institute.

Hughes, W and Parece, A (2002), *The Economics of Price Spikes in Deregulated Power Markets*, USA: Charles River Associates.

International Energy Agency (1999), *Electricity Reform, Power Generation Costs and Investment*, Paris: OECD.

International Energy Agency (2002a), *World Energy Outlook 2002*, Paris: OECD.

International Energy Agency (2002b), *Security of Supply in Electricity Markets, Evidence and Policy Issues*, Paris: OECD.

International Energy Agency (2002c), *Distributed Generation in Liberalised Electricity Markets*, Paris: OECD.

International Energy Agency (2002d), *Russia Energy Survey 2002*, Paris: OECD.

International Energy Agency (2003a), *Power Generation Investment in Electricity Markets*, Paris: OECD (forthcoming).

International Energy Agency (2003b), *South American Gas*, Paris: OECD.

Jamasb, T (2002), *Reform and Regulation of the Electricity Sectors in Developing Countries*, Working Paper CMI EP 08/DAE 0226, UK: University of Cambridge.

Joskow, P L (2003), *The Difficult Transition to Competitive Electricity Markets in the US*, US: MIT.

Karakezi, S (2002), *Options for Addressing the Nexus of Energy Poverty in the Framework of NEPAD*, Nairobi: AFREPREN.

Lamech, R and Saeed, K (2003), *What International Investors Look for when Investing in Developing Countries*, Washington D.C.: World Bank and the Energy and Mining Sector Board.

Mkhwanazi, X (2003), *Power Sector Development in Africa*, Dakar: United Nations.

Newbery, D M (2000), *Privatization, Restructuring, and Regulation of Network Utilities*, USA: MIT Press.

North American Electric Reliability Council (2002b), *Reliability Assessment 2002*, USA: NERC.

OECD Development Centre (2003), *African Economic Outlook 2002/2003*, Paris: OECD.

Official Journal of the European Union (2003), *Decision No 1229/2003/EC of the European Parliament and of the Council of June 26 2003*, Luxembourg: Office for Official Publications of the European Communities.

OSCAL (2001), *Energy for Sustainable Development of the Least Developed Countries in Africa*, New York: United Nations.

PLN (2002), *Annual Report 2001*, Indonesia: PLN.

Thomas, S (2003), The Seven Brothers, *Energy Policy 31* (2003)393-403.

Turkson, J (2000), *Power Sector Reform in Sub-Saharan Africa*, UNEP Collaborating Center on Energy and Environment.

US Department of Energy (2002a), *National Transmission Grid Study*, Washington, D.C.: US DOE.

US Department of Energy (2002b), *National Transmission Grid Study*, Washington, Issue Papers, D.C.: US DOE.

World Bank (2002a), *Strengthening of the Institutional and Regulatory Structure of the Brazilian Power Sector*, Washington, D.C.: World Bank.

World Bank (2002b), *Accounting for Poverty in Infrastructure Reform*, Washington, D.C.: World Bank

World Bank (2003a), *Indonesia, Maintaining Stability, Deepening Reforms*, Washington, D.C.: World Bank.

World Bank (2003c), *What International Investors Look for When Investing in Developing Countries*, Washington, D.C.: World Bank.

World Bank (2003d), *Private Participation in Infrastructure in Developing Countries, Trends, Impacts and Policy Lessons*, Washington, D.C.: World Bank.

Data sources

Canadian Electricity Association and Natural Resources Canada (2000), *Electric Power in Canada 1998-1999*, Ottawa: Canadian Electricity Association and Natural Resources Canada.

Commission of the European Communities (CEC) (2002), *First benchmarking report on the implementation of the internal electricity and gas market*, Commission Staff Working Paper, SEC(2001)1957, Brussels: CEC.

Direktorat Jenderal Listrik Dan Pengembangan Energi (2000), *Statistik Dan Informasi Tahun 1999/2000*, No. 13, Jakarta: Direktorat Jenderal Listrik Dan Pengembangan Energi.

Edison Electric Institute (2003), *Statistical Yearbook of the electric utility industry*, May 2003, Washington D.C.: Edison Electric Institute.

Endesa (2002), *Annual Report 2001*, Madrid: Endesa.

Electricity Supply Association of Australia (ESAA) (2002), *Electricity Australia 2002*, Sidney: Electricity Supply Association of Australia.

Energy Information Administration (EIA) (2001), *Electric Power Annual 2000*, Vol. I, August 2001, Washington D.C.: US DOE.

Energy Information Administration (EIA) (2002), *Electric Power Annual 2000*, Vol. II, November 2002, Washington D.C.: US DOE.

Energy Information Administration (EIA) (2003), *Existing Electric Generating Units in the United States by State, Company and Plant, 2002 (Database)*, Washington D.C.: US DOE.

Eurelectric (1999), *Programmes and Prospects for the European Electricity Sector*, Brussels: Eurelectric.

Eurelectric (2003), *Electricity Tariffs as of 1 January 2003 (Published Tariffs)*, May, Brussels: Eurelectric.

Federation of Electric Power Companies of Japan (2002), *Electricity Review Japan 2002*, Japan: FEPC.

Government of India, Planning Commission (2002), *Annual Report on The Work of State Electricity Boards & Electricity Departments*, May, New Delhi: Government of India.

Government of India, *Five-year Plan* (various years), India: GOI.

Government of the People's Republic of China, *China's Energy Development Report 2001*, Beijing: Government of the People's Republic of China.

IAEW and Consentec (2001), *Analysis of Electricity Network Capacities and Identification of Congestion*, December, Aachen: IAEW and Consentec.

Power Grid Corporation of India Ltd, *Annual Report 2000-2001*, New Delhi: Power Grid Corporation of India Ltd.

Japan Electric Power Information Center (2002), *Electric Power Industry in Japan 2002/2003*, Tokyo: Japan Electric Power Information Center.

Korean Electric Power Corporation (2000), *Biannual News & Statistics*, Seoul: KEPCO.

Korean Electric Power Corporation (2002), *2002 Annual Report*, Seoul: KEPCO.

Ministry of Mines and Energy, Secretariat for Energy (2001), *2001 Brazilian Energy Balance*, Brasilia: Federative Republic of Brazil.

National Grid (2001), *Statements of Charges for Use of the Transmission System and for Connection to the Transmission System*, NETA, UK: National Grid.

Nordel Planning Committee (2002), *Nordic Grid Master Plan 2002*, Denmark: Nordel.

North American Electric Reliability Council (NERC) (2002a), *Electricity Supply and Demand Database*, Princeton: NERC.

OECD (2003), *National Income Account Database*, Paris: OECD.

Oxera (2001), *Electricity Liberalisation Indicators in Europe*, October 2001: European Commission DG TREN.

Platts (2001), *World Electric Power Plant Database*: Platts.

Platts (2002), *International Private Power Quarterly, Second Quarter 2002*, New York: The McGraw-Hill Companies.

Standard & Poor's (2003), *Compustat Global Database*, London: MacGraw-Hill Companies.

State Statistical Bureau of the People's Republic of China, *China Statistical Yearbook* (various years), Beijing: China Statistical Information and Consultancy Service Center.

Tata Energy Research Institute (2002), *Teri Energy Data Directory & Yearbook 2001/02*, New Delhi: Tata Energy Research Institute.

Woodhill (2002), *Transmission of CO_2 and Energy, Report Number PH4/6*, March, UK: IEA Greenhouse Gas R&D Programme.

World Bank (2003b), *Global Development Indicator*, Washington D.C.: World Bank.

World Bank (2003e), *The Private Participation in Infrastructure Project Database*, Washington D.C.: World Bank.

Chapter 8

References

An, F and Santini, D J (2003), *Assessing tank-to-wheel efficiencies of advanced technology vehicles*, Society of Automotive Engineers (SAE) paper 2003-01-0412, USA: SAE.

Arthur, D Little (2002), *Guidance for transportation technologies: Fuel choice for fuel cell vehicles*, Cambridge, MA, USA.

Benato, R, Fellin L, Lorenzoni, A, Paolucci, A (2001), *Elettrodotti blindati nel territorio: connessioni dei nuovi impianti di generazione alla RTN*, in AEI, Vol.88, N° 3, March, pp. 28-37, Milano: Associazione Elettrotecnica ed Elettronica Italiana.

David, J and Herzog, H (2000), *The cost of carbon capture,* Paper presented at the 5th international meeting on greenhouse gas control technologies, 13-16 August, Cairns, Australia.

Dijkstra, J and Jansen, D (2002), *Novel concepts for CO_2 capture with SOFC*, Paper presented at the 6th international meeting on greenhouse gas control technologies, October, Kyoto, Japan.

Duffey, R *et al.* (2001), AECL, *Advanced Applications of Nuclear Reactors – Technology opportunities and their impacts on global energy, markets and emissions*, Paper presented at the World Energy Conference 2001.

European Copper Institute (1999), *The scope for energy saving in the EU through the use of energy-efficient electricity distribution transformers*, THERMIE PROJECT, December, Brussels: European Copper Institute.

Elvestad, O (2003), *Case study: CO_2 capture and storage from a 400 MW gas fired power plant at Kårstø, Norway*, Paper presented at IEA CO_2 capture workshop, 24th January, Paris: IEA.

Freund, P and Davison, J (2002), *General overview of costs*, Proceedings of the IPCC Workshop on Carbon Dioxide Capture and Storage, pp. 79-94, 18-21 November, Regina, Canada: Energy Research Centre of the Netherlands (ECN), Petten.

Gray, D and Tomlinson, G (2001), *Coproduction of ultra clean transportation fuels, hydrogen, and electric power from coal*, MTR 2001-43, Falls Church, Virginia, USA: Mitretek.

Herzog, H (2003), *Pathways for reducing the cost of CO_2 capture*, Paper presented at IEA CO_2 capture workshop, 24th January, Paris: IEA.

IEA (2001), *Saving Oil and Reducing CO_2 Emissions in Transport: Options & strategies*, Paris: OECD.

IEA GHG (2000), *Leading options for the capture of CO_2 emissions at power stations*, IEA GHG R&D programme, Orchard Road, Cheltenham, UK.

Klara, S (2003), *Systems analysis supporting the carbon sequestration technology roadmap.* Paper presented at the 2nd annual conference on carbon sequestration: Developing & Validating the Technology Base to Reduce Carbon Intensity, 5-8 May, Alexandria, VA, USA.

Koch, H (2003), *Experience with 2nd Generation Gas-Insulated Transmission Lines (GIL)*, Meudon (France): Wets Conference, CNRS.

Lange, T, de Beeldman, M, Kiel J, Uil, H and den Veenkamp, J (2001), *Co-production of fuels as an option for Demkolec?* ECN-C-01-004: Energy Research Centre of the Netherlands (ECN), Petten.

McDermott and Associates (2000), *Energy-Efficient Distribution Transformers – Utility Initiatives*, International Copper Association, December, Brussels: European Copper Institute.

McKee, B (2002), *Solutions for the 21st century. Zero Emissions Technologies for Fossil Fuels*, International Energy Agency, Working Party on Fossil Fuels, Paris: IEA.

Myers, D B, Ariff, G D, James, B D, Lettow, J S, Thomas, C E and Kuhn, R E (2002), *Cost and performance analysis of stationary hydrogen fueling appliances*, Arlington, VA, USA: Directed Technologies.

Ogden, J M (2002), *Review of small stationary reformers for hydrogen production.* IEA/H2/TR-02/002, USA: IEA Hydrogen Implementing Agreement/Princeton University.

Santini, D J, Vayas, A D, Kumar, R and Anderson, J L (2002), *Comparing estimates of fuel economy improvement via fuel-cell power trains*. Society of Automotive Engineers (SAE) paper 02FCC-125.USA: SAE.

Tsuchiya, H (2002), *Fuel cell cost study by learning curve*, Paper presented at the international energy workshop organised by EMF/IIASA, 18-20 June, Stanford, CA, USA.

US Department of Energy Nuclear Energy Research Advisory Committee (2002), *A Technology Roadmap for Generation IV Nuclear Energy Systems*, Washington, D.C.: US DOE.

Williams, R H (2002), *Decarbonised fossil energy carriers and their energy technology competitors*, Proceedings of the IPCC Workshop on Carbon Dioxide Capture and Storage, pp. 119-135, 18-21 November, Regina, Canada: Energy Research Centre of the Netherlands (ECN), Petten.

Annex 2

References

International Energy Agency (2000), *Experience Curves for Energy Technology Policy*, Paris: OECD.

International Energy Agency (2002), *World Energy Outlook 2002*, Paris: OECD.

ORDER FORM

IEA BOOKS

Fax: +33 (0)1 40 57 65 59
E-mail: books@iea.org
www.iea.org/books

INTERNATIONAL ENERGY AGENCY

9, rue de la Fédération
F-75739 Paris Cedex 15

I would like to order the following publications

PUBLICATIONS	ISBN	QTY	PRICE	TOTAL
☐ **World Energy Investment Outlook - *2003 Insights***	**92-64-01906-5**		**€150**	
☐ Power Generation Investment in Electricity Markets	92-64-10556-5		€75	
☐ The Power to Choose - *Demand Response in Liberalised Electricity Markets*	92-64-10503-4		€75	
☐ Renewables in Russia - *From Opportunity to Reality*	92-64-10544-1		€100	
☐ Energy Policies of IEA Countries - 2003 Review (Compendium)	92-64-01480-2		€120	
☐ Electricity Information - 2003 Edition	92-64-10219-8		€130	
☐ Natural Gas Information - 2003 Edition	92-64-10241-8		€150	
☐ Renewables for Power Generation - *Status and Prospects - 2003 edition*	92-64-01918-9		€75	
			TOTAL	

DELIVERY DETAILS

Name ____________________ Organisation ____________________

Address ____________________

____

Country ____________________ Postcode ____________________

Telephone ____________________ E-mail ____________________

PAYMENT DETAILS

☐ I enclose a cheque payable to IEA Publications for the sum of $ ____________ or € ____________

☐ Please debit my credit card (tick choice). ☐ Mastercard ☐ VISA ☐ American Express

Card no: ____________________

Expiry date: ____________

Signature:

OECD PARIS CENTRE
Tel: (+33-01) 45 24 81 67
Fax: (+33-01) 49 10 42 76
E-mail: distribution@oecd.org

OECD BONN CENTRE
Tel: (+49-228) 959 12 15
Fax: (+49-228) 959 12 18
E-mail: bonn.contact@oecd.org

OECD MEXICO CENTRE
Tel: (+52-5) 280 12 09
Fax: (+52-5) 280 04 80
E-mail: mexico.contact@oecd.org

You can also send your order to your nearest OECD sales point or through the OECD online services: www.oecd.org/bookshop

OECD TOKYO CENTRE
Tel: (+81-3) 3586 2016
Fax: (+81-3) 3584 7929
E-mail: center@oecdtokyo.org

OECD WASHINGTON CENTER
Tel: (+1-202) 785-6323
Toll-free number for orders:
(+1-800) 456-6323
Fax: (+1-202) 785-0350
E-mail: washington.contact@oecd.org

IEA PUBLICATIONS - 9, rue de la Fédération - 75739 PARIS CEDEX 15
PRINTED IN FRANCE BY STEDI
((61 2003 27 1 P1) ISBN 92-64-01906-5
Dépôt légal : 8148 - Novembre 2003